AF835342

Environmental Modelling with Contemporary Statistics

This book places an emphasis upon statistical methodology, data interpretation, and future-proofing with the intention of advancing statistics for the environment, ecology, and environmental health, in addition to environmental theory and practice, via the application of reliable statistics.

With a focus on advances in statistical methodology and application within the environmental sciences, the overarching purpose of this volume is to illuminate current trends, stimulate a focus on, and connect multidisciplinary domains originating from and within statistical analysis, development, and research. Given that the contributions consist of current improvements and new innovations in climate and environmental science research that are based on statistical theory, researchers can derive inspiration for future advancements or similar analyses on other environmental data.

Authored by internationally renowned scholars, this book is organized into three parts: Part I on Supervised and Unsupervised Learning, Part II on Directional Statistics, and Part III on Spatial and Temporal Modeling. Primarily intended as a reference book for academic researchers and graduate-level students in statistics as well as multidisciplinary domains, the chapters reflect a shared commitment to advancing methodological rigor while addressing real-world environmental concerns. They illustrate how environmental complexity drives the evolution of statistical thinking—and how statistical insight, in turn, informs meaningful action.

Key Features:

- Emphasizes the ongoing necessity to progress basic statistical theory and explores its relevance to environmental research.
- Employs multidisciplinary approaches and applications, appealing to a broader audience beyond solely theoretical statistics.
- Enhances the collective understanding of the ecosystem's diverse perspectives to ensure the welfare of present and future generations.
- Written by renowned subject matter experts and researchers, making it appealing to scholars from diverse fields.
- The statistical framework is not limited to a single methodology based on data complexity but promotes different techniques.

Environmental Modelling with Contemporary Statistics

Learning, Directionality, and Space-Time Dynamics

Edited by
Andriëtte Bekker, Priyanka Nagar, Johan Ferreira,
Barend Erasmus, and Abel Ramoelo

CRC Press
Taylor & Francis Group
Boca Raton London New York

CRC Press is an imprint of the
Taylor & Francis Group, an **informa** business
A CHAPMAN & HALL BOOK

First edition published 2026
by CRC Press
2385 NW Executive Center Drive, Suite 320, Boca Raton FL 33431

and by CRC Press
4 Park Square, Milton Park, Abingdon, Oxon, OX14 4RN

CRC Press is an imprint of Taylor & Francis Group, LLC

© 2026 selection and editorial matter, Andriëtte Bekker, Priyanka Nagar, Johan Ferreira, Barend Erasmus, and Abel Ramoelo; individual chapters, the contributors

Reasonable efforts have been made to publish reliable data and information, but the author and publisher cannot assume responsibility for the validity of all materials or the consequences of their use. The authors and publishers have attempted to trace the copyright holders of all material reproduced in this publication and apologize to copyright holders if permission to publish in this form has not been obtained. If any copyright material has not been acknowledged please write and let us know so we may rectify in any future reprint.

Except as permitted under U.S. Copyright Law, no part of this book may be reprinted, reproduced, transmitted, or utilized in any form by any electronic, mechanical, or other means, now known or hereafter invented, including photocopying, microfilming, and recording, or in any information storage or retrieval system, without written permission from the publishers.

For permission to photocopy or use material electronically from this work, access www.copyright.com or contact the Copyright Clearance Center, Inc. (CCC), 222 Rosewood Drive, Danvers, MA 01923, 978-750-8400. For works that are not available on CCC please contact mpkbookspermissions@tandf.co.uk

For Product Safety Concerns and Information please contact our EU representative GPSR@taylorandfrancis.com. Taylor & Francis Verlag GmbH, Kaufingerstraße 24, 80331 München, Germany.

Trademark notice: Product or corporate names may be trademarks or registered trademarks and are used only for identification and explanation without intent to infringe.

ISBN: 978-1-032-90391-0 (hbk)
ISBN: 978-1-032-91679-8 (pbk)
ISBN: 978-1-003-56442-3 (ebk)

DOI: 10.1201/9781003564423

Typeset in CMR10 font
by KnowledgeWorks Global Ltd.

Publisher's Note: This book has been prepared from camera-ready copy provided by the author.

Contents

Preface

The accelerating sustainability crisis—manifested in biodiversity loss, climate volatility, pollution, and resource depletion—poses profound and urgent challenges to both science and society. These challenges defy disciplinary boundaries. They demand integrative approaches that connect environmental science, ecology, public health, economics, and beyond. At the heart of this convergence lies data: vast, heterogeneous, and increasingly complex. From satellite imagery and climate records to genomic sequences and citizen-science platforms, environmental research now operates in a data-rich landscape that requires statistical ingenuity to navigate.

This edited volume responds to that call. It brings together 19 invited contributions from internationally recognized scholars, each rooted in both statistical and environmental applications. The chapters reflect a shared commitment to advancing methodological rigor while addressing real-world environmental concerns. They illustrate how environmental complexity drives the evolution of statistical thinking—and how statistical insight, in turn, informs meaningful action.

The volume facilitates synergies across domains of study and research concepts. The chapters that follow illustrate how environmental needs drive advances in fundamental statistics. The focus is on contemporary methodological development in three broad areas: (1) supervised and unsupervised learning, (2) directional statistics, and (3) spatial and temporal modeling. These themes were chosen because they capture both enduring challenges and emerging frontiers in environmental data analysis, from the complexity of air pollution to the far-reaching consequences of large-scale wildfires.

The book underscores the indispensable role of diverse statistical methodologies in addressing six critical environmental domains: air pollution, coastal dynamics, climate analysis, soil texture, biological complexity, and wildfire impact. These areas represent urgent global challenges, where data-driven approaches are essential for understanding, managing, and mitigating risk. Six critical domains anchor the environmental focus of this volume:

- **Air Pollution**, a multifaceted threat to public health and economic resilience, is examined in Chapter 1 by Tortora; Chapter 3 by Maruotti, Ricciotti, and Russo; Chapter 5 by Kleynhans, Kanfer, Millard, and du Plessis; and Chapter 12 by Borges, Ros Žario Ramos, Cordeiro, and Carvalho.

- **Coastal Dynamics**, essential for disaster preparedness and sustainable management, are explored in Chapter 7 by Lagona; Chapter 8 by Barbieri, Battaglia, and Cucina; and Chapter 10 by Alonso-Pena.

- **Climate Analysis**, central to policy and stakeholder engagement, is addressed in Chapter 6 by Nazari, Arashi, and Sadeghkhani; Chapter 9 by SenGupta and Laha; Chapter 11 by Jona Lasinio, Mastrantonio, and Pollice; Chapter 14 by van Niekerk, Teixeira Krainski, Rustand, and Rue; Chapter 16 by Banerjee; Chapter 17 by Gimeno-Sotelo, Richards, Hazra, Mhalla, and de Zea Bermudez; and Chapter 18 by Nadifar, Bekker, and Arashi.

- **Soil Texture analysis**, foundational to agriculture and land use, is investigated in Chapter 4 by Masoumifard, van der Westhuizen, and Gardner-Lubbe.

- **Biological Complexity,** which arises through intricate interactions and feedback that are challenging to quantify and predict, is illuminated in Chapter 2 by d'Angella and Hennig; Chapter 13 by Lyubchich, Kilbourne, and Nesslage; and Chapter 19 by Salehi, Mazza, and Punzo.

- **Wildfire Impact**, increasingly severe and far-reaching, is the focus of Chapter 15 by Yoo, Zhang, and Wikle.

Instead of endorsing a one-size-fits-all approach, this volume embraces the richness of methodological diversity, emphasizing the need to tailor statistical techniques to the intricate nature of environmental data and the specific scientific contexts in which they are applied. The goal is not only to anticipate future conditions but to identify pathways for constructive intervention—where statistical reasoning becomes a catalyst for sustainable change.

As global efforts intensify toward achieving the Sustainable Development Goals (United Nations 2015, Transforming Our World: The 2030 Agenda for Sustainable Development), reliable and forward-looking analytics have become indispensable for shaping effective policy and guiding strategic interventions. This book stands at the intersection of statistical science and environmental inquiry, offering readers both technical depth and a compelling vision of how data-driven approaches can illuminate and address the planet's most pressing challenges.

We invite readers—statisticians, environmental scientists, technical advisors, policy analysts, and students alike—to engage with the ideas presented here. May this volume inspire new collaborations, provoke thoughtful dialogue, and contribute meaningfully to a more sustainable and resilient world.

Acknowledgments

The editors wish to express their sincere gratitude to all those who contributed to the development of this volume. We thank the contributing authors for their enthusiastic participation and for sharing their professional expertise. We are also grateful to the reviewers for their careful, thorough, and constructive evaluations of the chapters included in this book.

We gratefully acknowledge the professional support of Lara Spieker, Jain Saveetha, and Mayank Sharma at Chapman & Hall/CRC, Taylor & Francis, whose guidance and assistance were instrumental in the successful publication of this volume.

This work was made possible through financial support from the following sources:

1. Centre of Excellence in Mathematical and Statistical Sciences, University of the Witwatersrand, South Africa (Ref. No. CoE-MaSS 2024-045-STA-Environmental Science).
2. Bekker UP cc: A0S937.
3. The Faculty of Natural and Agricultural Sciences at the University of Pretoria.

The views and conclusions expressed in this book are those of the author(s) and do not necessarily reflect those of the National Research Foundation (NRF), South Africa.

The editors welcome comments from readers, including notifications of typographical or other errors, and suggestions for improving future editions. Correspondence may be directed to any of the editors.

Editors

Professor Andriëtte Bekker is an Emeritus Professor and former Head of the Department of Statistics at the University of Pretoria (2012–2022), and an Associate of the National Institute for Theoretical and Computational Sciences (NITheCS). A recipient of the S2A3 Medal for scientific achievement, she is internationally recognized for her contributions to multivariate and matrix variate distribution theory, with expertise spanning directional statistics, model-based clustering, and graphical network modeling. She has authored over 140 peer-reviewed publications and edited volumes advancing statistical methodology and computation. Professor Bekker is an elected member of the International Statistical Institute and leads the Statistical Theory and Applied Statistics focus area within the DSTI-NRF Centre of Excellence in Mathematical and Statistical Sciences. Her recent accolades include the University of Pretoria's Exceptional Academic Achiever Award (2023), a fellowship from the South African Statistical Association (2024), and an NRF rating as a researcher of international standing.

Doctor Priyanka Nagar is a Senior Lecturer in Statistics and Actuarial Science at Stellenbosch University and an Associate of the National Institute for Theoretical and Computational Sciences (NITheCS). She holds a PhD in Mathematical Statistics from the University of Pretoria. Her research focuses on statistical learning theory, directional statistics, and copula-based modeling, with applications in environmental, biomechanical, and energy domains. Dr. Nagar's work advances statistical methodologies for complex environmental systems. Drawing on experience in both academia and industry, she brings a rigorous and applied perspective to statistical analysis. She is actively engaged in mentoring, supervision, and strengthening statistical capacity within environmental statistics research across South Africa.

Professor Johan Ferreira is a Professor in the School of Statistics and Actuarial Science at the University of the Witwatersrand, and previously served as the Assistant Focus Area Coordinator for the Statistical Theory and Applied Statistics focus area of the Centre of Excellence in Mathematical and Statistical Science, based at the University of the Witwatersrand in Johannesburg. He is an ASLP 4.1/4.2 fellow of Future Africa and was identified as one of the Top 200 South Africans under the age of 35 by the Mail & Guardian newspaper in the Education category. Johan regularly published in peer-reviewed, accredited journals, and his research interests include the probabilistic modeling of entropy, meaningful mixture modeling, directional statistics, and topics in educational statistics.

Professor Barend Erasmus is the previous Dean of the Faculty of Natural and Agricultural Sciences at the University of Pretoria, and currently holds a tenured professorship at the Gordon Institute of Business Science, University of Pretoria. He is an ecologist with broad experience in climate change risk and adaptation. His publication record reflects his interests in wide range of interdisciplinary topics. His doctoral degree at the University of Pretoria on assessing the impacts of climate change and its impacts on biodiversity in South Africa remains relevant in international literature. Over time, his research interests expanded from climate change impacts to broader sustainability issues across a wide range of sectors. His current academic work is on exploring the risks and opportunities of rapidly developing climate science for business and industry. He is passionate about postgraduate training, and students are deeply embedded in collaborative and interdisciplinary research programs.

Professor Abel Ramoelo is an Executive Director of the Earth Observation Programme at the South African National Space Agency (SANSA) and an Extraordinary Professor at the University of Pretoria. He has a PhD in Geoinformation Science and Earth Observation from the University of Twente, the Netherlands. He leads a dynamic team focused on developing earth intelligence to address the societal challenges we face today. He previously worked at the CSIR, advancing from junior to principal researcher, at SANParks as a Regional Ecologist/ Remote Sensing Specialist, and at the University of Pretoria as an Associate Professor and Director of the Centre for Environmental Studies in the Department of Geography, Geoinformatics, and Meteorology.

List of Contributors

María Alonso-Pena
Department of Statistics, Mathematical Analysis and Optimization
Universidade de Santiago de Compostela
Spain

Mohammad Arashi
Department of Statistics
Ferdowsi University of Mashhad
Iran

Department of Statistics
University of Pretoria
Republic of South Africa

Sudipto Banerjee
Department of Biostatistics
University of California Los Angeles
United States of America

Maria Maddalena Barbieri
Department of Economics
Roma Tre University
Italy

Francesco Battaglia
Roma Sapienza Foundation
Italy

Andriëtte Bekker
Department of Statistics
University of Pretoria
Republic of South Africa

National Institute for Theoretical and Computational Sciences (NITheCS)
Gauteng node
University of Pretoria
Republic of South Africa

Patrícia de Zea Bermudez
Departamento de Ciências Matemáticas
Universidade de Lisboa
Portugal

CEAUL – Centro de Estatística e Aplicações, Universidade de Lisboa
Portugal

Ana Borges
CIICESI, ESTG
Polytechnic of Porto
Portugal

Mariana Carvalho
CIICESI, ESTG
Polytechnic of Porto
Portugal

Clara Cordeiro
Center for Research and Development in Mathematics and Applications (CIDMA)
Department of Mathematics
University of Aveiro
Portugal

Faculdade de Ciências e Tecnologia da Universidade do Algarve
Portugal

Domenico Cucina
Department of Political Science
Roma Tre University
Italy

Gabriele d'Angella
Dipartimento di Scienze Statistiche "Paolo Fortunati"
ALMA MATER STUDIORUM – Università di Bologna
Italy

Salomi du Plessis
Department of Statistics
University of Pretoria
Republic of South Africa

Sugnet Gardner-Lubbe
MuViSU
Stellenbosch University
Republic of South Africa

Department of Statistics and Actuarial Science
Stellenbosch University
Republic of South Africa

Luis Gimeno-Sotelo
Departamento de Ciências Matemáticas
Universidade de Lisboa
Portugal

CEAUL Centro de Estatística e Aplicações
Universidade de Lisboa
Portugal

Arnab Hazra
Department of Mathematics and Statistics
Indian Institute of Technology Kanpur
India

Christian Hennig
Dipartimento di Scienze Statistiche "Paolo Fortunati"
ALMA MATER STUDIORUM – Università di Bologna
Italy

Frans Hendrik Jacobus Kanfer
Department of Statistics
University of Pretoria
Republic of South Africa

K. Halimeda Kilbourne
Chesapeake Biological Laboratory
University of Maryland Center for Environmental Science
United States of America

Andre Ruben Kleynhans
Department of Statistics
University of Pretoria
Republic of South Africa

Elias Teixeira Krainski
EMSE Division
King Abdullah University of Science and Technology
Kingdom of Saudi Arabia

Francesco Lagona
Department of Political Sciences
University of Roma Tre
Italy

Arnab K. Laha
Interdisciplinary Statistical Research Unit
Indian Statistical Institute
India

Operations and Decision Sciences
Indian Institute of Management
Ahmedabad
India

Giovanna Jona Lasinio
Department of Statistical Sciences
"Sapienza" University of Rome
Italy

Vyacheslav Lyubchich
Chesapeake Biological Laboratory
University of Maryland Center for Environmental Science
United States of America

Antonello Maruotti
Department of Public Health and Epidemiology
Khalifa University
United Arab Emirates

Center for Biotechnology
Khalifa University
United Arab Emirates

Department GEPLI
LUMSA University
Italy

Khaled Masoumifard
MuViSU
Stellenbosch University
Republic of South Africa

Department of Statistics and Actuarial Science
Stellenbosch University
Republic of South Africa

Gianluca Mastrantonio
Department of Mathematical Science
Polytechnic of Turin
Italy

Angelo Mazza
Department of Economics and Business
University of Catania
Italy

Linda Mhalla
Institute of Mathematics
EPFL – Swiss Federal Technology Institute of Lausanne
Switzerland

Solomon Marthinus Millard
Department of Statistics
University of Pretoria
Republic of South Africa

Mahsa Nadifar
Department of Statistics
University of Pretoria
Republic of South Africa

Samaneh Nazari
Department of Statistics
Ferdowsi University of Mashhad
Iran

Geneviève Nesslage
Chesapeake Biological Laboratory
University of Maryland Center for Environmental Science
United States of America

Alessio Pollice
Department of Economics and Finance
University of Bari Aldo Moro
Italy

Antonio Punzo
Department of Economics and Business
University of Catania
Italy

M. Rosário Ramos
CEAUL
Universidade de Lisboa
Portugal

LE@D
Universidade Aberta
Portugal

Lorena Ricciotti
Department of Economics, Statistics and Finance
"Giovanni Anania" – University of Calabria
Italy

Jordan Richards
School of Mathematics and Maxwell Institute for Mathematical Sciences
University of Edinburgh
United Kingdom

Håvard Rue
CEMSE Division
King Abdullah University of Science and Technology
Kingdom of Saudi Arabia

Alfonso Russo
Department of Economics, Statistics and Finance
"Giovanni Anania" – University of Calabria
Italy

Denis Rustand
Inserm UMR 1219 Bordeaux Population Health Research Center
University of Bordeaux
France

CEMSE
King Abdullah University of Science and Technology
Kingdom of Saudi Arabia

Abdolnasser Sadeghkhani
Department of Mathematics and Statistics
North Carolina Agricultural and Technical State University
United States of America

Mahdi Salehi
Department of Economics and Business
University of Catania
Italy

Ashis SenGupta
Department of Population Health Sciences
Augusta University
United States of America

Applied Statistics Unit
Indian Statistical Institute
India

Cristina Tortora
Department of Mathematics and Statistics
San José State University
United States of America

Stephan van der Westhuizen
MuViSU
Stellenbosch University
Republic of South Africa

Department of Statistics and Actuarial Science
Stellenbosch University
Republic of South Africa

Janet van Niekerk
Department of Statistics
University of Pretoria
Republic of South Africa

CEMSE
King Abdullah University of Science and Technology
Kingdom of Saudi Arabia

National Institute for Theoretical and Computational Sciences (NITheCS)
Gauteng node
University of Pretoria
Republic of South Africa

Christopher K. Wikle
Department of Statistics and Data Science
University of Missouri
United States of America

Myungsoo Yoo
Department of Statistics and Data Sciences
The University of Texas at Austin
United States of America

Likun Zhang
Department of Statistics and Data Science
University of Missouri
United States of America

Part I

Supervised and Unsupervised Learning

1

Mapping Emission Dynamics: A High-Dimensional Approach to Cluster Analysis with Outlier Detection

Cristina Tortora

This chapter examines the application of cluster analysis to the Emissions Database for Global Atmospheric Research (EDGAR), focusing on identifying patterns in greenhouse gas and air pollution emissions across various countries. Utilizing the comprehensive and regularly updated data provided by EDGAR, this study employs model-based clustering with finite mixture models to categorize nations according to their emissions profiles. The chapter includes an extension of model-based clustering algorithms based on the contaminated normal distribution to high-dimensional data, thanks to variable reduction and variable selection, allowing for outlier detection in high-dimensional data. The use of the multiple scaled contaminated normal distribution in this context has the extra advantage of directional outlier detection. The analysis provides significant insights into country homogeneity, outlier detection, and variable correlation. Ultimately, this research highlights the importance of employing advanced statistical techniques to enhance our understanding of global emissions dynamics and facilitate effective responses to air quality concerns.

1.1 Introduction

The Emissions Database for Global Atmospheric Research (EDGAR)[1]is a comprehensive global database that provides detailed estimates of anthropogenic emissions of greenhouse gases (GHGs) and air pollutants (EDGAR Team, 2022; Muntean et al., 2024). It is developed and maintained by the European Commission's Joint Research Centre (JRC) in collaboration with other institutions. EDGAR serves as a critical tool for policymakers, researchers, and environmental agencies, offering consistent and scientifically robust data to better understand the sources of emissions worldwide. The database plays a key role in supporting air quality studies and the development of global, regional, and national policies. It enables a comprehensive analysis of human activities contributing to environmental change by providing detailed emission estimates across various sectors, including energy production, industry, agriculture, waste management, and transportation.

[1]https://edgar.jrc.ec.europa.eu/data set_ap81#sources

DOI: 10.1201/9781003564423-1

One of EDGAR's notable strengths is its global scope and high resolution. These enable emissions data to be represented on fine spatial grids, allowing for comparisons across countries and regions and facilitating the monitoring of international commitments. Moreover, EDGAR is regularly updated to reflect the latest scientific and statistical developments, ensuring its relevance for environmental impact assessments and sustainable development planning. By providing open access to emissions data, EDGAR significantly contributes to transparency and accountability in addressing global air pollution challenges.

The database contains a lot of information, and exploratory data analysis tools can be essential for gaining valuable insights. Among those tools, cluster analysis is a valuable statistical technique used to identify groups of countries based on similarities in pollutant emissions data. It enables researchers and policymakers to understand patterns, relationships, and disparities in emissions across nations, facilitating more targeted and effective environmental strategies. By grouping countries with similar emissions profiles, cluster analysis provides insights into regional trends, emission drivers, and shared challenges, which can inform collaborative solutions. One key advantage of using cluster analysis on pollutant data is its ability to handle the complexity and heterogeneity of global emissions. Countries differ significantly in their economic activities, energy consumption, industrialization levels, population sizes, and environmental policies, all of which contribute to emissions. Cluster analysis can identify hidden patterns within these diverse data sets, grouping countries with similar emissions characteristics. The results of cluster analysis can also support policy design and international cooperation. Countries within the same cluster often face similar challenges or emission profiles, making it easier to design tailored policies that address their unique needs. For example, countries in a cluster characterized by high-energy-related emissions might benefit from shared strategies for renewable energy deployment or energy efficiency improvements.

Cluster analysis has indeed been used often across various fields (Everitt et al., 2011; Kaufman and Rousseeuw, 1990) and to study atmospheric pollution (Gramsch et al., 2006; Taubman et al., 2006; Zhang et al., 2016). Among various approaches to cluster analysis, model-based clustering using finite mixture models (McLachlan and Peel, 2000a) has gained increasing attention in recent years due to its flexibility and inferential advantages (Bouveyron et al., 2019). In this framework, a population is assumed to be distributed following a convex combination of a finite number of probability distributions, usually from the same family but with different parameters. Furthermore, a one-to-one correspondence is usually considered between each component distribution and a group. Pollution data are, however, characterized by some extreme observations; for example, big countries such as China or India have extremely high emission values for most pollutants. Small countries such as Guam, instead, have extremely low emissions. To apply model-based clustering to this type of data, the clusters can be modeled using a distribution with heavier tails than the normal distribution. Some examples of such distributions are t mixtures (Andrews and McNicholas, 2012; Peel and McLachlan, 2000) and contaminated Gaussian mixtures (Punzo and McNicholas, 2016). More recently, Forbes and Wraith (2014) and Punzo and Tortora (2021) proposed multiple scaled distributions to allow for directional outliers, i.e., when observations are outliers only for some dimensions, which can be the case for pollutant emissions.

A final issue observed on a database such as the EDGAR database is that many variables are also highly correlated; in this situation, two strategies are common: variable reduction and variable selection. Variable reduction refers to methods that transform the original variables into a smaller set of new variables. The goal is to reduce the data set's dimensionality while preserving as much information as possible. Techniques include principal component analysis (PCA), factor analysis, and multidimensional scaling. The new variables (like principal components [PCs]) are combinations of the original variables. Variable selection involves identifying and selecting a subset of the original variables most relevant to the outcome of interest. The aim is to retain only the most informative variables while excluding those that contribute little to the model's predictive power. The output of dimension reduction is a new set of variables that are typically uncorrelated with each other. For example, in PCA, the result is a set of PCs that capture the largest variance in the data. Understanding the contribution of individual original variables to these new components can be complex. The output in variable selection is a subset of the original variables. Selected variables retain their original interpretation and meaning, making it easier to understand their impact on the response variable. Analysts can directly assess the relationship between the selected variables and the outcome.

Both strategies have been used within model-based clustering literature. A dimension reduction solution is to use a mixture of factor analyzers (MFA), which does not entirely reduce the dimensionality of the problem but instead reduces the dimensionality of specific model parameters. (Ghahramani and Hinton, 1997; McLachlan and Peel, 2000b), some examples of extensions to non-Gaussian distributions can be found in Tortora et al. (2016) and Punzo et al. (2020). Variable selection for Gaussian mixture models was proposed by (Raftery and Dean's, 2006). Wallace et al. (2018) proposed extending variable selection to mixture models with skewed variables. For a review on clustering high-dimensional data, see Bouveyron and Brunet-Saumard (2014).

This study proposes clustering techniques that can detect clusters in data sets characterized by outliers, global or directional, and high-dimensional correlated variables. These techniques can then be used on the EDGAR database to identify groups of countries based on similarities in pollutant emissions data.

1.2 Background

As mentioned in Section 1.1, to apply model-based clustering on a data set characterized by outliers, the clusters can be modeled using a leptokurtic distribution, which is characterized by heavier tails than the normal distribution. Since, in this context, we are also interested in detecting outliers, the mixture of contaminated normal (MCN) distributions (Punzo and McNicholas, 2016) is preferred.

1.2.1 Mixtures of Contaminated Normal Distributions

As proposed by Tukey (1960), the contaminated normal (CN) distribution is a mixture of two normal distributions that share the same location but feature different covariance structures to account for mild outliers. The probability density function (pdf) for a d-dimensional random vector $\boldsymbol{X} = (X_1, \ldots, X_d)^\top$ that follows a CN distribution can be expressed as follows:

$$f_{\mathrm{CN}}\left(\boldsymbol{x} \mid \boldsymbol{\mu}, \boldsymbol{\Sigma}, \alpha, \eta\right) = \alpha f_{\mathrm{N}}\left(\boldsymbol{x} \mid \boldsymbol{\mu}, \boldsymbol{\Sigma}\right) + (1-\alpha) f_{\mathrm{N}}\left(\boldsymbol{x} \mid \boldsymbol{\mu}, \eta\boldsymbol{\Sigma}\right). \tag{1.1}$$

In this equation, $f_{\mathrm{N}}\left(\boldsymbol{x} \mid \boldsymbol{\mu}, \boldsymbol{\Sigma}\right)$ denotes the pdf of a multivariate normal distribution characterized by mean vector $\boldsymbol{\mu}$ and covariance matrix $\boldsymbol{\Sigma}$. Here, $\alpha \in (0,1)$ represents the proportion of good data points, and $\eta > 1$ indicates the extent of contamination, accounting for increased variability due to mild outliers. The condition $\eta > 1$ ensures the model remains identifiable. Additionally, both α and η serve as tailedness parameters, contributing to the leptokurtic nature of the distribution. More detailed discussion on the kurtosis of the CN distribution relative to α and η is available in Appendix G of Bagnato et al. (2017).

In practical scenarios, it is common to apply the constraint $\alpha \in (0.5, 1)$ to ensure that at least half of the observations are classified as 'good', facilitating the model effectiveness in outlier detection (Punzo and McNicholas, 2016).

Building on this, Punzo and McNicholas (2016) introduce the MCN distributions for model-based clustering. The pdf of MCN distributions is a convex combination of G multivariate CN density functions. This can be represented as follows:

$$f_{\text{MCN}}(\boldsymbol{x} \mid \boldsymbol{\vartheta}) = \sum_{g=1}^{G} \pi_g f_{\text{CN}}(\boldsymbol{x} \mid \boldsymbol{\mu}_g, \boldsymbol{\Sigma}_g, \alpha_g, \eta_g). \tag{1.2}$$

In this equation, $\boldsymbol{\vartheta}$ represents the vector of all model parameters, with $\boldsymbol{\pi} = (\pi_1, \ldots, \pi_G)$ indicating the mixing proportions (where $0 < \pi_g \leq 1$ and $\sum_{g=1}^{G} \pi_g = 1$). The term $f_{\text{CN}}(\boldsymbol{x} \mid \boldsymbol{\mu}_g, \boldsymbol{\Sigma}_g, \alpha_g, \eta_g)$ refers to the pdf of the CN distribution as described in equation (1.1), with other parameters defined for each $g = 1, \ldots, G$.

One of the limitations of the MCN is that the tail parameters, α_g and η_g, are the same across all dimensions. This implies that it is not possible to account for different tail behaviors in different dimensions; outliers are automatically down-weighted in the maximum likelihood (ML) estimation of the parameters but in the same way for each dimension, and the procedure to detect outliers could be defined as an omnibus in the sense that when a point is detected as an outlier, it is a global outlier (Punzo and Tortora, 2021). To overcome this issue, Punzo and Tortora (2021) proposed the multiple scaled contaminated normal (MSCN) distribution.

1.2.2 Mixtures of Multiple Scaled Contaminated Normal Distributions

To overcome specific limitations associated with MCN distributions, Punzo and Tortora (2021) propose a mixture of multiple scaled contaminated normal distributions (MMSCN). The pdf of the MMSCN can be expressed as follows:

$$\begin{aligned} f_{\text{MMSCN}}(\boldsymbol{x} \mid \boldsymbol{\vartheta}) &= \sum_{g=1}^{G} \pi_g f_{\text{MSCN}}(\boldsymbol{x} \mid \boldsymbol{\mu}_g, \boldsymbol{\Gamma}_g, \boldsymbol{\Lambda}_g, \boldsymbol{\alpha}_g, \boldsymbol{\eta}_g) \\ &= \sum_{g=1}^{G} \pi_g \prod_{h=1}^{d} f_{\text{CN}}\left(\left[\boldsymbol{\Gamma}_g^{\top}(\boldsymbol{x} - \boldsymbol{\mu}_g)\right]_h \mid 0, \Lambda_{hg}, \alpha_{hg}, \eta_{hg}\right). \end{aligned} \tag{1.3}$$

Here, d represents the number of PCs, while π_g and $\boldsymbol{\vartheta}$ align with the definitions provided in equation (1.2). The parameter $\boldsymbol{\mu}_g$ serves as the location parameter for the observed data set, and the matrix $\boldsymbol{\Gamma}_g$ comprises the eigenvectors that transform $\boldsymbol{x}$ into its PC representation. The diagonal matrix $\boldsymbol{\Lambda}_g$ includes the eigenvalues $\Lambda_{1g}, \ldots, \Lambda_{dg}$. Moreover, $\boldsymbol{\alpha}_g = (\alpha_{1g}, \ldots, \alpha_{dg})^{\top}$ and $\boldsymbol{\eta}_g = (\eta_{1g}, \ldots, \eta_{dg})^{\top}$ denote, respectively, the ratios of typical points and the extent of contamination across each PC. From equation (1.3), it can be inferred that the MSCN exhibits marginal CN distributions for every PC.

In contrast to the MCN, the MMSCN framework facilitates the detection of outliers in a specific direction spanned by each PC. This innovation allows for a more adaptable approach to modeling intricate data distributions, especially in cases where traditional elliptical models struggle to capture the inherent variability and extreme behaviors observed in the data.

Parameter estimates can be obtained using the Expectation-Maximization (EM) algorithm (Dempster et al., 1977); for details on the estimates, see Punzo and Tortora (2021). The software to use the MCN and MMSCN for clustering is freely available on `R` (R Core

Team, 2021) with the packages `ContaminatedMixt` (Punzo et al., 2018) and `MSclust` (Tortora et al., 2024b), respectively.

1.2.3 Variable Selection for Model-Based Clustering

The original data set often contains many variables, making it complex and time-intensive to execute clustering algorithms on the complete data set. As a result, effective variable selection becomes a crucial aspect of the clustering process. Raftery and Dean's (2006) present the challenge of identifying significant variables by comparing diverse models constructed from various combinations of the available variables. Two nested models are evaluated when assessing the potential exclusion of a specific set of candidate variables. The larger model encompasses all variables presumed to contribute information about the cluster memberships. In contrast, some variables that might be excluded in the smaller model are conditionally independent of the cluster memberships, assuming that the other variables retained in the smaller model contain relevant clustering information.

While Bayes factors are typically preferred for model selection, their precise computation can be difficult. Therefore, Raftery and Dean's (2006) suggests employing the Bayesian Information Criterion (BIC) as a substitute for model comparison. The BIC is an effective metric for assessing model performance, especially when clustering with the CN distribution, as highlighted by Tran and Tortora (2021). Specifically, the difference in BIC values between the competing models helps identify which model demonstrates superior performance. Letting $\boldsymbol{\vartheta}$ be the model parameters, the BIC is defined as follows:

$$BIC = 2\log(L(\boldsymbol{\vartheta})) - \rho\log(n),$$

where $L(\boldsymbol{\vartheta})$ is the likelihood obtained at convergence of the algorithm, ρ corresponds to the number of parameters, and n is the sample size.

More in detail, let's consider an original data set denoted as $\boldsymbol{X}$. This data set is partitioned into three distinct groups of variables: $\boldsymbol{X}_1$, $\boldsymbol{X}_2$, and $\boldsymbol{X}_3$. Here, $\boldsymbol{X}_1$ consists of variables that have already been verified to contribute information on cluster memberships; $\boldsymbol{X}_2$ contains candidate variables under evaluation for their potential inclusion or exclusion based on their relevance to cluster memberships, and $\boldsymbol{X}_3$ is the remainder of the variables that are not currently considered.

In this framework, two models are analyzed. Having already observed $\boldsymbol{X}_1$, model M_1 assumes that $\boldsymbol{X}_2$ adds no more information about cluster memberships. Conversely, model M_2 postulates that $\boldsymbol{X}_2$ does contribute additional insights into cluster membership determination. More specifically, defining with $\boldsymbol{z}$ the cluster memberships, given $\boldsymbol{X}_1$ already selected, $\boldsymbol{X}_2$ is conditionally independent of $\boldsymbol{z}$ in M_1, and $\boldsymbol{X}_2$ is not conditionally independent of $\boldsymbol{z}$ in M_2. However, $\boldsymbol{X}_3$ can potentially be associated with $\boldsymbol{X}_1$ and $\boldsymbol{X}_2$.

As Raftery and Dean's (2006) demonstrated, the integrated likelihood of models M_1 and M_2 can be expressed as:

$$p(\boldsymbol{X}|M_1) = p(\boldsymbol{X}_3|\boldsymbol{X}_2, \boldsymbol{X}_1, M_1)p(\boldsymbol{X}_2|\boldsymbol{X}_1, M_1)p(\boldsymbol{X}_1|M_1),$$
$$p(\boldsymbol{X}|M_2) = p(\boldsymbol{X}_3|\boldsymbol{X}_2, \boldsymbol{X}_1, M_2)p(\boldsymbol{X}_2, \boldsymbol{X}_1|M_2).$$

Since the conditional distribution of $\boldsymbol{X}_3$ given $\boldsymbol{X}_2$ and $\boldsymbol{X}_1$ is the same under M_1 and M_2, it follows that $p(\boldsymbol{X}_3|\boldsymbol{X}_2, \boldsymbol{X}_1, M_1) = p(\boldsymbol{X}_3|\boldsymbol{X}_2, \boldsymbol{X}_1, M_2)$. Therefore, comparing M_1 and M_2 is the same as comparing $p(\boldsymbol{X}_2|\boldsymbol{X}_1, M_1)p(\boldsymbol{X}_1|M_1)$ and $p(\boldsymbol{X}_2, \boldsymbol{X}_1|M_2)$.

Let d_1 be the number of variables in $\boldsymbol{X}_1$. Let RSS be the residual sum of squares obtained from the linear regression of $\boldsymbol{X}_2$ on $\boldsymbol{X}_1$, the BIC approximation to $p(\boldsymbol{X}_2|\boldsymbol{X}_1, M_1)$ is as follows:

$$BIC_{reg} = -nlog(2\pi) - nlog(RSS/n) - n - (d_1 + 2)log(n).$$

BIC approximation to $p(\boldsymbol{X}_1|M_1)$ is the BIC value of the model that consists of $\boldsymbol{X}_1$. BIC approximation to $p(\boldsymbol{X}_2, \boldsymbol{X}_1|M_2)$ is the model's BIC value consisting of $\boldsymbol{X}_1$ and $\boldsymbol{X}_2$.

Let BIC_1 be the sum of BIC_{reg} and the value of BIC approximation to $p(\boldsymbol{X}_1|M_1)$. Let BIC_2 be the value of BIC approximation to $p(\boldsymbol{X}_2, \boldsymbol{X}_1|M_2)$. To compare M_1 and M_2, $BIC_2 - BIC_1$ can be used. If the difference is greater than 0, then it is counted toward strong evidence that $\boldsymbol{X}_2$ contains additional cluster membership information, and therefore M_2 is preferred, according to Kass and Raftery (1995). In our code, we use a threshold of 0.1.

The steps of the variable selection algorithm proposed by Raftery and Dean's (2006) can be summarized as follows:

- Step 1: select the variable that provides the most cluster membership information among all variables.
- Step 2: select another variable that provides the most cluster membership information, including the variable chosen from the first step.
- Step 3: propose another variable that provides the most cluster membership information, including the previously selected variables; add this variable to the selected variable set if there is strong evidence that this variable contains additional cluster membership information.
- Step 4: propose a variable for removal from the selected variable set that shows the least cluster membership information; remove this variable from the selected variable set if removing this variable leads to a model that contains more cluster membership information.
- Execute steps 3 and 4 iteratively until (1) both steps do not change the selected variable set, or (2) the variable added in step 3 is the same as the variable removed in step 4.

To improve the algorithm performance above, the variable selection algorithm proposed by Maugis et al.'s (2009) is incorporated into the above algorithm.

1.3 Methodology

The MCN and MMSCN distributions allow for cluster analysis and outlier detection. However, they are not capable of handling high-dimensional data. In this section, we propose two ways of extending the proposed methods to handle highly correlated high-dimensional variables: variable reduction and variable selection.

1.3.1 Variable Reduction

The use of a variable reduction technique, such as PCA, followed by cluster analysis has been discouraged since the two steps optimize different criteria and lead to suboptimal results (Chang, 1983). To avoid this issue, subspace clustering methods have been proposed; they model each cluster in a group-specific low-dimensional subspace. One of the first proposed models, the mixture of probabilistic PC analyzers (Tipping and Bishop, 1999), models each cluster within the mixture model using a probabilistic PC analyzer; this model enables ML parameter estimates in the cluster-specific PC spaces. The structure of the MSCN naturally leads to a clusterwise dimension reduction. Specifically, instead of estimating all the PCs in all the clusters, as in equation (1.3), we can include only the first q PCs in each cluster, where $q < d$. The density of the reduced MMSCN (RMMSCN) becomes the following:

$$f_{\text{MMSCN}}(\boldsymbol{x} \mid \boldsymbol{\vartheta}) = \sum_{g=1}^{G} \pi_g \prod_{h=1}^{q} f_{\text{CN}}\left(\left[\tilde{\boldsymbol{\Gamma}}_g^\top (\boldsymbol{x} - \boldsymbol{\mu}_g)\right]_h \mid 0, \Lambda_{hg}, \alpha_{hg}, \eta_{hg}\right),$$

where $\tilde{\boldsymbol{\Gamma}}_g$ is a $d \times q$ matrix containing the first q column vectors of $\boldsymbol{\Gamma}_g$. It is important to note that the vectors $\boldsymbol{\mu}_g$ are still estimated in the original space and, therefore, are d-dimensional. The parameters $\boldsymbol{\alpha}_g = (\alpha_{1g}, \ldots, \alpha_{hg})^\top$ and $\boldsymbol{\eta}_g = (\eta_{1g}, \ldots, \eta_{hg})^\top$ are now q dimensional vectors. Parameter estimates can be obtained using the EM algorithm, the algorithm requires the choice of the number of clusters, G, and the number of PCs, q.

1.3.2 Variable Selection

We developed two variable selection methods, one for the MCN distributions illustrated in Section 1.2.1 and another for the MMSCN distributions explained in Section 1.2.2. In both cases, we followed the structure in Section 1.2.3. Specifically, the variable selection method proposed by Raftery and Dean's (2006) was combined with the method proposed by Maugis et al.'s (2009). To be precise, we replace the original steps 3 and 4 from Raftery and Dean's (2006) method with Maugis et al.'s (2009) exclusion-inclusion steps.

The resulting algorithms have been implemented in the `R` programming language. The algorithms do not require the selection of the number of variables since it is automatically obtained as a by-product of the algorithm; they only require the selection of the number of clusters.

1.3.3 Illustrative Examples

The proposed algorithms are shown in two illustrative examples. The first data set is obtained pseudo randomly, generating two clusters from an MMSCN. The number of variables is 6, and $n_1 = n_2 = 300$. For simplicity, we set $\boldsymbol{\Gamma}_1^T \boldsymbol{\Lambda}_1 \boldsymbol{\Gamma}_1 = \boldsymbol{\Gamma}_2^T \boldsymbol{\Lambda}_2 \boldsymbol{\Gamma}_2 = \mathbf{I}$, $\boldsymbol{\mu}_1 = \mathbf{0}$, $\boldsymbol{\mu}_2 = \mathbf{5}$, where $\mathbf{I}$ is the identity matrix and boldface numbers indicate a vector containing that value. The tails parameters are $\boldsymbol{\alpha}_1 = (0.8, 0.6, 0.8, 0.6, 0.8, 0.6)$, $\boldsymbol{\eta}_1 = (2, 4, 3, 2, 2, 6)$, $\boldsymbol{\alpha}_2 = (0.9, 0.9, 0.6, 0.6, 0.9, 0.7)$, and $\boldsymbol{\eta}_2 = (3, 2, 2, 6, 2, 4)$. The left panel of Figure 1.1 shows a scatter plot of the first data set with points colored based on the cluster to which they belong. The first three variables of the second data set are generated as the first three variables of data set one; the next three variables are obtained pseudo randomly, generating values from a uniform distribution between 0 and 1. The right panel of Figure 1.1 shows a scatter plot of the second data set with points colored based on the cluster to which they belong.

We run the MCN and MMSCN variable selection methods on the two illustrative data, setting $G = 2$. Moreover, we run RMMSCN setting $G = 2$ and q equal to 3, 4, and 5. Results are measured using the adjusted Rand index (ARI; Hubert and Arabie, 1985), which corrects the Rand index (Rand, 1971) for chance and has an expected value of 0 under random partitions and an expected value of 1 under perfect agreement (for more information on the interpretation of the ARI, see Steinley, 2004). Table 1.1 shows the obtained ARI per model and the selected variables when using variable selection.

The variable selection techniques give high ARI values with just two variables, with the MMSCN giving slightly better results. The RMMSCN achieves the best ARI in both cases, but it requires a q value of at least 4, which is 75% of d. Since the methods are based on different data sets, model selection indices cannot be used to compare them.

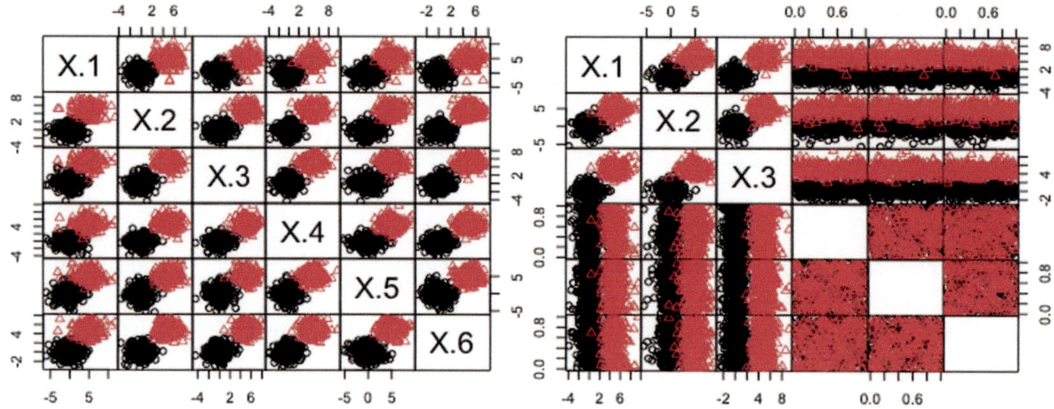

FIGURE 1.1
Toy data sets: On the left, the data is generated pseudo randomly by creating each cluster from an MSCN distribution with varying parameters. On the right, the first three variables are analogous to the data on the left, while the next three are uniformly generated within the range of 0 to 1.

TABLE 1.1
ARI and q or selected variables per method on the two toy data sets

	Data Set 1		**Data Set 2**	
	ARI	**q/Selected Variables**	**ARI**	**q/Selected Variables**
Variable selection MCN	0.94	1 and 5	0.99	2 and 3
Variable selection MMSCN	0.95	1 and 5	1.00	2 and 3
RMMSCN	0.01	$q = 3$	0.01	$q = 3$
RMMSCN	1.00	$q = 4$	1.00	$q = 4$
RMMSCN	1.00	$q = 5$	1.00	$q = 5$

1.4 Pollution Data

The data are available on the EDGARv8.1 air pollutant website (EDGAR Team, 2022). Specifically, we are using the EDGAR Community GHG emissions database (Crippa et al., 2024), which is the outcome of an agreement between the JRC and the International Energy Agency (IEA) to work more closely on gathering and providing consistent, harmonized CO_2 emissions from fossil fuel combustion. The data used refers to the total of the following air pollutants by country for the year 2022. The air pollutants include some ozone precursor gases: carbon monoxide (CO), nitrogen oxides (NOX), non-methane volatile organic compounds (NMVOC), and methane (CH_4); acidifying gases: ammonia (NH_3), nitrogen oxides (NOX), and sulfur dioxide (SO_2); primary particulates: fine particulate matter (PM10 and PM2.5) and carbonaceous speciation (BC, OC); carbon monoxide from the combustion of biofuels (CO_2Bio). We also included data on toxic pollutants: gaseous elemental mercury (Hg_G), gaseous oxidized mercury (Hg_D), and particle-bound mercury (Hg_P) (Muntean et al., 2024). We selected the most recent year available, 2022. In total, we obtained 14 variables with data available for 223 countries.

1.4.1 Pre-Treatment

The total number of variables is 14; four small countries had a lot of missing data: Marshall Islands (10/14), Northern Mariana Islands (11/14), Mayotte (10/14), and Nauru (8/14), and were therefore excluded from the analysis. The remaining sample size is 219 countries, the sample has a few missing values: 1 in CO, BC, and SO_2, 2 in OC, Hg_G, Hg_P, 4 in CO_2BIO, and 6 in Hg_D. Although MCN and MMSCN have been extended to work with values missing at random (Tong and Tortora, 2022, 2024), this extension cannot be easily included in a variable selection framework since it requires at least one observation per row. One non-missing observation per row is not guaranteed when all the univariate and bivariate models are fitted in the variable selection process. Therefore, the algorithms cannot be used, and a different imputation technique is required. Thus, considering the small number of missing values relative to the sample size, 19 in total over 11 observations, we proceeded with mean imputation. Moreover, the data set contains countries of very different sizes in terms of area and population. To mitigate this effect, we applied a log transformation to the data.

1.4.2 Results

We evaluated several indices to determine the optimal number of clusters, including the Akaike Information Criterion (AIC), the BIC, the silhouette score, and the Dunn index. However, these indices did not converge on the same value for G. This is not surprising since they focus on different aspects. The silhouette score and Dunn index are not specific for model-based clustering; they both tend to work better with fewer clusters and are negatively impacted by noise (Arbelaitz et al., 2013). Focusing on indices specific to model-based clustering, i.e., AIC and BIC, AIC remains relatively constant when G varies between 2 and 5. At the same time, BIC exhibits a gradual decline as the number of clusters increases. Consequently, given the discordance of the indices, we opted for a cluster count that enhances the interpretability of the results, setting $G = 4$. Using RMMSCN, the method only identifies two large clusters and two small ones, whereas variable selection with either MCN or MMSCN yields more balanced clusters. Furthermore, given the ease of interpretation provided by the variable selection method for the MCN and MMSCN, we present the results from these models herein. When $G = 4$, MCN and MMSCN select the same variables, allowing model selection indices to be used for comparison. For the MCN, the BIC is -2295.268, and for the MMSCN, the BIC is -2312.814. Among model selection methods, BIC usually tends to privilege MCN due to the lower number of parameters (Punzo and Bagnato, 2021; Punzo and Tortora, 2021; Tortora et al., 2024a); therefore, it should be looked at together with the AIC (Akaike, 1998), $AIC = 2\log(L(\boldsymbol{\vartheta})) - 2\rho$, which penalizes less. The AIC for the MCN is -2190.207, while for the MMSCN, it is -2180.641. Therefore, we selected the MMSCN. Interestingly, an ozone precursor gas and one toxic pollutant were selected, namely OC and Hg_D, confirming the high correlation between the pollutants. Figure 1.2 shows the scatter plot of the selected variables OC and Hg_D. The points are colored based on cluster membership, and those with a superimposed triangle or black cross are outliers. Figure 1.3 shows a world map with countries colored based on cluster membership. Countries in white have not been included in the analysis. Figure 1.4 shows a world map with outliers colored based on cluster membership; the countries in gray, also labeled with a 0, are not outliers. A simultaneous interpretation of the three figures leads to some interesting conclusions.

Cluster 1 (green) consists of 26% of the sample, 53 countries. Those countries have low emissions of OC and Hg_D and are, therefore, the least polluting countries. Among them are three outliers: Guam, Niue, and the Virgin Islands (USA), which pollute

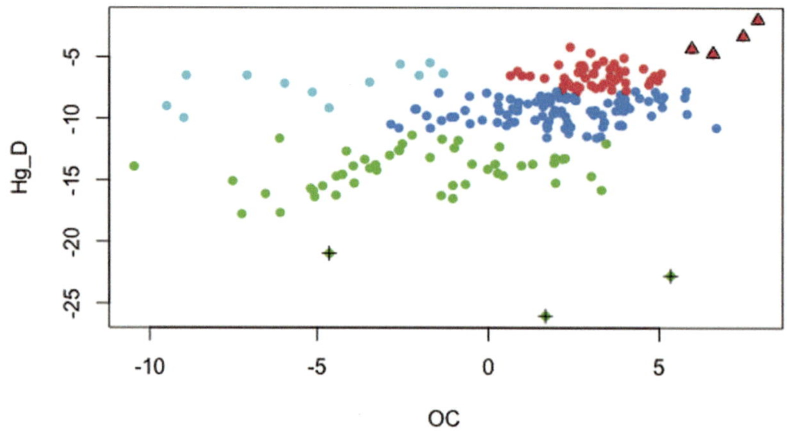

FIGURE 1.2
Scatter plot of the two selected variables OC and Hg_D, the points are colored based on the clusters, the points with superimposed black triangle and black cross are outliers.

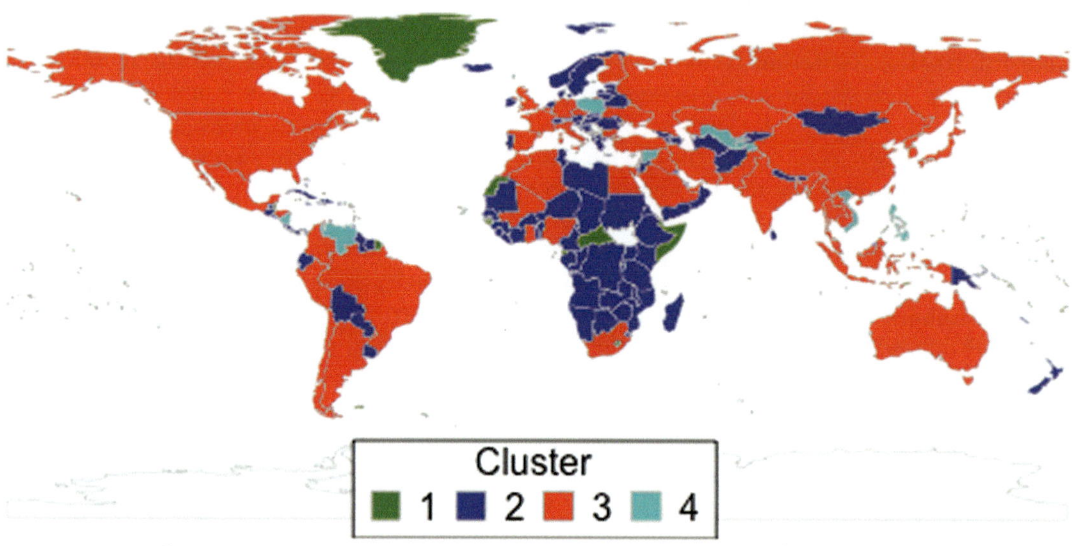

FIGURE 1.3
World map with country colored based on the clustering partition.

even less. The three outliers are small islands, so, unsurprisingly, they are among the least polluting countries, especially when looking at toxic pollutants. This group includes many islands, such as those in the Caribbean and Greenland.

Cluster 2 (blue) consists of 43% of the sample, 101 countries. This cluster can be considered average since both gas emissions are average. The average-polluting countries are mainly located in Africa; several European countries also fall into this group.

Cluster 3 (red) consists of 26% of the sample, 53 countries. Those countries pollute the most. This group includes the USA, Canada, Russia, South Africa, Australia, and several European and South American countries. There are four outliers, namely the countries that pollute the most: Brazil, China, Indonesia, and India.

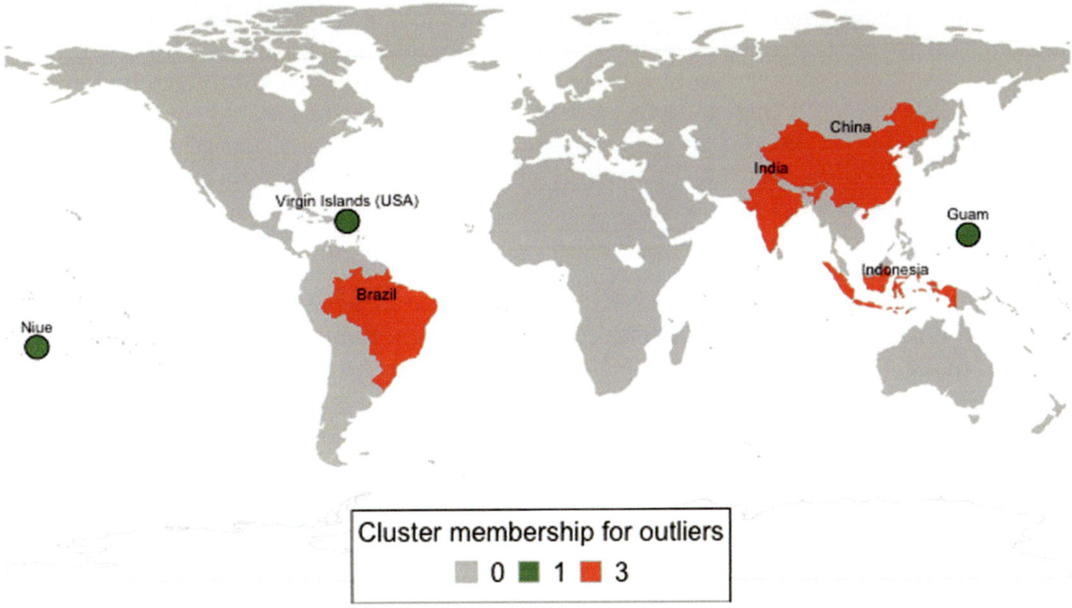

FIGURE 1.4
World map with outliers colored based on the clustering partition.

Cluster 4 (light blue) consists of 5% of the sample, 12 countries. These countries can be described as having high to average levels of pollutants in toxic gases and low levels of pollutants in ozone precursor gases. This group includes American Samoa, the Federated States of Micronesia, Nicaragua, the Philippines, Poland, Singapore, the Syrian Arab Republic, Tajikistan, Uzbekistan, Venezuela, Vietnam, and Wallis and Futuna.

Considering these results in conjunction with the characteristics of the gas yields interesting insights. OC emissions mainly stem from biomass burning (wildfires, deforestation, and agricultural practices) as well as fossil fuel combustion. Biomass burning releases substantial amounts of OC in tropical regions, such as Africa and South America, whereas fossil fuels contribute more in industrialized areas (Salma et al., 2020). Hg emissions, instead, are primarily attributed to coal combustion, small-scale artisanal gold mining, and deforestation. Coal combustion dominates in countries like China and India due to energy production. In contrast, mercury emissions from artisanal gold mining are prevalent in parts of South America, Africa, and Asia (Sundseth et al., 2017).

Therefore, the small countries in group 1 are not surprisingly the lowest polluters since they do not have intensive agricultural or industrial practices. On the other hand, the countries in group 3 are mainly bigger countries that have both agricultural and industrial emissions. The 12 countries in class 4 are different from the others and could be analyzed in a one-on-one case.

1.5 Conclusion

The EDGAR is a comprehensive global database that contains a wealth of valuable information. Utilizing appropriate unsupervised learning techniques is essential for extracting new insights from this data. One of the specific challenges we faced was identifying homogeneous groups of countries while managing highly correlated variables and outliers.

To address this challenge, we employed two existing models designed for clustering and outlier detection: the MCN distributions and the mixture of multiple scaled contaminated normal distributions (MMSCN). By implementing variable reduction and selection techniques, we were able to extend the applicability of these models to high-dimensional data.

Focusing on the selected air pollution metrics from the EDGAR database, we found that variable selection within the MMSCN model performed best in identifying four distinct clusters of countries: Cluster One: Countries with low pollution levels, primarily consisting of island nations. Cluster Two: Countries with average pollution levels, including many from Africa and several in Europe. Cluster Three: Countries with high pollution levels, such as the USA, Canada, Russia, South Africa, and Australia, as well as several countries in Europe and South America. Cluster Four: Countries that have high to average levels of toxic gas pollutants and low levels of pollutants from ozone precursor gases. Additionally, we identified three outliers—Guam, Niue, and the Virgin Islands (USA)—as the least polluting countries and four more outliers—Brazil, China, Indonesia, and India—as the highest polluters.

Funding

This material is based upon work supported by the National Science Foundation under Grant No. 2209974.

Bibliography

Akaike, H. (1998). Information theory and an extension of the maximum likelihood principle. In Parzen, E., Tanabe, K., and Kitagawa, G., editors, *Selected Papers of Hirotugu Akaike*, pages 199–213. Springer New York, New York, NY.

Andrews, J. L. and McNicholas, P. D. (2012). Model-based clustering, classification, and discriminant analysis via mixtures of multivariate t-distributions. *Statistics and Computing*, 22(5):1021–1029.

Arbelaitz, O., Gurrutxaga, I., Muguerza, J., Pérez, J. M., and Perona, I. (2013). An extensive comparative study of cluster validity indices. *Pattern Recognition*, 46(1):243–256.

Bagnato, L., Punzo, A., and Zoia, M. G. (2017). The multivariate leptokurtic-normal distribution and its application in model-based clustering. *Canadian Journal of Statistics*, 45(1):95–119.

Bouveyron, C. and Brunet-Saumard, C. (2014). Model-based clustering of high-dimensional data: A review. *Computational Statistics and Data Analysis*, 71:52–78.

Bouveyron, C., Celeux, G., Murphy, T. B., and Raftery, A. E. (2019). *Model-based Clustering and Classification for Data Science, with Applications in R.* Cambridge Series in Statistical and Probabilistic Mathematics. Cambridge University Press, Cambridge, UK.

Chang, W.-C. (1983). On using principal components before separating a mixture of two multivariate normal distributions. *Journal of the Royal Statistical Society: Series C (Applied Statistics)*, 32(3):267–275.

Crippa, M., Janssens-Maenhout, G., Dentener, F., Guizzardi, D., Sindelarova, K., Muntean, M., Van Dingenen, R., and Granier, C. (2016). Forty years of improvements in European air quality: Regional policy-industry interactions with global impacts. *Atmospheric Chemistry and Physics* 16: 3825–3841, doi:10.5194/acp-16-3825-2016.

Crippa, M., Guizzardi, D., Banja, M., Solazzo, E., Muntean, M., Schaaf, E., Pagani, F., Monforti-Ferrario, F., Olivier, J., Quadrelli, R., et al. (2024). GHG emissions of all world countries. *JRC Science for Policy Report, European Commission, EUR*, JRC138862.

Dempster, A. P., Laird, N. M., and Rubin, D. B. (1977). Maximum likelihood from incomplete data via the EM algorithm. *Journal of the Royal Statistical Society: Series B*, 39(1):1–38.

European Commission, Joint Research Centre (JRC)/Netherlands Environmental Assessment Agency (PBL). (2016). Emission Database for Global Atmospheric Research (EDGAR), release version 4.3.1. http://edgar.jrc.ec.europa.eu/overview.php?v=431.

Everitt, B., Landau, S., Leese, M., and Stahl, D. (2011). *Cluster Analysis*. Wiley Series in Probability and Statistics. John Wiley & Sons, Chichester, UK.

Forbes, F. and Wraith, D. (2014). A new family of multivariate heavy-tailed distributions with variable marginal amounts of tailweights: Application to robust clustering. *Statistics and Computing*, 24(6):971–984.

Ghahramani, Z. and Hinton, G. E. (1997). The EM algorithm for factor analyzers. Technical Report CRG-TR-96-1, University Of Toronto, Toronto.

Gramsch, E., Cereceda-Balic, F., Oyola, P., and Von Baer, D. (2006). Examination of pollution trends in Santiago de Chile with cluster analysis of PM10 and ozone data. *Atmospheric Environment*, 40(28):5464–5475.

Hubert, L. and Arabie, P. (1985). Comparing partitions. *Journal of Classification*, 2(1):193–218.

Kass, R. E. and Raftery, A. E. (1995). Bayes factors. *Journal of the American Statistical Association*, 90(430):773–795.

Kaufman, L. and Rousseeuw, P. J., editors (1990). *Finding Groups in Data: An Introduction to Cluster Analysis*. Wiley Series in Probability and Statistics. John Wiley & Sons, Hoboken, NJ.

Maugis, C., Celeux, G., and Martin-Magniette, M.-L. (2009). Variable selection for clustering with Gaussian mixture models. *Biometrics*, 65(3):701–709.

McLachlan, G. and Peel, D. (2000a). *Finite Mixture Models*. Wiley Series in Probability and Statistics. John Wiley & Sons, Hoboken, NJ.

McLachlan, G. J. and Peel, D. (2000b). Mixtures of factor analyzers. In *Proceedings of the Seventh International Conference on Machine Learning*, pages 599–606. Morgan Kaufmann, San Francisco, CA.

Muntean, M., M. Crippa, D. Guizzardi, F. Pagani, W. Becker, M. Banja, E. Schaaf, and A. Simonati. (2024). EDGAR v8. 1 Global Mercury Emissions. *European Commission, Joint Research*, 660.

Peel, D. and McLachlan, G. J. (2000). Robust mixture modelling using the t distribution. *Statistics and Computing*, 10(4):339–348.

Punzo, A. and Bagnato, L. (2021). The multivariate tail-inflated normal distribution and its application in finance. *Journal of Statistical Computation and Simulation*, 91(1):1–36.

Punzo, A., Blostein, M., and McNicholas., P. D. (2020). High-dimensional unsupervised classification via parsimonious contaminated mixtures. *Pattern Recognition*, 98:107031.

Punzo, A., Mazza, A., and McNicholas, P. D. (2018). ContaminatedMixt: An R package for fitting parsimonious mixtures of multivariate contaminated normal distributions. *Journal of Statistical Software*, 85(10):1–25.

Punzo, A. and McNicholas, P. D. (2016). Parsimonious mixtures of multivariate contaminated normal distributions. *Biometrical Journal*, 58(6):1506–1537.

Punzo, A. and Tortora, C. (2021). Multiple scaled contaminated normal distribution and its application in clustering. *Statistical Modelling*, 21(4):332–358.

R Core Team (2021). *R: A Language and Environment for Statistical Computing*. R Foundation for Statistical Computing, Vienna, Austria.

Raftery, A. E. and Dean, N. (2006). Variable selection for model-based clustering. *Journal of the American Statistical Association*, 101(473):168–178.

Rand, W. M. (1971). Objective criteria for the evaluation of clustering methods. *Journal of the American Statistical Association*, 66(336):846–850.

Salma, I., Vasanits-Zsigrai, A., Machon, A., Varga, T., Major, I., Gergely, V., and Molnár, M. (2020). Fossil fuel combustion, biomass burning and biogenic sources of fine carbonaceous aerosol in the carpathian basin. *Atmospheric Chemistry and Physics*, 20(7):4295–4312.

Steinley, D. (2004). Properties of the Hubert-Arabie adjusted Rand index. *Psychological Methods*, 9(3):386–396.

Sundseth, K., Pacyna, J. M., Pacyna, E. G., Pirrone, N., and Thorne, R. J. (2017). Global sources and pathways of mercury in the context of human health. *International Journal of Environmental Research and Public Health*, 14(1):105.

Taubman, B., Hains, J., Thompson, A., Marufu, L., Doddridge, B., Stehr, J., Piety, C., and Dickerson, R. (2006). Aircraft vertical profiles of trace gas and aerosol pollution over the mid-Atlantic United States: Statistics and meteorological cluster analysis. *Journal of Geophysical Research: Atmospheres*, 111(D10): D10S07.

Tipping, M. E. and Bishop, C. M. (1999). Mixtures of probabilistic principal component analyzers. *Neural Computation*, 11(2):443–482.

Tong, H. and Tortora, C. (2022). Model-based clustering and outlier detection with missing data. *Advances in Data Analysis and Classification*, 16(1):5–30.

Tong, H. and Tortora, C. (2024). Missing values and directional outlier detection in model-based clustering. *Journal of Classification*, 41(3):480–513.

Tortora, C., Franczak, B. C., Bagnato, L., and Punzo, A. (2024a). A Laplace-based model with flexible tail behavior. *Computational Statistics & Data Analysis*, 192:107909.

Tortora, C., McNicholas, P. D., and Browne, R. P. (2016). A mixture of generalized hyperbolic factor analyzers. *Advanced in Data Analysis and Classification*, 10(4):423–440.

Tortora, C., Punzo, A., and Tran, L. (2024b). *MSclust: Multiple-Scaled Clustering.* R package version 1.0.4.

Tran, L. and Tortora, C. (2021). How many clusters are best? investigating model selection in robust clustering. *JSM Proceedings, Statistical Learning and Data Science Section*, pages 1159–1180.

Tukey, J. W. (1960). A survey of sampling from contaminated distributions. In Olkin, I., Ghurye, S. G., Hoeffding, W., Madow, W. G., and Mann, H. B., editors, *Contributions to Probability and Statistics: Essays in Honor of Harold Hotelling*, pages 448–485. Stanford University Press, Stanford, CA.

Wallace, M. L., Buysse, D. J., Germain, A., Hall, M. H., and Iyengar, S. (2018). Variable selection for skewed model-based clustering: Application to the identification of novel sleep phenotypes. *Journal of the American Statistical Association*, 113(521):95–110.

Zhang, J., Zhang, L.-Y., Du, M., Zhang, W., Huang, X., Zhang, Y.-Q., Yang, Y.-Y., Zhang, J.-M., Deng, S.-H., Shen, F., et al. (2016). Indentifying the major air pollutants base on factor and cluster analysis, a case study in 74 Chinese cities. *Atmospheric Environment*, 144:37–46.

2

Determining the Number of Biological Species in the Presence of Spatial Patterns of Differentiation

Gabriele d'Angella and Christian Hennig

Various methods for the estimation of the number of biological species K are introduced that take into account geographic distances between individuals. Such distances can explain genetic differences between the individuals that can appear within the same species. For a given K, the TESS3 approach (Caye et al., 2018) is used for species delimitation. For estimating K, the ΔK statistic Evanno et al. (2005) is adapted to TESS3, and an additional modification (SOD; "second order differences") is also investigated. As, by definition, homogeneous data (i.e., $K = 1$) cannot be recognized in this way, Monte Carlo/parametric bootstrap tests of models for the null hypothesis $K = 1$ against $K > 1$ are introduced. The estimation of the influence of spatial distance within such null models is discussed. Systematic simulations on data sets generated by SLiM Haller and Messer (2023) compare the methods, including the cross-entropy originally proposed by Caye et al. (2016) for use with TESS3. The SOD method shows particularly promising results.

2.1 Introduction

For the delimitation of biological species, empirical data is used to determine which groups of individual organisms constitute different populations of a single species and which constitute different species Rannala and Yang (2020). Species delimitation is of paramount importance in systematics and has practical implications for conservation and biodiversity management (e.g., Pedraza-Marrón et al., 2019). Neglecting geographic information when delimiting species can lead to the misassessment of the genetic structure in the data Frantz et al. (2009). This is more likely to happen in the presence of spatial patterns of genetic differentiation, such as isolation by distance (IBD; Wright, 1943). Clustering methods applied to genetic data without incorporating geographic information might wrongly separate individuals to different species that belong to the same one. This is because genetic dissimilarity increases with geographic separation even within the same species Bradburd et al. (2018). Geographic patterns may violate the usual assumption of random mating within the population (panmixia).

Most existing literature and software for species delimitation do not use geographic information. Tools like MEGA5 Tamura et al. (2011) allow for sequence-based phylogenetic

DOI: 10.1201/9781003564423-2

inference including the reconstruction of phylogenetic trees, but they are not suitable for spatially explicit assignment of the individuals. BPP Yang (2015) delimits species only using genetic information and may overestimate their number in the presence of IBD Mason et al. (2020). There exist statistical frameworks where geographic information is taken into account when clustering individuals, as implemented in the software packages TESS3 Caye et al. (2016, 2018) or conStruct Bradburd et al. (2018). These methodologies require the user to specify a number of ancestral populations K from which the individuals in the sample inherit their genetic information. In many situations, this number can be interpreted as the number of species.

In the current chapter, we investigate the estimation of K. In TESS3, this is done by a cross-entropy criterion based on cross-validation Caye et al. (2016). Our results indicate that this criterion has a strong tendency to overestimate the number of species in the data. We propose instead an adaptation of the ΔK statistic by Evanno et al. (2005) for TESS3. This approach cannot distinguish $K = 1$ from larger K. For doing this, we also discuss null models for homogeneous data from a single species, which we use to define a test for the null hypothesis $K = 1$ against the alternative $K > 1$.

We then employ spatially-explicit genetic simulations based on the SLiM software Haller and Messer (2023) to explore the efficacy of these criteria when using TESS3 for species delimitation purposes.

When discussing the conspecificity criterion adopted for this study, it is important to acknowledge that species delimitation is closely connected to species conceptualization De Queiroz (1998), which can be controversial Hausdorf (2011). Indeed, there exists biological differentiation between populations at different levels: some species are more or less closely related, and there can be heterogeneity also within species. The information in the data does not always suffice to identify whether observed heterogeneity is due to differences between species or occurs within species for reasons that may go beyond the analyzed data. The methodologies developed here do therefore not allow conclusive verdicts regarding species, and additional criteria and information should be considered as well De Queiroz (2007). Nonetheless, we will use the term "species" for the resulting groups of the methodology under investigation here.

In this work, all analyses are based on artificial situations, in which the truth can be controlled; while this allows for objective assessment of the methodology, real data analysis is of course also relevant.

The chapter is arranged as follows: the geographic and genetic data and the dissimilarities based on them are introduced in Section 2.2, whereas the methodologies explored are described in Section 2.3. In particular, after the treatment of the optimization carried out in TESS3, that section includes the definition of the ΔK method, the construction of null models, and the development of a test for checking whether the dataset contains more than one species. In Section 2.4, a simulation study based on SLiM is introduced, with scenarios where the number of species ranges from 1 to 6. Section 2.5 reports the performance of the methodologies introduced in Section 2.3 on data simulated with SLiM. Section 2.6 concludes the chapter.

2.2 Data

Spatially explicit genetic data consists of individuals carrying information about their location and genetic make-up. For each of the n individuals in the dataset, a pair of geographic coordinates and P loci (locations on the DNA) are available. Individual-level co-dominant data, such as single-nucleotide polymorphism or microsatellites, with diploid genotypes is considered Waits and Storfer (2015); this means that each locus contains two alleles. Following Hausdorf and Hennig (2019), alleles are represented by single characters, although elsewhere in the literature more elaborate coding is used (see, e.g., Rousset, 2008). The resulting $n \times (P+2)$ data frame is denoted by

$$\mathbf{Z} = \begin{pmatrix} \mathbf{z}_1 \\ \vdots \\ \mathbf{z}_n \end{pmatrix} = \begin{pmatrix} z_1^{(x)} & z_1^{(y)} & Z_1^1 & \cdots & Z_1^P \\ z_2^{(x)} & z_2^{(y)} & Z_2^1 & \cdots & Z_2^P \\ \vdots & \vdots & \vdots & \ddots & \vdots \\ z_n^{(x)} & z_n^{(y)} & Z_n^1 & \cdots & Z_n^P \end{pmatrix},$$

where each observed locus Z_i^p, $p = 1, \ldots, P$, is a set of characters, like $\{A, B\}$ for heterozygous loci ("BA") or $\{B\}$ for homozygous ones ("BB"); also note that alleles are arranged in lexicographical order in the sets because a meaningful order is not normally observable. Each $\mathbf{z}_i$ is a $1 \times (P+2)$ vector representing the ith individual, $i = 1, \ldots, n$, with geographic coordinates $(z_i^{(x)}, z_i^{(y)})$.

In this study, the Euclidean distance is employed as geographic distance (subscript x):

$$d_x(\mathbf{z}_i, \mathbf{z}_j) = \sqrt{\left(z_i^{(x)} - z_j^{(x)}\right)^2 + \left(z_i^{(y)} - z_j^{(y)}\right)^2}. \tag{2.1}$$

As genetic dissimilarity (subscript y), the shared allele dissimilarity Bowcock et al. (1994) is used:

$$d_y(\mathbf{z}_i, \mathbf{z}_j) = 1 - \frac{1}{2P} \sum_{p=1}^{P} \left| Z_i^p \cap Z_j^p \right| \cdot \left[1 + \mathbb{1}\left(|Z_i^p| + |Z_j^p| = 2 \right) \right], \tag{2.2}$$

where $\mathbb{1}(\text{condition}) = 1$ if the condition is true and zero otherwise (the indicator makes sure that identical homozygous loci contribute 2 to the sum, as do identical heterozygous ones).

In the literature, a logarithmic transformation of the geographic distances is sometimes applied to achieve a more linear relationship between genetic and geographic distances, as done for example in Rousset (1997), Vekemans and Hardy (2004), and Hausdorf and Hennig (2020). Zero geographical distances can occur if two individuals are observed in the same location. Therefore, based on Hausdorf and Hennig (2020), the following transformation is considered:

$$f(d_x(\mathbf{z}_r, \mathbf{z}_c)) = \ln(d_x(\mathbf{z}_r, \mathbf{z}_c) + F_x^{-1}(0.25)), \tag{2.3}$$

where

$$\begin{aligned} F_x^{-1}(q) &= \inf\{d : F_x(d) \geq q\}, \quad \text{with} \\ F_x(d) &= \frac{1}{w} \sum_{i \neq j} \mathbb{1}(d_x(\mathbf{z}_i, \mathbf{z}_j) \leq d) \end{aligned}$$

being the empirical cumulative distribution function of all geographic distances in the dataset, and $w = n(n-1)/2$.

2.3 Methodology

The data presented in the previous section can be fed into the TESS3 program, which describes the genetic information of the individuals as proportionally inherited from K ancestral populations. The resulting proportions can be used to assign individuals to groups, which can provide an indication for species delimitation. In TESS3, K is a hyperparameter to be specified before the execution of the algorithm. To decide what is the best K, a built-in cross-validation scheme is available in TESS3, but elbow methods can also be adopted. Since the latter are only useful when the dataset truly contains more than one species, null models tailored to TESS3 are conceived that can help construct a test for $K = 1$ versus $K > 1$ (homogeneity test). In this section, the optimization in TESS3 and all the mentioned strategies to choose the best K are introduced, before defining a test statistic for a homogeneity test.

2.3.1 TESS3

The TESS3 program provides statistical estimates of ancestry proportions using multilocus genotype data and geographic information Caye et al. (2016). It can be seen as a spatially explicit version of the methods available in the LEA R-package Frichot and François (2015). In admixture models, the alleles that make up the genetic material of the individuals are supposed to be proportionally inherited from K ancestral populations François and Waits (2015). Ancestry coefficients quantify to what extent each individual's genetic material originated from these populations. In practice, individuals in the dataset are assigned to K groups according to their largest estimated ancestry proportion.

Consider a genotypic matrix $\mathbf{X}$, whose entries contain the number of instances of each allele for all loci and all individuals. In this work, simulations will be based on nucleotidic models, where four allelic stata are possible: adenine ("A"), cytosine ("C"), guanine ("G"), and thymine ("T"). Thus, by encoding the genetic information stored in $\mathbf{Z}$, matrix $\mathbf{X}$ will have n rows and $4P$ columns. For instance, if Z_i^1 is "CG", the first four elements on the ith row of $\mathbf{X}$ will be "0110", whereas locus "TT" will correspond to "0002".

Denote by $\widetilde{\mathbf{X}} = (\widetilde{X}_{il})$ the expanded version of $\mathbf{X}$, containing three times its number of columns: its entries are either zero or one and indicate for each allele count in $\mathbf{X}$ (0, 1 or 2 with diploid loci) whether an individual has that many instances of a given allele in a given locus.

By using an allelic status indicator m, equal to 1 for adenine, 2 for cytosine, 3 for guanine, and 4 for thymine, the probability that individual i displays c instances of the mth allelic status at locus p can be expressed as:

$$\mathbb{P}(\widetilde{X}_{il} = c) = \sum_{k=1}^{K} Q_{ik} G_{lk}\,, \qquad c = 0, 1, 2\,, \tag{2.4}$$

where Q_{ik} is the share of genetic information that individual i inherited from the ancestral population k, G_{lk}—with $l = 3(\widetilde{p} + m) + c - 2$—quantifies how often c instances of the mth allelic status are observed at locus p in population k and $\widetilde{p} = 4(p - 1)$.

Estimates of Q_{ik} and G_{lk} (which in line with much literature are not distinguished by notation from the estimands) for every i, k, and l are found by minimizing

$$||\widetilde{\mathbf{X}} - \mathbf{Q}\mathbf{G}^{\intercal}||_F^2 + \frac{1}{2} \sum_{i,j=1}^{n} \omega_{ij} ||\mathbf{Q}_i - \mathbf{Q}_j||^2\,, \tag{2.5}$$

where:

- $\mathbf{Q} = (\mathbf{Q}_1, \ldots, \mathbf{Q}_n)^{\intercal} = (Q_{ik})$ is the $n \times K$ matrix of admixture proportions.
- $\mathbf{G} = (G_{lk})$ is the $12P \times K$ matrix of allele count frequencies.
- $|| \cdot ||_F$ denotes the Frobenius norm and $|| \cdot ||$ the Euclidean norm.

- $\omega_{ij} = \exp\left(-d_x(\mathbf{z}_i, \mathbf{z}_j)^2/(\rho\bar{d}_x)^2\right)$ is the weight associated with individuals i and j, which is large if the individuals are geographically close.

- ρ is a tuning parameter, whose default value is 0.05.

- $\bar{d}_x = w^{-1}\sum_{i\neq j} d_x(\mathbf{z}_i, \mathbf{z}_j)$ is the average geographic distance across all pairs of individuals.

This optimization is subject to the following constraints:

$$\begin{aligned} Q_{ik}, G_{lk} &> 0 \quad \forall i,k,l\,; \\ \sum_{k=1}^{K} Q_{ik} &= 1 \quad \forall i\,; \\ \sum_{c=0,1,2} G_{lk} &= 1 \quad \forall p,m\,. \end{aligned} \tag{2.6}$$

The second addend in (2.5) enforces a geographically-informed regularity condition, such that individuals close in space tend to display similar admixture coefficients. Note that this does not model co-specific individuals that can be located far away, a scenario that will be explored in the simulation study. Two numerical methods were proposed to carry out the optimization in (2.5), see Caye et al. (2018).

In order to select a value for K, the authors suggest adopting a cross-validation scheme, also detailed in Frichot et al. (2014). A share (e.g., 5%) of all entries in $\widetilde{\mathbf{X}}$ is masked (artificially set as missing) and their value is predicted with (2.4). Note that this exclusion is not row-wise, but rather cell-wise. The optimization in (2.5) is carried out using the masked version of $\widetilde{\mathbf{X}}$ and the resulting matrices $\mathbf{Q}$ and $\mathbf{G}$ are employed to predict the probability of each masked genotype. The actual values of the masked entries that were excluded from the optimization are then used to compute a cross-entropy criterion equal to the average negative log-probability of the masked allele counts. The best K corresponds to the lowest value of the cross-validation cross-entropy.

2.3.2 The ΔK Method

In molecular ecology, a widely used statistic for determining K is the ΔK proposed by Evanno et al. (2005) Janes et al. (2017). This statistic is connected with the search for elbows on charts where some clustering error measure is plotted against K. Here, ΔK is adapted to TESS3, using the root mean squared error:

$$\text{RMSE}_K = \sqrt{\left(\frac{\sum_{i=1}^{n}\sum_{l=1}^{12P}\left(\widetilde{X}_{il} - \sum_{k=1}^{K} Q_{ik}G_{lk}\right)^2}{12Pn}\right)}\,, \tag{2.7}$$

which quantifies how accurate the estimation of allele counts for each individual and each locus is when K ancestral populations are assumed. Unlike the cross-validation cross-entropy criterion, which is supposed to hit its minimum at the best K, (2.7) tends to decrease as K grows. Theoretically, given an intermediate configuration with $K < n$, it is always possible to introduce an additional ancestral population and split the individuals in one of the K starting groups so as to better describe their genotypes, up to the point where $K = n$ and each column in $\mathbf{G}$ reports the allele count frequencies observed for each individual (zero RMSE). In practice, due to the numerical optimization in TESS3, a random TESS3 run with $K + 1$ can actually display larger RMSE than one with K groups, but on average, a decreasing trend emerges upon the execution of a sufficiently large number of runs for each K.

Given that measures like the RMSE are expected to decrease with K, the rationale of the elbow search is that adding one more group to the solution brings about stronger improvements when the best K is yet to be reached, while, once it is overcome, additional groups are not that useful anymore. The change in the rate of improvement will result in a flattened trend, generating an elbow. This heuristic dates back at least to Thorndike (1953). In the field of molecular ecology, it was translated into mathematical terms by Evanno et al. (2005). They tailored it to the built-in error measure in the STRUCTURE program Pritchard et al. (2000), the posterior probability of the data for a given K, but here its use is extended to (2.7). The visual inspection of the elbow plot is replaced by the maximization of a function of the second-order rate of change of the average RMSE:

$$\begin{aligned} \mathrm{RMSE}'_K &= \overline{\mathrm{RMSE}}_K - \overline{\mathrm{RMSE}}_{K-1}\,, \\ \Delta K &= \frac{|\mathrm{RMSE}'_{K+1} - \mathrm{RMSE}'_K|}{s(\mathrm{RMSE}_K)}\,, \end{aligned} \tag{2.8}$$

where $\overline{\mathrm{RMSE}}_K$ and $s(\mathrm{RMSE}_K)$ are the mean and standard deviation of the RMSE, respectively, across all TESS3 runs executed for a given K; these runs rely on random initializations. The chosen K is the one that maximizes ΔK. The division by $s(\mathrm{RMSE}_K)$ relates the "elbow" to the variation in the computation of RMSE_K. Evanno et al. (2005) observed that the variance of the posterior probability in STRUCTURE tends to increase with K, and the division by the standard deviation can help in that case. This does not necessarily hold for TESS3. For this reason, as an alternative, we also investigate maximizing the non-standardized second-order differences (SOD),

$$\mathrm{SOD} = |\mathrm{RMSE}'_{K+1} - \mathrm{RMSE}'_K|. \tag{2.9}$$

2.3.3 Null Models

By construction, elbow rules like the ΔK method can never suggest that the best K to be chosen is 1. We test null models modeling $K = 1$ by means of a Monte Carlo or parametric bootstrap test in order to decide whether such a model can fit the data well. The null models aim at capturing all features of the data that cannot be interpreted as information distinguishing species; Huang et al. (2015) review tests of homogeneity null hypotheses in cluster analysis.

When relying upon algorithms like TESS3 for species delimitation, the use of null models is motivated as follows: given the requested number of ancestral populations K, TESS3 will return a clustering configuration with K groups even if the dataset contains only one species. Model-based clustering algorithms like this assume that K ancestral populations are to be found, and will describe the genotypes as a mixture of the contributions of these populations—regardless of species delimitation considerations. Discrepancies between the K co-specific groups may be explained by spatial patterns of differentiation, if present. However, even with panmictic species displaying no geographic segregation, TESS3 might capture noise patterns. Therefore, the objective is to discriminate clustering solutions where distinct species are being delimited from those where the assignment is guided by within-species heterogeneity or just noise.

Ideally, for species delimitation as tackled in this study, datasets simulated from the null model should have the same number of individuals, number of loci, geographic locations, and IBD patterns of the original dataset. Indeed, any strictly geographic grouping is not necessarily associated with species distinctness. Moreover, mimicking IBD patterns in the data is crucial because a dataset containing only one species might display heterogeneity due to spatial patterns: data from a null model neglecting this may then look significantly different from the original dataset, leading to the selection of configurations with $K > 1$ in a setup where the goal is to conclude that $K = 1$. On the other hand, the recovery of these data features should not cause any relevant clustering information to enter the null model; in this case any test relying upon it is likely to have little power when the data truly contains more than one species.

Null models can be used to construct tests for the presence of more than one species in the dataset ($K = 1$ versus $K > 1$), but also to estimate the number of species. If TESS3 is used to cluster individuals, it is crucial to apply it both on the original and the null data, a minimum requirement for the comparability of the resulting test statistics.

In the following, two strategies to construct null models for this applications are delineated.

Uniform null model

One of the simplest ways to construct a null model in this context is to only use the information about the area inhabited by all individuals and the genotypes observed in the dataset. For simulated data, where generic eastings and northings on a two-dimensional map are considered, individuals under this null model can be uniformly sampled from the rectangular area covered by the original ones. Their genetic make-up can be re-created via a random resampling with replacement of the diploid loci observed in the dataset. This null model discards information about any geographic clustering and generates completely panmictic datasets with no grouping whatsoever. It only reproduces the original number of individuals and loci, and merely preserves the range of the original geographical coordinates.

By neglecting information on geographic structure and spatial patterns of differentiation, this null model can be expected to lead to liberal tests of homogeneity (i.e., where $H_0 : K = 1$). For instance, genetic discrepancies between geographically segregated groups from a species with IBD behavior are likely to constitute a deviation from what is expected to be observed under this uniform null model.

Weighted null model

The weighted null model takes into account that genetic exchange is easier between geographically close individuals, which is required to potentially classify populations as homogeneous where there is larger genetic variety in bigger geographic distance. This model is based on resampling individuals at their original locations in the dataset. Based on these positions, spatial weights are computed for each pair of individuals. These weights are used to inform a resampling of the genetic make-up at each position. An appropriate choice of the weights may help to reconstruct the spatial patterns of differentiation in the datasets.

Given the locations of the n individuals, the ith row in the original genotypic matrix $\mathbf{X}$ is resampled based on spatially-weighted allele count frequencies:

$$\boldsymbol{\xi}_i = \frac{\boldsymbol{\omega}_i \mathbf{X}}{2\boldsymbol{\omega}_i \mathbf{1}_n}, \tag{2.10}$$

where $\boldsymbol{\omega}_i = \{\omega_{ij}\}_{j=1,\ldots,n}$ is the $1 \times n$ vector of weights related to individual i and $\mathbf{1}_n$ is a $n \times 1$ vector of 1s. These weights should be large for geographically close pairs of individuals. They can be chosen as in (2.5) for TESS3 optimization. Assuming once again that $\mathbf{X}$ has $4P$ columns, each of the P sub-vectors in $\boldsymbol{\xi}_i$ will contain four elements that sum to one. These can be used as event probabilities in a multinomial experiment. For example, the vector $\{\xi_{3,5}, \xi_{3,6}, \xi_{3,7}, \xi_{3,8}\}$ is used to resample the second locus of the third individual in

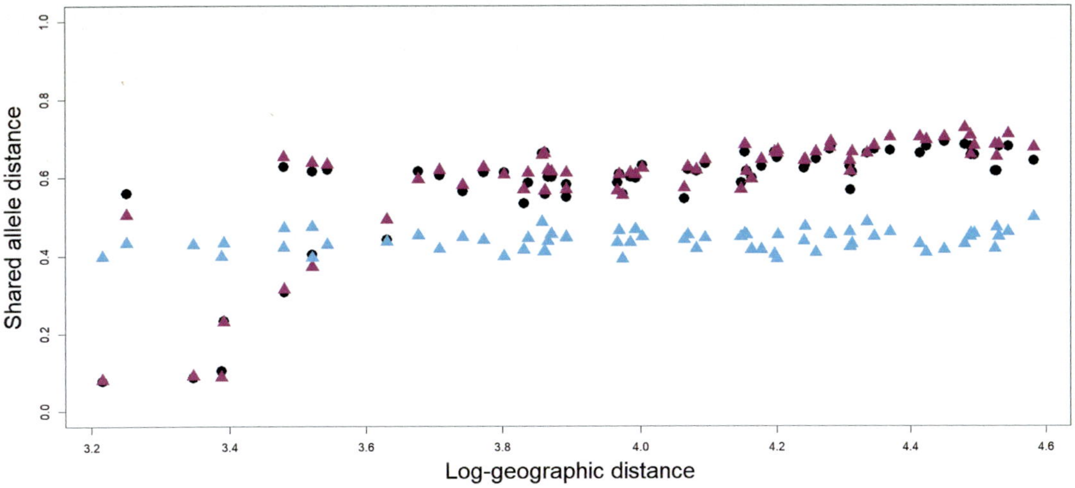

(a)

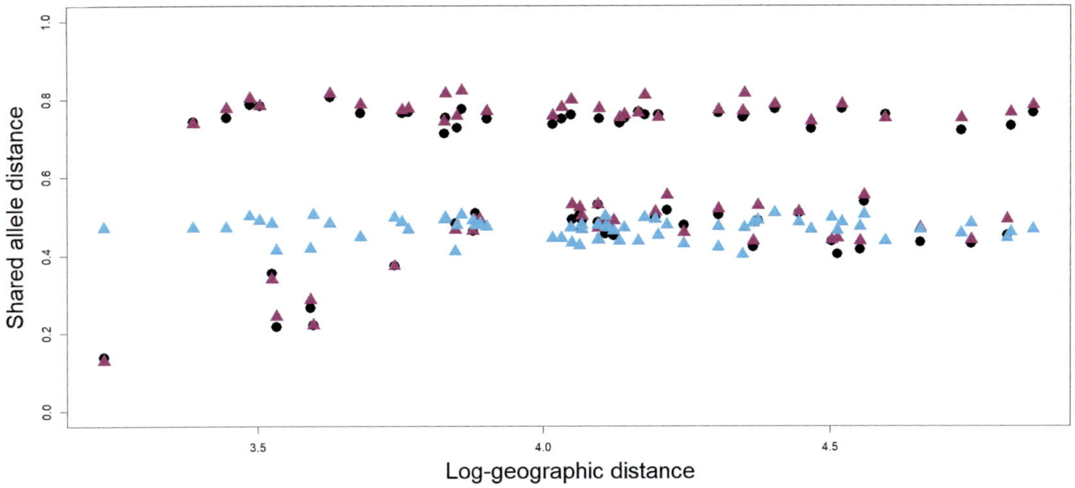

(b)

FIGURE 2.1
Distance-distance plots from simulated datasets with one species (top) and two species (bottm). Both datasets have $n = 12$ and $P = 134$. Black circles are distances based on original data, cyan triangles are from a weighted null model with $\rho = 5$ and magenta triangles are from a weighted null model with $\rho = 0.05$.

the dataset. In general, the draw of the two alleles for individual i at locus p is modeled as a multinomial random variable with probability mass function equal to

$$P(X_{i,\tilde{p}+1} = x_{i,\tilde{p}+1}, \ldots, X_{i,\tilde{p}+4} = x_{i,\tilde{p}+4}) = \frac{2}{x_{i,\tilde{p}+1}! \cdots x_{i,\tilde{p}+4}!} \xi_{i,\tilde{p}+1}^{x_{i,\tilde{p}+1}} \cdots \xi_{i,\tilde{p}+4}^{x_{i,\tilde{p}+4}} \qquad (2.11)$$

when $x_{i,\tilde{p}+1} + x_{i,\tilde{p}+2} + x_{i,\tilde{p}+3} + x_{i,\tilde{p}+4} = 2$ and to 0 otherwise. Recall that $\tilde{p} = 4(p-1)$. This drawing mechanism assumes that loci are independent.

The choice of ρ in the weights has a strong impact on the datasets generated from the model. In Figure 2.1, dissimilarities from datasets obtained from this null model are superimposed on the original dissimilarities from two exemplary datasets (simulated by SLiM, see Section 2.4), one containing one species and the other containing two species:

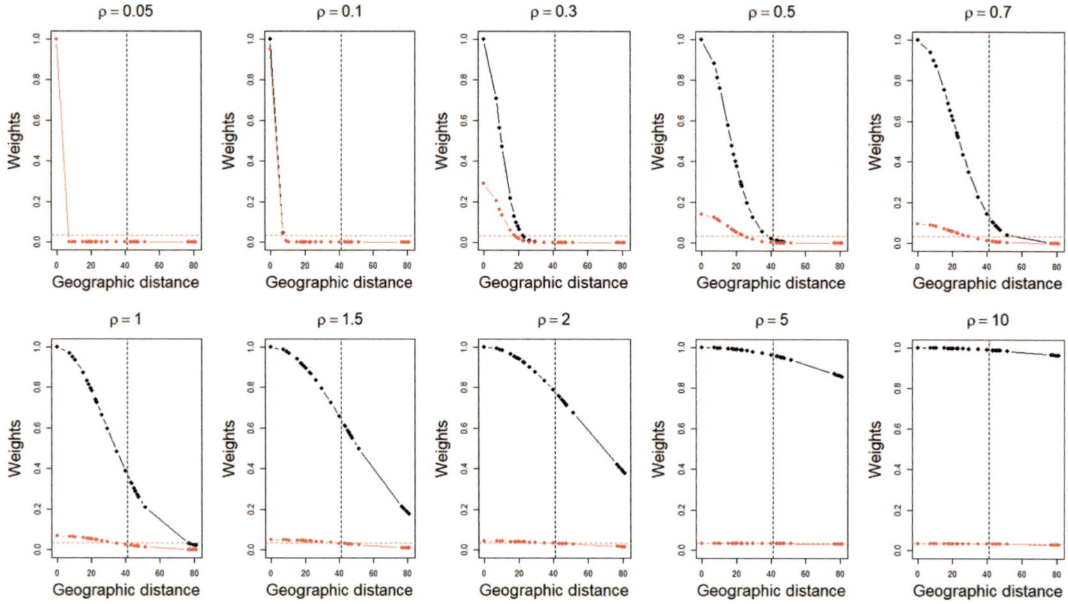

FIGURE 2.2
Weights $\boldsymbol{\omega}_i = \{\omega_{ij}\}_{j=1,\dots,n}$ (black) and their scaled version $\frac{\boldsymbol{\omega}_i}{2\boldsymbol{\omega}_i \mathbf{1}_n}$ (red) against geographic distance, for a random individual i from a simulated dataset with $n = 30$ individuals. Panels by value of parameter ρ. ω_{ij} is defined in (2.5), scaled weights in (2.10). The vertical dotted black line indicates the average geographic distance $\bar{d}_x$, whereas the horizontal dotted red line indicates the uniform weight $1/n$.

- When the parameter $\rho = 0.05$ (its default value in TESS3 optimization), the resulting dataset is very similar to the original one, therefore, it possesses the same IBD pattern and variability of the genetic dissimilarities. However, this comes at the cost of absorbing clustering information, since also in the null dataset it is possible to spot an upward shift in the genetic dissimilarities that perfectly matches species membership (bottom plot). As a consequence, tests based on such null datasets will be very conservative and hardly lead to the rejection of the homogeneity hypothesis, as null datasets will be too similar to the original one.

- When $\rho = 5$, the dissimilarities look more homogeneous (a uniform null model will produce a similar pattern). The potential problem here is that the geographic structure hardly has any influence on the distances anymore and IBD patterns will not be reproduced, so that a test of such a null model may reject it too easily. A compromise between these two choices is required.

The factor ρ is a bandwidth parameter that determines the speed of decrease of weights with growing geographic distance relative to the overall average geographic distance. Figure 2.2 shows how the weights used to resample the loci of a random individual i in a SLiM dataset with $n = 30$ vary with ρ. For instance, an individual found one $\bar{d}_x$ away and another individual two $\bar{d}_x$s away from individual i are given weights approximately equal to 0.37 and 0.02, respectively, when $\rho = 1$; when $\rho = 2$, instead, these weights would be approximately equal to 0.78 and 0.37, respectively. Thus, the closer of the two individuals goes from having a weight ~18 times larger than that of the farther individual to only having twice its weight, approximately. With values as low as 0.05, only the individuals closest to the ith location will have a non-negligible impact on the re-sampling of its genetic material. More precisely, since ω_{ii} is included in $\boldsymbol{\omega}_i$, too, the original genetic make-up at that location will have the largest

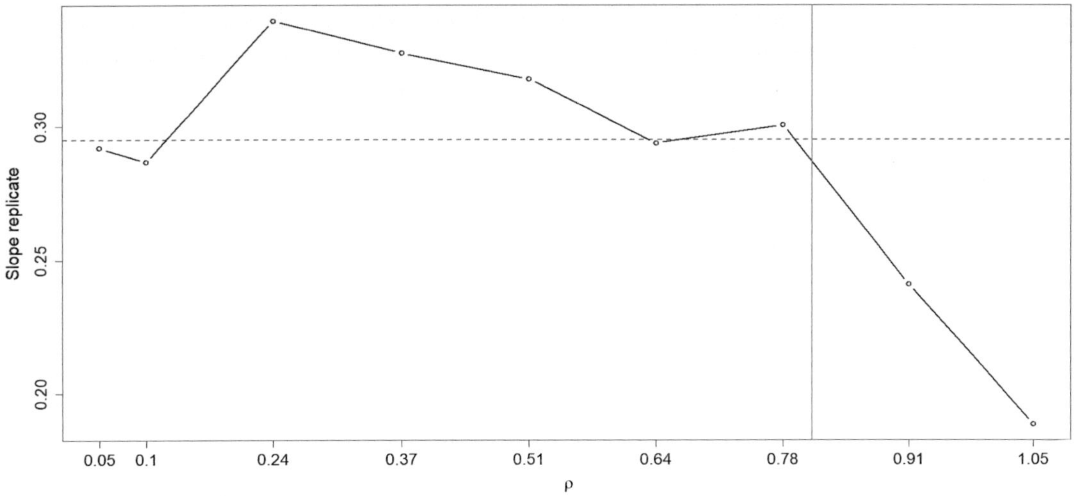

FIGURE 2.3
Illustration of the procedure to find a good value of ρ for the weighted null model. The dotted line is the slope of the regression between genetic and log-transformed geographic dissimilarities in the observed dataset of interest, $\hat{b}_{obs}$. The points are the slopes $\hat{b}$ for datasets generated from the weighted null model with different ρ-values. The blue line is the largest ρ-value so that the resulting $\hat{b}$ fulfills $|\hat{b} - \hat{b}_{obs}| < 0.01$.

weight in its resampling, leading to null datasets very close the original one. One could enforce $\omega_{ii} = 0$, but this does not solve the problem either, because then the ith location is filled with genetic information mostly based on its closest neighbor. As ρ increases, more and more individuals are assigned a positive weight and scaled weights converge to the reciprocal of the sample size n. In practice, with values as large as 5, all individuals contribute almost equally to the resampling of the genetic material at the ith location. This smoothes genetic discrepancies and causes the loss of spatial patterns of differentiation.

An optimal value of ρ between those considered in Figure 2.1 needs to be found that hits a good balance between recovering key IBD patterns and preserving the conspecificity requirement for null datasets. In order to choose ρ in a data-driven way, a possibility is to explore what values of ρ reproduce a key data feature. The slope coefficient estimate $\hat{b}$ in the simple linear regression between genetic and log-transformed geographic dissimilarities (as depicted in Figure 2.1) can be used as proxy to describe the IBD pattern that the null model needs to reproduce.

The relationship between ρ and $\hat{b}$ can be investigated by simulating datasets from the weighted null model over a set of ρ-values. $\hat{b}$ is computed for all these simulated data sets. Linear interpolation is used between the ρ-values. The resulting polygonal chain is used to map any value of ρ within the covered interval to a value of $\hat{b}$. The largest ρ associated to the slope $\hat{b}$ from null data close enough (with distance of at most 0.01, say) to the one computed on the original dataset ($\hat{b}_{obs}$) is then selected for the null model. It is mathematically possible to choose the largest ρ-value so that $\hat{b} = \hat{b}_{obs}$, but we found that due to the conservativity of the resulting test procedure, choosing ρ slightly larger is often better.

In practice, the search for the value of ρ has to be constrained to some interval. When $\rho \to 0$, null datasets become very close to the original one because all individuals $j \neq i$ have approximately zero weight in the resampling of the genetic information at location i. Very small values of ρ will normally reproduce $\hat{b}_{obs}$ well for this reason, but this is not very useful. With growing ρ, our experience is that $\hat{b}$ grows first, but then converges to zero (this is because a uniform model has zero expected slope). This means that there usually is a ρ much larger than 0 or 0.05 that can reproduce $\hat{b}_{obs}$ well, and this can be expected to be a good choice for the null model. The process is illustrated in Figure 2.3. Here, the typical pattern can be seen that the null model with very small ρ delivers $\hat{b} \approx \hat{b}_{obs}$. Then $\hat{b}$ increases, and for large enough ρ it decreases again, reaching $\hat{b} \approx 0$ for very large ρ. Larger ρ-values are preferred because with large ρ the null model produces more variation, which is useful as long as ρ is "realistic" in the sense of leading to an impact of the geography that is similar to what happens in the observed data, i.e., $\hat{b} \approx \hat{b}_{obs}$.

As far as the upper boundary of the search space is concerned, heuristic rules can be adopted by observing how the weights in Figure 2.2 tend to vary with ρ. Since the scaled weights converge to $1/n$ as ρ grows, beyond some value of ρ, null datasets will no longer change the estimated slope much, and one can check for what value of ρ all scaled weights for all individuals are sufficiently close to their limit value $1/n$. We set the upper boundary for the search of ρ, denoted $\rho_{\max}$, so that

$$\left| \frac{\omega_{ij}}{2\boldsymbol{\omega}_i \mathbf{1}_n} - \frac{1}{n} \right| < \frac{2}{n} \quad \forall i, j\,.$$

On simulated data, this criterion was often fulfilled with $\rho < 2$.

Once $\rho_{\max}$ is set, the following algorithm can be adopted to inform weighted null models with a data-driven estimate of parameter ρ, cp. Figure 2.3. Given a dataset to be clustered:

1. obtain $\hat{b}_{obs}$, the OLS estimate of the slope of the regression between genetic and log-transformed geographic distances;
2. simulate datasets based on the weighted null model with $\rho = 0.05$, $\rho = \rho_{\max}$, and, say, eight equally spaced values in between, and interpolate the resulting estimates of the slope;
3. find the largest value of ρ so that the polygonal chain defined on the estimated slopes from the null models with the different ρ-values returns a slope that is at most by 0.01 different from the slope estimated from the observed data (in case such a value does not exist, tolerance is multiplied by 10 and the process repeated until convergence).

This allows us to use a dataset-specific value of ρ. In the literature, tests based on simulation from a null model with an estimated parameter are usually called "parametric bootstrap".

2.3.4 Test Statistic

The test statistic used for the Monte Carlo/parametric bootstrap test is obtained from the RMSE values from TESS3 which is a modified version of the second-order differences connected to the ΔK method.

Denote with $\text{RMSE}_{K,b}$ the minimum RMSE returned by TESS3 runs with K groups on the bth dataset generated under the null model, with $b = 1, \ldots, B$. TESS3 is defined by minimizing a penalized RMSE, see (2.5), and therefore, we use the minimum RMSE here, different from the average in ΔK. Define

$$\begin{aligned}
\text{RMSE}'_{K,b} &= \text{RMSE}_{K,b} - \text{RMSE}_{K-1,b}\,, \\
\text{RMSE}''_{K,b} &= \text{RMSE}'_{K+1,b} - \text{RMSE}'_{K,b}\,, \\
\overline{\text{RMSE}}''_K &= \frac{\sum_{b=1}^{B} \text{RMSE}''_{K,b}}{B}\,, \\
s(\text{RMSE}''_K) &= \sqrt{\frac{\sum_{b=1}^{B} (\text{RMSE}''_{K,b} - \overline{\text{RMSE}}''_K)^2}{(B-1)}}\,.
\end{aligned} \tag{2.12}$$

Denote with RMSE''_K the analogue of $\text{RMSE}''_{K,b}$ based on the minimum RMSE values across TESS3 runs with K groups on the original dataset. With B sufficiently large, RMSE''_K can be compared with the quantiles of its empirical distribution under the null hypothesis of interest here, which is that the null model holds. It can be rejected when RMSE''_K computed on the observed data is larger than, e.g., 95% of its values on data from the null model. The one-sided alternative is justified by the fact that positive RMSE$''$s are expected when an elbow occurs.

With lower values of B, other options for significance assessment might be preferable. In general, for any real-valued random variable Y and for $t > 0$, the Chebyshev-Cantelli inequality prescribes that $\mathbb{P}(Y - \mathbb{E}(Y) \geq t) \leq \mathbb{V}(Y)/(\mathbb{V}(Y) + t^2)$ (see, e.g., Boucheron et al., 2013). If, given K, the expected value $\mathbb{E}$ and variance $\mathbb{V}$ of the RMSE$''$s under the null model are estimated with $\overline{\text{RMSE}}''_K$ and $s(\text{RMSE}''_K)^2$, respectively, this result can provide an indication for the upper boundary of the probability to observe deviations equal to $\text{RMSE}''_K - \overline{\text{RMSE}}''_K$ or larger. A rejection of the null hypothesis can follow if this upper boundary is lower than some threshold, e.g., 5%. Anyway, given that the true moments of the statistic remain unknown, the indication provided by this inequality is not guaranteed to be accurate.

Regardless of how the tail probability is computed in the test for every K, since $K_{\max}-2$ tests are carried out (where $K_{\max}$ is the largest K for which TESS3 solutions are obtained), multiple testing issues can arise. We will apply the Bonferroni correction, see Section 2.6 for a discussion. For a test level of 5%, the Bonferroni corrected level of $0.05/(K_{\max} - 2)$ will be used. If no tail probability falls below this threshold across the $K_{\max} - 2$ tests, there is not enough evidence to reject the hypothesis that $K = 1$, assuming that the best K is not larger than $K_{\max}$.

To recapitulate, each RMSE''_K, with $K = 2, \ldots, K_{\max} - 1$, obtained on the original data will be compared with its null replicates using the Chebyshev-Cantelli inequality. Based on quantities defined in (2.12), the tail probability

$$\frac{s(\text{RMSE}''_K)^2}{s(\text{RMSE}''_K)^2 + (\text{RMSE}''_K - \overline{\text{RMSE}}''_K)^2} \tag{2.13}$$

is computed for $K = 2, \ldots K_{\max} - 1$ and compared with a Bonferroni-corrected threshold equal to $0.05/(K_{\max}-2)$. If any of the $K_{\max}-2$ tail probabilities is lower than this threshold, the null hypothesis that $K = 1$ is rejected. That is, RMSE$''$ is too large to be compatible with the null hypothesis that only one species is represented in the dataset.

2.4 Simulations

The methodologies discussed in this study will be assessed based on datasets simulated via the SLiM software Haller and Messer (2023). This is a simulation package that allows to generate the genetic material of the individuals in a spatially explicit framework. Its flexibility helps to investigate the impact of a plethora of factors on species delimitation routines, including sample size, number of available loci P, and spatial patterns of differentiation. The technical details and code related to the simulations employed here can be found in d'Angella and Hennig (2026).

Eight data scenarios were conceived. For each of them, a version with quasi-panmictic species and one with species showing the same IBD behavior were generated. The genetic and geographic dissimilarities between quasi-panmictic individuals display no relationship, whereas the genetic dissimilarities between individuals isolated by distance grow with geographic separation. Additionally, three sample size options were explored. Each combination of scenario, sample size and IBD behavior was simulated 100 times, returning geographic positions of the individuals and 1000 diploid loci with four possible allelic stata (the four aforementioned nucleobases). A random selection of 134 of these loci would then be recorded, and divided into a subset with 5, one with 40 and one with 89 loci, which allows to investigate the impact of data richness on methods' performance.

In Figure 2.4, the locations of the individuals from exemplary datasets are shown for each combination of scenario and IBD behavior. Eight different SLiM scripts were used to create:

- a scenario with only one species covering the whole map;
- a scenario with two species inhabiting the whole map, one having twice the number of members of the other;
- a scenario with two unequally sized species inhabiting separate areas of the map;
- a scenario with two distinct species inhabiting the whole map, but where one of them has two subpopulations, each having as many members as those from the other species;
- a scenario with three equally sized species inhabiting the whole map;
- a scenario with three equally sized species segregated in three disjoint areas of the map;
- a scenario with six species inhabiting the whole map, where three of them have a number of individuals twice as big as the number of individuals from the other three species;
- a scenario with six species as the previous one, but where now each species inhabits its exclusive niche on the map.

Scenarios with three or fewer species were simulated with a total number of individuals equal to either 12, 30, or 90, while for the two scenarios with six species, there were either 36, 90, or 270 individuals. These choices for n and P are by and large aligned with the scale observed in empirical and simulation studies in the field—see, e.g., Aurelle and Ledoux (2013); Janes et al. (2017); Wang (2017).

As can be seen in Figure 2.4, individuals from quasi-panmictic species tended to be geographically close, even in scenarios where they were allowed to populate the whole map. On the contrary, individuals from species isolated by distance could well be separated by very large geographic distances. As a result, in absence of spatial patterns of differentiation, the amount of geographic segregation in scenarios with overlapping species often ended up being not that different from that observed with separate species, because distinct species would occasionally populate disjoint areas of the map. This was most evident in scenarios with two species and less impactful in scenarios with six species. In these simulations all species from a dataset exhibited the same IBD behavior.

All methodological assessments discussed in this study will be based on the data from these eight SLiM scenarios.

2.5 Results

In the following, the performance of the methodologies treated in this study is evaluated using the simulated data presented in the previous section.

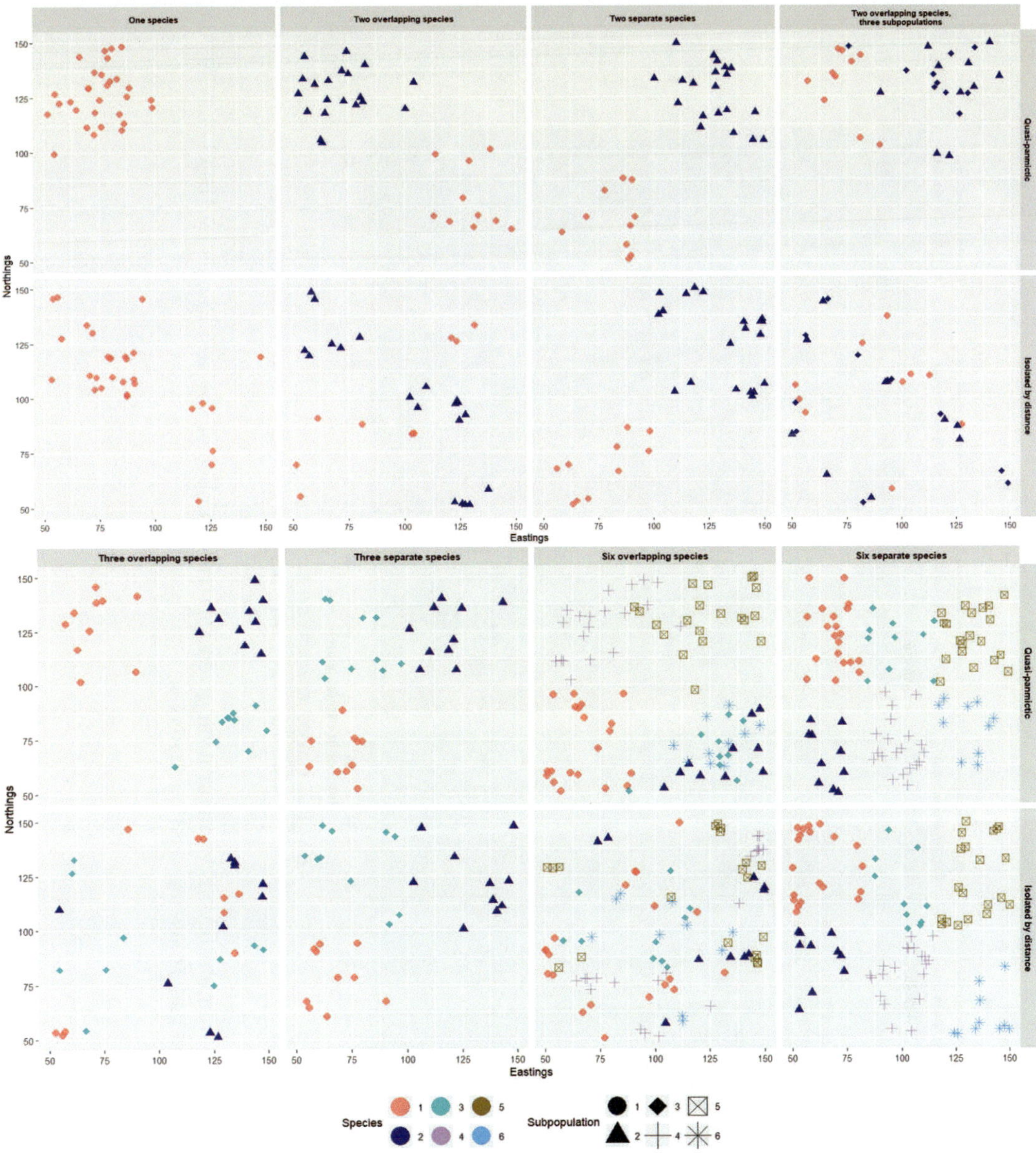

FIGURE 2.4
Maps of the geographic positions of individuals from exemplary SLiM datasets with average sample sizes. Panels by scenario and IBD behavior. Colors map species membership, while shapes map subpopulations: the latter only differ from species membership in the scenario where one of the two species has two subpopulations. All visual cues are only comparable within a scenario, as the individuals from different scenarios are not related.

Species retrieval

Although this contribution focuses on investigating strategies to estimate the number of species in a dataset, the wide simulation study presented in the previous section lends itself to the assessment of the accuracy of TESS3 when it comes to the allocation of the individuals to a known number of species. In Figure 2.5, the average adjusted Rand index (ARI; Hubert

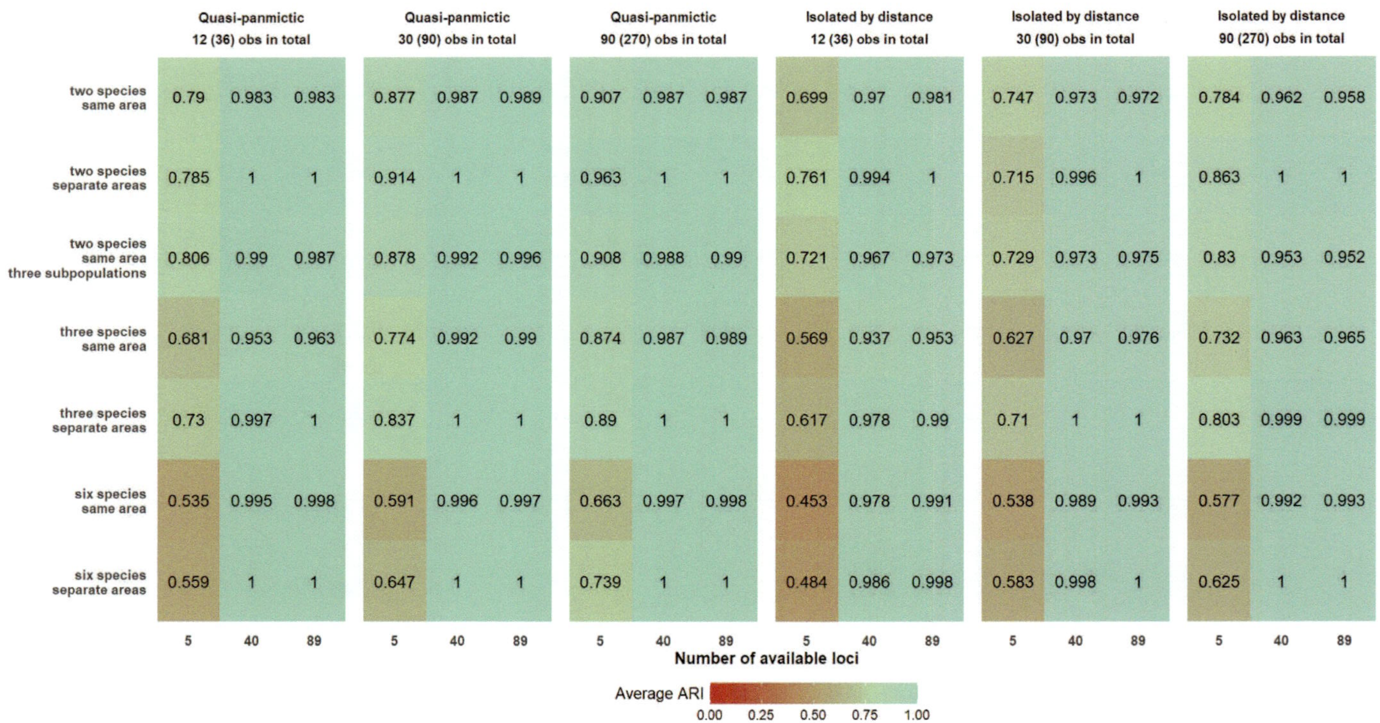

FIGURE 2.5
Average ARI over 100 datasets of the TESS3 minimum RMSE configuration given the true number of species. Panels by IBD behavior and sample size n; scenarios and number of available loci P are mapped on the axes.

and Arabie, 1985) scored by the TESS3 solutions with known K over 100 datasets from the same scenario is reported. This was obtained for all combinations of scenario, sample size, value of P, and IBD behavior explored in SLiM simulations. TESS3 was quite accurate at allocating individuals to the species when their true number was known, as long as the genetic information available was sufficient ($P \geq 40$). Recall that TESS3 imposes a spatial prior on the admixture coefficients, implying that geographically close individuals are more likely to be assigned to the same species. For this reason, the average ARI was larger in scenarios with separate species with respect to scenarios with species inhabiting the same area. Analogously, larger ARIs were scored when grouping quasi-panmictic species with respect to isolated by distance ones. As expected, larger sample sizes also led to better results, whereas the presence of a larger number of species made the allocation more difficult.

Cross-entropy

The cross-entropy criterion used in Caye et al. (2018) for choosing K was found to severely overestimate the true number of species K for the SLiM datasets presented in the previous section. As it can be seen in Figure 2.6, excluding SLiM scenarios where only 5 loci were available, the most common selected value for K according to this criterion was the maximum explored one, regardless of the true number of species in the dataset. The maximum K explored with the algorithm, K_{max}, changed according to the sample size: it was equal to 6 when $n = 12$, to 10 when $n \in \{30, 36\}$, and to 20 when $n \in \{90, 270\}$. These findings suggest that this criterion might not be reliable when using TESS3 for species delimitation, regardless of the presence of isolation by distance patterns.

ΔK and SOD

Figure 2.7 helps assess the efficacy of the ΔK rule in recovering the number of species in the SLiM datasets with more than one species along with SOD, see (2.9).

In scenarios with two or three species, these heuristic rules showed almost perfect results, provided that at least 40 loci were available. In these SLiM scenarios, SOD did slightly worse than ΔK. However, the overall performance strongly deteriorated in scenarios with six species, where SOD was much better than ΔK. Maximizing SOD led to success rates equal or larger than those obtained using ΔK in 100 out of 126 setups shown in Figure 2.7. In addition, when ΔK's success rate was lower, its gap to SODs' rate was on average four times larger than that observed in the cases where ΔK's rates were better.

All in all, these simulations suggest that it might be better to rely upon SOD than upon ΔK when using TESS3 to determine the best value for K.

As far as other data features are concerned, larger sample sizes and longer available genetic sequences helped to improve the accuracy of the methods, in line with the conclusions in Evanno et al. (2005). The presence of within-species IBD patterns, instead, was associated with worse results.

Null model tests

Out of the methods considered so far, the cross-entropy criterion provides an indication for whether the dataset is homogeneous, i.e., there is only one species. However, empirically this criterion proved to be extremely liberal as a test for homogeneity.

The following analysis, regarding the uniform and weighted null models from Section 2.3.4, only considers datasets with $P = 40$. Four null models were employed: a uniform null model; two weighted null models with $\rho = 0.05$ and $\rho = 1$; and a weighted null model where the value of ρ is estimated for each specific dataset with the algorithm described in Section 2.3.3. From each null model, $B = 30$ null datasets were generated in order to simulate

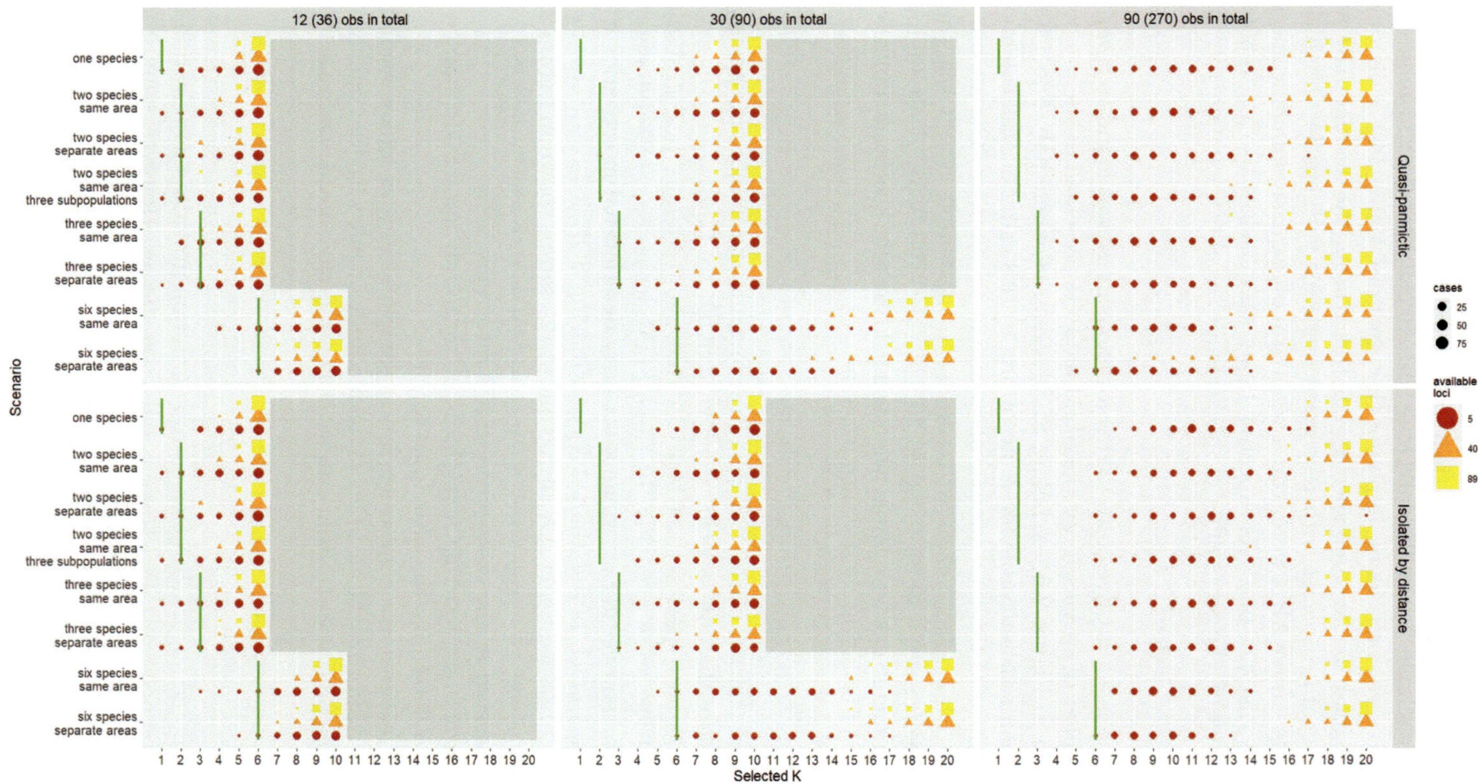

FIGURE 2.6
Counts of times each K was selected by the cross-entropy criterion in TESS3 (symbol size proportional to count). Panels by sample size and IBD behavior, color, and shape by number of loci. The green line indicates the true K for the specific scenario; a gray rectangle covers unexplored values of K given the sample size. Cross-entropy based on 20 runs with 5% masked genotypes.

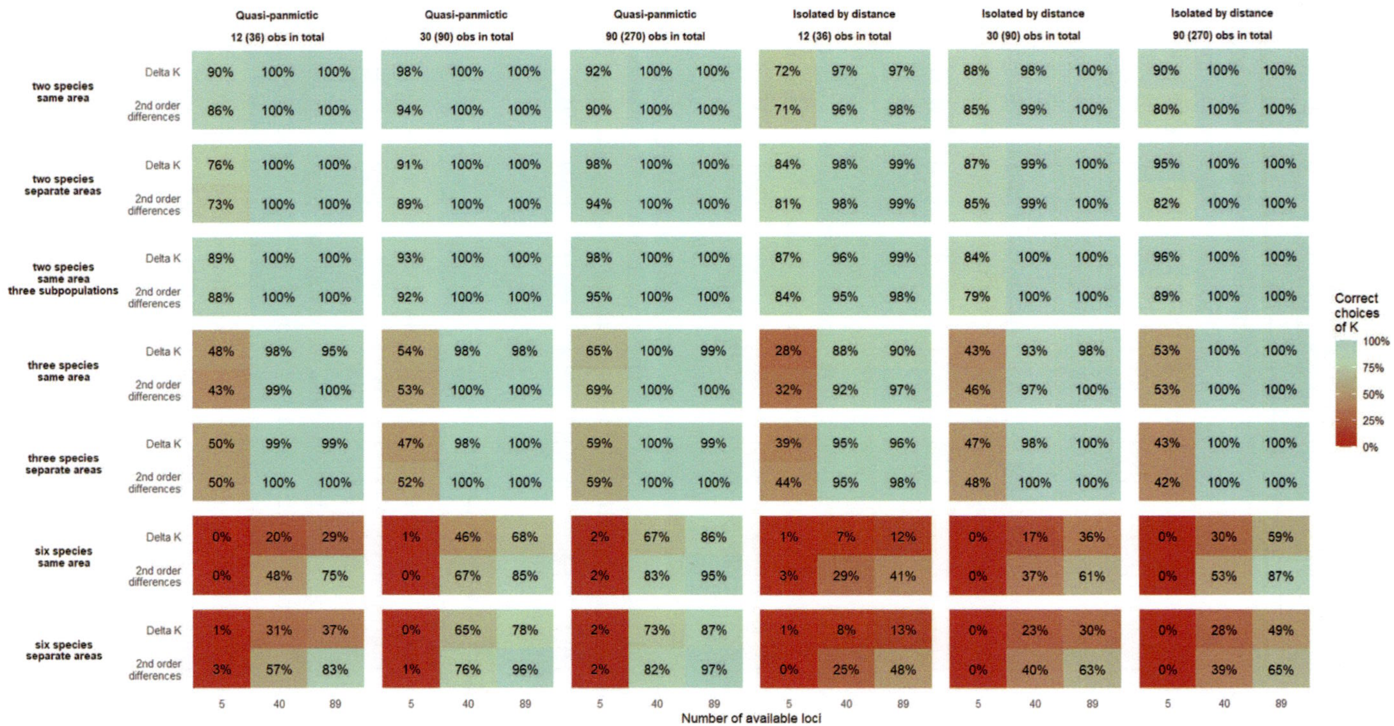

FIGURE 2.7
Proportion of datasets for which the heuristics to choose K selected the number of species. Each tile of the heatmap refers to a combination of heuristic rule and number of available loci. Panels by scenario, IBD pattern, and sample size. The number of TESS3 runs for each value of K was 20. Maximum K per scenario chosen as in Figure 2.6.

the null distribution of the test statistic - a choice informed by reasons of computing time. Whenever TESS3 was applied to the original or to a null dataset, 10 runs for each K were executed and the minimum RMSE across them was recorded. The value of $K_{\max}$ for a given combination of scenario and sample size was set as in Figure 2.6. All tests were carried out at a 5% significance level.

In Figure 2.8, the number of rejections of the hypothesis that the dataset contains only one species according to the test described in Section 2.3.4 is reported for all investigated combinations of null model, SLiM scenario, IBD behavior, and sample size.

Testing the uniform null model using RMSE$''$ (green line) for deciding whether $K = 1$ did surprisingly well in most scenarios, with excellent performance when quasi-panmictic species were involved. However, too many rejections occurred in scenarios with one species displaying IBD behavior. This is because uniform null datasets show no association between genetic and geographic distances. Single species showing heterogeneity due to spatial patterns of differentiation thus constituted an anomaly with respect to what was generated under the null.

As regards the weighted null model, setting $\rho = 0.05$ (pink line) implied that the test for homogeneity had no power in scenarios with true $K > 1$. The null datasets generated with this value of ρ were too similar to the original ones, therefore the observed value of the test statistic could hardly deviate from what reproduced under the null model, yielding extremely conservative tests.

A better compromise between type I error rate and power was achieved using the adaptive choice of ρ (blue line), which led to decent power in scenarios where the species inhabited the same geographical area, with n sufficiently large. Given the value of K, going from scenarios with overlapping species to scenarios with separated ones, larger regression slopes were targeted for the search of ρ; this often caused the selected ρ to be smaller, jeopardizing the power of the test. Across all scenarios, the vast majority of the selected values of ρ lay below 1, explaining why the best overall performance was recorded when ρ was fixed to 1 (brown line). Indeed, this had negligible impact on the type I error rate, while ensuring more power especially in scenarios with three species or three subpopulations. However, from Figure 2.8 it is apparent how none of the explored ρs was very successful in scenarios with two species inhabiting separate areas.

The fact that empirical power was at a minimum with true $K = 2$, grew sharply with $K = 3$ and deteriorated again with six species might be due to the interplay between geographic separation and sample size. As it is visible in Figure 2.4, in scenarios with two separate species the ratio between between-group geographic distances and within-group ones was usually larger than that observed in scenarios with three species: in this situation, $\rho = 1$ was still not sufficiently large in order for the resampling delineated in Section 2.3.3 to smooth the genetic differences between the species, allowing for clustering information to enter the null model. This aligns with the lower power observed in the scenario with two quasi-panmictic species inhabiting the same area compared to IBD species: also in this case they often ended up covering disjoint areas of the map. The presence of a third species located in a somewhat intermediate position probably helped reshuffle the genetic material, given ρ, increasing the power of the test.

While this would be the case also in scenarios with six species, a lower power was observed probably because in these scenarios the trend in the RMSE values decreases more steadily over K, making the presence of elbows less evident. Put differently, the larger the true K, the more similar is the trend in the RMSE values to what is generated by a uniform null model, which explains also the lower power testing uniform null models in these scenarios. Having more individuals (larger n) helped alleviate this issue.

All in all, these results suggest that no fixed value of ρ may be able to ensure the best type I error and power performance of the test for homogeneity introduced in Section 2.3.4.

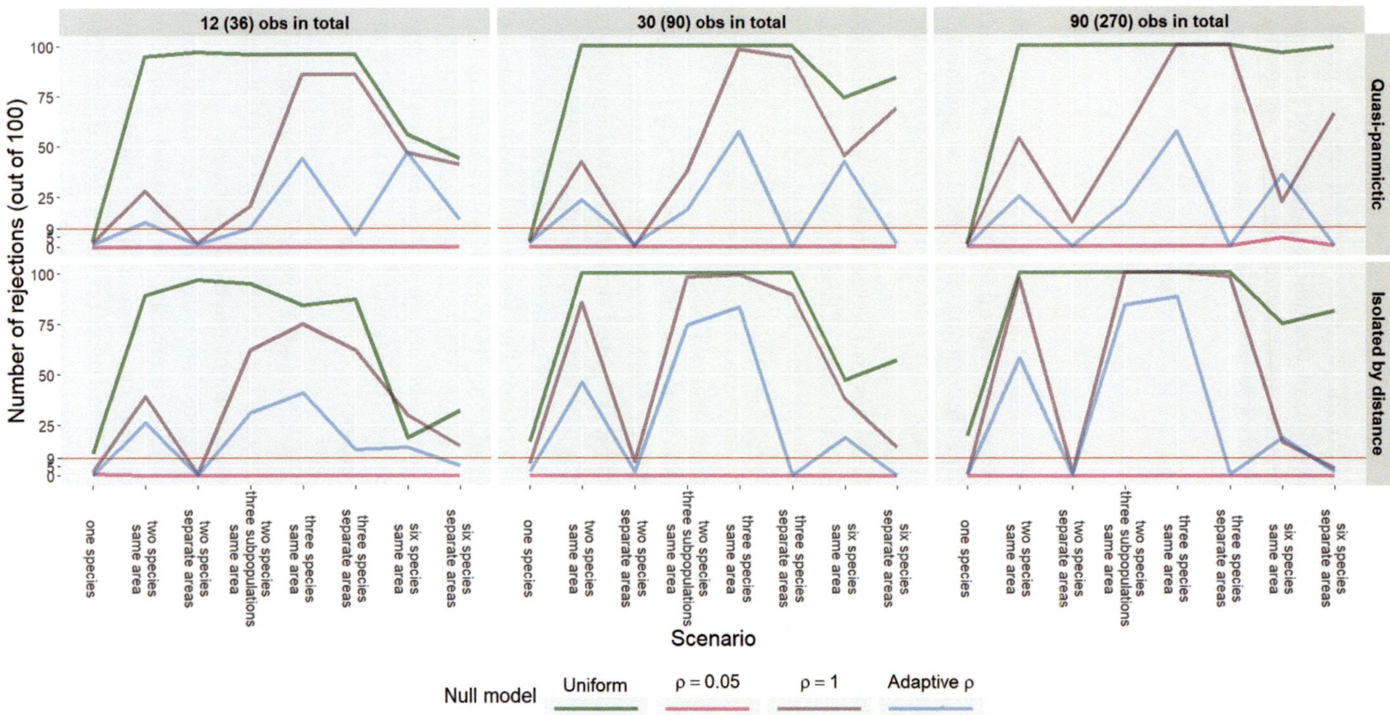

FIGURE 2.8

Number of rejections (out of 100) of the null hypothesis that $K = 1$ tested on SLiM data. Colors by null model employed. Panels by total sample size n and IBD behavior. All datasets with $P = 40$. A red horizontal line is superimposed at 9 rejections because $\mathbb{P}(Y \geq 9) < 0.05$, where Y is distributed as a Binomial RV with size 100 and success probability 0.05.

Furthermore, the adaptive choice of ρ might be improved by targeting a different statistic than the slope between genetic distance and log-transformed geographic one. Indeed, this target was misleading in scenarios with quasi-panmictic separate species, where the IBD behavior within the species was often overestimated, leading to smaller values for the adaptively chosen ρ.

Despite these shortcomings, this test can provide some indication on the presence of more than one species in the dataset: upon rejection of the null hypothesis of homogeneity, statistics like those discussed in Figure 2.7 can be used to get an estimate of the most supported value of K in the dataset.

2.6 Discussion

In this work, the performance of methods to determine the number of species in a dataset containing geographic and genetic information at an individual level was assessed via a broad simulation study. All techniques revolve around the output of TESS3, a spatial ancestry estimation program that can assign the individuals to K ancestral populations Caye et al. (2018).

Given the correct number of species, TESS3 showed rather accurate allocations provided sufficient genetic information was available. On SLiM data, TESS3's built-in cross-validation scheme was found to systematically overestimate the true number of species. More reliable estimates were returned by the ΔK adapted to the TESS3 root mean squared error measure. Simulations show how it might be preferable to only maximize its numerator (second-order difference) to determine the best value of K when using TESS3.

In order to deal with the impossibility to determine whether the dataset is homogeneous (i.e., it contains only one species) using ΔK, Monte Carlo/parametric bootstrap tests were applied to RMSE$''$ from TESS3. A uniform and a weighted null model were proposed, the latter endowed with an algorithm for the selection of a critical hyperparameter, ρ. The test based on the uniform null model showed excellent performance in scenarios where there was no isolation by distance, but was too liberal otherwise. The test based on the weighted null model with $\rho = 1$ turned out to be more powerful than its version with the dataset-specific estimation of ρ, while keeping its same type I error. While achieving rather high power in scenarios with three species, this test did not have acceptable power in the simulated scenarios with two species, which is disappointing. It suggests that no fixed value of ρ fits all datasets, and that the algorithm for the adaptive choice of ρ leaves something to be desired.

Despite these shortcomings, the tests discussed here still provide some guidance for the identification of the number of species in a dataset where spatial patterns of differentiation are present and can be used together with the aforementioned elbow rules.

For further research, other assignment methods, such as conStruct Bradburd et al. (2018), can be evaluated through simulations and potentially be adapted along the lines explored here. Experiments with real data with an agreed number of species are also desirable. A major issue is the bad power of the tests based on the weighted null model in some situations. There is some scope to run the tests in a somewhat less conservative way (conservativity is caused by use of the Bonferroni correction among other things), but this needs to be done in such a way that the type I error behavior is still satisfactory. In our study, the Bonferroni method helped in keeping the type I error rate under control. The Holm-Bonferroni correction Holm (1979) does not make a difference here, as only the smallest p-value is relevant. The method by Benjamini and Hochberg (1995) may lead to too many rejections but would be worth a try. We tried the strategy proposed by Hennig and Lin (2015), but it did not achieve a satisfactory type I error rate.

Bibliography

Aurelle, D. and Ledoux, J.-B. (2013). Interplay between isolation by distance and genetic clusters in the red coral corallium rubrum: insights from simulated and empirical data. *Conservation Genetics*, 14(3):705–716.

Benjamini, Y. and Hochberg, Y. (1995). Controlling the false discovery rate: a practical and powerful approach to multiple testing. *Journal of the Royal Statistical Society. Series B (Methodological)*, 57(1):289–300.

Boucheron, S., Lugosi, G., and Massart, P. (2013). *Concentration inequalities: a nonasymptotic theory of independence*. OUP Oxford.

Bowcock, A. M., Ruiz-Linares, A., Tomfohrde, J., Minch, E., Kidd, J. R., and Cavalli-Sforza, L. L. (1994). High resolution of human evolutionary trees with polymorphic microsatellites. *Nature*, 368(6470):455–457.

Bradburd, G. S., Coop, G. M., and Ralph, P. L. (2018). Inferring continuous and discrete population genetic structure across space. *Genetics*, 210(1):33–52.

Caye, K., Deist, T. M., Martins, H., Michel, O., and François, O. (2016). Tess3: fast inference of spatial population structure and genome scans for selection. *Molecular Ecology Resources*, 16(2):540–548.

Caye, K., Jay, F., Michel, O., and François, O. (2018). Fast inference of individual admixture coefficients using geographic data. *The Annals of Applied Statistics*, 12(1):586–608.

d'Angella, G., Hennig, C. (2026). Approaches to biological species delimitation based on genetic and spatial dissimilarity. *Adv Data Anal Classif.* https://doi.org/10.1007/s11634-026-00669-6

De Queiroz, K. (1998). The general lineage concept of species, species criteria, and the process of speciation. *Endless Forms: Species and Speciation.* 57–75. https://doi.org/10.1093/oso/9780195109009.003.0005

De Queiroz, K. (2007). Species concepts and species delimitation. *Systematic Biology*, 56(6):879–886.

Evanno, G., Regnaut, S., and Goudet, J. (2005). Detecting the number of clusters of individuals using the software structure: a simulation study. *Molecular Ecology*, 14(8):2611–2620.

Frantz, A. C., Cellina, S., Krier, A., Schley, L., and Burke, T. (2009). Using spatial bayesian methods to determine the genetic structure of a continuously distributed population: clusters or isolation by distance? *Journal of Applied Ecology*, 46(2):493–505.

François, O. and Waits, L. P. (2015). *Clustering and assignment methods in landscape genetics*, chapter 7, pages 114–128. John Wiley & Sons, Ltd.

Frichot, E. and François, O. (2015). Lea: An r package for landscape and ecological association studies. *Methods in Ecology and Evolution*, 6(8):925–929.

Frichot, E., Mathieu, F., Trouillon, T., Bouchard, G., and François, O. (2014). Fast and efficient estimation of individual ancestry coefficients. *Genetics*, 196(4):973–983.

Haller, B. C. and Messer, P. W. (2023). Slim 4: multispecies eco-evolutionary modeling. *The American Naturalist*, 201(5):E127–E139.

Hausdorf, B. (2011). Progress toward a general species concept. *Evolution*, 65(4):923–931.

Hausdorf, B. and Hennig, C. (2019). Package 'prabclus', version 2.3-2.

Hausdorf, B. and Hennig, C. (2020). Species delimitation and geography. *Molecular Ecology Resources*, 20(4):950–960.

Hennig, C. and Lin, C. (2015). Flexible parametric bootstrap for testing homogeneity against clustering and assessing the number of clusters. *Statistics and Computing*, 25:821–833.

Holm, S. (1979). A simple sequentially rejective multiple test procedure. *Scandinavian Journal of Statistics*, 6(2):65–70.

Huang, H., Liu, Y., Hayes, D. N., Nobel, A., Marron, J. S., and Hennig, C. (2015). Significance testing in clustering. In Hennig, C., Meila, M., Murtagh, F., and Rocci, R., editors, *Handbook of cluster analysis*, pages 315–336.

Hubert, L. and Arabie, P. (1985). Comparing partitions. *Journal of Classification*, 2(1):193–218.

Janes, J. K., Miller, J. M., Dupuis, J. R., Malenfant, R. M., Gorrell, J. C., Cullingham, C. I., and Andrew, R. L. (2017). The k = 2 conundrum. *Molecular Ecology*, 26(14):3594–3602.

Mason, N. A., Fletcher, N. K., Gill, B. A., Funk, W. C., and Zamudio, K. R. (2020). Coalescent-based species delimitation is sensitive to geographic sampling and isolation by distance. *Systematics and Biodiversity*, 18(3):269–280.

Pedraza-Marrón, C., Silva, R., Deeds, J., Van Belleghem, S. M., Mastretta-Yanes, A., Domínguez-Domínguez, O., Rivero-Vega, R. A., Lutackas, L., Murie, D., Parkyn, D., Bullock, L. H., Foss, K., Ortiz-Zuazaga, H., Narváez-Barandica, J., Acero, A., Gomes, G., and Betancur, R. R. (2019). Genomics overrules mitochondrial dna, siding with morphology on a controversial case of species delimitation. *Proceedings of the Royal Society B: Biological Sciences*, 286(1900):20182924.

Pritchard, J. K., Stephens, M., and Donnelly, P. (2000). Inference of population structure using multilocus genotype data. *Genetics (Austin)*, 155(2):945–959.

Rannala, B. and Yang, Z. (2020). Species delimitation. In Scornavacca, C., Delsuc, F., and Galtier, N., editors, *Phylogenetics in the genomic era*, pages 5.5:1–5.5:18. Self published.

Rousset, F. (1997). Genetic differentiation and estimation of gene flow from F-statistics under isolation by distance. *Genetics*, 145(4):1219–1228.

Rousset, F. (2008). GENEPOP'007: a complete re-implementation of the GENEPOP software for windows and linux. *Molecular Ecology Resources*, 8(1):103–106.

Tamura, K., Peterson, D., Peterson, N., Stecher, G., Nei, M., and Kumar, S. (2011). Mega5: Molecular evolutionary genetics analysis using maximum likelihood, evolutionary distance, and maximum parsimony methods. *Molecular Biology and Evolution*, 28(10):2731–2739.

Thorndike, R. L. (1953). Who belongs in the family? *Psychometrika*, 18(4):267–276.

Vekemans, X. and Hardy, O. J. (2004). New insights from fine-scale spatial genetic structure analyses in plant populations. *Molecular Ecology*, 13(4):921–935.

Waits, L. P. and Storfer, A. (2015). *Basics of population genetics: quantifying neutral and adaptive genetic variation for landscape genetic studies*, chapter 3, pages 35–57. John Wiley & Sons, Ltd.

Wang, J. (2017). The computer program structure for assigning individuals to populations: easy to use but easier to misuse. *Molecular Ecology Resources*, 17(5):981–990.

Wright, S. (1943). Isolation by distance. *Genetics*, 28(2):114.

Yang, Z. (2015). The bpp program for species tree estimation and species delimitation. *Current Zoology*, 61(5):854–865.

3

Multivariate Longitudinal Latent Markov Models to Characterize Pollutant Exposures

Antonello Maruotti, Lorena Ricciotti, and Alfonso Russo

Motivated by the analysis of pollutant's concentrations collected by 35 environmental monitoring stations in the United Kingdom in December 2024, we introduce a family of parsimonious latent Markov models (LMMs) aiming at clustering stations over time, identifying risk-states of pollutant/environmental exposure. The proposal gives an alternative to existing methods to model-based clustering of multivariate longitudinal data. As parsimony is desirable under a multivariate setting, we assume a reparameterization of the covariance matrix based on a factor model representation, obtaining a family of eight models. An Alternating Expectation Conditional Maximization (AECM) algorithm is presented and used to perform parameters estimation. Our results reveal a substantial degree of heterogeneity among monitoring stations and across times, as indicated by the four clusters estimated, and strong correlations among pollutants as summarized by the two latent factors estimated.

3.1 Introduction

Understanding pollution exposure in urban areas requires analyzing complex, high-dimensional data that capture both temporal and spatial heterogeneity in air quality. In particular, the concentration levels of multiple pollutants are often measured at different monitoring stations, which exhibit heterogeneous exposure patterns.

Although it is known that air pollution is a complex mixture of multiple pollutants, the pollution analyzes are in general conducted on each pollutant separately (see, e.g., Draper et al. (2023); Gianquintieri et al. (2024); Martínez-Silva et al. (2016); Martinez-Zarzoso and Maruotti (2013); Uluocak et al. (2025); Yang et al. (2022)). Hence, alternative multivariate approaches have been overlooked so far, although the investigation of exposures to multiple pollutants would be useful because it could provide a better understanding of potential combined effects of individual pollutants and their interactions, with a few exceptions Boaz et al. (2019); Bose et al. (2018); He et al. (2022); Lagona et al. (2011).

Many existing models rely on oversimplified assumptions that do not represent the real-world complexity of pollution exposures. Several statistical challenges arise in the analysis of multivariate longitudinal pollutants concentrations, including:

DOI: 10.1201/9781003564423-3

- High correlation among pollutants, leading to redundant information and potentially biased inference.
- Confounding effects arising from time-varying dependencies, making it difficult to isolate the impact of each pollutant.
- The presence of heterogeneous monitoring stations, as well as time periods characterized by different pollutant concentrations.
- Computational difficulties are particularly present in high-dimensional settings where standard multivariate approaches are broken down.

To address these challenges, we develop a multivariate longitudinal LMM. In the considered setting, pollution is understood as levels of the various pollutants, i.e., pollutants concentrations, and is measured at different locations, leading to a three-way data matrix (monitorning stations, times, pollutant concentrations). To address these challenges, multivariate longitudinal LMMs constitute a natural candidate for studying their joint temporal dynamics, simultaneously clustering network monitoring stations, reducing the curse of dimensionality, and segmenting exposure periods. More precisely, the latent process of the LMM approach allows for inferring a finite number of concentration profiles, each of which is characterized by a specific combination of pollutant concentrations and, at the same time, shows differences across monitoring stations. These profiles are a direct result of clustering similarities occurring in both the variable space (i.e., between pollutants) and in a temporal neighborhood. By integrating station-level clustering within the Markov framework, we identify groups of monitoring stations that share similar exposure dynamics while capturing distinct pollution exposures. Each profile is linked to an easily interpretable multivariate distribution, which provides an accurate representation of the respective concentration of air pollutants, illustrating the potential of this type of modeling for defining exposure conditions.

In the context of understanding exposure to multiple pollutants, dimensionality-related aspects present a challenge because these pollutants could be potentially highly correlated. Therefore, estimation and interpretation of the parameters of interest may become nontrivial. In order to examine the interrelationships between pollutant concentrations to perform dimensionality reduction in the variable space simultaneously (allowing for an easy interpretation of model parameters), pollutant concentrations are modeled using mixtures of factor analyzers, where the latent factors evolve according to a latent Markov process Maruotti et al. (2017). Accordingly, we define a class of parsimonious LMMs, inspired to the notable work of [Tomarchio et al. (2022) and Neal et al. (2024)] for the case of Hidden Markov models in empirical time series analysis, by imposing a factor decomposition on state-specific covariance matrices. The loadings and noise terms of the covariance matrix may be constrained to be equal or unequal across latent states. In addition, the noise term may be subject to further restrictions, resulting in a set of eight parsimonious covariance structures. This model structure allows to account for dependence between pollutants, and provides a clear interpretation of the (latent) association structure between pollutants.

In this framework, model parameters can be estimated by a full maximum likelihood method based on the AECM algorithms, and recursions widely used in the LMM literature.

By integrating clustering, longitudinal modeling, and dimensionality reduction within a unified probabilistic framework, our approach provides a powerful tool for understanding pollution exposure dynamics and supporting targeted environmental policies.

3.2 Data Description

3.2.1 Data Source

The empirical analysis presented in this study is based on air pollution data collected from a comprehensive network of environmental monitoring stations across the United Kingdom. Specifically, data belong to the UK Automatic Urban and Rural Network (AURN) archive and are obtained using the `importAURN()` function in the `openair` R package Carslaw and Ropkins (2012). This package is a well-established tool in the environmental sciences community that facilitates access to air quality data from a variety of UK-based monitoring systems. The package serves as a valuable interface for researchers, providing standardized and harmonized pollutant measurements from multiple official sources.

The UK monitoring infrastructure includes several national and regional networks coordinated through public environmental agencies. Among these, the most prominent is the AURN, operated under the supervision of the Department for Environment, Food and Rural Affairs (DEFRA). In addition to AURN, the dataset encompasses information from the Air Quality England (AQE) network, the Scottish Air Quality Network (SAQN), the Welsh Air Quality Network (WAQN), and the Northern Ireland (NI) Air Quality Network. Furthermore, the dataset incorporates contributions from a number of locally managed monitoring networks, which play an essential role in providing fine-scale information on pollutant concentrations, particularly in urban and peri-urban areas.

The data infrastructure has been developed and maintained to support a broad set of goals, ranging from regulatory compliance with national and European environmental directives to the facilitation of epidemiological research and the design of mitigation strategies. This system has grown in response to the increasing demand for high-quality, accessible environmental data, both for public transparency and for the scientific assessment of pollution exposure and its health effects.

3.2.2 Data Description

In the present study, we focus on a curated sample of 35 monitoring stations distributed across the United Kingdom. These stations have been selected based on the availability of complete time series data for the month of December 2024, allowing us to perform a detailed temporal analysis of air pollution dynamics during a typical winter period, often characterized by atmospheric stagnation and elevated pollution levels.

Several pollutants are monitored by the network such as ozone, carbon monoxide, and sulfur dioxide, although here we focus on a core set of five indicators commonly associated with urban air quality and adverse health outcomes: nitric oxide (NO), nitrogen dioxide (NO_2), oxides of nitrogen (NO_x) and particulate matter with a diameter less than 10 micrometers (PM_{10}) and less than 2.5 micrometers ($\mathrm{PM}_{2.5}$). The data provide an opportunity to investigate both the marginal behavior of pollutant concentrations and their joint temporal evolution. In particular, the use of multiple pollutants measured simultaneously across several monitoring locations enables a multivariate perspective on pollution exposure. Rather than relying solely on conventional threshold exceedance metrics or aggregated indices, the methodology adopted in this study seeks to capture the underlying dynamics of environmental exposure through a model-based framework. In doing so, we aim to identify latent structures in the data, represented by latent exposure states, that summarize pollution conditions. By modeling the temporal evolution of these latent states, we are able to construct a data-driven classification of environmental exposure that is inherently dynamic. These latent states, coupled with their respective emission distributions, serve as a reduced-dimensionality summary of pollution exposure. While the analysis in this study focuses on a subset of monitoring stations with complete data for the month of December 2024, it is important to note that hidden Markov models can be effectively applied even when time series are partially observed (see Zucchini et al. (2016)). Under the assumption of missing at random, the model's probabilistic structure allows for handling of missing values via likelihood-based inference, without requiring ad hoc imputation (see e.g. Pennoni et al. (2024)). This feature supports the broader applicability of the proposed methodology in real-world settings, where data completeness cannot always be guaranteed. If the missing data mechanism is not at random, Marino et al. (2018) and Maruotti (2015) introduced straightforward method to handle non-ignorable dropouts in longitudinal data.

Figure 3.1 offers a preliminary visual summary of the pollutant concentration data across both time and space, revealing two prominent characteristics of the dataset. First, the temporal evolution of pollutant levels appears highly irregular and nonstationary. Rather than following smooth or seasonal trends, the series exhibits pronounced bursts of elevated concentrations, often followed by sudden drops to relatively low levels. These dynamics suggest the influence of episodic emission events and rapid changes in atmospheric conditions, such as temperature inversions, wind shifts, or changes in traffic and heating patterns, particularly relevant during the winter month considered. This temporal volatility underscores the need for a modeling approach that can accommodate abrupt regime changes, motivating the use of state-switching models such as the latent Markov framework developed in this paper. Second, considerable spatial heterogeneity is evident across the 35 monitoring stations included in the study. While some stations report concentration levels broadly in line with others, a number of sites display substantially wider ranges, as well as values that deviate markedly from the rest—indicating the presence of atypical or extreme measurements. The boxplots highlight this variability, reflecting the challenges of generalizing risk across space without a model that accounts for station-specific behavior and joint pollutant dynamics.

Another important feature of the data considered in this study lies in the distinct atmospheric lifetimes of the pollutants under analysis. Gaseous compounds such as NO and NO_2 tend to react and disperse more rapidly, while others—such as PM_{10}—can persist in the atmosphere for extended periods, depending on meteorological conditions and emission sources. These differences have a direct impact on the temporal dependence structure of the observed concentrations. As a result, a pronounced daily and intra-day dependence is expected across pollutant time series, which underscores the importance of modeling not only their marginal behavior but also their joint dynamic evolution. In this context, the inter-pollutant associations play a pivotal role. Rather than treating correlations among pollutants as nuisance artifacts or controlling for them in a marginal framework, we explicitly regard their joint dependence structure as a central component of the overall environmental risk landscape. Disregarding these associations may lead to misleading conclusions, particularly under conditions of atmospheric instability, where multiple pollutants are likely to exhibit simultaneous concentration spikes due to shared or interacting emission sources and meteorological influences. To better understand the dependence structure among pollutants, Figure 3.2 presents scatter plots for each pair of variables considered in our analysis. The fitted black lines in each panel represent simple linear correlations; however, the visual patterns reveal much richer and more nuanced relationships. In particular, clear signs of nonlinear dependencies emerge, along with evidence of heterogeneous correlation structures across different pollutant pairs. For some combinations, the relationships appear nearly linear, while for others, more complex, nonlinear patterns are evident. Additionally, the presence of multiple clusters and heavier-tailed distributions is suggested by the shape of the scatter plots, especially in the upper extremes of the concentration ranges. These features hint at the existence of multimodality in the joint distribution, possibly corresponding to distinct environmental regimes or exposure conditions—an observation that further

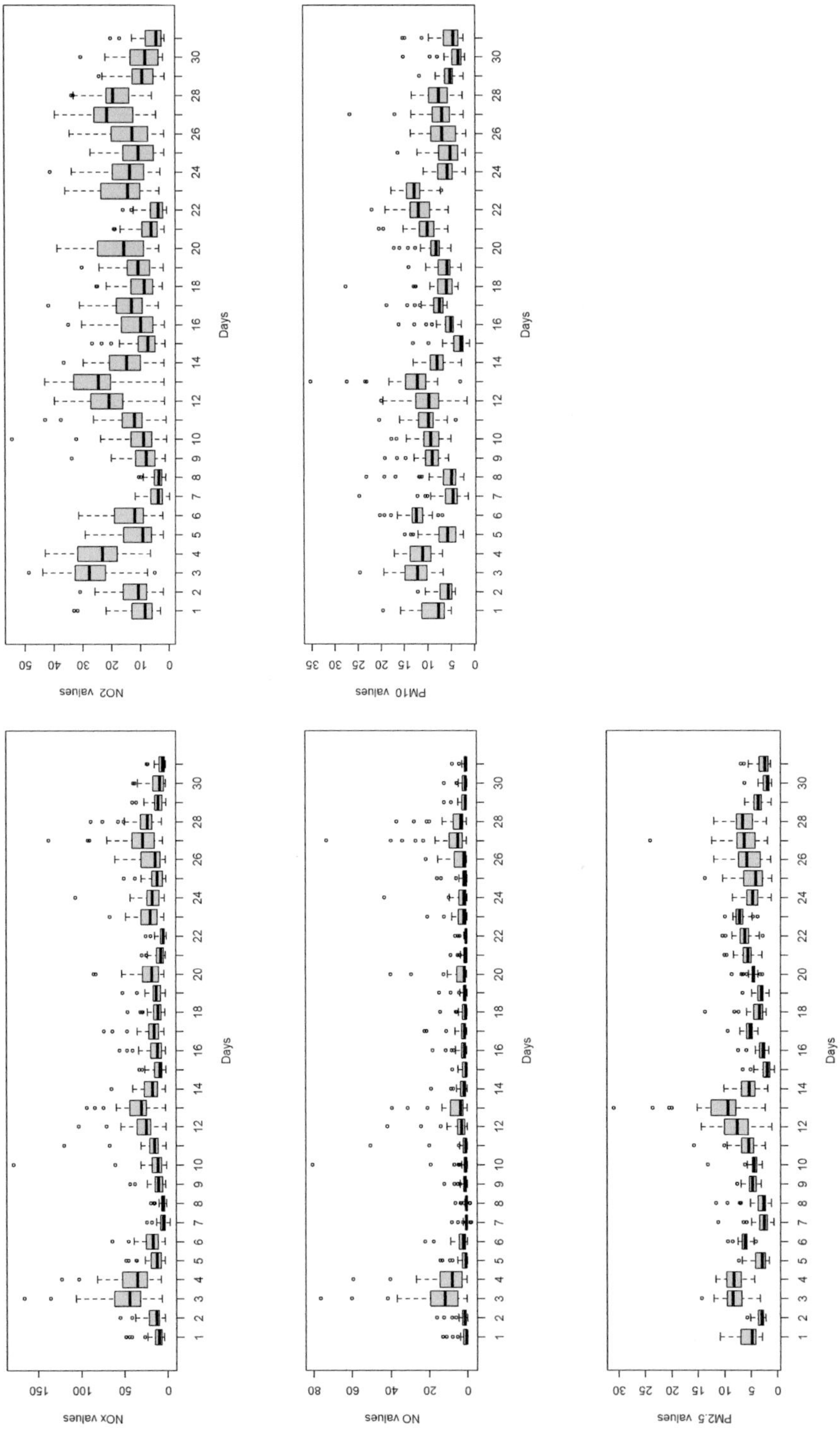

FIGURE 3.1
Summary information of pollutant concentrations over time, collapsed at the day level.

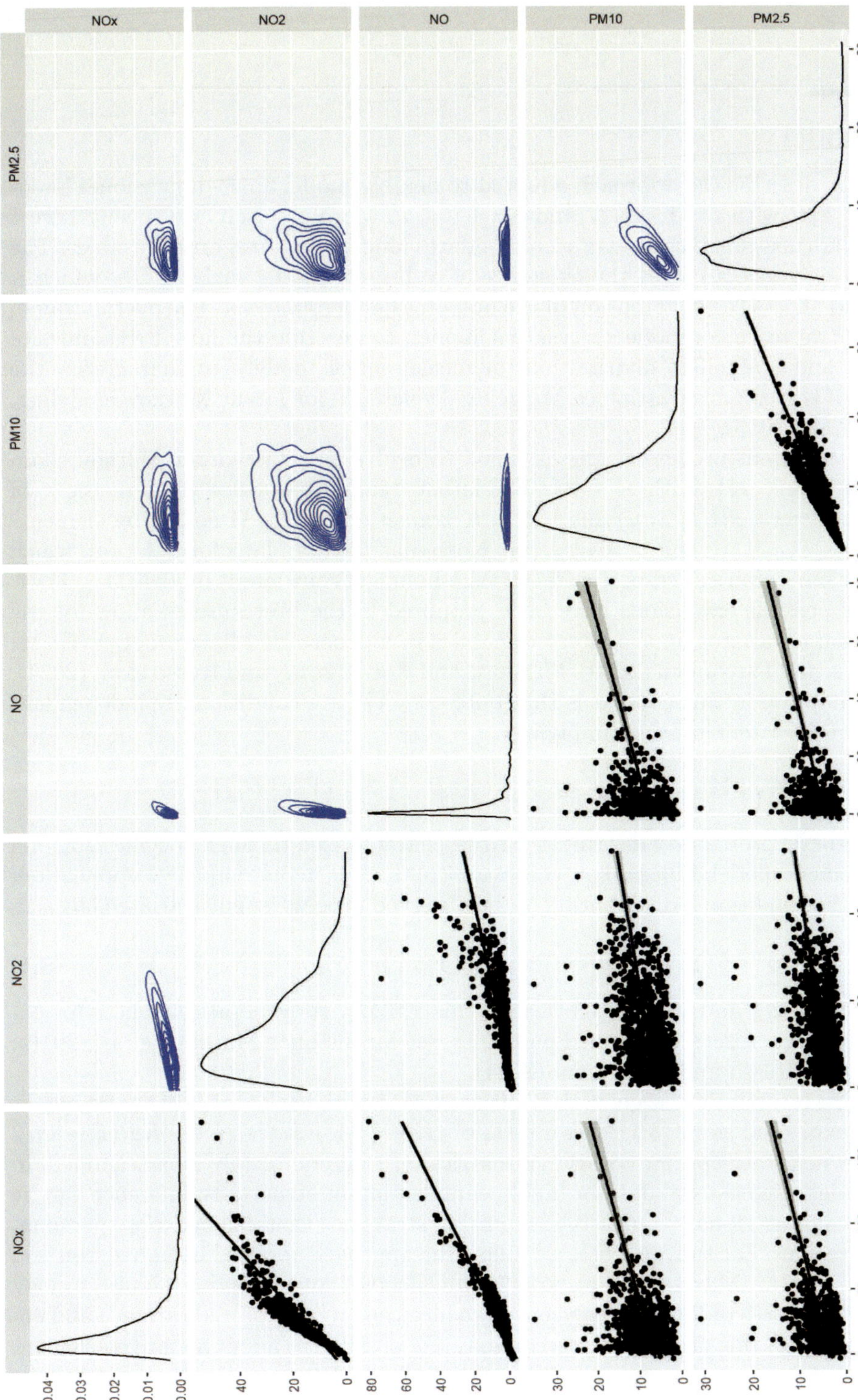

FIGURE 3.2
Pollutant concentrations: marginal distributions, bivariate correlations, and pairwise scatter plots.

motivates the use of latent state models, such as the hidden semi-Markov framework proposed in this study.

3.3 The Model

In this section, we describe the approach adopted to flexibly model an R-dimensional longitudinal outcome $\mathbf{Y}_{i,t}$, with $i = 1, \ldots, N$ and $t = 1, \ldots, T$. More precisely, we specify here a LMM, a particular class of mixture that accommodates longitudinal response variables. The general formulation conceptualizes the existence of a discrete latent variable $U_{i,t}$ which is time-varying, and affects the distribution of the outcomes under analysis. The overall framework consists of two main components: a latent model, to describe the latent dynamics of the process $U_{i,t}$, and a manifest distribution (or measurement model), to characterize the probability law of the response variables. As general reference for latent Markov modeling, see Bartolucci et al. (2013).

Specifically, the latent process $U_{i,t}$ is assumed to evolve as a first-order Markov chain with state space $\{1, \ldots, K\}$, with $K \in \mathcal{N}$. The latent dynamics is ruled by a vector $\boldsymbol{\pi}$ of initial probabilities $\pi_u = \mathbb{P}(U_{i,1} = u)$ and a $K \times K$ transition matrix $\mathbf{\Pi}$ collecting elements $\pi_{u,v} = \mathbb{P}(U_{i,t} = u \mid U_{i,t-1} = v)$, $u, v = 1, \ldots, K$. Note that the standard formulation, which we will adopt for our empirical case, specifies a time-homogeneous latent model ${}_t\mathbf{\Pi} = \mathbf{\Pi} \, \forall t$, although several possible extensions have been proposed in the literature to allow for the effect of individual covariates on the latent dynamics in a generalized linear model (GLM) fashion Bartolucci and Farcomeni (2009); Maruotti and Rocci (2012) and Russo et al. (2024).

One key assumption in this context is that outcomes $\mathbf{Y}_{i,t}$ are conditionally independent on the past given the latent process, implying

$$f\left(\mathbf{Y}_{i,t} \mid \mathbf{Y}_{i,1}, \ldots, \mathbf{Y}_{i,t-1} \,;\, U_{i,1}, \ldots, U_{i,T}\right) = f\left(\mathbf{Y}_{i,t} \mid U_{i,t}\right)$$

where $f(\cdot)$ is a generic probability distribution to be characterized through specification of a suitable measurement model. Since all outcomes are continuous in our empirical application, we assume that the conditional distribution of the observed process is multivariate Gaussian of order R

$$f\left(\mathbf{Y}_{i,t} \mid U_{i,t} = u\right) = \boldsymbol{\mathcal{N}}_R\left(\mathbf{Y}_{i,t} \mid \boldsymbol{\mu}_u, \mathbf{\Sigma}_u\right) \tag{3.1}$$

with mean vector $\boldsymbol{\mu}_u$ and variance-covariance matrix $\mathbf{\Sigma}_u$. As general references for extensions to multivariate discrete or mixed outcomes, see for instance Maruotti et al. (2016); Punzo et al. (2021); Russo and Farcomeni (2024).

In its standard formulation, a general R-variate LMM with K latent states and conditional densities expressed as in (3.1), has a total of $(K-1)(1+K) + KR + KR(R+1)/2$ free parameters when state-specific covariance matrices are left completely unconstrained. In the presence of high dimensional data or highly complex latent structures, inference can be cumbersome or computationally unfeasible, requiring strategies that achieve more parsimonious parametrizations. For this purpose, we follow the approach adopted in Bouveyron and Brunet-Saumard (2014); Maruotti et al. (2017); McNicholas and Brendan (2008), among others, based on mixtures of factors analyzers to reduce the number of free variance and covariance parameters. In this factor analysis context, it is possible to adopt a more convenient representation of the observed process that, conditionally on $U_{i,t} = u$, can be expressed as:

$$\mathbf{Y}_{i,t} = \boldsymbol{\mu}_u + \mathbf{\Lambda}_u \boldsymbol{\eta}_{i,t,u} + \boldsymbol{\epsilon}_{i,t,u} \tag{3.2}$$

TABLE 3.1
Eight possible covariance structures, characterized by different degrees of parsimony, resulting from mixtures of factor analyzers.

Model Identifier	Loading Matrix $\mathbf{\Lambda}_u$	Error Variance $\mathbf{\Psi}_u$	Isotropic $\mathbf{\Psi}_u = \psi_u \mathbf{I}_R$	Covariance Structure	Covariance Parameters
CCC	Constrained	Constrained	Constrained	$\mathbf{\Sigma}_u = \mathbf{\Lambda}\mathbf{\Lambda}' + \psi \mathbf{I}_R$	$[RQ - Q(Q-1)/2] + 1$
CCU	Constrained	Constrained	Unconstrained	$\mathbf{\Sigma}_u = \mathbf{\Lambda}\mathbf{\Lambda}' + \mathbf{\Psi}$	$[RQ - Q(Q-1)/2] + R$
CUC	Constrained	Unconstrained	Constrained	$\mathbf{\Sigma}_u = \mathbf{\Lambda}\mathbf{\Lambda}' + \psi_u \mathbf{I}_R$	$[RQ - Q(Q-1)/2] + K$
CUU	Constrained	Unconstrained	Unconstrained	$\mathbf{\Sigma}_u = \mathbf{\Lambda}\mathbf{\Lambda}' + \mathbf{\Psi}_u$	$[RQ - Q(Q-1)/2] + KR$
UCC	Unconstrained	Constrained	Constrained	$\mathbf{\Sigma}_u = \mathbf{\Lambda}_u\mathbf{\Lambda}_u' + \psi \mathbf{I}_R$	$K[RQ - Q(Q-1)/2] + 1$
UCU	Unconstrained	Constrained	Unconstrained	$\mathbf{\Sigma}_u = \mathbf{\Lambda}_u\mathbf{\Lambda}_u' + \mathbf{\Psi}$	$K[RQ - Q(Q-1)/2] + R$
UUC	Unconstrained	Unconstrained	Constrained	$\mathbf{\Sigma}_u = \mathbf{\Lambda}_u\mathbf{\Lambda}_u' + \psi_u \mathbf{I}_R$	$K[RQ - Q(Q-1)/2] + K$
UUU	Unconstrained	Unconstrained	Unconstrained	$\mathbf{\Sigma}_u = \mathbf{\Lambda}_u\mathbf{\Lambda}_u' + \mathbf{\Psi}_u$	$K[RQ - Q(Q-1)/2] + KR$

where $\boldsymbol{\eta}_{i,t,u} \sim \boldsymbol{\mathcal{N}}_Q(\mathbf{0}, \mathbf{I}_Q)$ is a Q-dimensional vector of conditional latent factors, $\mathbf{\Lambda}_u$ is a $R \times Q$ real matrix of state-specific factor loadings (typically with $Q \ll R$) and $\mathbf{I}_Q$ denotes the identity matrix of rank Q. Additionally, errors $\boldsymbol{\epsilon}_{i,t,u}$ are assumed to be zero-mean multivariate Gaussian with covariance matrix $\mathbf{\Psi}_u = \{\psi_{u,1}, \ldots, \psi_{u,R}\}$.

This method allows for a variety of association structures resulting from the following decomposition of each state-specific variance covariance matrix:

$$\mathbf{\Sigma}_u = \mathbf{\Lambda}_u \mathbf{\Lambda}_u' + \mathbf{\Psi}_u \tag{3.3}$$

It can be shown that, conditionally on the latent Markov chain, outcomes have marginal distribution taking the following form:

$$\mathbf{Y}_{i,t} \mid U_{i,t} = u \sim \boldsymbol{\mathcal{N}}_R \left(\boldsymbol{\mu}_u, \mathbf{\Lambda}_u \mathbf{\Lambda}_u' + \mathbf{\Psi}_u \right). \tag{3.4}$$

Eight distinct parsimonious parametrizations for our described multivariate Gaussian LMM can be derived by imposing constraints on either $\mathbf{\Lambda}_u$ and/or the free elements in $\mathbf{\Psi}_u$, in combination with whether or not $\mathbf{\Psi}_u = \psi_u \mathbf{I}_R$ ($\psi_u \in \mathbb{R}^+$) and the further isotropic special case implying $\psi_u = \psi$, $\forall\, u = 1, \ldots, K$. Table 3.1 presents a schematic description of the possible formulations.

3.4 Inference

This section describes computational and analytical details of the inferential scheme we develop to estimate all the parameters of the model described in Section 3.3.

3.4.1 The Likelihood

Maximum likelihood estimation of all model's parameters hinges upon computation of the observed likelihood, taking the following form:

$$\boldsymbol{\mathcal{L}}(\boldsymbol{\theta}) = \prod_{i=1}^{N} \mathcal{L}_i(\boldsymbol{\theta}) = \prod_{i=1}^{N} \boldsymbol{\pi}'\boldsymbol{\Phi}(\boldsymbol{Y}_{i,1}) \cdot \boldsymbol{\Pi}\boldsymbol{\Phi}(\boldsymbol{Y}_{i,2})\boldsymbol{\Pi} \cdots \boldsymbol{\Phi}(\boldsymbol{Y}_{i,T-1})\boldsymbol{\Pi}\boldsymbol{\Phi}(\boldsymbol{Y}_{i,T})\mathbf{1}_K \tag{3.5}$$

where $\boldsymbol{\theta} = \{\boldsymbol{\mu}_u, \boldsymbol{\Lambda}_u, \boldsymbol{\Psi}_u, \boldsymbol{\pi}, \boldsymbol{\Pi}\}$ is a short-hand notation for all model parameters, $\boldsymbol{\Phi}(\boldsymbol{Y}_{i,t})$ denotes a $K \times K$ diagonal matrix collecting conditional densities $f(\boldsymbol{Y}_{i,t} \mid U_{i,t} = u)$ while $\mathbf{1}_K$ is a K-dimensional vector of ones. Computation of (3.5) cannot be performed directly, as it would require a telescopic sum over all possible configurations of the latent Markov chain. Therefore, we derive a custom recursive algorithm Baum et al. (1970); Welch (2003), with complexity linearly increasing in $N \cdot T$, in order to efficiently compute $\boldsymbol{\mathcal{L}}(\boldsymbol{\theta})$. The procedure is based on forward variables $a_{i,t}(u) = f(\boldsymbol{Y}_{i,1}, \ldots, \boldsymbol{Y}_{i,t}, U_{i,t} = u \mid \boldsymbol{\theta})$, for $u = 1, \ldots. K$, $t = 1, \ldots, T$, and $i = 1, \ldots, N$. The recursion is initialized as

$$a_{i,1}(u) = \pi_u \, f(\boldsymbol{Y}_{i,1} \mid U_{i,1} = u) \tag{3.6}$$

and then, for $t = 2, \ldots, T$, it can be shown that

$$a_{i,t}(u) = \sum_{v=1}^{K} \pi_{u,v} \, a_{i,t-1}(v) \, f(\boldsymbol{Y}_{\boldsymbol{i,t}} \mid U_{i,t} = u) \tag{3.7}$$

Since $a_{i,T}(u) = f(\boldsymbol{Y}_{i,1}, \ldots, \boldsymbol{Y}_{i,T}, U_{i,T} = u \mid \boldsymbol{\theta})$ by definition, the observed log-likelihood $\boldsymbol{\ell}(\boldsymbol{\theta}) = \log[\boldsymbol{\mathcal{L}}(\boldsymbol{\theta})]$ is therefore

$$\boldsymbol{\ell}(\boldsymbol{\theta}) = \sum_{i=1}^{N} \log\left[\sum_{u=1}^{K} a_{i,T}(u)\right] \tag{3.8}$$

Employing forward variables addresses the computational infeasibility of directly calculating the likelihood. However, a major numerical limitation persists. Since forward variables involve repeated products of probabilities and densities, their values can rapidly diminish or grow as the time index increases, leading to underflow or overflow in practical computation. The recursion for computing the forward variables is replaced by a scaled version, as e.g., described in Visser (2011) and Zucchini et al. (2016).

We also introduce backward variables $b_{i,t}(u) = f(\boldsymbol{Y}_{i,t+1}, \ldots, \boldsymbol{Y}_{i,T} \mid U_{i,t} = u)$. Initial conditions are given by $b_{i,T}(u) = 1 \, \forall \, u$, and then backward filtering is performed for $t = T-1, \ldots, 1$ as

$$b_{i,t}(u) = \sum_{v=1}^{K} \pi_{u,v} \, b_{i,t+1}(v) \, f(\boldsymbol{Y}_{\boldsymbol{i,t}} \mid U_{i,t} = u) \tag{3.9}$$

3.4.2 Maximum Likelihood Estimation

In order to maximize $\boldsymbol{\ell}(\boldsymbol{\theta})$, it is possible to adapt an Alternating Expectation Conditional Maximization (ECMA) algorithm Field et al. (2016); Maruotti et al. (2017); Yao et al. (2005) and break the overall model into smaller subparts to be individually optimized. Indeed, we divide all model's parameters into two disjoint sets $\{\boldsymbol{\theta}_M, \boldsymbol{\theta}_\Sigma\}$ defined as follows: $\boldsymbol{\theta}_M$ collects the parameters ruling the latent dynamics $(\boldsymbol{\pi}, \boldsymbol{\Pi})$ as well as state-specific means $\boldsymbol{\mu} = \{\boldsymbol{\mu}_1, \ldots, \boldsymbol{\mu}_K\}$; quantities $\widetilde{\boldsymbol{\Lambda}} = \{\boldsymbol{\Lambda}_u\}$ and $\widetilde{\boldsymbol{\Psi}} = \{\boldsymbol{\Psi}_u\}$, for each $u = 1, \ldots, K$, are stored in $\boldsymbol{\theta}_\Sigma$.

To illustrate the iterative procedure, we first introduce the complete-data log-likelihood, which takes form

$$\begin{aligned}\boldsymbol{\ell}^*(\boldsymbol{\theta}) &= \sum_{i=1}^{N}\sum_{u=1}^{K} w_{i,1,u}\, \log(\pi_u) + \sum_{i=1}^{N}\sum_{t=2}^{T}\sum_{u=1}^{K}\sum_{v=1}^{K} z_{i,t,u,v}\, \log\left(\pi_{u,v}\right) \\ &+ \sum_{i=1}^{N}\sum_{t=1}^{T}\sum_{u=1}^{K} w_{i,t,u}\, \boldsymbol{f}(\boldsymbol{Y}_{i,t} \mid U_{i,t} = u)\end{aligned} \tag{3.10}$$

where $w_{i,t,u} = 1$ if unit i is in state u at time t, and 0 otherwise, $z_{i,t,u,v} = 1$ if unit i transitions from state v, held at $t-1$, to state u at time t, and again is zero otherwise; while

$$\begin{aligned}\boldsymbol{f}(\boldsymbol{Y}_{i,t} \mid U_{i,t} = u) &= -\frac{R}{2}\log(2\pi) - \frac{1}{2}\log\left(|\boldsymbol{\Lambda}_u\boldsymbol{\Lambda}_u' + \boldsymbol{\Psi}|\right) \\ &- \frac{1}{2}\mathrm{tr}\left\{(\boldsymbol{Y}_{i,t} - \boldsymbol{\mu}_u)(\boldsymbol{Y}_{i,t} - \boldsymbol{\mu}_u)'\left(\boldsymbol{\Lambda}_u\boldsymbol{\Lambda}_u' + \boldsymbol{\Psi}\right)^{-1}\right\}\end{aligned} \tag{3.11}$$

Then, the following steps are iterated until convergence:

- *First stage*:
  - E-step: compute the conditional expectation of the complete-data-log-likelihood given the data and the current estimates for the parameters $(\boldsymbol{\pi}, \boldsymbol{\Pi}, \boldsymbol{\mu})$ while keeping $\boldsymbol{\theta}_\Sigma$ fixed
  - M-step: Maximise the value of this expected complete-data log-likelihood with respect to $(\boldsymbol{\pi}, \boldsymbol{\Pi}, \boldsymbol{\mu})$
- *Second stage*:
  - E-step: Compute the conditional expectation of the complete-data log-likelihood conditionally on the data and parameters in $\boldsymbol{\theta}_\Sigma$ while values for $(\boldsymbol{\pi}, \boldsymbol{\Pi}, \boldsymbol{\mu})$ are held fixed from the previous stage
  - CM-step: Maximize this latter expected value with respect to $\left(\widetilde{\boldsymbol{\Lambda}}, \widetilde{\boldsymbol{\Psi}}\right)$.

At the E-step in the first-stage, the conditional expectation of $\boldsymbol{\ell}^*(\boldsymbol{\theta})$, given the data and current estimates of model's parameters, can be computed as

$$\begin{aligned}\mathcal{H}(\boldsymbol{\theta}_M) &= \sum_{i=1}^{N}\sum_{u=1}^{K} \widehat{w}_{i,1,u}\, \log\left(\pi_u\right) \\ &+ \sum_{i=1}^{N}\sum_{t=2}^{T}\sum_{u=1}^{K}\sum_{v=1}^{K} \widehat{z}_{i,t,u,v}\, \log\left(\pi_{u,v}\right) \\ &- \frac{1}{2}\sum_{i=1}^{N}\sum_{t=1}^{T}\sum_{u=1}^{K} \widehat{w}_{i,t,u}\, \log\left(|\boldsymbol{\Lambda}_u\boldsymbol{\Lambda}' + \boldsymbol{\Psi}_u|\right) \\ &- \frac{1}{2}\sum_{i=1}^{N}\sum_{t=1}^{T}\sum_{u=1}^{K} \widehat{w}_{i,t,u}\, \mathrm{tr}\left\{\left[\widetilde{\boldsymbol{\Sigma}}_u\left(\boldsymbol{\Lambda}_{\boldsymbol{u}}\boldsymbol{\Lambda}_u' + \boldsymbol{\Psi}_u\right)^{-1}\right]\right\}\end{aligned} \tag{3.12}$$

where the conditional expected value of the labels $w_{i,t,u}$ and $z_{i,t,u,v}$ are available in closed form as:

$$\widehat{w}_{i,t,u} = \mathbb{E}\left[w_{i,t,u} \mid \boldsymbol{Y}_{i,1},\ldots,\boldsymbol{Y}_{i,T}\right] = \frac{a_{i,t}(u)\, b_{i,t}(u)}{\sum_u a_{i,T}(u)}$$

$$\widehat{z}_{i,t,u,v} = \mathbb{E}\left[z_{i,t,u,v} \mid \boldsymbol{Y}_{i,1},\ldots,\boldsymbol{Y}_{i,T}\right] = \frac{a_{i,t}(u)\, b_{i,t+1}(u)}{\sum_{u=1}^{K} a_{i,t}(u)\, b_{i,t}(v)} f\left(\boldsymbol{Y}_{i,t+1} \mid U_{i,t}=v\right) \tag{3.13}$$

while the state-specific variance-covariance matrix $\widetilde{\boldsymbol{\Sigma}}_u$ is kept fixed at

$$\widetilde{\boldsymbol{\Sigma}}_u = \frac{\sum_{i=1}^{N}\sum_{t=1}^{T} \widehat{w}_{i,t,u}\left(\boldsymbol{Y}_{i,t}-\boldsymbol{\mu}_u\right)\left(\boldsymbol{Y}_{i,t}-\boldsymbol{\mu}_u\right)'}{\sum_{i=1}^{N}\sum_{t=1}^{T} \widehat{w}_{i,t,u}} \qquad u=1,\ldots,K. \tag{3.14}$$

Then, in the first-stage M-step, we maximize the expected value of $\boldsymbol{\ell}^*(\boldsymbol{\theta})$ with respect to $\boldsymbol{\theta}_M$. Updates for latent parameters and state-specific means are available in closed form:

$$\widehat{\pi}_u = \frac{\sum_{i=1}^{N} \mathbb{E}\left[w_{i,1,u} \mid \boldsymbol{Y}_{i,1},\ldots,\boldsymbol{Y}_{i,T}\right]}{\sum_{i=1}^{N}\sum_{v=1}^{K} \mathbb{E}\left[w_{i,1,v} \mid \boldsymbol{Y}_{i,1},\ldots,\boldsymbol{Y}_{i,T}\right]}$$

$$\widehat{\pi}_{u,v} = \frac{\sum_{i=1}^{N}\sum_{t=1}^{T-1} \mathbb{E}\left[z_{i,t,u,v} \mid \boldsymbol{Y}_{i,1},\ldots,\boldsymbol{Y}_{i,T}\right]}{\sum_{i=1}^{N}\sum_{t=1}^{T-1}\sum_{v=1}^{K} \mathbb{E}\left[z_{i,t,u,v} \mid \boldsymbol{Y}_{i,1},\ldots,\boldsymbol{Y}_{i,T}\right]} \tag{3.15}$$

$$\widehat{\mu}_u = \frac{\sum_{i=1}^{N}\sum_{t=1}^{T} \mathbb{E}\left[w_{i,t,u} \mid \boldsymbol{Y}_{i,1},\ldots,\boldsymbol{Y}_{i,T}\right]\boldsymbol{Y}_{i,t}}{\sum_{i=1}^{N}\sum_{t=1}^{T} \mathbb{E}\left[w_{i,t,u} \mid \boldsymbol{Y}_{i,1},\ldots,\boldsymbol{Y}_{i,T}\right]}$$

Concerning the second state of the algorithm, we focus on the estimation of $\boldsymbol{\Lambda}_u$ and $\boldsymbol{\Psi}_u$ by first computing the following conditional expectation

$$\begin{aligned}
\mathcal{H}(\boldsymbol{\theta}_\Sigma) = C &+ \sum_{i=1}^{N}\sum_{t=1}^{T}\sum_{u=1}^{K} \frac{\widehat{w}_{i,t,u}}{2} \log\left(|\boldsymbol{\Psi}_u^{-1}|\right) \\
&- \sum_{i=1}^{N}\sum_{t=1}^{T}\sum_{u=1}^{K} \frac{\widehat{w}_{i,t,u}}{2} \operatorname{tr}\left(\boldsymbol{\Psi}_u^{-1}\widetilde{\boldsymbol{\Sigma}}_u\right) \\
&+ \sum_{i=1}^{N}\sum_{t=1}^{T} \widehat{w}_{i,t,u}\left(\boldsymbol{Y}_{i,t}-\boldsymbol{\mu}_u\right)\boldsymbol{\Psi}_u^{-1}\boldsymbol{\Lambda}_u\, \mathbb{E}\left[\boldsymbol{\eta}_{i,t,u} \mid \boldsymbol{Y}_{i,t},\boldsymbol{\mu}_u,\boldsymbol{\Lambda}_u,\boldsymbol{\Psi}_u\right] \\
&- \frac{1}{2}\operatorname{tr}\left\{\boldsymbol{\Lambda}_u'\boldsymbol{\Psi}_u\boldsymbol{\Lambda}_u \sum_{i=1}^{N}\sum_{u=1}^{K} \mathbb{E}\left[\boldsymbol{\eta}_{i,t,u}\boldsymbol{\eta}_{i,t,u}' \mid \boldsymbol{Y}_{i,t},\boldsymbol{\mu}_u,\boldsymbol{\Lambda}_u,\boldsymbol{\Psi}_u\right]\right\}
\end{aligned} \tag{3.16}$$

where the $\widehat{w}_{i,t,u}$ are obtained from the previous stage and C is a constant.

Fixing the quantities

$$\boldsymbol{\Omega} = \boldsymbol{I}_R - \boldsymbol{\beta}\boldsymbol{\Lambda} + \boldsymbol{\beta}\widetilde{\boldsymbol{\Sigma}}\boldsymbol{\beta}' \qquad \text{and} \qquad \boldsymbol{\Omega}_u = \boldsymbol{I}_R - \boldsymbol{\beta}_u\boldsymbol{\Lambda} + \boldsymbol{\beta}_u\widetilde{\boldsymbol{\Sigma}}_u\boldsymbol{\beta}_u',$$

it is possible to derive model-specific updates for $\boldsymbol{\Lambda}_u$ and $\boldsymbol{\Psi}_u$, denoted henceforth as $\left(\widehat{\boldsymbol{\Lambda}}_u, \widehat{\boldsymbol{\Psi}}_u\right)$, as follows:

- Model CCC:

$$\widehat{\boldsymbol{\Lambda}} = \widetilde{\boldsymbol{\Sigma}}\boldsymbol{\beta}'\boldsymbol{\Omega}^{-1} \qquad\qquad \widehat{\psi} = \frac{1}{R}\operatorname{tr}\left(\widetilde{\boldsymbol{\Sigma}}' - \widehat{\boldsymbol{\Lambda}}\boldsymbol{\beta}\widetilde{\boldsymbol{\Sigma}}\right)$$

  where $\boldsymbol{\beta} = \boldsymbol{\Lambda}'\left(\boldsymbol{\Lambda}\boldsymbol{\Lambda}' + \psi\boldsymbol{I}_R\right)^{-1}$.

- Model CCU:

$$\widehat{\boldsymbol{\Lambda}} = \widetilde{\boldsymbol{\Sigma}}\boldsymbol{\beta}'\boldsymbol{\Omega}^{-1} \qquad \widehat{\boldsymbol{\Psi}} = \text{diag}\left\{\widetilde{\boldsymbol{\Sigma}} - \widehat{\boldsymbol{\Lambda}}\boldsymbol{\beta}\widetilde{\boldsymbol{\Sigma}}\right\}$$

with $\boldsymbol{\beta} = \boldsymbol{\Lambda}'\left(\boldsymbol{\Lambda}\boldsymbol{\Lambda}' + \boldsymbol{\Psi}\right)^{-1}$.

- Model CUC:

$$\widehat{\boldsymbol{\Lambda}} = \left[\sum_{u=1}^{K} \frac{\sum_{i=1}^{N}\sum_{t=1}^{T} \widehat{w}_{i,t,u}}{\psi_u} \widetilde{\boldsymbol{\Sigma}}\,\boldsymbol{\beta}'_u\right]\left[\sum_{u=1}^{K} \frac{\sum_{i=1}^{N}\sum_{t=1}^{T} \widehat{w}_{i,t,u}}{\psi_u}\boldsymbol{\Omega}_u\right]^{-1}$$

and

$$\widehat{\psi}_u = \frac{1}{R}\text{tr}\left\{\widetilde{\boldsymbol{\Sigma}} - 2\widehat{\boldsymbol{\Lambda}}\boldsymbol{\beta}_u\widetilde{\boldsymbol{\Sigma}} + \widehat{\boldsymbol{\Lambda}}\boldsymbol{\Omega}_u\widehat{\boldsymbol{\Lambda}}'\right\}$$

where $\boldsymbol{\beta}_u = \boldsymbol{\Lambda}'\left(\boldsymbol{\Lambda}\boldsymbol{\Lambda}' + \psi_u\boldsymbol{I}_R\right)^{-1}$.

- Model CUU:

$$\widehat{\boldsymbol{\lambda}}_j = \boldsymbol{r}_j\left(\sum_{u=1}^{K} \frac{\sum_{i=1}^{N}\sum_{t=1}^{T} \widehat{w}_{i,t,u}}{\psi_{u,j}}\boldsymbol{\Omega}_u\right) \qquad \widehat{\boldsymbol{\Psi}}_u = \text{diag}\left\{\widetilde{\boldsymbol{\Sigma}} - 2\widehat{\boldsymbol{\Lambda}}\boldsymbol{\beta}_u\widetilde{\boldsymbol{\Sigma}} + \widehat{\boldsymbol{\Lambda}}\boldsymbol{\Omega}_u\widehat{\boldsymbol{\Lambda}}'\right\}$$

where $\boldsymbol{\beta}_u = \boldsymbol{\Lambda}'\left(\boldsymbol{\Lambda}\boldsymbol{\Lambda}' + \boldsymbol{\Psi}_u\right)^{-1}$, $\widehat{\boldsymbol{\lambda}}_j$ denotes the j-th row of the matrix $\widehat{\boldsymbol{\Lambda}}$, $\psi_{u,j}$ is the j-th element along the diagonal of $\boldsymbol{\Psi}_u$, and $\boldsymbol{r}_j$ denotes the j-th row of the matrix $\sum_{u=1}^{K}\left(\sum_{i=1}^{N}\sum_{t=1}^{T}\widehat{w}_{i,t,u}\boldsymbol{\Psi}_u^{-1}\widetilde{\boldsymbol{\Sigma}}_u\boldsymbol{\beta}'_u\right)$ for $j = 1, \ldots, R$.

- Model UCC:

$$\widehat{\boldsymbol{\Lambda}}_u = \widetilde{\boldsymbol{\Sigma}}_u\boldsymbol{\beta}'_u\boldsymbol{\Omega}_u^{-1} \qquad \widehat{\psi} = \frac{1}{R}\sum_{u=1}^{K}\widehat{\pi}_u\text{tr}\left\{\widetilde{\boldsymbol{\Sigma}}_u - \widehat{\boldsymbol{\Lambda}}_u\boldsymbol{\beta}_u\widetilde{\boldsymbol{\Sigma}}_u\right\}$$

with $\boldsymbol{\beta}_u = \boldsymbol{\Lambda}'_u\left(\boldsymbol{\Lambda}_u\boldsymbol{\Lambda}'_u + \psi\boldsymbol{I}_R\right)^{-1}$.

- Model UCU:

$$\widehat{\boldsymbol{\Lambda}}_u = \widetilde{\boldsymbol{\Sigma}}_u\boldsymbol{\beta}'_u\boldsymbol{\Omega}_u^{-1} \qquad \widehat{\boldsymbol{\Psi}} = \sum_{u=1}^{K}\widehat{\pi}_u\text{diag}\left\{\widetilde{\boldsymbol{\Sigma}}_u - \widehat{\boldsymbol{\Lambda}}_u\boldsymbol{\beta}_u\widetilde{\boldsymbol{\Sigma}}_u\right\}$$

with $\boldsymbol{\beta}_u = \boldsymbol{\Lambda}'_u\left(\boldsymbol{\Lambda}_u\boldsymbol{\Lambda}'_u + \boldsymbol{\Psi}\right)^{-1}$.

- Model UUC:

$$\widehat{\boldsymbol{\Lambda}}_u = \widetilde{\boldsymbol{\Sigma}}_u\boldsymbol{\beta}'_u\boldsymbol{\Omega}_u^{-1} \qquad \widehat{\psi}_u = \frac{1}{R}\text{tr}\left\{\widetilde{\boldsymbol{\Sigma}}_u - \widehat{\boldsymbol{\Lambda}}_u\boldsymbol{\beta}_u\widetilde{\boldsymbol{\Sigma}}_u\right\}$$

where $\boldsymbol{\beta}_u = \boldsymbol{\Lambda}'_u\left(\boldsymbol{\Lambda}_u\boldsymbol{\Lambda}'_u + \psi_u\boldsymbol{I}_R\right)^{-1}$.

- Model UUU:

$$\widehat{\boldsymbol{\Lambda}}_u = \widetilde{\boldsymbol{\Sigma}}_u\boldsymbol{\beta}'_u\boldsymbol{\Omega}_u^{-1} \qquad \widehat{\boldsymbol{\Psi}}_u = \text{diag}\left\{\widetilde{\boldsymbol{\Sigma}}_u - \widehat{\boldsymbol{\Lambda}}_u\boldsymbol{\beta}_u\widetilde{\boldsymbol{\Sigma}}_u\right\}$$

with $\boldsymbol{\beta}_u = \boldsymbol{\Lambda}'_u\left(\boldsymbol{\Lambda}_u\boldsymbol{\Lambda}'_u + \boldsymbol{\Psi}_u\right)^{-1}$.

The approach described above does not yield standard errors for the parameter estimates, as approximations relying on the observed information matrix typically require very large sample sizes to be reliable. To address this limitation in the context of Hidden Markov Models (HMMs), Visser et al. (2000) and Bulla et al. (2012) examined the estimation of confidence intervals and recommended the use of bootstrap-based methods. Following their guidance, a parametric bootstrap procedure was adopted. Specifically, the model was re-estimated on 100 bootstrap samples, each simulated from the originally estimated parameters. This approach provides an empirical distribution of the estimates, allowing for the construction of confidence intervals that are robust to small-sample issues and the potential non-regularity of the likelihood surface commonly encountered in HMMs. An alternative to parametric bootstrap is given by Bartolucci and Farcomeni (2015). The selection of starting values for implementing the algorithm to obtain maximum likelihood estimates is not an element of the model specification itself. Nonetheless, their proper choice is critical, as poor initialization can lead to convergence failures or misleading inference. In practice, the goal is to identify plausible starting values that are informed by the observed data. The underlying rationale we follow here is straightforward: the closer the initial values are to the global maximum of the likelihood function, the higher the probability that the optimization algorithm will converge to that maximum. This reduces the risk of the algorithm becoming trapped in local optima or failing to converge at all, particularly when the initial values lie far from the true maximum in the parameter space. In this sense, initializing the parameters with values that are reasonably close to the true estimates enhances both the stability and the efficiency of the estimation process. In practice, we follow Maruotti and Punzo (2021) and initialize the parameters starting from partitional clustering methods, assuming independence across observations over time.

3.4.3 Path Prediction

One of the primary objectives in the context of LMMs is the reconstruction of the latent cluster sequence underlying the observed data. Specifically, given the full sequence of observations $\{\mathbf{y}_{i,1}, \mathbf{y}_{i,2}, \ldots, \mathbf{y}_{i,T}\}$, one may wish to determine the most probable sequence of latent clusters $\{u_{i,1}, u_{i,2}, \ldots, u_{i,T}\}$ that has generated the data. This is formally obtained by maximizing the posterior probability of the latent path:

$$\Pr(U_{i,1} = u_{i,1}, U_{i,2} = u_{i,2}, \ldots, U_{i,T} = u_{i,T} \mid \mathbf{y}_{i,1}, \ldots, \mathbf{y}_{i,T}),$$

with respect to the sequence $\{u_{i,1}, u_{i,2}, \ldots, u_{i,T}\}$. Unlike local decoding of clusters, which may lead to inconsistent state transitions, this approach—known as global decoding—ensures that the most likely entire path is selected, taking into account the joint structure of the latent sequence.

The most widely used method to perform global decoding is the Viterbi algorithm Viterbi (1967), which provides an efficient recursive procedure to obtain the maximum a posteriori sequence of clusters. The algorithm consists of a forward recursion to compute the optimal path probabilities up to each time point, followed by a backward recursion to reconstruct the corresponding latent clusters. Let $\rho_{i,t}(k)$ denote the highest probability of any cluster sequence ending in state k at time t for unit i, jointly with the observed data up to time t:

$$\rho_{i,t}(v) = \max_{u_{i,1},\ldots,u_{i,t-1}} \Pr(U_{i,1} = u_{i,1}, \ldots, U_{i,t-1} = u_{i,t-1}, U_{i,t} = v, \mathbf{y}_{i,1}, \ldots, \mathbf{y}_{i,t}).$$

The algorithm proceeds as follows:

1. **Initialization (at $t = 1$):**

$$\rho_{i,1}(u) = \pi_u \cdot f(\mathbf{y}_{i,1} \mid U_{i,1} = u), \quad \text{for } u = 1, \ldots, K.$$

2. **Forward recursion (for $t = 2, \ldots, T$):**

$$\rho_{i,t}(u) = f(\mathbf{y}_{i,t} \mid U_{i,t} = u) \cdot \max_v \left[\rho_{i,t-1}(v) \cdot \pi_{u,v}\right], \quad \text{for each } u, v = 1, \ldots, K.$$

3. **Termination:** Identify the final state in the most likely cluster at time T as:

$$u_{i,T} = \arg\max_k \rho_{i,T}(u).$$

4. **Backward recursion (for $t = T-1, \ldots, 1$):** Reconstruct the optimal path by:

$$u_{i,t} = \arg\max_v \left[\rho_{i,t}(v) \cdot \pi_{u_{i,t+1},v}\right].$$

This dynamic programming procedure ensures that the jointly most probable sequence is selected efficiently, avoiding the exponential complexity of brute-force enumeration over all possible state sequences.

All quantities involved in the algorithm are evaluated using the maximum likelihood estimates of the model parameters. The resulting sequence $\{u_{i,t}\}$ offers a clear and interpretable classification of the time series into latent regimes, which can be used for further risk stratification, clustering, or diagnostic purposes.

3.5 Data Analysis

We fitted all eight models (see Table 3.1) to the data for$K = 1, \ldots, 8$ and $q = \{1, 2\}$, as the number of factors q must respect $(P - q)^2 > P + q$ (see Lawley and Maxwell (1962)), and computed the akaike information criterion (AIC), the Bayesian information criterion, and integrated completed likelihood (ICL) values for each model. The ICL essentially penalizes the BIC for estimated mean entropy, and it is given by

$$ICL = BIC + \sum_{i=1}^{N} \sum_{t=1}^{T} \sum_{k=1}^{K} \hat{w}_{i,t,k} \log \hat{w}_{i,t,k}.$$

All criteria involve penalization terms depending on the number of redundant parameters, which is equal to the sum of the following:

- $(K - 1)$ initial probabilities,
- $K(K - 1)$ transition probabilities,
- KR state-specific means, collected in $\boldsymbol{\mu}_k, k = 1, \ldots, K$,
- covariance parameters, as defined in Table 3.1.

Figure 3.3 shows the minimum values over the eight models of each pair (q, K) for all the criteria. We would like to remark that all the criteria select the UUU model with $K = 4$ and $Q = 2$.

It should be noted that the selection of the number of hidden states in HMMs remains an open and complex problem, despite the substantial literature on parameter estimation. This challenge arises due to the intricate dependence structure of HMMs and the latent nature of the hidden states. Empirical evaluations have become increasingly important for

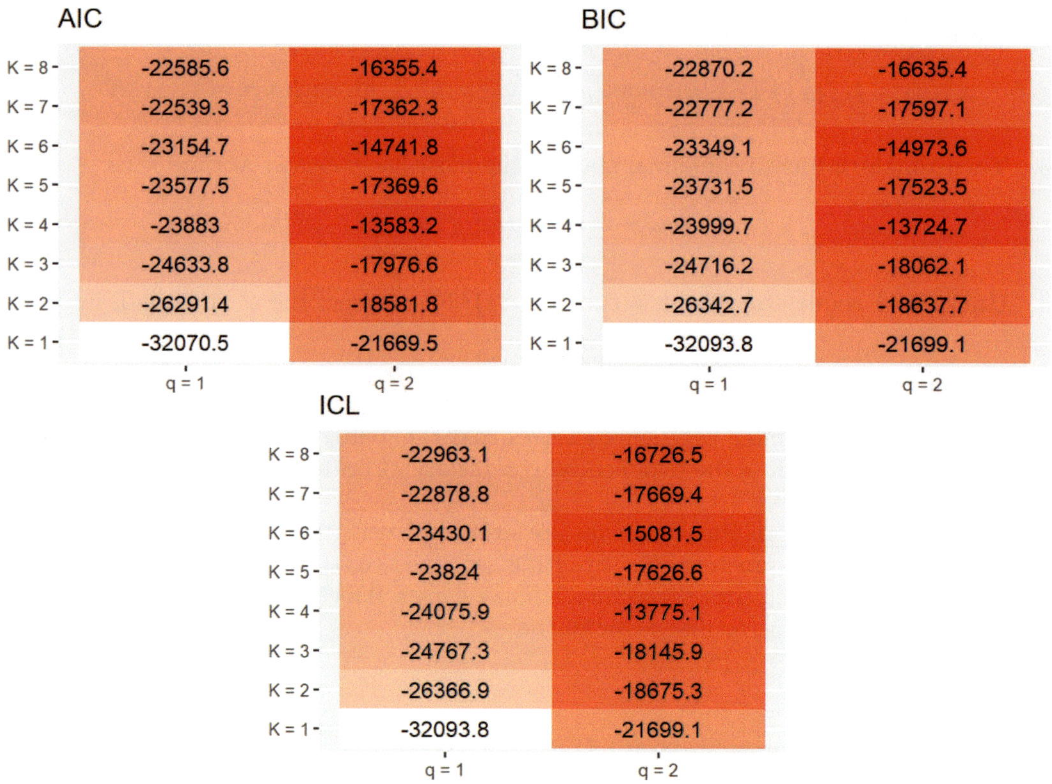

FIGURE 3.3
Heat map of model selection criteria. Heat maps of AIC, BIC, ICL values for each (K, q). For fixed (K, q), the minimum value is determined from all eight models.

understanding the practical behavior of model selection criteria and several simulation-based investigation are available in the literature Buckby et al. (2023); Pohle et al. (2017); Zou et al. (2024).

It is possible to quantify the uncertainty surrounding the obtained classification by looking at histograms or rootograms of the posterior state probabilities. In fact, both are suitable to assess the hidden structure. As a more informative alternative, a scatter plot of the state-specific posterior probabilities may be used: a good classification results in the distribution to be concentrated near the vertices (0,0), (0,1), and (1,0). In our analysis, States are somehow ordered, with State 1 being the one with the *best* environmental conditions. This State is well separated then from States 3 and 4, the worse states, as expected. State 2 is the *normal* conditions state, and it overlaps with States 1 and 4. Similarly, in a few cases States 3 and 4 overlaps (see Figure 3.4). Overall, the uncertainty surrounding the classification is limited and we can use a maximum-a-posteriori allocation without many concerns.

This allows us to describe the overall pollution exposure in terms of percentages of monitoring networks clustered in each state at different times. From Figure 3.5, it is rather clear that State 2, the normal state, is the most visited over the entire period. There are however day where the pollutants exposure is higher, and accordingly the probability of States 3 and 4 increases. State 1, instead, is rarely visited and, as we will discuss later, in a few monitoring stations only.

FIGURE 3.4
Uncertainty surrounding the classification.

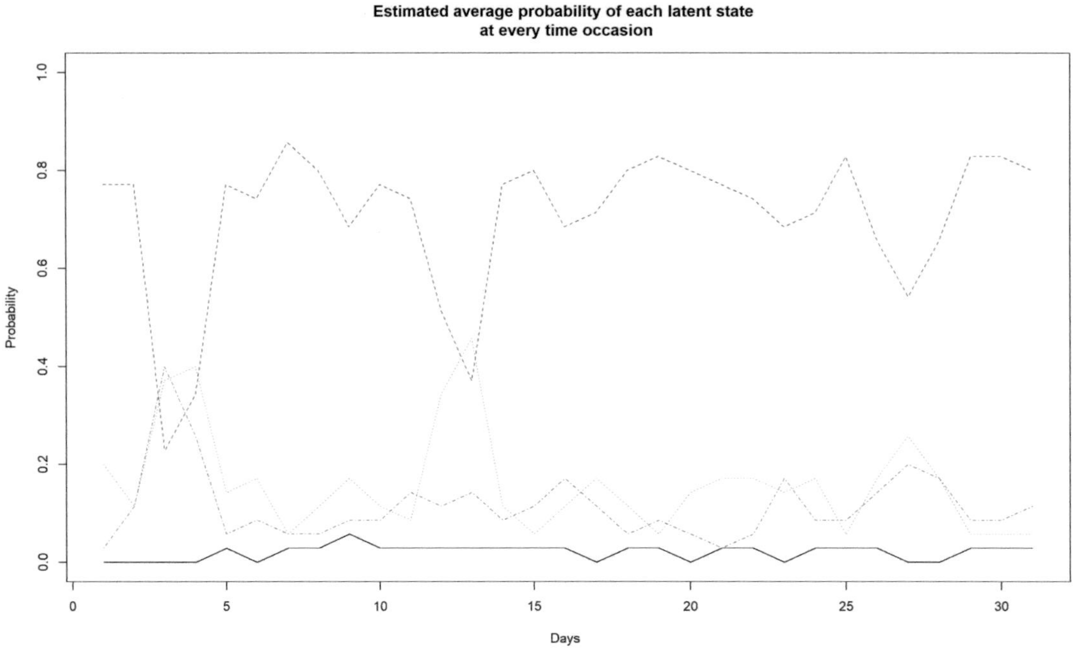

FIGURE 3.5
Estimated average probability of each latent state at every time occasion. State 1: black line; State 2: red line; State 3: green line; State 4: blue line.

To further illustrate the clustering capabilities of the proposed model-based framework, we present in Figure 3.6 the inferred hidden state allocations across monitoring stations and days. The classification into latent environmental states reveals clear heterogeneity in pollution exposure profiles, both across space and over time. The analysis shows that only a minority of monitoring stations—seven in total—were assigned to State 1, which is associated with good air quality conditions. Moreover, the assignment to State 1 is typically limited to just a few days during the observation period. Among all sites, the High Muffles monitoring station stands out as the location with the most prolonged exposure to clean air, being classified in State 1 for 11 days. This result is consistent with expectations based on the site's physical characteristics: located in a remote clearing within a forest plantation, approximately 100 metres from the nearest habitation and 200 metres from a minor rural road, the station is officially designated as a rural background site. The surrounding area is open and sparsely populated, making High Muffles an ideal benchmark location for understanding pollution exposure in the absence of significant anthropogenic influence.

In contrast, several stations exhibit persistent exposure to high pollution conditions, as indicated by frequent assignment to States 3 and 4, which are associated with poor air quality. These include:

- Birmingham Watery Lane Middleway, situated adjacent to the A4540 ring road and classified as urban traffic. The monitoring station is just 6.5 meters from the roadside, contributing to high pollutant concentrations and 26 days spent in one of the two most polluted states.
- Salford Eccles, officially classified as urban background, but located near both residential developments and industrial infrastructure. Although surrounded by suburban properties, the proximity to the M602 motorway, the A56/A567 arterial roads, and a chemical plant

situated only 350 meters to the south contribute to frequent episodes of elevated pollution, with 18 days in the worst states.

- London Harlington, a station located on the grounds of Imperial College's Sports Facility in Harlington, approximately 1 kilometer north of Heathrow Airport. Despite being surrounded by open fields and residential areas, its proximity to the M4 motorway (800 metres away) and its classification as urban industrial explains the 18 days of assignment to highly polluted states.

- Stoke-on-Trent Centre, located in the heart of Hanley's central business district, is only 5 metres from a road leading to a heavily trafficked multi-storey car park, and approximately 130 metres from the A50 Potteries Way ring road. These characteristics place the site in an area with substantial vehicular emissions, accounting for 25 days in States 3 or 4.

These findings not only validate the ability of the proposed model to capture pollution exposure consistent with known site characteristics, but also emphasize the value of the latent state structure in providing a categorical summary of environmental exposure conditions. This approach enables a concise and interpretable stratification of monitoring locations based on empirical concentration profiles, offering a tool for both scientific understanding and policy-relevant risk communication.

The temporal dynamics of the inferred environmental states are governed by the estimated transition probability matrix, which describes how the system evolves from one latent state to another over time. The matrix (with standard errors in brackets) is expressed as follows.

$$\begin{bmatrix} 0.486 & 0.514 & 0.000 & 0.000 \\ (0.066) & (0.074) & (0.027) & (0.032) \\ 0.019 & 0.810 & 0.061 & 0.110 \\ (0.042) & (0.111) & (0.038) & (0.033) \\ 0.000 & 0.259 & 0.560 & 0.181 \\ (0.039) & (0.055) & (0.057) & (0.036) \\ 0.000 & 0.519 & 0.077 & 0.404 \\ (0.040) & (0.097) & (0.039) & (0.121) \end{bmatrix}$$

From this matrix, several key features of the hidden process emerge. Most notably, State 2 plays a central role in the system's dynamics: it is the only state that communicates with all others, acting as a hub or intermediary between more extreme pollution regimes. Furthermore, it exhibits the highest self-transition probability (0.810), confirming its persistence and stabilizing effect within the latent structure. In contrast, States 1, 3, and 4 display more asymmetrical transitions, suggesting that once the system enters State 2, it is more likely to remain there or shift to another state through it, rather than transitioning directly between peripheral states.

To better understand the environmental meaning of each latent state, Table 3.2 reports the state-specific mean pollutant concentrations across the five measured pollutants. These empirical summaries offer a concise interpretation of the latent structure uncovered by the model.

These values clearly reflect an increasing gradient of pollution intensity from State 1 (lowest exposure) to State 4 (highest exposure), supporting the interpretability of the hidden states in terms of air quality regimes. Specifically:

- State 1 represents periods of clean air conditions, with all pollutants at minimal levels, aligning well with rural background areas.

- State 2, though still relatively low in pollutant concentration, exhibits elevated NO_x and particulate levels compared to State 1, representing a mildly polluted but persistent state.

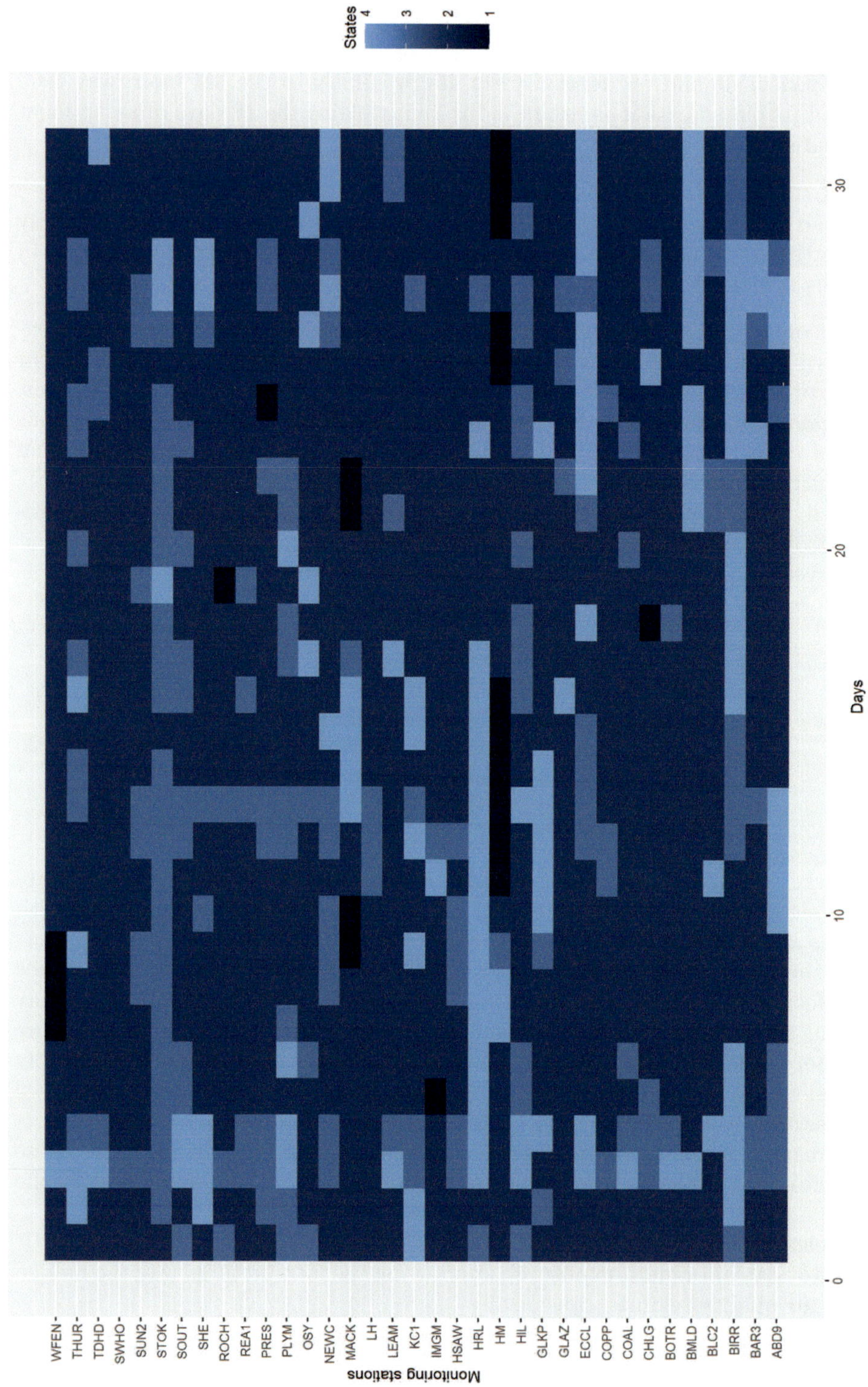

FIGURE 3.6
Most probable latent state for each monitoring station at each time.

TABLE 3.2
Cluster-specific mean concentrations (μg/m³ for NO, NO_2, NO_x, PM_{10}; $PM_{2.5}$), with standard errors in brackets.

	State 1	**State 2**	**State 3**	**State 4**
NO_x	3.824	12.386	31.546	44.144
	(1.393)	(1.907)	(1.905)	(4.855)
NO_2	3.145	10.141	21.125	22.063
	(1.165)	(1.368)	(1.240)	(1.112)
NO	0.445	1.464	6.796	14.406
	(0.339)	(0.455)	(0.547)	(1.230)
PM_{10}	5.030	7.334	11.802	9.576
	(1.207)	(0.299)	(0.541)	(0.529)
$PM_{2.5}$	2.996	4.452	7.534	6.205
	(0.720)	(0.188)	(0.384)	(0.395)

- States 3 and 4 indicate severe pollution episodes, with high NO_x concentrations and notably elevated NO values in particular—pointing to primary traffic emissions. The distinction between them lies in the relative levels of PM_{10} and $PM_{2.5}$, which are slightly lower in State 4 despite extremely high NO_x levels, possibly reflecting variations in source type or meteorological dispersion.

Taken together, the transition dynamics and the state-specific mean profiles provide a rich and interpretable latent structure for dynamic environmental exposure classification, enabling not only risk stratification but also early-warning identification of transitions into high-risk pollution regimes.

A key advantage of our modeling framework lies in its ability to uncover and further characterize distinct latent exposures by cluster-specific correlation structure among pollutants. These structures ($\widetilde{\Sigma}_u$, displayed below) not only reflect the intensity of pollution exposure, but also reveal the interdependencies and divergence between various pollutant types under different atmospheric and emission clusters. To this end, we examine the correlation matrices derived from the estimated state-specific covariance matrices across the four latent clusters identified by the LMM. The pollutants under analysis are known to have diverse atmospheric behaviors and sources, ranging from direct traffic emissions to secondary particle formation, and from short-lived reactive gases to more persistent particulates. The varying patterns of their co-movements thus provide insight into the nature and origin of pollution episodes. The first latent state, associated with the cleanest air quality profile, displays a well-structured and moderate-to-high correlation pattern among pollutants, especially between NO_x and its components. These strong associations are consistent with coherent emissions from a common source, most likely vehicular combustion, under conditions of low background concentrations. Particulate matter shows moderate positive correlations with NO_x and NO_2, suggesting some degree of source overlap, possibly due to secondary formation or combustion residues. Overall, State 1 reflects stable and uniform environmental conditions, typical of rural or low-traffic urban settings. In contrast, State 2 exhibits a dramatic breakdown in the correlation structure between gaseous and particulate pollutants. While NO_x and its components maintain an extremely high correlation, their relationships with PM components are weak or even slightly negative. These patterns point to a decoupling between combustion gases and particulate concentrations, possibly due to e.g., meteorological dispersion events, localized non-overlapping sources, or periods following precipitation, which may remove PM from the atmosphere more effectively than gases. This is the most common condition in the observational period. State 3 captures a markedly different exposure, characterized by widespread and simultaneous pollution across all variables. The correlation matrix is dominated by extremely high pairwise associations. These patterns suggest a scenario in which all pollutants—both gaseous and particulate—are being emitted or accumulated together, possibly due to e.g. atmospheric stagnation, temperature inversion layers, or peak-hour traffic and heating emissions during winter. The strong synchronization of pollutants implies that State 3 represents severe pollution episodes, warranting targeted environmental alerts. The high coupling across pollutants also supports the necessity of multivariate modeling approaches, as the impact of one pollutant cannot be disentangled from others under such conditions. Finally, State 4 presents a scenario of pollutant-family-specific correlations between NO-types and the particulate components. Each cluster encapsulates a unique interaction exposure. The variability in the strength and direction of correlations across states provides strong empirical support for the latent classification derived from the LMM. Moreover, it highlights the critical need for dynamic, cluster-aware multivariate modeling in environmental risk assessment, as fixed correlation assumptions may fail to capture the true complexity of air pollution dynamics.

$$\widetilde{\Sigma}_1 = \begin{bmatrix} 1.000 & 0.802 & 0.982 & 0.599 & 0.643 \\ 0.802 & 1.000 & 0.675 & 0.440 & 0.445 \\ 0.982 & 0.675 & 1.000 & 0.601 & 0.654 \\ 0.599 & 0.440 & 0.601 & 1.000 & 0.393 \\ 0.643 & 0.445 & 0.654 & 0.393 & 1.000 \end{bmatrix} \quad \widetilde{\Sigma}_2 = \begin{bmatrix} 1.000 & 0.963 & 0.897 & -0.186 & -0.016 \\ 0.963 & 1.000 & 0.745 & -0.123 & 0.054 \\ 0.897 & 0.745 & 1.000 & -0.259 & -0.129 \\ -0.186 & -0.123 & -0.259 & 1.000 & 0.057 \\ -0.016 & 0.054 & -0.129 & 0.057 & 1.000 \end{bmatrix}$$

$$\widetilde{\Sigma}_3 = \begin{bmatrix} 1.000 & 0.999 & 0.963 & 0.303 & 0.358 \\ 0.999 & 1.000 & 0.956 & 0.275 & 0.330 \\ 0.963 & 0.956 & 1.000 & 0.512 & 0.560 \\ 0.303 & 0.275 & 0.512 & 1.000 & 0.990 \\ 0.358 & 0.330 & 0.560 & 0.990 & 1.000 \end{bmatrix} \quad \widetilde{\Sigma}_4 = \begin{bmatrix} 1.000 & 0.979 & 0.751 & 0.236 & 0.314 \\ 0.979 & 1.000 & 0.599 & 0.286 & 0.374 \\ 0.751 & 0.599 & 1.000 & 0.003 & 0.023 \\ 0.236 & 0.286 & 0.003 & 1.000 & 0.159 \\ 0.314 & 0.374 & 0.023 & 0.159 & 1.000 \end{bmatrix}$$

3.6 Conclusion

In this chapter, we have proposed a flexible and parsimonious class of multivariate LMMs for the dynamic clustering of environmental monitoring stations, with the goal of identifying evolving risk profiles of pollutant exposure. Our approach is motivated by the analysis of air pollution data and addresses key challenges in modeling high-dimensional, temporally dependent data.

By integrating a latent Markov chain with a factor-analytic decomposition of the covariance structure, the proposed model accommodates both temporal dependence and cross-sectional heterogeneity in a coherent probabilistic framework. The factor model structure allows for dimension reduction and enhances model interpretability, while the latent Markov chain captures regime-switching behavior and time-varying risk patterns. The adoption of an AECM algorithm facilitates efficient parameter estimation under the proposed model structure.

Our empirical results highlight the heterogeneity in pollution exposure across both monitoring stations and time. The inferred latent states correspond to interpretable pollution exposures, ranging from clean-air conditions to severe multi-pollutant episodes. The clustering dynamics reflect known urban features of the monitoring stations and underscore the importance of accounting for both latent state transitions and pollutant interdependence.

Moreover, the model's capacity to jointly model multiple pollutants makes it a valuable tool for exposure assessment, policy evaluation, and early-warning systems in environmental monitoring. Future research could explore extensions of the proposed framework, including the incorporation of meteorological covariates, spatial dependence, or the transition to continuous-time specifications.

Overall, the proposed multivariate latent Markov framework contributes to a growing class of model-based approaches for environmental risk assessment, offering both methodological innovation and practical utility in characterizing complex pollution patterns over time.

Bibliography

Bartolucci, F. and Farcomeni, A. (2009). A multivariate extension of the dynamic logit model for longitudinal data based on a latent Markov heterogeneity structure. *Journal of the American Statistical Association*, 104(486):816–831.

Bartolucci, F. and Farcomeni, A. (2015). Information matrix for hidden Markov models with covariates. *Statistics and Computing*, 25:515–526.

Bartolucci, F., Farcomeni, A., and Pennoni, F. (2013). *Latent Markov Models for Longitudinal Data.* Chapman & Hall/CRC Press, Boca Raton, FL.

Baum, L., Petrie, T., Soules, G., and Weiss, N. (1970). A maximization technique occurring in the statistical analysis of probabilistic functions of Markov chains. *Annals of Mathematical Statistics*, 41:164–171.

Boaz, R., Lawson, A., and Pearce, J. (2019). Multivariate air pollution prediction modeling with partial missingness. *Environmetrics*, 30(7):e2592.

Bose, M., Larson, T., and Szpiro, A. A. (2018). Adaptive predictive principal components for modeling multivariate air pollution. *Environmetrics*, 29(8):e2525.

Bouveyron, C. and Brunet-Saumard, C. (2014). Model-based clustering of high-dimensional data: A review. *Computational Statistics & Data Analysis*, 71:52–78.

Buckby, J., Wang, T., Fletcher, D., Zhuang, J., Takeo, A., and Obara, K. (2023). Finding the number of latent states in hidden Markov models using information criteria. *Environmental and Ecological Statistics*, 30(4):797–825.

Bulla, J., Lagona, F., Maruotti, A., and Picone, M. (2012). A multivariate hidden markov model for the identification of sea regimes from incomplete skewed and circular time series. *Journal of Agricultural, Biological, and Environmental Statistics*, 17:544–567.

Carslaw, D. C. and Ropkins, K. (2012). Openair—an R package for air quality data analysis. *Environmental Modelling & Software*, 27:52–61.

Draper, E. L., Whyatt, J. D., Taylor, R. S., and Metcalfe, S. E. (2023). Estimating background concentrations of PM2. 5 for urban air quality modelling in a data poor environment. *Atmospheric Environment*, 314:120107.

Field, M., Stirling, D., Pan, Z., and Naghdy, F. (2016). Learning trajectories for robot programing by demonstration using a coordinated mixture of factor analyzers. *IEEE Trans Cybern*, 46:706–717.

Gianquintieri, L., Oxoli, D., Caiani, E. G., and Brovelli, M. A. (2024). State-of-art in modelling particulate matter (PM) concentration: A scoping review of aims and methods. *Environment, Development and Sustainability*, 27:25889–25911.

He, K., Wang, Y., Su, W., and Yang, H. (2022). A varying-coefficient regression approach to modeling the effects of wind speed on the dispersion of pollutants. *Environmental and Ecological Statistics*, 29(2):433–452.

Lagona, F., Maruotti, A., and Picone, M. (2011). A non-homogeneous hidden Markov model for the analysis of multi-pollutant exceedances data. *Hidden Markov Models, Theory and Applications*, 10:207–222. https://doi.org/10.5772/14749

Lawley, D. N. and Maxwell, A. E. (1962). Factor analysis as a statistical method. *Journal of the Royal Statistical Society. Series D (The Statistician)*, 12(3):209–229.

Marino, M. F., Tzavidis, N., and Alfò, M. (2018). Mixed hidden Markov quantile regression models for longitudinal data with possibly incomplete sequences. *Statistical Methods in Medical Research*, 27(7):2231–2246.

Martínez-Silva, I., Roca-Pardiñas, J., and Ordóñez, C. (2016). Forecasting SO2 pollution incidents by means of quantile curves based on additive models. *Environmetrics*, 27(3):147–157.

Martinez-Zarzoso, I. and Maruotti, A. (2013). The environmental Kuznets curve: functional form, time-varying heterogeneity and outliers in a panel setting. *Environmetrics*, 24(7):461–475.

Maruotti, A. (2015). Handling non-ignorable dropouts in longitudinal data: a conditional model based on a latent Markov heterogeneity structure. *Test*, 24(1):84–109.

Maruotti, A., Bulla, J., Lagona, F., Picone, M., and Martella, F. (2017). Dynamic mixtures of factor analyzers to characterize multivariate air pollutant exposures. *The Annals of Applied Statistics*, 11(3):1617 – 1648.

Maruotti, A. and Punzo, A. (2021). Initialization of hidden markov and semi-markov models: A critical evaluation of several strategies. *International Statistical Review*, 89(3):447–480.

Maruotti, A., Punzo, A., Mastrantonio, G., and Lagona, F. (2016). A time-dependent extension of the projected normal regression model for longitudinal circular data based on a hidden Markov heterogeneity structure. *Stochastic Environmental Research and Risk Assessment*, 30:1725–1740.

Maruotti, A. and Rocci, R. (2012). A mixed non-homogeneous hidden Markov model for categorical data, with application to alcohol consumption. *Statistics in Medicine*, 31:871–886.

McNicholas, P. D. and Brendan, M. T. (2008). Parsimonious Gaussian mixture models. *Stat Compt*, 18:285–296.

Neal, M. R., Sochaniwsky, A. A., and McNicholas, P. D. (2024). Hidden Markov models for multivariate panel data. *Statistics and Computing*, 34(6):182.

Pennoni, F., Bartolucci, F., and Pandolfi, S. (2024). Variable selection for hidden markov models with continuous variables and missing data. *Journal of Classification*, 41:568–589.

Pohle, J., Langrock, R., Van Beest, F. M., and Schmidt, N. M. (2017). Selecting the number of states in hidden Markov models: Pragmatic solutions illustrated using animal movement. *Journal of Agricultural, Biological and Environmental Statistics*, 22:270–293.

Punzo, A., Ingrassia, S., and Maruotti, A. (2021). Multivariate hidden Markov regression models: Random covariates and heavy-tailed distributions. *Statistical Papers*, 62:1519–1555.

Russo, A. and Farcomeni, A. (2024). A copula formulation for multivariate latent Markov models. *TEST*, 33:731–751.

Russo, A., Farcomeni, A., Pittau, M. G., and Zelli, R. (2024). Covariate-modulated rectangular latent Markov models with an unknown number of regime profiles. *Statistical Modeling*, 24:368–388.

Tomarchio, S. D., Punzo, A., and Maruotti, A. (2022). Parsimonious hidden Markov models for matrix-variate longitudinal data. *Statistics and Computing*, 32:53.

Uluocak, I., Pinar, E., and Bilgili, M. (2025). Atmospheric NO2 concentration prediction with statistical and hybrid deep learning methods. *Environmental and Ecological Statistics*, pages 1–30.

Visser, I. (2011). Seven things to remember about hidden Markov models: A tutorial on Markovian models for time series. *Journal of Mathematical Psychology*, 55(6):403–415.

Visser, I., Raijmakers, M. E., and Molenaar, P. C. (2000). Confidence intervals for hidden Markov model parameters. *British Journal of Mathematical and Statistical Psychology*, 53(2):317–327.

Viterbi, A. (1967). Error bounds for convolutional codes and an asymptotically optimum decoding algorithm. *IEEE Transactions on Information Theory*, 13(2):260–269.

Welch, L. R. (2003). Hidden Markov models and the Baum-Welch algorithm. *IEEE Information Theory Society Newsletter*, 53:1–13.

Yang, R., Yin, L., Hao, X., Liu, L., Wang, C., Li, X., and Liu, Q. (2022). Identifying a suitable model for predicting hourly pollutant concentrations by using low-cost microstation data and machine learning. *Scientific Reports*, 12(1):19949.

Yao, K., Paliwal, K. K., and Lee, T.-W. (2005). Generative factor analyzed hmm for automatic speech recognition. *Speech Communication*, 45:435–454.

Zou, Y., Lin, Y., and Song, X. (2024). Bayesian heterogeneous hidden markov models with an unknown number of states. *Journal of Computational and Graphical Statistics*, 33(1):15–24.

Zucchini, W., MacDonald, I. L., and Langrock, R. (2016). *Hidden Markov models for time series: An introduction using R*. Springer-Verlag, New York.

4

Dirichlet-Random Forest for Predicting Compositional Data

Khaled Masoumifard, Stephan van der Westhuizen, and Sugnet Gardner-Lubbe

Compositions are observations that are unit-sum constrained and frequently encountered in environmental studies, such as species abundance in ecological research, the relative abundance of greenhouse gases in climate studies, and soil texture in soil science. The primary goal of compositional data analysis is to account for the composite nature of observations and to ensure that model outputs adhere to the same constraints as the original data. However, conventional machine learning models may not optimally handle compositional data and may therefore require additional transformations, such as the isometric log-ratio (ILR) transformation. Another modeling approach involves utilizing Dirichlet regression to model compositional data directly in the simplex space. Nevertheless, this approach, like many other parametric models, may not typically achieve the predictive accuracy associated with machine learning models. Consequently, there is a need to explore machine learning approaches specifically designed for compositional data. In this research, we investigate the use of a Dirichlet-adapted random forest (RF) capable of training directly in the simplex space. In a classical RF model, trees are split based on the mean-squared-error (MSE; in regression). We propose splitting nodes by maximizing the log-likelihood of the Dirichlet distribution. This approach extends the beta-RF, which is used for modeling bounded outcomes. The Dirichlet-RF ensures that the resulting predictions maintain the unit-sum constraint while achieving competitive predictive accuracy.

4.1 Introduction

Machine learning methods have significantly advanced environmental modeling by improving the ability to predict spatial response variables from environmental covariates, outperforming traditional models such as linear regression in fields such as ecology [1], forestry [2, 3], hydrology [4], and soil science [5]. Among these methods, RF, introduced by [6], have become widely used for spatial-environmental predictive tasks [7]. RF effectively captures complex, nonlinear relationships between environmental attributes (e.g., soil properties or species abundance) and spatially continuous covariates while providing interpretability through variable importance measures and uncertainty estimates via quantile regression forests [8]. Its versatility is evident in applications such as digital soil mapping (DSM), where RF has been employed, for example, to predict soil texture [9, 10], organic carbon content [11], and thickness of soil [12].

DOI: 10.1201/9781003564423-4

A significant challenge in environmental studies is the analysis of compositional data, where observations are constrained to sum to a constant (typically 1 or 100%). Such data are ubiquitous in environmental sciences, appearing as species abundance in ecological studies, relative concentrations of greenhouse gases in climate research, and particle-size fractions in soil science. Although conventional machine learning approaches, including RFs, have proven highly effective for many prediction tasks, they are not inherently designed to handle compositional data. The main challenge lies in ensuring predictions respect the constant-sum constraint while preserving the relationships between components [13]. Current approaches typically rely on data transformations, such as the ILR transformation [14], to address these constraints. However, these transformations can complicate interpretation and may not fully preserve the compositional structure of the data [15]. Alternative approaches using parametric models like Dirichlet regression can directly model compositional data but often fail to achieve the prediction accuracy associated with machine learning methods, particularly when dealing with complex, high-dimensional data. This highlights the need for machine learning models that inherently respect compositional constraints without requiring data transformations.

To address this limitation, we propose Dirichlet-RF, a novel adaptation of the RF algorithm that incorporates maximum log-likelihood splitting based on the Dirichlet distribution. This approach extends prior work on beta-RFs for bounded response variables [16] to multivariate compositional data. Unlike the conventional RF, which relies on the MSE as the splitting criterion, Dirichlet-RF selects splits by maximizing the Dirichlet log-likelihood. This ensures that predictions inherently satisfy the unit-sum constraint while also accounting for heteroscedastic variance patterns—an aspect not addressed by MSE-based methods. The programming language that we used is Julia because of computational performance. Experimental results demonstrate that Dirichlet-RF achieves competitive predictive accuracy compared to existing approaches, particularly for high-dimensional datasets.

The remainder of this chapter is structured as follows: Section 4.2 introduces the Dirichlet-RF algorithm, detailing its theoretical foundation and implementation. Section 4.3 compares Dirichlet-RF with existing models. Section 4.4 presents the proposed methodology to real data of modeling soil texture. Finally, in sections 4.5 and 4.6, we discuss the broader implications of this research and potential future directions, and present the final conclusion.

4.2 Model Formulation

4.2.1 The Dirichlet Distribution

The Dirichlet distribution represents a multivariate probability generalization of the beta distribution. Consider the multivariate variable $\mathbf{Z} = (Z_1, \ldots, Z_k)$ that follows the Dirichlet distribution with parameters α_i, $i = 1, \ldots, k$, where k is the number of composites. The distribution function at point $\mathbf{z} = (z_1, \ldots, z_k)$ is given by

$$f_{\mathbf{Z}}(\mathbf{z}; \boldsymbol{\alpha}) = \frac{\Gamma(\alpha_0)}{\prod_{i=1}^{k} \Gamma(\alpha_i)} \prod_{i=1}^{k} z_i^{\alpha_i - 1}, \quad \text{subject to } z_i \in [0, 1], \quad \sum_{i=1}^{k} z_i = 1,$$

where $\alpha_0 = \sum_{i=1}^{k} \alpha_i$ is the sum of the concentration parameters. The expected value and variance for each component, z_i, of the Dirichlet-distributed random vector are respectively

$$E(Z_i) = \frac{\alpha_i}{\alpha_0},$$

and

$$\mathrm{Var}(Z_i) = \frac{\alpha_i \alpha_0 - \alpha_i}{\alpha_0^2(\alpha_0 + 1)}.$$

For further details on the Dirichlet distribution, we invite the reader to refer to [17].

4.2.2 Dirichlet-Random Forest

The proposed `Dirichlet-RF` model, based on maximum likelihood estimation (MLE), eliminates the need for data transformations and addresses concerns about heteroscedasticity in the response variable. We now present a step-by-step procedure of the Dirichlet decision tree model which recursively partitions the feature space using binary splits that maximize the Dirichlet log-likelihood.

For a sample size, n, and for m features, let $\mathbf{X} \in \mathbb{R}^{n\times m}$ denote the feature matrix and $\mathbf{Y} \in \mathbb{R}^{n\times k}$ denote the response matrix. The response matrix $\mathbf{Y}$ represents k-dimensional compositional data, where $Y_{i,j} > 0$ for all i, j and $\sum_{j=1}^{k} Y_{i,j} = 1$. The compositional responses are assumed to follow a Dirichlet distribution.

The proposed algorithm recursively partitions the predictor space $\mathbf{X}$ into non-overlapping regions (nodes) via binary splits. At each step, the splitting criterion is determined by maximizing the Dirichlet log-likelihood. Consider a node containing a subset of observations indexed by $\mathcal{M} \subset \{1, 2, \ldots, n\}$, denoted as

$$\{\mathbf{Y}_i \mid i \in \mathcal{M}\}.$$

where $\mathbf{Y}_i = (Y_{i1}, Y_{i2}, \cdots, Y_{ik})$. For a candidate split based on the s-th feature and threshold c, let $I_s(c) = \{i \mid X_{si} \leq c, i \in \mathcal{M}\}$ denote the set of observations where the s-th feature is less than or equal to c. The Dirichlet log-likelihood for observations in $I_s(c)$ is given by

$$\mathcal{L}(I_s(c)) = \sum_{i \in I_s(c)} \left[\log\Gamma(\alpha_0) - \sum_{j=1}^{k} \log\Gamma(\alpha_j) + \sum_{j=1}^{k} (\alpha_j - 1)\log Y_{i,j} \right].$$

The optimal feature s^* and threshold c^* are selected to maximize the combined log-likelihood of the two resulting child nodes

$$(s^*, c^*) = \mathrm{argmax}_{s,c} \left(\max_{\boldsymbol{\alpha}} \mathcal{L}(I_s(c)) + \max_{\boldsymbol{\alpha}} \mathcal{L}(\mathcal{M} \setminus I_s(c)) \right),$$

where $\mathcal{M} \setminus I_s(c)$ denotes the relative complement of $I_s(c)$ in $\mathcal{M}$. The recursive splitting process continues until one of the following termination criteria is met. If the number of observations in a node falls below a predefined threshold, further splitting is halted. The process stops when the tree reaches the maximum allowable depth. Finally, if no candidate split (s, c) results in an improvement in the Dirichlet log-likelihood function, the splitting procedure terminates. Mathematically, this condition is expressed as

$$\max_{(s,c)} \left(\max_{\boldsymbol{\alpha}} \mathcal{L}(I_s(c)) + \max_{\boldsymbol{\alpha}} \mathcal{L}(\mathcal{M} \setminus I_s(c)) \right) - \max_{\boldsymbol{\alpha}} \mathcal{L}(\mathcal{M}) \leq \epsilon,$$

where ϵ is a predefined threshold for log-likelihood improvement. At terminal nodes, the estimated Dirichlet parameters $\hat{\alpha}_j$ are stored for prediction.

For a new observation, predictions are made by traversing the tree according to the splitting rules to reach a terminal node. The Dirichlet parameters at the terminal node are used to predict the distribution

$$\hat{Y}_{\mathrm{new},j} = \frac{\hat{\alpha}_j}{\sum_{j=1}^{k} \hat{\alpha}_j}, \quad j = 1, \ldots, k.$$

The Dirichlet regression tree is then extended into a `Dirichlet-RF`. This approach constructs an ensemble of Dirichlet trees where in each tree and at each node, a randomly selected subset of features is considered, and the optimal split is determined as explained above.

For a new observation, the final prediction is obtained by averaging the predictions from all individual trees in the ensemble. Specifically, let B denote the total number of trees in the forest. The predicted value for the jth component of a new observation is computed as

$$\hat{Y}_{\text{new},j} = \frac{1}{B}\sum_{t=1}^{B}\hat{Y}_{\text{new},j}^{(t)}, \quad j = 1,\ldots,k,$$

where $\hat{Y}_{\text{new},j}^{(t)}$ represents the prediction from the t-th tree. The `Dirichlet-RF` algorithm is detailed in Algorithm 1.

4.3 Simulation Study

We evaluated the performance of our proposed methodology against alternative approaches, including transformation-based RFs and a Dirichlet regression model, in a simulation study.

Specifically, we compared `Dirichlet-RF` with the following methods:

1. `CLR-RF`: In this model, we first apply the Centered Log-Ratio (CLR) transformation to map the compositional data into Euclidean space. For a compositional vector $\mathbf{Z} = (Z_1,\ldots,Z_k)$, the CLR transformation is defined as

$$\text{CLR}(\mathbf{Z}) = \left[\log\frac{Z_1}{g(\mathbf{Z})},\ldots,\log\frac{Z_k}{g(\mathbf{Z})}\right], \quad g(\mathbf{Z}) = \left(\prod_{i=1}^{k} Z_i\right)^{1/k},$$

where $g(\mathbf{Z})$ is the geometric mean of the components of $\mathbf{Z}$. A RF model is then trained on the transformed data, and the resulting predictions are subsequently mapped back to the original compositional space using the inverse CLR transformation.

2. `ILR-RF`: The ILR transformation is defined as $\text{ILR}(\mathbf{Z}) = \text{CLR}(\mathbf{Z})\mathbf{V} \in \mathbb{R}^{k-1}$, where $\mathbf{V} \in \mathbb{R}^{k\times(k-1)}$ is a matrix whose columns form an orthonormal basis of the CLR-plane $\{\mathbf{v} = (v_1,\cdots,v_k) \in \mathbb{R}^k : \sum_{i=1}^{k} v_i = 0\}$. The choice of $\mathbf{V}$ is not unique; a commonly used construction is

$$\mathbf{V} = [\mathbf{v}_1\,\mathbf{v}_2\,\ldots\,\mathbf{v}_{k-1}] \in \mathbb{R}^{k\times(k-1)},$$

where

$$\mathbf{v}_j = \left(\underbrace{\frac{1}{\sqrt{j(j+1)}},\ldots,\frac{1}{\sqrt{j(j+1)}}}_{j\text{ times}}, -\frac{j}{\sqrt{j(j+1)}}, 0,\ldots,0\right)^{\top}, \quad j = 1,\ldots,k-1.$$

For more details, see [18].

Algorithm 1 Dirichlet Random Forest

Require:
$X \in \mathbb{R}^{n \times m}$ ▷ Feature matrix
$Y \in \mathbb{R}^{n \times k}$ ▷ Compositional response matrix
B ▷ Number of trees
m_{try} ▷ Number of features to consider at each split
n_{min} ▷ Minimum node size
d_{max} ▷ Maximum tree depth
procedure DIRICHLETFOREST($X, Y, B, m_{try}, n_{min}, d_{max}$)
    $\mathcal{F} \leftarrow \emptyset$ ▷ Initialize empty forest
    for $b = 1$ to B **do**
        $\mathcal{I}_b \leftarrow$ Bootstrap sample from $\{1, \ldots, n\}$
        $T_b \leftarrow$ GROWTREE($X[\mathcal{I}_b], Y[\mathcal{I}_b], m_{try}, n_{min}, d_{max}$)
        $\mathcal{F} \leftarrow \mathcal{F} \cup \{T_b\}$
    end forreturn $\mathcal{F}$
end procedure
procedure GROWTREE($X, Y, m_{try}, n_{min}, d_{max}$)
    if $|Y| < n_{min}$ **or** depth $= d_{max}$ **then**
        return FITTERMINALNODE(Y)
    end if
    $\mathcal{M} \leftarrow$ Random subset of m_{try} features
    $(s^*, c^*) \leftarrow$ FINDBESTSPLIT($X, Y, \mathcal{M}$)
    if no valid split found **then**
        return FITTERMINALNODE(Y)
    end if
    $I_L \leftarrow \{i : x_{i,s^*} \leq c^*\}$ ▷ Left child indices
    $I_R \leftarrow \{i : x_{i,s^*} > c^*\}$ ▷ Right child indices
    left_child $\leftarrow$ GROWTREE($X[I_L], Y[I_L], m_{try}, n_{min}, d_{max}$)
    right_child $\leftarrow$ GROWTREE($X[I_R], Y[I_R], m_{try}, n_{min}, d_{max}$)
     return Node(s^*, c^*, left_child, right_child)
end procedure
procedure FINDBESTSPLIT($X, Y, \mathcal{M}$)
    best_gain $\leftarrow -\infty$
    for each feature s in $\mathcal{M}$ **do**
        for each potential split point c in $X_{:,s}$ **do**
            $I_L \leftarrow \{i : x_{i,s} \leq c\}$
            $I_R \leftarrow \{i : x_{i,s} > c\}$
            if $|I_L| \geq n_{min}$ **and** $|I_R| \geq n_{min}$ **then**
                $\alpha_L \leftarrow$ ESTIMATEDIRICHLET($Y[I_L]$)
                $\alpha_R \leftarrow$ ESTIMATEDIRICHLET($Y[I_R]$)
                gain $\leftarrow$ LOGLIK($Y[I_L], \alpha_L$) + LOGLIK($Y[I_R], \alpha_R$)
                if gain > best_gain **then**
                    best_gain $\leftarrow$ gain
                    $(s^*, c^*) \leftarrow (s, c)$
                end if
            end if
        end for
    end forreturn (s^*, c^*)
end procedure
procedure PREDICT($\mathcal{F}, X_{new}$)
    for each tree T_b in $\mathcal{F}$ **do**
        $\hat{Y}_b \leftarrow$ TREEPREDICT(T_b, X_{new})
    end for
    $\hat{Y} \leftarrow \frac{1}{B} \sum_{b=1}^{B} \hat{Y}_b$ ▷ Average predictions **return** $\hat{Y}$
end procedure

3. `Dirichlet regression`: A parametric model specifically designed for compositional data that follows a Dirichlet distribution. Unlike standard regression models, Dirichlet regression accounts for the constraints of compositional data without requiring prior transformation [19].

The CLR model and the ILR transformation used in this section for transforming compositional data are two well-known approaches [14, 20]. Furthermore, alternative transformation methods exist, each with distinct mathematical properties and characteristics and we refer readers to [21] for an overview. The choice of transformation is non-trivial, as it can influence the results of subsequent analyses, including RF models [22, 23]. However, a key advantage arises when employing the maximum likelihood criterion for splitting in RF, as it eliminates the need for any preliminary transformation of the compositional data.

To assess the models, we employ two distance metrics: the Aitchison distance [24], and the root mean-square-error (RMSE). Below, we define these metrics using notation consistent with the response matrix $\mathbf{Y} \in \mathbb{R}^{n \times k}$ (defined in Subsection 4.2.2) and its predicted counterpart $\hat{\mathbf{Y}} \in \mathbb{R}^{n \times k}$.
The Aitchison distance is computed as

$$\frac{1}{n} \sum_{i=1}^{n} \sqrt{\sum_{j=1}^{k} \left(\log \frac{Y_{ij}}{g(\mathbf{Y}_i)} - \log \frac{\hat{Y}_{ij}}{g(\hat{\mathbf{Y}}_i)} \right)^2},$$

where $g(\mathbf{Y}_i) = \left(\prod_{j=1}^{k} Y_{ij} \right)^{1/k}$. The RMSE is used to evaluate overall predictive performance. For one of the j composites

$$\text{RMSE}_j = \sqrt{\frac{1}{n} \sum_{i=1}^{n} (Y_{ij} - \hat{Y}_{ij})^2}.$$

For both metrics, smaller values indicate better model performance.
Our research involved simulations across two distinct scenarios: simple and complex data structures. Datasets with compositional responses were created, varying the sample sizes to include 100, 200, 500, and 2000 observations. Each dataset was divided into training (80% of the data) and testing (20% of the data) subsets. The experiments were repeated 40 times to ensure robustness, and results were aggregated to compute mean values and standard deviations for the metrics.

4.3.1 Simple Structure Synthetic Simulations

In this section, we used simulated datasets containing three feature variables, X_1, X_2, X_3, and a three-dimensional Dirichlet-distributed response vector $\mathbf{Y} = (Y_1, Y_2, Y_3)$, where $Y_1 + Y_2 + Y_3 = 1$. The response vector $\mathbf{Y}$ is generated using Dirichlet-distributed random variables, where the parameters $\alpha_1, \alpha_2, \alpha_3$ depend on the corresponding feature variables. The full details of the model parameters can be found in the supplementary materials.

In Table 4.1 we present the results. Although `Dirichlet-RF` did not outperform the other models, it maintained a competitive standing. Across all sample sizes, Dirichlet regression exhibited better performance, with smaller Aitchison distances and RMSE-averaged values, although these differences were relatively modest. This is perhaps because the feature variables are low-dimensional and generated from a Dirichlet distribution with a simple relationship between the response and feature variables. As a result, Dirichlet regression is naturally well-suited to the data, allowing it to achieve a good fit.

TABLE 4.1
Simulation results for synthetic data with a simple structure. Values represent the means and (standard deviations) of the metrics for test datasets. RMSE-averaged is calculated as the mean of RMSE across all compositional components. Train-Test Diff represents the absolute difference between train and test Aitchison Distance, indicating model generalization capability. Bold values indicate best performance for each metric (smaller is better for Aitchison Distance, RMSE-averaged, and Train-Test Diff).

Sample Size	Model	Aitchison Distance	RMSE-averaged	Train-Test Diff
n = 100	CLR-RF	1.327 (0.187)	0.699 (0.059)	0.701
	ILR-RF	1.326 (0.191)	0.697 (0.059)	0.699
	Dirichlet Regression	**1.266** (0.190)	**0.736** (0.047)	**0.072**
	`Dirichlet-RF`	1.458 (0.223)	0.622 (0.077)	0.603
n = 200	CLR-RF	1.299 (0.141)	0.703 (0.039)	0.684
	ILR-RF	1.303 (0.143)	0.700 (0.040)	0.686
	Dirichlet Regression	**1.233** (0.139)	**0.745** (0.033)	**0.030**
	`Dirichlet-RF`	1.451 (0.153)	0.620 (0.052)	0.600
n = 500	CLR-RF	1.282 (0.077)	0.716 (0.019)	0.684
	ILR-RF	1.284 (0.077)	0.716 (0.020)	0.685
	Dirichlet Regression	**1.234** (0.078)	**0.752** (0.016)	**0.035**
	`Dirichlet-RF`	1.450 (0.077)	0.627 (0.031)	0.616
n = 2000	CLR-RF	1.249 (0.032)	0.726 (0.010)	0.652
	ILR-RF	1.252 (0.032)	0.725 (0.010)	0.654
	Dirichlet Regression	**1.209** (0.032)	**0.757** (0.008)	**0.006**
	`Dirichlet-RF`	1.431 (0.041)	0.635 (0.016)	0.596

A noteworthy observation is that the CLR-RF and ILR-RF models exhibited substantial gaps between their training and testing performance, indicating significant overfitting tendencies (results shown only for Aitchison Distance, with similar patterns observed across other evaluation metrics). In contrast, `Dirichlet-RF` demonstrated more moderate differences between training and testing performance, suggesting superior generalization capabilities relative to alternative RF implementations.

4.3.2 Complex Structure Synthetic Simulations

To assess model performance in a more complex setting involving feature selection challenges, we created a dataset that included both informative and non-informative predictor variables, following a method similar to that used by [16]. The dataset consisted of 10 predictor variables ($X_1, \ldots, X_{10}$), with only the first three variables (X_1, X_2, X_3) being truly informative for predicting the compositional response. The three-component compositional response vector $\mathbf{Y} = (Y_1, Y_2, Y_3)$ was constructed through complex non-linear transformations of the informative features. Detailed specifications of the model parameters and transformation functions are provided in the supplementary materials.

The primary distinction between the simple and complex datasets lies in their structural design: the simple scenario is based on straightforward covariate–response relationships (linear), which can be modeled easily, while the complex scenario incorporates substantial non-linearity that obscures these associations, resulting in a more challenging modeling task.

In Table 4.2 the results of the `Dirichlet-RF` model against CLR-RF, ILR-RF, and Dirichlet regression are presented for the Aitchison distance and RMSE-averaged metrics.

On the test datasets, the `Dirichlet-RF` model consistently outperformed alternative approaches across all evaluation metrics. For Aitchison distance, `Dirichlet-RF` achieved the

TABLE 4.2
Simulation results for synthetic data with a complex structure. Values represent the means and (standard deviations) of the metrics for test datasets. RMSE-averaged is calculated as the mean of RMSE across all compositional components. Train-Test Diff represents the absolute difference between train and test Aitchison Distance, indicating model generalization capability. Bold values indicate best performance for each metric (smaller is better for Aitchison Distance, RMSE-averaged, and Train-Test Diff).

Sample Size	Model	Aitchison Distance	RMSE-averaged	Train-Test Diff
n = 100	CLR-RF	0.734 (0.207)	0.881 (0.034)	0.409
	ILR-RF	0.734 (0.208)	0.881 (0.035)	0.408
	Dirichlet Regression	0.871 (0.206)	0.825 (0.041)	**0.139**
	Dirichlet-RF	**0.717** (0.207)	**0.886** (0.029)	0.245
n = 200	CLR-RF	0.671 (0.152)	0.898 (0.018)	0.366
	ILR-RF	0.670 (0.151)	0.899 (0.018)	0.364
	Dirichlet Regression	0.826 (0.155)	0.844 (0.026)	**0.068**
	Dirichlet-RF	**0.659** (0.152)	**0.903** (0.017)	0.221
n = 500	CLR-RF	0.617 (0.072)	0.912 (0.011)	0.334
	ILR-RF	0.618 (0.072)	0.911 (0.012)	0.334
	Dirichlet Regression	0.790 (0.075)	0.855 (0.012)	**0.029**
	Dirichlet-RF	**0.606** (0.068)	**0.917** (0.010)	0.201
n = 2000	CLR-RF	0.557 (0.045)	0.931 (0.006)	0.302
	ILR-RF	0.557 (0.044)	0.931 (0.006)	0.301
	Dirichlet Regression	0.785 (0.044)	0.858 (0.006)	**0.016**
	Dirichlet-RF	**0.544** (0.044)	**0.936** (0.004)	0.178

lowest values across all sample sizes, with values of 0.717, 0.659, 0.606, and 0.544 for $n = 100$, 200, 500, and 2000, respectively. This represents an improvement of 2.3% over CLR-RF and ILR-RF and 17.7% over standard Dirichlet regression at $n = 100$, with the advantage increasing to 30.7% over Dirichlet regression at $n = 2000$. Similarly, `Dirichlet-RF` exhibited superior performance in terms of RMSE-averaged.

In regard to overfitting, as expected, Dirichlet-RF performs better in all examples shown in Tables 4.1 and 4.2. However, overfitting is always a concern in RF-based models. Based on our analysis, we observe that while CLR-RF and ILR-RF demonstrate superior performance on training datasets across all metrics, the `Dirichlet-RF` model achieves better results on test datasets. This balanced performance indicates that `Dirichlet-RF` effectively mitigates overfitting while maintaining superior predictive accuracy. To further examine overfitting behavior, Figure 4.1 shows Aitchison distance as a function of training sample size, comparing `Dirichlet-RF`, CLR-RF, and ILR-RF across different dataset structures. `Dirichlet-RF` consistently exhibits a smaller gap between training and test performance compared to CLR-RF and ILR-RF, indicating better generalization and reduced overfitting.

A noteworthy observation is that our model demonstrates effective feature selection capabilities through two complementary variable importance measures: log-likelihood importance, which quantifies each variable's contribution to model fit improvement, and frequency-based importance, which measures how often each variable is selected for splitting nodes. Both metrics are averaged across all trees in the RF. Analysis shows that X_1, X_2, and X_3 receive the highest importance scores across both metrics, as shown in Table 4.3, identifying them as the informative predictors in our simulation. The remaining variables (X_4 through X_{10}) receive substantially lower importance scores, confirming the model's ability to distinguish signal from noise.

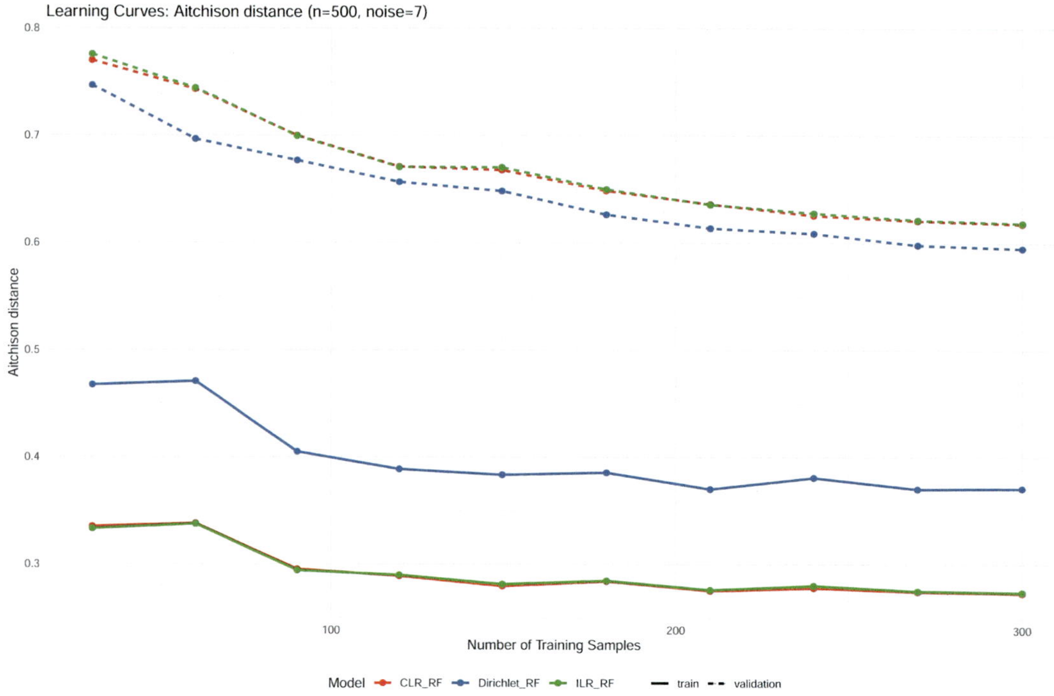

FIGURE 4.1
Aitchison distance performance for Dirichlet-RF, CLR-RF, and ILR-RF as a function of training sample size ($n = 500$). The curves show both training and test performance, demonstrating Dirichlet-RF's superior generalization and minimal overfitting.

TABLE 4.3
Variable importance scores based on log-likelihood improvement and selection frequency, averaged across all trees in the `Dirichlet-RF`.

Feature	Log-Likelihood Importance	Frequency-Based Importance
X_1	225.039	14
X_2	303.618	21
X_3	272.036	19
X_4	16.839	2
X_5	9.972	2
X_6	15.649	3
X_7	0.000	0
X_8	40.240	6
X_9	5.369	1
X_{10}	23.295	3

4.4 Real-World Case Study

Digital soil maps are important inputs in a variety of fields and provide information on the spatial distribution of soil properties. A widely used tool to generate these maps is DSM [25] which uses statistical models to quantify the relationship between a soil property observed at various locations within a given region of interest and environmental covariates, which are available as raster maps covering the entire area of the region. This relationship is then used to predict the soil property at all locations in the region of interest. The covariates should ideally represent soil-forming factors [25, 26]. It is useful to conceptualize the observations of the soil property across a region as a realization of a stochastic process. Spatial autocorrelation in the soil property is commonly used in modeling. However, the inherent spatial variation in the covariates provides significant information about the spatial variation of the soil property. Therefore, by incorporating a sufficiently complex model (i.e., machine learning model), this spatial autocorrelation can be effectively captured through the covariates [27].

In this case study, soil texture (i.e., composites of sand, silt, and clay adding to 100%) is mapped in the province of Kwazulu-Natal in South Africa. The dataset is a subset from the South African Profile Database of the Agricultural Research Council, Institute of Soil Climate and Water [28]. The dataset used in this study consisted of 685 observations from the A horizon (i.e., topsoil). In Figure 4.2(a), the spatial distribution of the observations is shown. In Figure 4.2(b) the composites are shown in a ternary plot which indicates a skewed distribution toward the sand composite.

We used 120 covariates (at $250\,m \times 250\,m$) which included rasters of vegetation indices, elevation variables, precipitation, temperatures, and other climate variables. These covariates are mainly the same as those used in SoilGrids 2.0 [11].

In this case study, the models CLR-RF, ILR-RF, `Dirichlet-RF`, and RF without the use of transformations were evaluated when modeling soil texture. We did not include the parametric Dirichlet regression as it was not feasible with the large number of covariates. The models were evaluated in a standard 10-fold cross-validation with the same metrics as in the simulation study. We used the Ranger package [29] in R to implement the RF models except for the `Dirichlet-RF` for which code is available in the Supplementary Materials. We did not calibrate the models and only used default values for the hyper-parameters of the RF models implemented with Ranger (i.e., m-try $= \sqrt{m}$, num.trees $= 500$, and min.node.size $= 5$). For the `Dirichlet-RF`, we used the same hyper-parameters except m-try which was set to $m/3$.

The results of the test set predictions of the cross-validation are shown in Table 4.4. The proposed model `Dirichlet-RF` outperformed the other RF models. Interestingly, RF without the use of transformations produced comparable performance in comparison to RF-CLR and RF-ILR. These findings emphasize `Dirichlet-RF` as an effective approach for modeling high-dimensional compositional data. Beyond its competitive predictive accuracy, it offers advantages in generalization and robustness, making it a superior alternative to existing methods. It should be pointed out that the overall performance for all models was relatively poor as the respective standard deviations for clay, silt, and sand were 16.86, 11.77, and 24.03, respectively. Two possible reasons for this could be that the models were not calibrated, and that the covariates do not explain much of the variation of soil texture. However, we want to point out that the purpose of this research was to compare `Dirichlet-RF` to other related methods, and not necessarily obtain the best optimal calibrated version of each model. The reason for this was the considerable time consumption of fitting the `Dirichlet-RF` model.

The superior performance of `Dirichlet-RF` came at a significant computational cost which is a great disadvantage. One of the main challenges lies in the absence of closed-form solutions for the MLE of the parameters, which requires iterative numerical optimization techniques and consequently results in low computational performance. There are some efficient methods in the literature for estimating Dirichlet parameters, including fixed-point iteration and Newton-Raphson techniques (for further details, see [30]). In our implementation, we employ the Newton-Raphson algorithm. In addition to these methods, recent

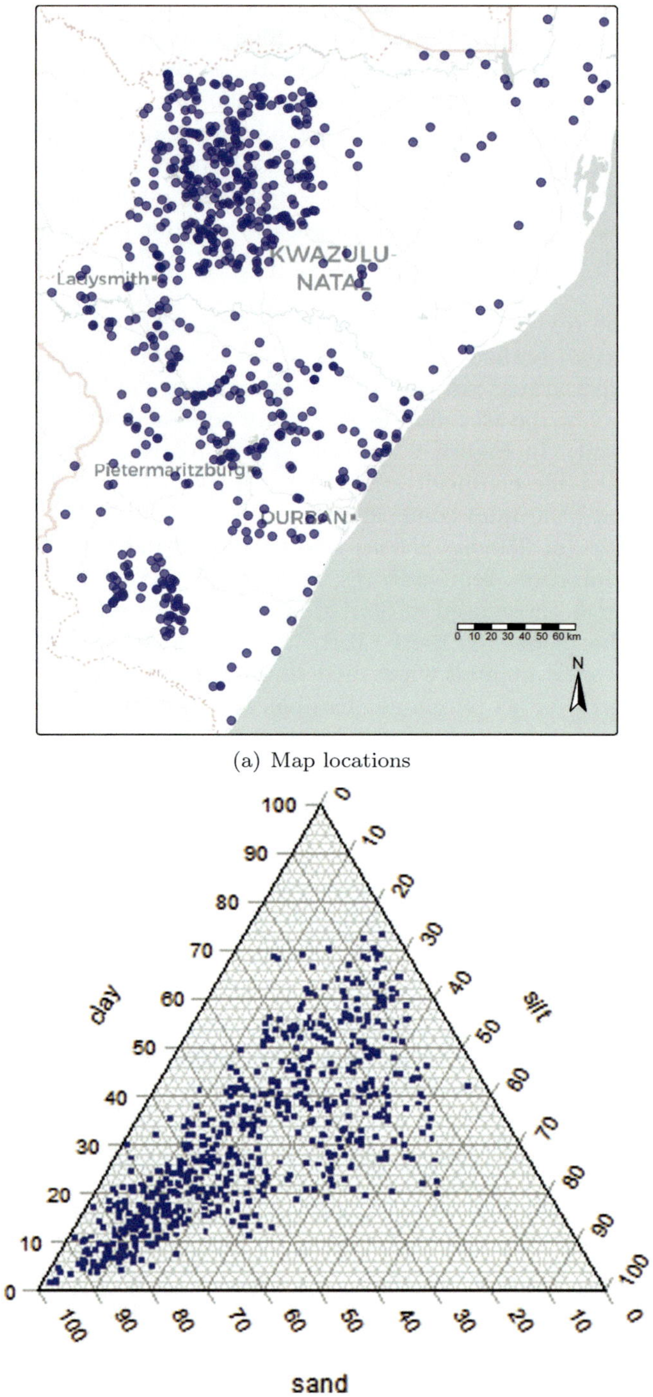

(a) Map locations

(b) Ternary plot

FIGURE 4.2
Locations of observations in South Africa with the corresponding distribution as shown in the ternary plot.

TABLE 4.4
Results of `Dirichlet-RF`, `ILR-RF`, `CLR-RF`, and an RF without the use of a transformation on the soil texture data in KwaZulu-Natal in South Africa.

	RF-CLR	**RF-ILR**	**RF**	**Dirichlet-RF**
Aitchison distance	1.08	1.08	1.05	0.98
clayRMSE	20.09	20.14	19.44	17.85
siltRMSE	13.94	13.91	13.46	12.41
sandRMSE	29.79	29.83	28.66	25.93
meanRMSE	22.25	22.28	21.45	19.54

studies have investigated alternative estimation techniques, including the method of moments (MoM), and entropy-based approaches [31]. For further details, see [32] and [33].

4.5 Discussion

In this chapter, we proposed the `Dirichlet-RF` algorithm which incorporates a log-likelihood-based splitting criterion. Our empirical analyses demonstrate the competitive performance of this approach compared to existing methodologies. The discussion of computational performance remains a significant issue for big datasets in our methodology. However, we have identified several promising directions for future research and performance improvements. For example:

- We are exploring closed-form approximations to MLE, or alternative closed-form estimators such as MoM, which can provide reasonable approximations while offering significant computational advantages. A hybrid approach could be implemented by strategically substituting MLE with MoM at particular nodes according to predetermined criteria, allowing us to balance estimation accuracy with computational efficiency.
- Given that our results showed very similar performance between `Dirichlet-RF` and ILR-RF methods on test data, another promising direction would be combining these approaches. Specifically, the development of a hybrid splitting criterion that integrates both `Dirichlet-RF` and ILR-RF approaches warrant investigation. One possible strategy is a two-stage procedure: first, candidate splits are selected using ILR-RF criteria, and then the optimal split is determined using a log-likelihood-based criterion. This hybrid approach has the potential to improve both computational efficiency and statistical robustness by reducing the complexity of parameter estimation for the Dirichlet distribution.

One issue that remains open is that sometimes the use of the log-likelihood criterion for splitting and the mean of the distribution in a leaf for prediction does not work well. This mostly occurs when the Dirichlet distribution is not sufficiently sensitive to permutations of some components of the data, especially when the values of those components are not close. For example, as illustrated in Figure 4.3 , for the case where $k = 2$ (which follows a beta distribution), a simulated sample from a beta distribution is shown along the x-axis, along with its mean. In this scenario, the mean does not provide a good prediction. Investigating effective constraints or alternative approaches to mitigate this issue presents an interesting direction for future work.

It is important to emphasize that, when applying `Dirichlet-RF` to compositional data, careful attention must be paid to the inherent properties of such data. In particular,

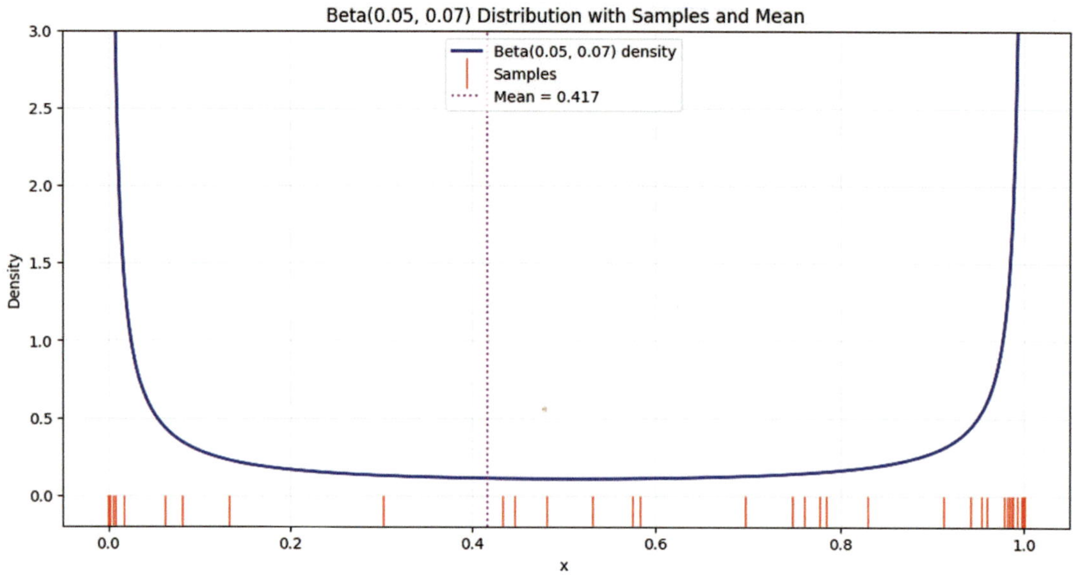

FIGURE 4.3
Simulated beta distribution and its mean.

`Dirichlet-RF` is not appropriate for subcompositional data and should be applied only to vectors of strictly positive components that either sum to one or can be normalized to do so. For more details on applying the Dirichlet distribution to compositional data, see [34].

4.6 Conclusion

`Dirichlet-RF` is introduced to model compositional data. This methodology makes use of a maximum log-likelihood approach to split trees based on the Dirichlet distribution. The proposed model was compared to other approaches for compositional data in a synthetic simulation study as well as in a real-world case of soil texture in KwaZulu-Natal in South Africa. The proposed model demonstrated superior performance, especially for larger and more complex datasets. However, more research is needed to address the computational challenges of fitting Dirichlet trees to improve the applicability of `Dirichlet-RF`.

Author Contributions

Khaled Masoumifard: Conceptualization; Investigation; Methodology; Software; Visualization; Writing—Original draft; Writing—Review and editing. **Stephan van der Westhuizen:** Conceptualization; Data curation; Funding acquisition; Investigation; Methodology; Project Administration; Supervision; Validation; Visualization; Writing—Original draft; Writing—Review and editing. **Sugnet Gardner-Lubbe:** Conceptualization; Project administration; Supervision; Writing—Review and Editing.

Acknowledgments

This work is based on research supported by the Rector's Postdoctoral Fellowship of Stellenbosch University, and by the National Research Foundation of South Africa (Grant reference: TTK240319209815). We would like to thank the Agricultural Research Council—Institute for Soil Climate and Water for access to parts of the South African National Soil Database. Finally, we also want to thank ISRIC—World Soil Information (`isric.org`) for providing the environmental covariates.

Supplementary Data

The supplementary materials is available at `https://github.com/Xaleed/DirichletRandomForestJulia`.

Bibliography

[1] M. Pichler and F. Hartig. Machine learning and deep learning—A review for ecologists. *Methods in Ecology and Evolution*, 14(4):994–1016, 2023.

[2] J. Bastin, Y. Finegold, C. Garcia, D. Mollicone, M. Rezende, D. Routh, C. M. Zohner, and T. W. Crowther. The global tree restoration potential. *Science*, 365(6448):76–79, 2019.

[3] J. S. Estrada, A. Fuentes, P. Reszka, and F. A. Cheein. Machine learning assisted remote forestry health assessment: a comprehensive state of the art review. *Frontiers in Plant Science*, 14:1139232, 2023.

[4] N. A. Povak, P. F. Hessburg, T. C. McDonnell, K. M. Reynolds, T. J. Sullivan, R. B. Salter, and B. J. Cosby. Machine learning and linear regression models to predict catchment-level base cation weathering rates across the southern Appalachian mountain region, USA. *Water Resources Research*, 50(4):2798–2814, 2014.

[5] A. M. J.-C. Wadoux, B. Minasny, and A. B. McBratney. Machine learning for digital soil mapping: Applications, challenges and suggested solutions. *Earth-Science Reviews*, 210:103359, 2020.

[6] L. Breiman. Random forests. *Machine Learning*, 45:5–32, 2001.

[7] T. Hengl, M. Nussbaum, M. N. Wright, G. B. M. Heuvelink, and B. Gräler. Random forest as a generic framework for predictive modeling of spatial and spatio-temporal variables. *PeerJ*, 6:e5518, 2018.

[8] N. Meinshausen and G. Ridgeway. Quantile regression forests. *Journal of Machine Learning Research*, 7(6), 2006.

[9] M. Ließ, B. Glaser, and B. Huwe. Uncertainty in the spatial prediction of soil texture: Comparison of regression tree and random forest models. *Geoderma*, 170:70–79, 2012.

[10] F. Kaya, L. Başayiğit, A. Keshavarzi, and R. Francaviglia. Digital mapping for soil texture class prediction in northwestern Türkiye by different machine learning algorithms. *Geoderma Regional*, 31:e00584, 2022.

[11] L. Poggio, L. M. De Sousa, N. H. Batjes, G. B. M. Heuvelink, B. Kempen, E. Ribeiro, and D. Rossiter. Soilgrids 2.0: Producing soil information for the globe with quantified spatial uncertainty. *Soil*, 7(1):217–240, 2021.

[12] S. van der Westhuizen, G. B. M. Heuvelink, D. P. Hofmeyr, L. Poggio, M. Nussbaum, and C. Brungard. Mapping soil thickness by accounting for right-censored data with survival probabilities and machine learning. *European Journal of Soil Science*, 75(5):e13589, 2024.

[13] I. O. A. Odeh, A. J. Todd, and J. Triantafilis. Spatial prediction of soil particle-size fractions as compositional data. *Soil Science*, 168(8):501–515, 2003.

[14] J. J. Egozcue, V. Pawlowsky-Glahn, G. Mateu-Figueras, and C. Barcelo-Vidal. Isometric logratio transformations for compositional data analysis. *Mathematical Geology*, 35(3):279–300, 2003.

[15] Z. Wang and W. Shi. Mapping soil particle-size fractions: A comparison of compositional kriging and log-ratio kriging. *Journal of Hydrology*, 546:526–541, 2017.

[16] L. Weinhold, M. Schmid, R. Mitchell, K. O. Maloney, M. N. Wright, and M. Berger. A random forest approach for bounded outcome variables. *Journal of Computational and Graphical Statistics*, 29(3):639–658, 2020.

[17] S. Kotz, N. Balakrishnan, and N. L. Johnson. *Continuous Multivariate Distributions, Volume 1: Models and Applications*. Wiley Series in Probability and Statistics. John Wiley & Sons, 2nd edition, 2000.

[18] J. J. Egozcue and V. Pawlowsky-Glahn. Groups of parts and their balances in compositional data analysis. *Mathematical Geology*, 37:795–828, 2005.

[19] R. H. Hijazi and R. W. Jernigan. Modelling compositional data using Dirichlet regression. *Journal of Applied Probability and Statistics*, 4(1):77–91, 2009.

[20] C. Barcelo-Vidal, J. Martin-Fernandez, and V. Pawlowsky-Glahn. Mathematical foundations of compositional data analysis. 1, 2001.

[21] A. Abdulaziz. A review of compositional data analysis and recent advances. *Communications in Statistics-Theory and Methods*, 52(16):5535–5567, 2023.

[22] M. Greenacre. Log-ratio analysis is a limiting case of correspondence analysis. *Mathematical Geosciences*, 42(1):129–134, 2010.

[23] M. Tsagris, S. Preston, and A. Wood. A data-based power transformation for compositional data. 2011.

[24] J. Aitchison. The statistical analysis of compositional data. *Journal of the Royal Statistical Society: Series B*, 44(2):139–177, 1982.

[25] A. B. McBratney, M. L. Mendonça Santos, and B. Minasny. On digital soil mapping. *Geoderma*, 117(1–2):3–52, 2003.

[26] H. Jenny. *Factors of Soil Formation: A System of Quantitative Pedology*. McGraw-Hill, New York, 1941.

[27] S. van der Westhuizen, G. B. M. Heuvelink, D. P. Hofmeyr, and L. Poggio. Measurement error-filtered machine learning in digital soil mapping. *Spatial Statistics*, 47:100572, 2022.

[28] G. Paterson, D. Turner, L. Wiese, G. Van Zijl, C. Clarke, and J. Van Tol. Spatial soil information in South Africa: Situational analysis, limitations and challenges. *South African Journal of Science*, 111(5–6):1–7, 2015.

[29] M. N. Wright and A. Ziegler. ranger: A fast implementation of random forests for high dimensional data in C++ and R. *Journal of Statistical Software*, 77(1):1–17, 2017.

[30] Thomas Minka. Estimating a Dirichlet distribution, 2000.

[31] B. Şahin, A. A. Evren, E. Tuna, Z. Z. Şahinbaşoğlu, and E. Ustaoğlu. Parameter estimation of the Dirichlet distribution based on entropy. *Axioms*, 12(10):947, 2023.

[32] N. Wicker, J. Muller, R. K. R. Kalathur, and O. Poch. A maximum likelihood approximation method for Dirichlet's parameter estimation. *Computational Statistics & Data Analysis*, 52(3):1315–1322, 2008.

[33] J. Huang. Maximum likelihood estimation of Dirichlet distribution parameters. *CMU Technique Report*, 76, 2005.

[34] Vera Pawlowsky-Glahn and Antonella Buccianti. *Compositional data analysis*. Wiley Online Library, 2011.

5

Robust Model Selection in Mixture Regression with Application on CO_2 Emissions Data

Andre Ruben Kleynhans, Frans Hendrik Jacobus Kanfer, Solomon Marthinus Millard, and Salomi du Plessis

5.1 Introduction

The finite mixture of regression (FMR) model is often used in environmental statistical analysis due to its ability to study the relationship between the response and predictor variables across several latent groups present in the data [1, 2]. These models allow for flexible modeling, enabling researchers to extract more nuanced information from the data. When fitting an FMR model, the number of components must be specified beforehand. This presents a limitation, especially when the number of components is not known in advance [3]. The task of identifying the correct number of components based on observed data is known as order selection. Order selection forms part of the broader model selection problem in mixture models. Extensive work has been done in this area, and consistency has been shown for order selection in FMRs [4].

A common method for determining the number of components is the use of information criteria such as the Bayesian information criterion (BIC) and Akaike information criterion (AIC) [3]. An alternative approach is to use a penalized likelihood function for order selection. By introducing penalty terms specifically on the mixing proportions into the likelihood function, uninformative components can be shrunk to zero during maximum likelihood estimation. Examples of such techniques include penalization via Shannon entropy and non-concave penalties such as the smoothly clipped absolute deviation (SCAD) [5, 6].

Order selection becomes more challenging in the presence of non-typical observations [7]. Such observations generally impair accurate parameter recovery [8]. Common solutions include fitting mixtures of heavier-tailed distributions or augmenting the model with additional components that serve to cluster outliers separately [8, 9]. Another approach is to fit the FMR model within a self-paced learning (SPL) framework [10]. We refer to such models as SPLFMR. SPL alters the order in which observations are introduced during the estimation process, initially fitting the model to typical observations and gradually incorporating non-typical observations [11, 12]. The robustness of SPL has been well documented [13].

In this chapter, we introduce a robust model selection framework for FMR based on a SPL penalized likelihood approach. Section 5.2 provides a brief overview of related work, including essential concepts and notation. Section 5.3 introduces the component

DOI: 10.1201/9781003564423-5

penalized self-paced learning (CPSPL) FMR model, detailing its formulation and estimation algorithm. Section 5.4 presents a simulation study evaluating the model's performance under various contamination scenarios. Finally, Section 5.5 demonstrates the method's practical utility by applying it to real-world CO_2 emissions data. The goal of this analysis is to explore the different latent patterns that exist in the relationship between economic indicators and CO_2 emissions between countries. This helps inform policy by identifying groups of countries where economic growth may or may not increase emissions, an important issue in sustainable development and climate change mitigation.

5.2 Related Work

This section, we provide a brief introduction to the related work and notation used. We will first consider FMRs and the component mixing proportion penalized (CPFMR) variant. We then provide the expectation maximization (EM) algorithm to fit an FMR, followed by a description of SPL. Fitting FMR using SPL is also considered.

5.2.1 Finite Mixture of Regressions

Consider a univariate response variable Y with associated explanatory variables $\boldsymbol{x}$, the FMRs model is

$$Y = \begin{cases} \boldsymbol{x}^T\boldsymbol{\beta}_1 + \epsilon_1 \text{ with probability } \pi_1 \\ \boldsymbol{x}^T\boldsymbol{\beta}_2 + \epsilon_2 \text{ with probability } \pi_2 \\ \vdots \\ \boldsymbol{x}^T\boldsymbol{\beta}_K + \epsilon_k \text{ with probability } \pi_K \end{cases}$$

The response variable Y follows a FMRs model with probability density function

$$f_Y(y|\boldsymbol{\Theta}) = \sum_{k=1}^{K} \pi_k f_{Y_k}(y|\boldsymbol{x}, \boldsymbol{\theta}_k), \tag{5.1}$$

where π_k are the mixing probabilities, $0 < \pi_k < 1$, $\sum_{j=1}^{K} \pi_k = 1$ and $E(Y_k|\boldsymbol{x}, \boldsymbol{\beta}_k) = \boldsymbol{x}^T\boldsymbol{\beta}_k$ with $\boldsymbol{\beta}_k \in \mathbb{R}^p$ the regression coefficients and $\sigma_k^2 \in \mathbb{R}$ the scale parameter of component k, $k = 1, \dots, K$, $\boldsymbol{\Theta} = (\boldsymbol{\theta}_1, \boldsymbol{\theta}_2, \dots, \boldsymbol{\theta}_K)$, and $\boldsymbol{\theta}_k = (\pi_k, \boldsymbol{\beta}_k, \sigma_k^2)$.

The component labels are usually not known beforehand. The unknown component membership can be seen as missing data. We define the random vectors $\boldsymbol{z}_i = (z_{i1}, \dots, z_{iK})$ to be the unobserved component membership for the ith observation with

$$z_{ik} = \begin{cases} 0 & \text{if observation } i \text{ is not from component } k \\ 1 & \text{if observation } i \text{ is from component } k \end{cases},$$

and $P(z_{ik} = 1|\boldsymbol{\Theta}) = \pi_k$.

Consider the random sample $\boldsymbol{y} = (y_1, \dots, y_n)$, $\boldsymbol{X} = (\boldsymbol{x}_1, \dots, \boldsymbol{x}_n)$, and $\boldsymbol{Z} = (\boldsymbol{z}_1, \dots, \boldsymbol{z}_n)$ all of size n, the complete data are $(\boldsymbol{y}, \boldsymbol{X}, \boldsymbol{Z})$. The complete data likelihood function using the residual formulation of the finite mixture of regression model is

$$l_c(\boldsymbol{\Theta}|\boldsymbol{X}, \boldsymbol{y}, \boldsymbol{Z}) = \sum_{i=1}^{n}\sum_{k=1}^{K} z_{ik} \log\left[\pi_k f_{\epsilon_k}(y_i - \boldsymbol{x}_i^T\boldsymbol{\beta}_k|z_{ik}, \boldsymbol{x}_i, \boldsymbol{\theta}_k)\right]. \tag{5.2}$$

In order to fit (5.2), we need to pre-specify the number of components. This too is often unknown beforehand. The number of components can be determined by using the CPFMR approach. Similar to the approach in (5.2) but with a penalty on the number of components resulting in the penalized complete data log-likelihood,

$$l_c^p(\mathbf{\Theta}|\boldsymbol{X}, \boldsymbol{y}, \boldsymbol{Z}) = l_c(\mathbf{\Theta}|\boldsymbol{X}, \boldsymbol{y}, \boldsymbol{Z}) - P_1(\boldsymbol{\pi}), \tag{5.3}$$

where $P_1(\boldsymbol{\pi}) = n\lambda \sum_k \left[log(C_0 + p_{\lambda(\pi_k)}) - log(C_0)\right]$ with $p_{\lambda(\pi_k)}$ a non-negative and non-decreasing function in π_k and C_0 a small positive number. The choice of penalty function is flexible; however, for this chapter, we use the SCAD penalty. The SCAD penalty has good properties, such as the oracle property, which makes it attractive for this task [14].

5.2.1.1 Computing Algorithm for CPFMR

The EM-algorithm is regularly used to find maximum likelihood estimates. The EM algorithm consists of two iterative steps, the expectation (E) step and the maximization (M) step. We calculate the expectation of the latent variables in the E-step. We then substitute the latent variables with these expectations to obtain the Q function. In the M-step, we find the parameter values that maximize the Q function. Consider the following updating equations:

E-Step: Calculate $E(z_{ik}|\boldsymbol{y}, \boldsymbol{X}, \boldsymbol{\theta}_k)$,

$$\gamma_{ik}^{new} = \frac{\pi_k^{old} N(y_i - \boldsymbol{x}_i^T \boldsymbol{\beta}_k^{old} | \boldsymbol{x}_i^T \boldsymbol{\beta}_k^{old}, \sigma_k^{2\ old})}{\sum_{j=1}^K \pi_k^{old} N(y_i - \boldsymbol{x}_i^T \boldsymbol{\beta}_k^{old} | \boldsymbol{x}_i^T \boldsymbol{\beta}_k^{old} \sigma_k^{2\ old})}, \tag{5.4}$$

γ_{ik} is called the responsibilities.
By substituting z_{ik} in (5.3) with γ_{ik} yields

$$Q^P(\mathbf{\Theta}) = Q(\mathbf{\Theta}) - P_1(\boldsymbol{\pi}).$$

During the M-step, we calculate the following parameter updates:

In order to update π_k, we first substitute $log(C_0 + p_\lambda(\pi_k))$ with a linear approximation and maximize $Q^P(\mathbf{\Theta})$ with respect to π_k

$$\begin{aligned} \pi_k^{new} &= \arg\max_{\pi_k} Q^P(\mathbf{\Theta}) \\ &= \frac{\sum_{i=1}^{n_k} \gamma_{ik}^{new}}{n_c} \\ &= \frac{n_k}{n_c}, \end{aligned} \tag{5.5}$$

with

$$n_c = n\left[1 - \lambda \sum_k \frac{p_\lambda'(\pi_k^{old})\pi_k^{old}}{C_0 + p_\lambda(\pi_k^{old})} + \lambda \frac{p_\lambda'(\pi_k^{old})}{C_0 + p_\lambda(\pi_k^{old})}\right].$$

$$\begin{aligned} \boldsymbol{\beta}_k^{new} &= \arg\max_{\beta_k} Q^P(\mathbf{\Theta}) \\ &= \left(\boldsymbol{X}^T \boldsymbol{W}_k \boldsymbol{X}\right)^{-1} \boldsymbol{X}^T \boldsymbol{W}_k \boldsymbol{y}, \end{aligned} \tag{5.6}$$

with $\boldsymbol{X} = (\boldsymbol{x}_1, \ldots, \boldsymbol{x}_n)^T$, $\boldsymbol{y} = (y_1, \ldots, y_n)^T$, and $\boldsymbol{W}_k = diag(\gamma_{1k}^{new}, \ldots, \gamma_{nk}^{new})$.

$$\begin{aligned} \sigma_k^{2\ new} &= \arg\max_{\sigma_k^2} Q^P(\mathbf{\Theta}) \\ &= \frac{\sum_{i=1}^n \gamma_{ik}^{new} \left(y_i - \boldsymbol{x}_i^T \boldsymbol{\beta}_k^{new}\right)^2}{n_k}. \end{aligned} \tag{5.7}$$

This approach is summarized in Algorithm 1.

Algorithm 1 Algorithm for fitting a CPFMR model.

1. Choose a set of initial values of the parameters $\boldsymbol{\Theta}^{old}$, that is π_k^{old}, $\boldsymbol{\beta}_k^{old}$, and $\sigma_k^{2^{old}}$, for $k = 1, 2, \ldots, K$.

while (*parameter estimates have not converged*) **do**
  E-Step:
  Update γ_{ik}^{new} using equation (5.4)
  M-Step:
  Update π_k^{new} using equation (5.5)
  Update $\boldsymbol{\beta}_k^{new}$ using equation (5.6)
  Update σ_k^{new} using equation (5.7)

  Set $\pi_k^{old} \leftarrow \pi_k^{new}$
  Set $\boldsymbol{\beta}_k^{old} \leftarrow \boldsymbol{\beta}_k^{new}$
  Set $\sigma_k^{old} \leftarrow \sigma_k^{new}$

  If π_k is smaller than a pre-specified threshold value **then**:
  $\pi_k \leftarrow 0$ and the estimation algorithm re-starts with $K - 1$ components.
end while

5.2.2 Self-paced Learning

In this subsection, we discuss SPLFMR and define key concepts and notation. By introducing observations in a meaningful order, SPL mimics a learning strategy that starts with easy examples and gradually includes more difficult examples. Observations with a lower contribution to the complete log-likelihood are considered to be less typical and are gradually incorporated during the later stages of the estimation process.
Consider the SPL weights $\boldsymbol{V} = (v_{ik})$

$$v_{ik} = \begin{cases} 0 & \text{if observation } i \text{ is non-typical for component } k \\ 1 & \text{if observation } i \text{ is typical for component } k \end{cases}.$$

For a random sample of size n, SPLFMR finds the parameter values that maximize the penalized weighted complete data log-likelihood

$$\begin{aligned} l_c^{SPL}(\boldsymbol{\Theta}|\boldsymbol{X}, \boldsymbol{y}, \boldsymbol{Z}, \boldsymbol{V}) &= \sum_{i=1}^{n}\sum_{k=1}^{K}\{v_{ik} z_{ik} \times \\ &\quad log\left[\pi_k f_{\epsilon_k}(y_i - \boldsymbol{x}_i^T \boldsymbol{\beta}_k | z_{ik}, \boldsymbol{x}_i, \boldsymbol{\theta}_k)\right]\} \\ &\quad -\sum_{i=1}^{n}\sum_{k=1}^{K} \omega_k |v_{ik}|, \end{aligned}$$

where ω_k is the age parameter of the kth component. The age parameter controls the number of non-typical observations introduced in the fitting process. This is achieved by systematically decreasing ω_k by some constant factor $a > 1$. The value of a is determined using a grid search methodology.

5.2.2.1 Computing Algorithm for SPLFMR

In each iteration, the following equations are updated:
E-Step: Calculate the responsibilities $E(z_{ik}|\boldsymbol{y}, \boldsymbol{X}, \boldsymbol{\theta}_k v_{ik})$ to obtain the responsibilities,

$$\gamma_{ik}^{new} = \frac{v_{ik}^{old}\pi_k^{old}N(y_i - \boldsymbol{x}_i^T\boldsymbol{\beta}_k^{old}|\boldsymbol{x}_i^T\boldsymbol{\beta}_k^{2\ old}, \sigma_k^{2\ old})}{\sum_{j=1}^K v_{ik}^{old}\pi_k^{old}N(y_i - \boldsymbol{x}_i^T\boldsymbol{\beta}_k^{old}|\boldsymbol{x}_i^T\boldsymbol{\beta}_k^{2\ old}, \sigma_k^{2\ old})}. \tag{5.8}$$

In the first iteration of SPL, we set $v_{ik}^{old} = 1$ for all $i = 1, 2, \ldots, n$ and $j = 1, 2, \ldots, K$. Substituting z_{ik} in the complete data log-likelihood with γ_{ik}^{new} yields

$$\begin{aligned} Q^{SPL}(\boldsymbol{\Theta}) &= \sum_{i=1}^n\sum_{k=1}^K v_{ik}\gamma_{ik}^{new}\, log\left[\pi_k f_{\epsilon_k}(y_i - \boldsymbol{x}_i^T\boldsymbol{\beta}_k|z_{ik}, \boldsymbol{x}_i, \boldsymbol{\theta}_k)\right] \\ &\quad -\sum_{i=1}^n\sum_{k=1}^K \omega_k|v_{ik}| \\ &= \sum_{i=1}^n\sum_{k=1}^K v_{ik}q_{ik}(\boldsymbol{\Theta}) - \sum_{i=1}^n\sum_{k=1}^K \omega_k|v_{ik}|. \end{aligned}$$

M-Step equation updates:

$$v_{ik}^{new} = \begin{cases} 0 & \text{if } q_{ik}(\boldsymbol{\Theta}^{old}) < \omega_k^{old} \\ 1 & \text{if } q_{ik}(\boldsymbol{\Theta}^{old}) \geq \omega_k^{old} \end{cases}, \tag{5.9}$$

with $\boldsymbol{\Theta}^{old} = (\boldsymbol{\theta}_1^{old}, \ldots, \boldsymbol{\theta}_K^{old})$ and $\boldsymbol{\theta}_k^{old} = (\pi_k^{old}, \boldsymbol{\beta}_k^{old}, \sigma_k^{2\ old})$.

$$\begin{aligned} \pi_k^{new} &= \arg\max_{\pi_k} Q^{SPL}(\boldsymbol{\Theta}) \\ &= \frac{\sum_{i=1}^n v_{ik}^{new}\gamma_{ik}^{new}}{\sum_{i=1}^n\sum_{k=1}^K v_{ik}^{new}\gamma_{ik}^{new}} \\ &= \frac{n_k}{\sum_{i=1}^n\sum_{k=1}^K v_{ik}^{new}\gamma_{ik}^{new}}, \end{aligned} \tag{5.10}$$

$$\begin{aligned} \boldsymbol{\beta}_k^{new} &= \arg\max_{\beta_k} Q^{SPL}(\boldsymbol{\Theta}) \\ &= \left(\boldsymbol{X}^T\boldsymbol{W}_k\boldsymbol{X}\right)^{-1}\boldsymbol{X}^T\boldsymbol{W}_k\boldsymbol{y}, \end{aligned} \tag{5.11}$$

with $\boldsymbol{X} = (\boldsymbol{x}_1, \ldots, \boldsymbol{x}_n)^T, \boldsymbol{y} = (y_1, \ldots, y_n)^T$, and $\boldsymbol{W}_k = diag(v_{1k}^{new}\gamma_{1k}^{new}, \ldots, v_{nk}^{new}\gamma_{nk}^{new})$, and

$$\begin{aligned} \sigma_k^{2\ new} &= \arg\max_{\sigma_k^2} Q^{SPL}(\boldsymbol{\Theta}) \\ &= \frac{\sum_{i=1}^n v_{ik}^{new}\gamma_{ik}^{new}\left(y_i - \boldsymbol{x}_i^T\boldsymbol{\beta}_k^{new}\right)^2}{n_k}. \end{aligned} \tag{5.12}$$

The age parameter is updated with

$$\omega_k^{new} = \omega_k^{old} - a, \tag{5.13}$$

with $a > 1$.

The process is outlined in Algorithm 2:

Algorithm 2 Algorithm for fitting a SPLFMR model.

1. Choose a set of initial values of the parameters $\boldsymbol{\Theta}^{old}$, that is π_k^{old}, $\boldsymbol{\beta}_k^{old}$, and $\sigma_k^{2^{old}}$, for $k = 1, 2, \ldots, K$.

2. Determine an initial set values of the belongings γ_{ik}^{old} using a standard Gaussian mixture regression model, for $i = 1, 2, \ldots, n$ and $k = 1, 2, \ldots, K$.

3. Determine initial v_{ik}^{new} using equation (5.9)

while (*while at least one* $v_{ik}^{old} = 0$) or (*parameter estimates have not converged*) **do**
  while (*while at least one* $v_{ik}^{old} = 0$) or (*parameter estimates have not converged*) **do**
    E-Step:
    Update γ_{ik}^{new} using equation (5.8)
    M-Step:
    Update v_{ik}^{new} using equation (5.9)
    Update π_k^{new} using equation (5.10)
    Update $\boldsymbol{\beta}_k^{new}$ using equation (5.11)
    Update σ_k^{new} using equation (5.12)

    Set $v_{ik}^{old} \leftarrow v_{ik}^{new}$
    Set $\pi_k^{old} \leftarrow \pi_k^{new}$
    Set $\boldsymbol{\beta}_k^{old} \leftarrow \boldsymbol{\beta}_k^{new}$
    Set $\sigma_k^{old} \leftarrow \sigma_k^{new}$

    If π_k is smaller than a pre-specified threshold value **then**:
    $\pi_k \leftarrow 0$ and the estimation algorithm re-starts with $K - 1$ components.
  end while
  Update ω_k^{new} using (5.13)

  Set $\omega_k^{old} \leftarrow \omega_k^{new}$
end while

5.3 Component Penalized Self-paced Learning

In this section, we introduce the CPSPL model. This method fits an FMR with automatic order selection whilst mitigating the influence of non-typical observations.

Consider the complete data log-likelihood

$$l_c^{CPSPL}(\boldsymbol{\Theta}|\boldsymbol{X}, \boldsymbol{y}, \boldsymbol{Z}, \boldsymbol{V}) \quad = \quad l_c^{SPL}(\boldsymbol{\Theta}|\boldsymbol{X}, \boldsymbol{y}, \boldsymbol{Z}, \boldsymbol{V}) - P_1(\boldsymbol{\pi}),$$

$P_1(\boldsymbol{\pi}) = n\lambda \sum_k [log(C_0 + p_\lambda(\pi_k)) - log(C_0)]$ with $p_\lambda(\pi_k)$ a non-negative and non-decreasing function in π_k and C_0 a small positive number and $p_\lambda(\pi_k)$ the SCAD penalty.

5.3.1 Computing Algorithm for the CPSPL Model

The CPSPL estimates parameters by iteratively maximizing (5.14) with respect to v_{ik} and $\boldsymbol{\Theta}$ resulting in the following updating equations
E-Step: Calculate γ_{ik}^{new} using (5.8).

Substituting z_{ik} in the complete data log-likelihood with γ_{ik}^{new} yields

$$Q^{CPSPL}(\boldsymbol{\Theta}) = Q^{SPL}(\boldsymbol{\Theta}) - P_1(\boldsymbol{\pi}).$$

M-Step update equations:

$$v_{ik}^{new} = \begin{cases} 0 & \text{if } q_{ik}(\boldsymbol{\Theta}^{old}) < \omega_k^{old} \\ 1 & \text{if } q_{ik}(\boldsymbol{\Theta}^{old}) \geq \omega_k^{old} \end{cases}. \tag{5.14}$$

Substitute $log(C_0 + p_\lambda(\pi_k))$ by its linear approximation and maximize $Q^{CPSPL}(\boldsymbol{\Theta})$

$$\begin{aligned} \pi_k^{new} &= \arg\max_{\pi_k} Q^{CPSPL}(\boldsymbol{\Theta}) \\ &= \frac{\sum_{i=1}^{n_k} v_{ik}^{new}\gamma_{ik}^{new}}{n_c} = \frac{n_k}{n_c}, \end{aligned} \tag{5.15}$$

with

$$n_c = \sum_{i=1}^{n}\sum_{k=1}^{K} v_{ik}^{new}\gamma_{ik}^{new}\left[1 - \lambda\sum_k \frac{p'_\lambda(\pi_k^{old})\pi_k^{old}}{C_0 + p_\lambda(\pi_k^{old})} + \lambda\frac{p'_\lambda(\pi_k^{old})}{C_0 + p_\lambda(\pi_k^{old})}\right],$$

$$\begin{aligned} \boldsymbol{\beta}_k^{new} &= \arg\max_{\beta_k} Q^{CPSPL}(\boldsymbol{\Theta}) \\ &= \left(\boldsymbol{X}^T\boldsymbol{W}_k\boldsymbol{X}\right)^{-1}\boldsymbol{X}^T\boldsymbol{W}_k\boldsymbol{y}, \end{aligned} \tag{5.16}$$

with $\boldsymbol{X} = (\boldsymbol{x}_1, \ldots, \boldsymbol{x}_n)^T$, $\boldsymbol{y} = (y_1, \ldots, y_n)^T$, and $\boldsymbol{W}_k = diag(v_{1k}^{new}\gamma_{1k}^{new}, \ldots, v_{nk}^{new}\gamma_{nk}^{new})$,

$$\begin{aligned} \sigma_k^{2\ new} &= \arg\max_{\sigma_k^2} Q^{CPSPL}(\boldsymbol{\Theta}) \\ &= \frac{\sum_{i=1}^{n} v_{ik}^{new}\gamma_{ik}^{new}\left(y_i - \boldsymbol{x}_i^T\boldsymbol{\beta}_k^{new}\right)^2}{n_k}. \end{aligned} \tag{5.17}$$

Update ω_k with

$$\omega_k^{new} = \omega_k^{new} - a, \tag{5.18}$$

with $a > 1$.

Algorithm 3 summarizes this process:

Algorithm 3 Fitting a CPSPL model.

1. Choose a set of initial values of the parameters $\boldsymbol{\Theta}^{old}$, that is π_k^{old}, $\boldsymbol{\beta}_k^{old}$, and $\sigma_k^{2^{old}}$, for $k = 1, 2, \ldots, K$.

2. Determine an initial set values of the belongings γ_{ik}^{old} using a standard Gaussian mixture regression model, for $i = 1, 2, \ldots, n$ and $k = 1, 2, \ldots, K$.

3. Determine initial v_{ik}^{new} using equation (5.14)

while (*while at least one* $v_{ik}^{old} = 0$) or (*parameter estimates have not converged*) **do**
  while (*while at least one* $v_{ik}^{old} = 0$) or (*parameter estimates have not converged*) **do**
    E-Step:
    Update γ_{ik}^{new} using equation (5.8)
    M-Step:
    Update v_{ik}^{new} using equation (5.14)
    Update π_k^{new} using equation (5.15)
    Update $\boldsymbol{\beta}_k^{new}$ using equation (5.16)
    Update σ_k^{new} using equation (5.17)

    Set $v_{ik}^{old} \leftarrow v_{ik}^{new}$
    Set $\pi_k^{old} \leftarrow \pi_k^{new}$
    Set $\boldsymbol{\beta}_k^{old} \leftarrow \boldsymbol{\beta}_k^{new}$
    Set $\sigma_k^{old} \leftarrow \sigma_k^{new}$

    If π_k is smaller than a pre-specified threshold value **then**:
    $\pi_k \leftarrow 0$ and the estimation algorithm re-starts with $K - 1$ components.
  end while
  Update ω_k^{new} using (5.18)

  Set $\omega_k^{old} \leftarrow \omega_k^{new}$
end while

5.4 Simulation Study

In this section, we discuss the results from a simulation study comparing the model estimation abilities of the CPFMR and CPSPL models. Each method's ability to correctly identify the number of components and its capacity to mitigate the impact of non-typical observations are evaluated through simulation.

5.4.1 Data Generation

Two cases are considered for generating data:

- Easy case:

$$Y_i = \begin{cases} 1 + 3x_i + \epsilon_{1i} & \text{with probability } \pi_1 = 0.5 \\ 1 - 3x_i + \epsilon_{2i} & \text{with probability } \pi_2 = 0.5 \end{cases}$$

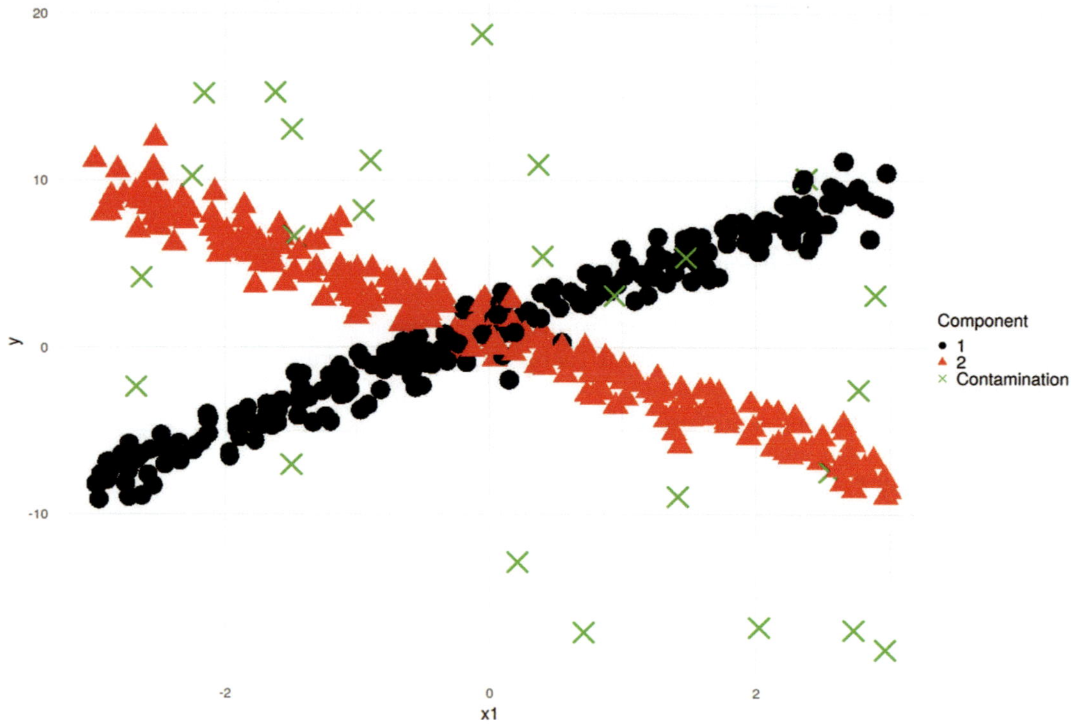

FIGURE 5.1
Easy case with 5% contamination.

- Difficult case:

$$Y_i = \begin{cases} 1 + 1x_i + \epsilon_{1i} & \text{with probability } \pi_1 = 0.5 \\ 1 - 1x_i + \epsilon_{2i} & \text{with probability } \pi_2 = 0.5 \end{cases}$$

with $\epsilon_{ji} \sim N(0, 1)$ for each $j = 1, 2$ and $i = 1, \ldots, n$.

A scatter plot of an easy case and a difficult case with 5% contamination can be found in Figures 5.1 and 5.2, respectively. The contaminated observations are plotted as green crosses. We generate data from each case and contaminate the data with non-typical observations. Four contamination levels are considered: $1\%, 5\%, 10\%, 20\%$.

To generate the contaminated data, we randomly selected observations to be contaminated. For the contamination, we keep the X values fixed and selected new random Y values uniformly within the interval $(-20, 20)$. We consider this type of contamination as it simulates gross outliers that are not generated by the underlying distribution of interest. We consider $B = 500$ replicates from each case, combined with the contamination levels. Sample sizes $100, 200, 500$, and 1000 are considered.

5.4.2 Performance Assessment

Bias and mean squared error (MSE) are used to quantify the accuracy of the parameter estimates. The bias and MSE for a parameter $\boldsymbol{\theta}$ are calculated using

$$Bias_{\hat{\boldsymbol{\theta}}} = \frac{\sum_{i=1}^{B} (\hat{\boldsymbol{\theta}}_i - \boldsymbol{\theta})}{B}$$

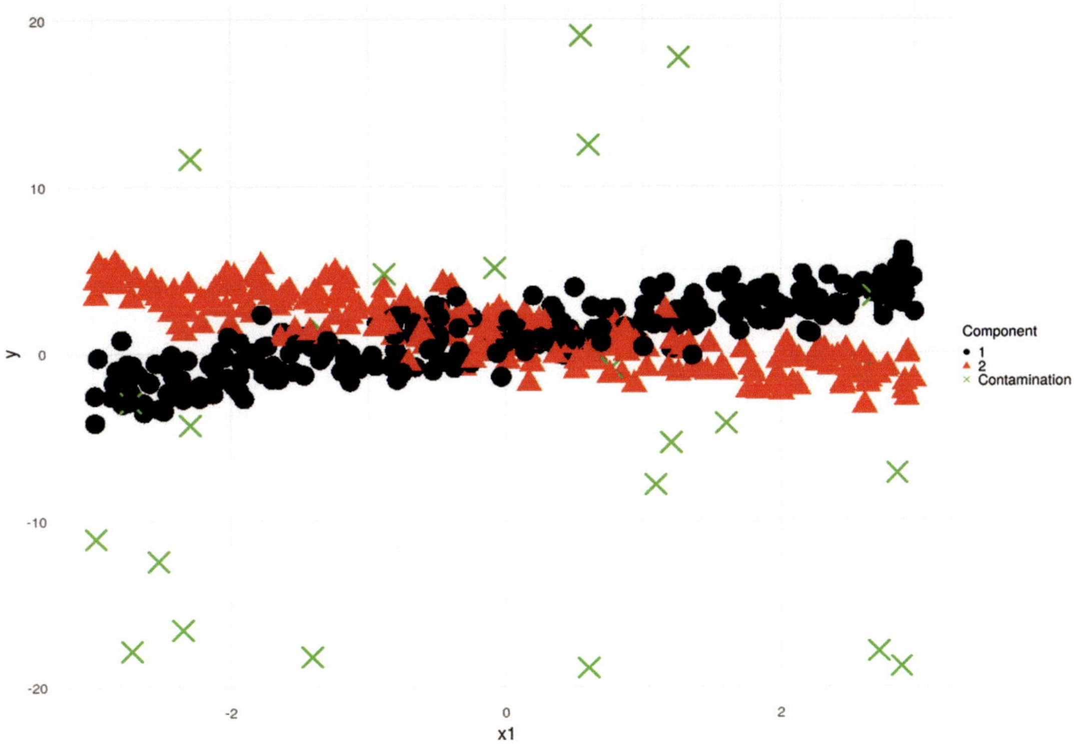

FIGURE 5.2
Difficult case with 5% contamination.

and

$$MSE_{\hat{\boldsymbol{\theta}}} = \frac{\sum_{i=1}^{B}(\hat{\boldsymbol{\theta}}_i - \boldsymbol{\theta})^2}{B},$$

where $\hat{\boldsymbol{\theta}} = (\hat{\theta}_1, ..., \hat{\theta}_B)$ is the set of parameter estimates for the simulation i. Bias and MSE values closer to zero indicate better estimation performance.

5.4.3 Model Fitting

It is well known that the initial values in FMR have a profound influence on the final estimated results. Although the EM algorithm guarantees convergence, it does not necessarily converge to the global optimum [15, 16]. It is therefore recommended to fit multiple FMR models with different initial values and select the model with the highest likelihood [3]. To mitigate the potential random effects that different initial values may have on the final simulation results, we use a fixed set of initial values for each data set. These include the true underlying distribution and two additional components with parameters $\boldsymbol{\beta}_3^T = \begin{pmatrix}1 & -4\end{pmatrix}$ and $\boldsymbol{\beta}_4^T = \begin{pmatrix}1 & 4\end{pmatrix}$. A grid search is performed to identify the hyper-parameters, λ and ω_k, that yield the highest BIC value.

5.4.3.1 Order Selection Results

For each contamination case, we kept track of the number of components fitted by each model. We recorded how many times each method identified under fitted ($k < 2$), correctly fitted ($k = 2$), or over fitted components ($k > 2$). We find each method are more prone to

TABLE 5.1
The proportion of easy cases where the order selection was under fitted, correctly fitted and over fitted for different contamination levels.

Method	**k estimate**	**No contamination**	**1%**	**5%**	**10%**	**20%**
CPSPL	under fitted	0.000	0.000	0.000	0.000	0.000
CPSPL	correctly fitted	**1.000**	**1.000**	**1.000**	**0.984**	**0.944**
CPSPL	over fitted	0.000	0.000	0.000	0.016	0.056
CPFMR	under fitted	0.000	0.000	0.004	0.206	0.246
CPFMR	correctly fitted	**1.000**	**1.000**	0.996	0.794	0.750
CPFMR	over fitted	0.000	0.000	0.000	0.000	0.004

TABLE 5.2
The proportion of difficult cases where the order selection was under fitted, correctly fitted, and over fitted for different contamination levels.

Method	**k estimate**	**No Contamination**	**1%**	**5%**	**10%**	**20%**
CPSPL	under fitted	0.004	0.002	0.002	0.004	0.006
CPSPL	correctly fitted	0.996	**0.998**	**0.998**	**0.996**	**0.992**
CPSPL	over fitted	0.000	0.000	0.000	0.000	0.002
CPFMR	under fitted	0.002	0.002	0.024	0.004	0.414
CPFMR	correctly fitted	**0.998**	**0.998**	0.976	**0.996**	0.586
CPFMR	over fitted	0.000	0.000	0.000	0.000	0.000

under fitting. This issue worsens as the number of non-typical observations increases. This is however not surprising, given the type of contamination considered. The contamination masks the distribution of interest uniformly, making it resemble a single component.

When considering an easy case and with no contamination as reported in Table 5.1, we find that both methods correctly estimated the number of components for all data sets for a sample size of $n = 500$. Table 5.2 shows a slight decrease in accuracy for both methods when applied to a difficult case without contamination. In this setting, CPFMR estimated k correctly more frequently than the CPSPL method.

When considering an easy case with 1% contamination, Table 5.1 indicates that both methods still accurately identify the correct number of components. This demonstrates their capability to select the correct order under mild contamination. However, we observe a decline in performance for both methods under the difficult case with 1% contamination, as shown in Table 5.2.

Cases with higher contamination, Table 5.1 shows that both methods tend to under fit more frequently. This issue becomes more pronounced in the difficult case, as seen in Table 5.2. In both cases, the CPSPL method fares better in selecting the correct number of components under contamination.

Overall, both methods are capable of accurately estimating the correct number of components. However, when non-typical observations are present in the data, the robust CPSPL method yields the best results, successfully mitigating the impact of such observations.

5.4.3.2 Parameter Estimates Results

In this subsection, we investigate each method's ability to estimate the correct model parameters under different contamination levels.

TABLE 5.3
Bias and MSE for an easy case with no contamination.

	CPFMR			CPSPL		
	Mean estimate	**Bias**	**MSE**	**Mean estimate**	**Bias**	**MSE**
β_{11}	0.999	**−0.001**	**0.013**	1.001	**0.001**	0.014
β_{12}	−2.989	0.011	**0.012**	−2.998	**0.002**	**0.012**
β_{21}	0.999	**−0.000**	**0.013**	1.000	**−0.000**	**0.013**
β_{22}	2.982	−0.018	**0.011**	2.992	**−0.008**	0.012
π_1	0.500	**−0.000**	**0.002**	0.498	−0.002	**0.002**

TABLE 5.4
Bias and MSE for an easy case with 5% contamination.

	CPFMR			CPSPL		
	Mean estimate	**Bias**	**MSE**	**Mean estimate**	**Bias**	**MSE**
β_{11}	1.001	**0.001**	**0.012**	1.007	0.007	0.013
β_{12}	−2.988	0.012	**0.011**	−3.004	**−0.004**	0.013
β_{21}	1.004	**0.004**	**0.013**	1.004	**0.004**	0.015
β_{22}	2.987	−0.013	**0.011**	2.997	**−0.003**	0.013
π_1	0.499	**−0.001**	**0.002**	0.497	−0.003	0.002

TABLE 5.5
Bias and MSE for a difficult case with 5% contamination.

	CPFMR			CPSPL		
	Mean estimate	**Bias**	**MSE**	**Mean estimate**	**Bias**	**MSE**
β_{11}	0.989	−0.011	0.014	1.006	**0.006**	**0.008**
β_{12}	0.001	1.001	0.008	−1.003	**−0.003**	**0.008**
β_{21}	0.989	−0.011	0.014	0.998	**−0.002**	**0.009**
β_{22}	0.012	−0.988	0.008	1.006	**0.006**	**0.009**
π_1	0.500	**−0.000**	**0.001**	0.501	0.001	0.003

Table 5.3 reports the bias and MSE of the parameter estimates for the CPSPL and CPFMR methods in an easy case with no contamination. Both methods produced similarly low biases and MSE values for their respective estimates.

When considering the 5% contamination, reported in Tables 5.4 and 5.5, we observe that the CPSPL method yields lower bias in the regression parameter estimates compared to the CPFMR model. Although both methods show high accuracy in selecting the correct number of components for the difficult case with 5% contamination, the CPFMR method produces parameter estimates with higher bias and MSE. In contrast, CPSPL achieves lower bias and MSE in this scenario.

For the 20% contamination level for both the easy and difficult cases, as shown in Tables 5.6 and 5.7 respectively, we clearly see that both methods yield poorer parameter estimates. This is evidenced by the higher bias and MSE values for each parameter compared to the cases with no contamination. However, the CPSPL method consistently demonstrates better bias and MSE values across these highly contaminated cases.

Both methods see reduced accuracy of parameter estimates as the contamination increases. The CPSPL method had overall the lowest bias and MSE values across the simulation study, indeed showing its robust properties.

TABLE 5.6
Bias and MSE for an easy case with 20% contamination.

	CPFMR			CPSPL		
	Mean estimate	**Bias**	**MSE**	**Mean estimate**	**Bias**	**MSE**
β_{11}	0.693	−0.307	0.421	0.956	**−0.044**	**0.047**
β_{12}	−1.619	1.381	0.681	−2.828	**0.172**	**0.234**
β_{21}	0.706	−0.294	0.395	0.955	**−0.045**	**0.040**
β_{22}	1.636	−1.364	0.668	2.832	**−0.168**	**0.236**
π_1	0.495	−0.005	0.026	0.499	**−0.001**	**0.003**

TABLE 5.7
Bias and MSE for a difficult case with 20% contamination.

	CPFMR			CPSPL		
	Mean estimate	**Bias**	**MSE**	**Mean estimate**	**Bias**	**MSE**
β_{11}	0.428	−0.572	2.948	0.996	**−0.004**	**0.033**
β_{12}	−0.914	0.086	2.288	−1.011	**−0.011**	**0.043**
β_{21}	0.156	−0.844	3.631	1.007	**0.007**	**0.034**
β_{22}	1.014	**0.014**	2.084	1.033	0.033	**0.041**
π_1	0.517	0.017	0.158	0.502	**0.002**	**0.012**

5.4.3.3 Simulation Study Conclusion

The simulation study confirms that both methods can find the correct number of components on clean and mildly contaminated data. However, when the data contains a high proportion of non-typical data, then the CPFMR method shows a decline in order selection and parameter estimation accuracy. The CPSPL method has shown to provide the most resilience to the impact of these non-typical observations, showing its robust properties.

5.5 Application on CO_2 Data

In this chapter, we study the relationship between the total annual CO_2 emissions, measured in metric tonnes, an indicator of environmental degradation, and GDP per capita measured in US dollar, a monetary value produced by a country over the period of 2023. This type of analysis has been studied before and allows to answer questions such as, do countries have differing development paths which allow to prosper economically whilst protecting the environment? Ideally, a country would have a high GDP while maintaining low CO_2 emissions. This type of analysis can help achieve the sustainable development goals (SDG) targets such as Target 13, integration of climate change measures into national policies, strategies, and planning. It can also help with understanding the factors to help achieve SDG Target 8.1, sustain per capita economic growth in accordance with national circumstances [17, 18].

For this application, we specifically focus on BRICS, an intergovernmental organization and Annex II countries as classified under the United Nations Framework Convention on Climate Change. Annex II countries are considered to be developed countries with an obligation to aid developing countries through financial resources and facilitation of technology to developing countries [19]. BRICS consisted of five country members at the time, but we

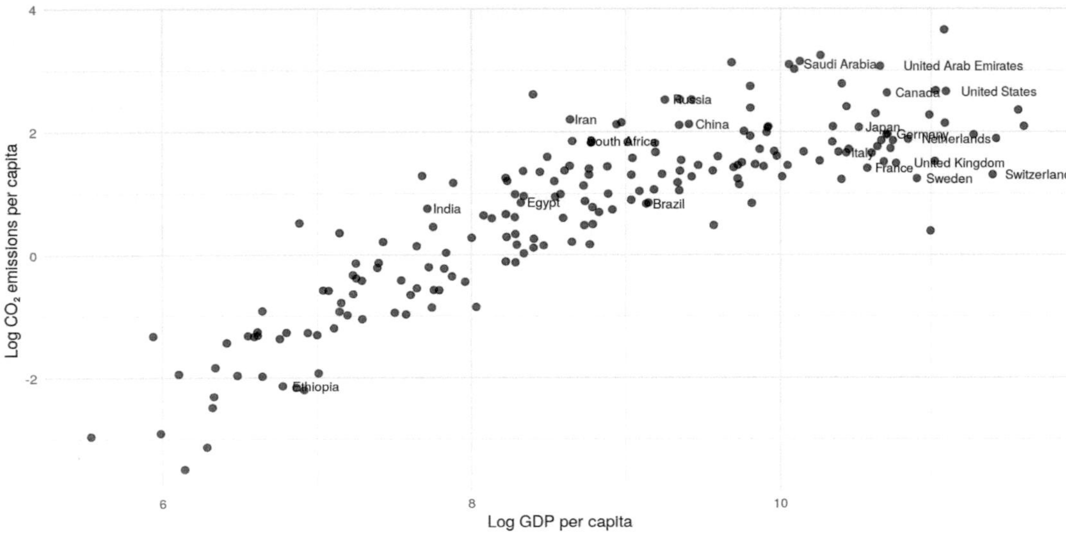

FIGURE 5.3
Scatter plot of the log of total annual CO_2 emissions versus log of GDP per capita for each country over the period of 2023.

will analyze the current members of BRICS as of January 2025 as well. All BRICS members consist of non-Annex I countries with the exception of the Russian federation which is considered an Annex I country. Non-Annex I countries consist of primarily developing nations with limited CO_2 emission restrictions placed on them [19]. The data used in this study can be freely found here [20, 21].

5.5.1 Preprocessing of Data

We consider the log of the CO_2 emissions per year and the GDP per capita. A scatter plot of the 2023 data is presented in Figure 5.3.

5.5.2 Methodology

All analyzes were performed using `R` version 4.2.0. We employ two methods to fit the data. The first method applies the CPSPL method for this analysis. A grid search is performed to identify the optimal set of hyper-parameters, using the configuration that results in the lowest BIC value. We also fit a FMR model with between 2 and 10 components, selecting the model order based on the lowest BIC value. To fit the standard FMR model, we use the `flexmix()` function from the `flexmix` package, version 2.3-18 [22].

5.5.3 Application Results

The models fitted using CPSPL and FMR respectively are:

$$Y_i^{CPSPL} = \begin{cases} -8.14 + 0.99x_i + \epsilon_{1i} \text{ with probability } \pi_1 = 0.54 \\ -9.64 + 1.28x_i + \epsilon_{2i} \text{ with probability } \pi_2 = 0.20 \\ -1.27 + 0.28x_i + \epsilon_{3i} \text{ with probability } \pi_3 = 0.26 \end{cases},$$

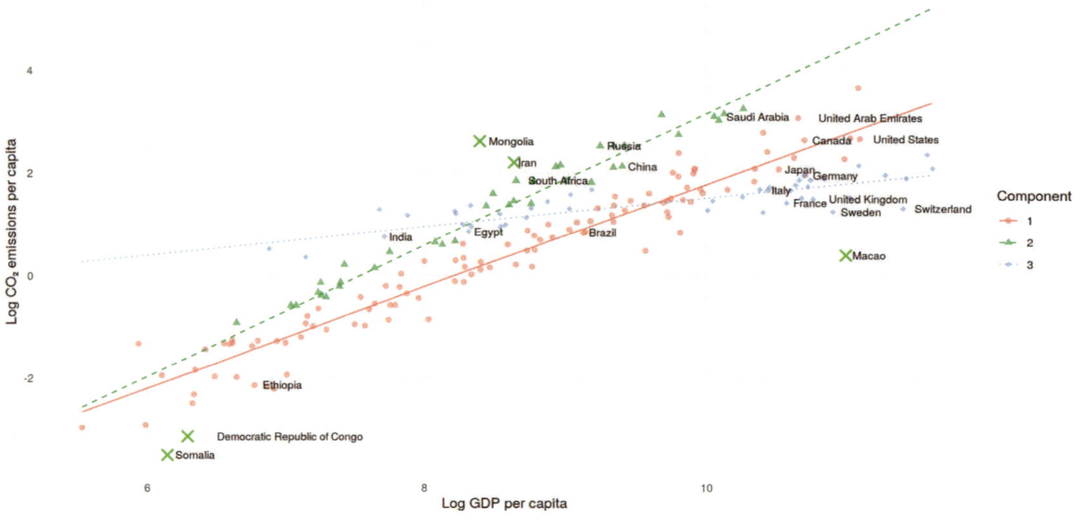

FIGURE 5.4
Scatter plot with the fitted CPSL model.

$\epsilon_{1i} \sim N(0, 0.43^2)$, $\epsilon_{2i} \sim N(0, 0.24^2)$, and $\epsilon_{3i} \sim N(0, 0.23^2)$.

$$Y_i^{FMR} = \begin{cases} -10.68 + 1.34x_i + \epsilon_{1i} \text{ with probability } \pi_1 = 0.36 \\ -6.30 + 0.78x_i + \epsilon_{2i} \text{ with probability } \pi_2 = 0.38 \\ -0.85 + 0.24x_i + \epsilon_{3i} \text{ with probability } \pi_3 = 0.26 \end{cases},$$

$\epsilon_{1i} \sim N(0, 0.36^2)$, $\epsilon_{2i} \sim N(0, 0.38^2)$, and $\epsilon_{3i} \sim N(0, 0.26^2)$.

A scatter plot with the fitted CPSL and FMR models can be seen respectively in Figures 5.4 and 5.5.

Both methods identified three components; one with higher levels of CO_2 emissions for a given GDP per capita, one intermediary group, and a group with the lowest amount of CO_2 emissions per GDP per capita. However, the methods differ in terms of slope parameters and country classification.

The CPSPL method classified countries such as South Africa, China, Russia, and Saudi Arabia in the high emissions group per unit of GDP per capita. This classification aligns with expectations, as these countries are Non-Annex I countries, with the exception of the Russian Federation. However, using the traditional FMR method, countries such as Russia, China, Saudi Arabia, and Egypt remain in the high emissions group, whereas South Africa and Iran are classified in the lower emissions group. This may be surprising, as Iran has a lower GDP per capita than China, but also higher emissions per capita. The CPSPL method identified Iran as a non-typical observation, with a higher level of emissions relative to its GDP per capita.

When examining the intermediary groups, we observe that the CPSPL method placed Canada, the United States of America, Ethiopia, the United Arab Emirates, and Brazil in this category. The latter three are Non-Annex I countries, yet they perform better in terms of emissions per GDP per capita than most BRICS countries like China, South Africa, Iran, Saudi Arabia and the Russian Federation. They are on par with some Annex II countries such as Canada, the United States of America and Japan. The FMR method classified countries such as Japan, Germany, the United states of America, Italy, and Brazil as part of the intermediary group containing more Annex II countries overall.

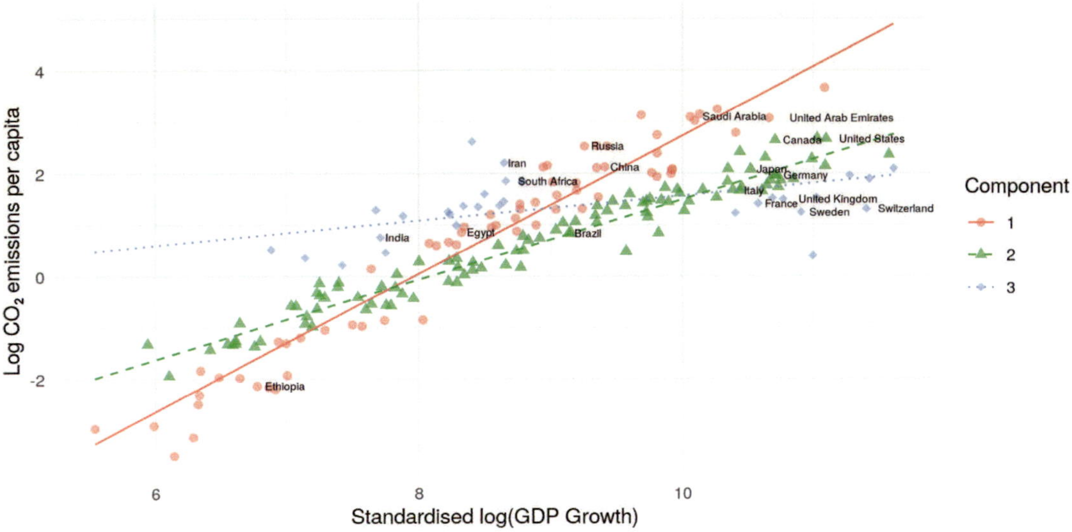

FIGURE 5.5
Scatter plot with the fitted FMR model.

Finally, in the group with the lowest emissions trend, the CPSPL model clustered most Annex II countries like Germany, Austria, Italy, Spain, Portugal, France, the Netherlands, Sweden, Denmark, the United Kingdom, Norway, Ireland, Switzerland and Luxembourg together. Interestingly, Egypt and India were also grouped with these countries, suggesting that their CO_2 emissions follow the same trend line relative to their GDP per capita levels. This is surprising, as both are Non-Annex I countries. The model also identified Macao as an outlier having lower CO_2 emissions per capita for their GDP per capita. The FMR model classified France, the United Kingdom, Denmark, Sweden, Switzerland, Norway, and Ireland in this group as well. It also classified India to be part of this group. The FMR method did however, place countries such as South Africa and Iran as part of the lower emission trend group.

It is clear that both methods uncovered three distinct trends in the data: countries with high emissions, intermediate emissions, and low emissions relative to GDP per capita. The CPSPL method did however identified five non-typical observations marked as green crosses in Figure 5.4. These non-typical observations include Iran, Mongolia, Somalia, the Democratic Republic of the Congo, and Macao. These outliers appear to have skewed the results in the regular FMR model. By plotting both models on the same plot in Figure 5.6, we can see the impact of the non-typical observations on the fitted FMR model. By mitigating the influence of these non-typical observations, the CPSPL method showed that the slope for the high-emission group is less steep as compared to that of the regular FMR. Similarly, the slope for countries in the low-emissions group are greater than in the FMR model. The same holds true for the intermediary group.

5.6 Conclusion and Future Work

This chapter has explored the problem of order selection in finite mixture regression models, with a particular focus on the challenges posed by non-typical observations. We proposed a

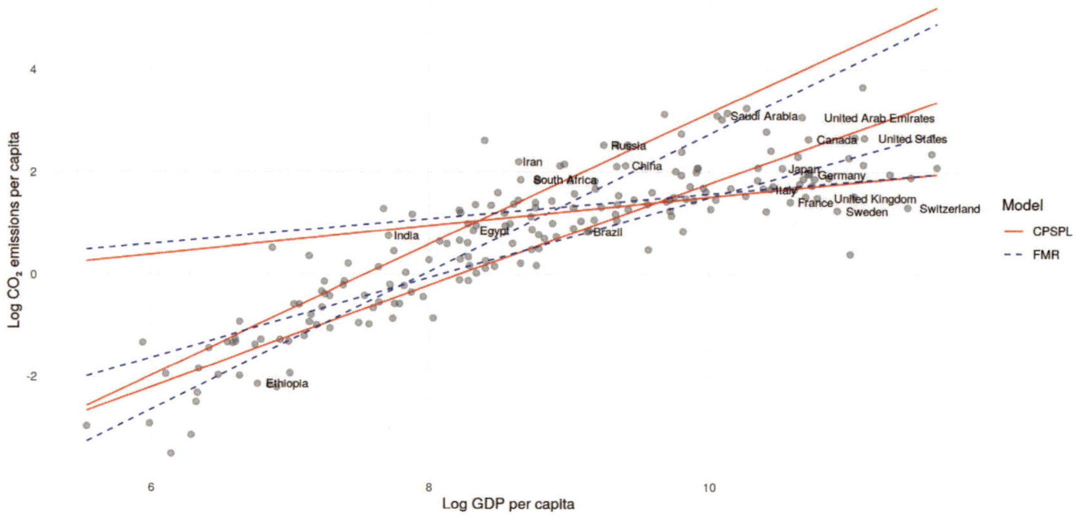

FIGURE 5.6
Combined plots of CPSPL and FMR methods.

robust methodology that integrates SPL and penalized likelihood to improve the accuracy and reliability of model selection under contamination.

Through extensive simulation studies, we demonstrated that the proposed approach effectively mitigates the influence of non-typical observations, consistently identifying the correct number of components even under increasing levels of data contamination. In our real-world application to climate data, the model successfully uncovered meaningful groupings based on CO_2 emissions per capita. Notably, it identified three distinct clusters, with most Annex I and II countries falling into the lower-emission groups, while some prominent non-Annex I nations appeared predominantly in the higher-emission cluster.

These results suggest that the proposed method is not only robust but also practically insightful, offering a valuable tool for researchers working with complex, contaminated datasets in environmental and socio-economic contexts.

Bibliography

[1] P. K. Dunstan, S.D. Foster, F. K. C. Hui, and D.I. Warton. Finite mixture of regression modeling for high-dimensional count and biomass data in ecology. *Journal of Agricultural, Biological, and Environmental Statistics*, 18:357–375, 2013.

[2] J. He. Mixture model based multivariate statistical analysis of multiply censored environmental data. *Advances in Water Resources*, 59:15–24, 2013.

[3] G. McLachlan and D. Peel. *Finite Mixture Models.* Wiley, 2000.

[4] C. Keribin. Consistent estimation of the order of mixture models. *Sankhyā: The Indian Journal of Statistics, Series A (1961-2002)*, 62(1):49–66, 2000.

[5] J. Ma and T. Wang. Entropy penalized automated model selection on Gaussian mixture. *International Journal of Pattern Recognition and Artificial Intelligence*, 18(8):1501–1512, 2004.

[6] J. Chen and A. Khalili. Order selection in finite mixture models with a nonsmooth penalty. *Journal of the American Statistical Association*, 104(485):187–196, 2009.

[7] Q. Tang and R. J. Karunamuni. Robust variable selection for finite mixture regression models. *Annals of the Institute of Statistical Mathematics*, 70:489–521, 2018.

[8] A. Mazza and A. Punzo. Mixtures of multivariate contaminated normal regression models. *Statistical Papers*, 61(3):787–822, 2020. Published online: 13 November 2017.

[9] D. Peel and G.J. McLachlan. Robust mixture modelling using the t distribution. *Statistics and Computing*, 10(4):339–348, 2000.

[10] L Han, D. Zhang, D. Huang, X. Chang, J. Ren, S. Luo, and J. Han. Self-paced mixture of regressions. In Carles Sierra, editor, *Proceedings of the 26th International Joint Conference on Artificial Intelligence*, pages 1816–1822, United States of America, 2017. Association for the Advancement of Artificial Intelligence (AAAI). International Joint Conference on Artificial Intelligence 2017, IJCAI 2017 ; Conference date: 19-08-2017 Through 25-08-2017.

[11] W. Li, H. Chen, T. Li, J. Wan, and B. Sang. Unsupervised feature selection via self-paced learning and low-redundant regularization. *Knowledge-Based Systems*, 240:108150, 2022.

[12] M. Kumar, B. Packer, and D. Koller. Self-paced learning for latent variable models. In J. Lafferty, C. Williams, J. Shawe-Taylor, R. Zemel, and A. Culotta, editors, *Advances in Neural Information Processing Systems*, volume 23. Curran Associates, Inc., 2010.

[13] D. Meng, Q. Zhao, and L. Jiang. A theoretical understanding of self-paced learning. *Information Sciences*, 414:319–328, 2017.

[14] Fan J. and Li R. Variable selection via nonconcave penalized likelihood and its oracle properties. *Journal of the American Statistical Association*, 96(456):1348–1360, 2001.

[15] C. M. Bishop. *Pattern Recognition and Machine Learning.* Springer, New York, 10th edition, 2006.

[16] A. P. Dempster, N. M. Laird, and D. B. Rubin. Maximum likelihood from incomplete data via the EM algorithm. *Journal of the Royal Statistical Society. Series B (Methodological)*, 39(1):1–22, 1977.

[17] United Nations. Sustainable development goal 13: Climate action `https://sdgs.un.org/goals/goal13#targets_and_indicators`.

[18] United Nations. Sustainable development goal 8: Decent work and economic growth `https://sdgs.un.org/goals/goal8#targets_and_indicators`.

[19] UNFCCC. Parties and Observers `https://unfccc.int/parties-observers#::text=Non%2DAnnex%20I%20Parties%20are,of%20climate%20change%20response%20measures`.

[20] H. Ritchie, P. Rosado, and M. Roser. Data page: Per capita CO_2 emissions. `https://ourworldindata.org/grapher/co-emissions-per-capita`, 2023. CO_2 and Greenhouse Gas Emissions.

[21] M. Roser, B. Rohenkohl, P. Arriagada, J. Hasell, H. Ritchie, and E. Ortiz-Ospina. Data page: GDP per capita. `https://ourworldindata.org/grapher/gdp-per-capita-worldbank`, 2023. Economic Growth.

[22] B. Grun and F. Leisch. FlexMix: An R package for finite mixture modelling. *R News*, 2007.

6

Bayesian Structure Learning of Directed Acyclic Graphs for Identifying Causal Effects of Weather Elements in South Africa

Samaneh Nazari, Mohammad Arashi, and Abdolnasser Sadeghkhani

In our endeavor to decode the intricate interplay of weather elements in South Africa, we embark on an exploration utilizing Bayesian structure learning techniques. By leveraging directed acyclic graphs (DAGs), we systematically unravel the complexities of causal relationships, establishing a robust framework for identifying and quantifying the effects of various meteorological factors. To this end, our analysis focuses on precise estimation, where the nodes of the DAG represent critical weather variables such as temperature, wind speed, precipitation, cloud cover, etc. DAGs serve as powerful structures for delineating and inferring dependencies among variables within multivariate scenarios. A thorough understanding of the underlying DAG-generating model is essential for effective causal discovery and accurate estimation. In our approach, we employ normal-gamma and normal-inverse gamma priors for Gaussian DAG structure learning, aimed at evaluating the posterior probabilities. Through meticulous Markov Chain Monte Carlo (MCMC) computations, we strive to derive a more precise causal graph. Our methodologies undergo rigorous evaluation through extensive simulation studies and are subsequently applied to analyze historical weather data from South Africa. This comprehensive framework not only enhances our understanding of the dynamics of weather elements but also contributes valuable insights to the field of meteorological research, paving the way for informed decision-making and effective climate strategies.

6.1 Introduction

Meteorological conditions play a crucial role in daily life, environmental sustainability, and public health. For instance, weather conditions play a critical role in shaping patterns of many diseases such as the tick-borne disease, Theileriosis, known as the east coast fever ([1], [2] and [3]). Also, climate change poses serious threats to agriculture, reducing water availability and shifting farming patterns, and it leads to job losses and economic inequality, see [4] for details. Monitoring and predicting these conditions are essential for disaster

DOI: 10.1201/9781003564423-6

preparedness, agriculture, pollution management, and energy planning. A comprehensive understanding of weather patterns facilitates informed decision-making across diverse sectors, including aviation and urban development. The climate is a combination of atmospheric phenomena, including temperature, precipitation, wind speed, solar radiation, humidity, etc., that characterizes a location for a long time. The elements that form climate are very important for developing prediction models. Understanding the complex interactions among these meteorological elements is essential for improving weather forecasts, climate modeling, and environmental surveillance. Traditional statistical approaches often fail to capture the complex probabilistic relationships among atmospheric variables, leading to inaccuracies in predictions ([5] and [6]). While traditional methods have laid the groundwork for atmospheric analysis, the integration of machine learning and advanced statistical techniques is essential for addressing the complexities of meteorological data and enhancing predictive accuracy. We refer to the couple of studies [7], [8], [9], and [10] to mention a few. Recent advancements in machine learning and probabilistic modeling have enabled the application of Bayesian networks in meteorology, facilitating data-driven discovery of interdependencies between climate variables ([11] and [12]). Bayesian networks offer a powerful probabilistic framework for representing and analyzing these dependencies. As discussed by [13] and [14], the Bayesian network models are powerful tools for analyzing meteorological data because they capture uncertainty, allow reasoning with incomplete data, and provide insights into causal relationships between meteorological factors. For example, [15], [16], and [17] used Bayesian networks to model the relationships between air pollution, climate change, and their impacts on human health, predict air quality and monthly rainfall forecast, respectively. A Bayesian network is a DAG where nodes represent random variables, and edges define conditional dependencies among them. This study aims to construct and analyze a Bayesian network model using key meteorological variables, including temperature, relative humidity, dew point, and precipitation, to uncover their probabilistic dependencies. Identifying these relationships is essential for enhancing forecasting accuracy, understanding climate dynamics, and supporting decision-making in weather-sensitive sectors such as agriculture, disaster management, and urban planning.

The rest of the paper is organized as follows: In Section 6.2, we provide an overview of the dataset utilized in this study. Section 6.3 presents the methodologies used to uncover the inherent dependencies among variables in a Bayesian network model, while in Section 6.4, we analyze the weather data using the proposed method to investigate causal effects of weather elements. Section 6.5 provides the conclusion.

6.2 The Descriptions of Weather Data

This study examines the complex relationships between meteorological variables to improve the precision of predictive models and reveals causal effects of the weather elements in some sites of South Africa. The dataset utilized includes a comprehensive set of meteorological factors, such as temperature, precipitation, dew point, humidity, wind speed, wind direction, cloud cover, solar radiation, and sea-level pressure. These variables have a fundamental role in characterizing atmospheric conditions and exhibit complex interdependencies that form weather patterns and climate dynamics. A comprehensive understanding of these interactions is essential for improving predictive models, identifying causal mechanisms, and gaining more profound insights into atmospheric dynamics. To accomplish this objective, Bayesian networks are utilized as a probabilistic graphical model that depicts causal relationships and conditional dependencies among variables. This methodology offers a structured framework for modeling uncertainty and effectively captures both direct and indirect interactions among meteorological factors. Moreover, the methodology allows the integration of prior knowledge with observational data, establishing it as a powerful tool for considering complexities in climate modeling and environmental research. This analysis presents valuable information that can enhance the weather forecasting models, contribute to a more precise understanding of climate patterns, and support decision-making processes in meteorology. In this study, weather data was obtained using the Visual Crossing Weather Query Builder. The Visual Crossing Weather Data platform provides the ability to easily access weather data sets including weather forecast data, weather history data, and historical weather summary data. Given its associated challenges and impacts, climate change in South Africa could be reframed as predominately a important issue, one necessitating an urgent review. The growing impact of climate change has major implications for South Africa, especially for the numerous vulnerable groups in the country. For our purpose, the hourly dataset for South Africa was acquired and utilized for subsequent analysis. The main goal of utilizing this dataset is to discover how the various weather variables in South Africa affect each other by leveraging the DAGs. Table 6.1 provides an overview of the dataset description. The hourly dataset is available at `https://www.visualcrossing.com/weather-query-builder`.

This dataset was collected for South Africa with the geographical and meteorological information, as presented in Table 6.2. The dataset comprises observations from three stations, whose details are detailed in Table 6.3. Figure 6.1 provides a comprehensive overview of South Africa, emphasizing the locations of these stations.

6.3 Bayesian Structure Learning

In the last few decades, probabilistic graphical models have been powerful and useful tools for displaying and inferring dependent relationships in complicated settings ([18]). Among these models, DAGs, also known as Bayesian networks, have been extensively utilized across diverse disciplines such as meteorology, biology, and economics, owing to their capacity to represent conditional dependencies among variables. One of the key challenges in this field is structural learning in these graphs. The structural learning of DAGs is important not only for understanding the dependencies between variables but also plays a prominent role in causal analysis and prediction. The problem of causal discovery has been considered through different methodological perspectives, with significant contributions from both frequentist and Bayesian frameworks. A key differentiation among frequentist approaches lies in the distinction between constraint-based and score-based algorithms. On the Bayesian side, learning the DAG structure is considered as a Bayesian model selection problem in which the objective is the posterior distribution in the space of all possible graphs, which is usually estimated via MCMC methods. Bayesian structure learning for DAGs is a powerful method for analyzing complex dependencies between variables, owing to its flexibility and capability to model uncertainty. One of the underlying challenges in Bayesian structure learning for DAGs is choosing the appropriate prior distribution for each model-dependent parameter. Prior distributions can reflect the researcher's prior knowledge about the graph structure and significantly effect the outcomes of the learning process. From this perspective, prior distribution specification plays an important role in the structure learning of DAGs. We employ the normal-inverse gamma and normal-gamma distributions as starting points in this study for Bayesian structure learning of DAGs. Before introducing the prior distributions utilized in this study, we first present preliminary concepts on DAGs and graph theory.

TABLE 6.1
The summary of South Africa weather data and the associated descriptions.

Variable	Description	Abbreviation
Temperature	Average (mean) temperature for the time period, for example, for each day.	temp
Feelslike	Temperature, heat index and wind chill are combined into a single 'feelslike' element for clarity and ease of use. Where heat Index is a measure of how hot it feels combining the actual air temperature with the relative humidity and wind chill is the measure of how cold it feels combining the actual air temperature with the wind speed.	feelslike
Dew point	The dew point is the temperature the air needs to be cooled to (at constant pressure) in order to produce a relative humidity of 100%.	dew
Relative humidity	Relative humidity is the amount of water vapor present in the air compared the maximum amount possible for a given temperature, expressed as a mean percentage.	humidity
Precipitation	The amount of precipitation that fell or is predicted to fall in the specified time period. The values are indicated in inches or mm.	precip
Wind speed and wind direction	The wind speed and wind direction indicate the wind speed for the location and time period requested.	windspeed and winddir
Sea level pressure	The atmospheric pressure at a location that removes reduction in pressure due to the altitude of the location. This is expressed in millibars.	sealevelpressure
Cloud cover	Cloud cover is the amount of sky that is covered by cloud expressed as a percentage. The cloud coverage is for all altitudes. Daily values include the mean of the hourly cloud coverage values.	cloudcover
Solar radiation	The solar radiation measures the power at the instantaneous moment of the observation (or forecast prediction).	solar radiation

6.3.1 Preliminaries on DAGs

This section concentrates on the basic concepts related to DAGs, particularly emphasizing Gaussian DAG models. For a comprehensive understanding of this topic, we refer to [18] and [19].

Consider a DAG denoted by $\mathcal{D}$=$(\mathbf{V}, \mathbf{E})$, where the vertices (or nodes) set is given by $\mathbf{V} = \{1, \ldots, q\}$, while the edge set $\mathbf{E} \subseteq \mathbf{V} \times \mathbf{V}$ consists of directed edges without forming any cycles. In this framework, if a directed edge exists from node u to node v (i.e., $u \rightarrow v$),

TABLE 6.2
The geographical and meteorological information of South Africa.

queryCost	1
latitude	−25.7459
longitude	28.1876
resolvedAddress	South Africa
address	South Africa
timezone	Africa/Johannesburg
tzoffset	2
description	Cooling down with a chanc...

TABLE 6.3
The geographical information of the stations considered in our weather data analysis of South Africa.

Station	Id	Distance (mi)	Latitude	Longitude
FAWK	FAWK	6	-25.83	28.22
FW8815 Centurion ZA	F8815	9	-25.871	28.143
FAGM	FAGM	34	-26.23	28.15

then u is referred to as the *parent* of v, and conversely, v is considered a *child* of u. The set of all parent nodes of v within $\mathcal{D}$ is denoted by $pa_{\mathcal{D}}(v)$. Furthermore, let $X_1, \ldots, Xq$ be a random vector characterized by the probability density function (pdf) $f(x_1, \ldots, x_q)$, where each random variable X_j is associated with a corresponding node in $\mathcal{D}$. The structure of the DAG encodes a set of conditional independence relationships among these variables through the Markov property. Consequently, $\mathcal{D}$ serves as a graphical representation of the underlying conditional independencies among the nodes (or variables) in the system.

As per modeling, assume the random vector $(X_1, \ldots, X_q)$ given the precision matrix $\boldsymbol{\Omega}_{\mathcal{D}}$, follows a multivariate normal distribution $\mathcal{N}_q\left(\mathbf{0}, \boldsymbol{\Omega}_{\mathcal{D}}^{-1}\right)$.

Here, the precision matrix is defined as $\boldsymbol{\Omega} = \boldsymbol{\Omega}_{\mathcal{D}} = \boldsymbol{\Sigma}_{\mathcal{D}}^{-1} \in \mathcal{P}_{\mathcal{D}}$, where $\mathcal{P}_{\mathcal{D}}$ denotes the space of all symmetric positive definite (s.p.d.) precision matrices that are Markov with respect to the DAG $\mathcal{D}$. Consequently, the structure of $\boldsymbol{\Omega}_{\mathcal{D}}$ encodes the conditional independence relationships dictated by the Markov property associated with $\mathcal{D}$.

As it is traditional in Bayesian modeling, one adopts (inverse) Wishart prior distribution. See [20] and [21] and references therein. However, to develop multi-shape parameter, following [22], we use the modified Cholesky decomposition of the precision matrix as $\boldsymbol{\Omega} = \boldsymbol{L}\boldsymbol{D}^{-1}\boldsymbol{L}^T$, where $\boldsymbol{L}$ is a $q \times q$ matrix whose off-diagonal elements, $\boldsymbol{L}_{uv}$, are nonzero if and only if u is a parent of v in $\mathcal{D}$, i.e., $u \in pa_{\mathcal{D}}(v)$. By construction, the diagonal elements of $\boldsymbol{L}$ are fixed at $\boldsymbol{L}_{uu} = 1$ for all $u = 1, \ldots, q$. Additionally, $\boldsymbol{D}$ is a $q \times q$ diagonal matrix,

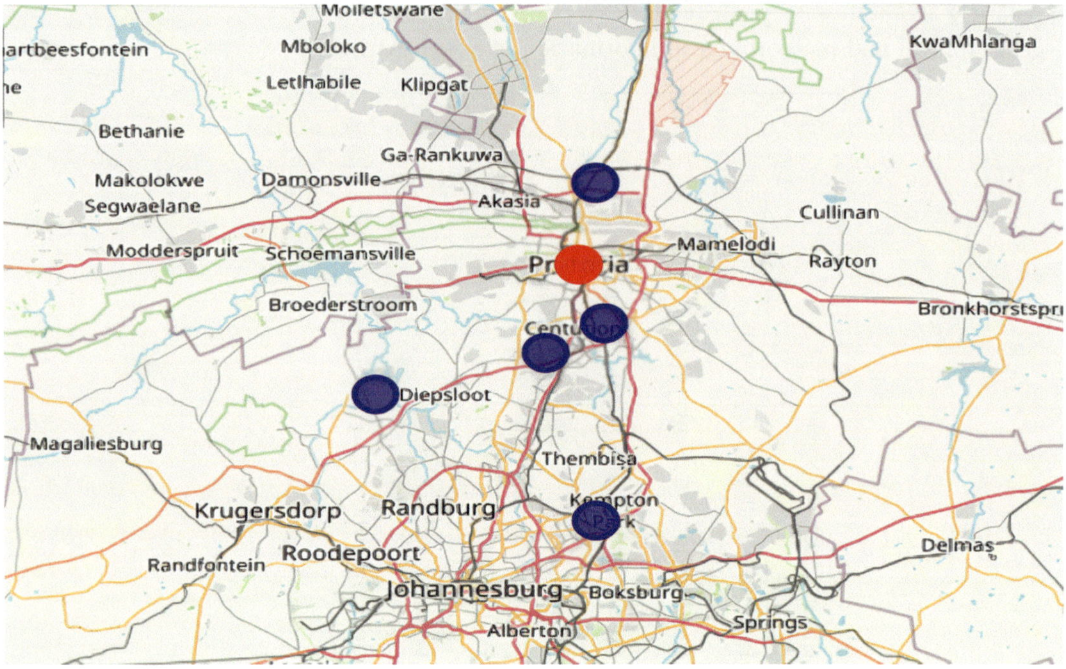

FIGURE 6.1
Map of the stations considered in our weather data analysis of South Africa.

where each diagonal element $\boldsymbol{D}_{uu}$ corresponds to the conditional variance of the respective node.

6.3.2 Prior Specification

Let us consider a collection of n independent observations, denoted as $\boldsymbol{x}_i = (x_{i,1}, \ldots, x_{i,q})^T \in \mathbb{R}^q$ for $i = 1, \ldots, n$, each drawn from $\mathcal{N}_q\left(\mathbf{0}, \boldsymbol{\Omega}_{\mathcal{D}}^{-1}\right)$, and the $n \times q$ data matrix $\boldsymbol{X}$, constructed by stacking the row vectors $\boldsymbol{x}_1, \ldots, \boldsymbol{x}_n$. The corresponding likelihood function is the product of $d\mathcal{N}_n\left(\boldsymbol{X}_j \mid -\boldsymbol{X}_{pa_{\mathcal{D}}(j)}\boldsymbol{L}_{\prec j]}, \boldsymbol{D}_{jj}\boldsymbol{I}_n\right)$, for $j = 1, \ldots, q$, that is

$$f\left(\boldsymbol{X}|\boldsymbol{D}, \boldsymbol{L}\right) = \prod_{j=1}^{q} d\mathcal{N}_n\left(\boldsymbol{X}_j \mid -\boldsymbol{X}_{pa_{\mathcal{D}}(j)}\boldsymbol{L}_{\prec j]}, \boldsymbol{D}_{jj}\boldsymbol{I}_n\right), \tag{6.1}$$

such that

$$\begin{aligned}
\boldsymbol{L}_{\prec j]} &= -\boldsymbol{\Sigma}_{\prec j \succ}^{-1}\boldsymbol{\Sigma}_{\prec j]}, \\
\boldsymbol{D}_{jj} &= \boldsymbol{\Sigma}_{jj|pa_{\mathcal{D}}(j)} = \boldsymbol{\Sigma}_{jj} - \boldsymbol{\Sigma}_{[j\succ}\boldsymbol{\Sigma}_{\prec j \succ}^{-1}\boldsymbol{\Sigma}_{\prec j]}, \\
\prec j] &= pa_{\mathcal{D}}(j) \times j,\ [j \succ = j \times pa_{\mathcal{D}}(j), \prec j \succ = pa_{\mathcal{D}}(j) \times pa_{\mathcal{D}}(j).
\end{aligned}$$

The parameter $\boldsymbol{D}_{jj}$ represents the conditional variance of X_j, denoted as $\mathbb{V}ar\left(X_j \mid \boldsymbol{x}_{pa_{\mathcal{D}}(j)}\right)$.

Also, $d\mathcal{N}(. \mid \mu, \sigma^2)$ denotes the pdf of a normal distribution with mean μ and variance σ^2. The symbol $\boldsymbol{X}_A$ refers to the sub-matrix of $\boldsymbol{X}$ formed by selecting the columns indexed by the set A, and $\boldsymbol{I}_n$ represents the $n \times n$ identity matrix.

In order to learn DAGs structures using Bayesian methods, it is necessary to specify appropriate prior distributions for the parameters associated with matrices $\mathbf{D}$ and $\mathbf{L}$, as well as for the DAGs structure itself. These prior distributions play a critical role in Bayesian inference and the resulting structure estimation. In the following, we will first start by choosing the prior distribution on DAG $\mathcal{D}$. Then, we will use normal-inverse-gamma (NIG) as the prior distribution on Cholesky parameters $(\mathbf{D}, \mathbf{L})$. Finally, we will show the posterior distribution of the model parameters $(\mathbf{D}, \mathbf{L})$ and the DAG marginal likelihood. To compare the effect of different prior specifications, we also carry out the same steps—including the derivation of the posterior distribution and marginal likelihood—using the normal-gamma (NG) prior distribution.

For a given prior probability of edge inclusion $\pi \in [0,1]$, we assume that for each pair (u,v) with $u > v$, the elements $\boldsymbol{S}^{\mathcal{D}}_{u,v}$ are independently and identically distributed according to a Bernoulli distribution with parameter π, i.e., $\boldsymbol{S}^{\mathcal{D}}_{u,v} \mid \pi \overset{iid}{\sim} \text{Ber}(\pi)$. Here, $\boldsymbol{S}^{\mathcal{D}}_{u,v} = 1$ if and only if $(u,v) \in E$ or $(v,u) \in E$, and $\boldsymbol{S}^{\mathcal{D}}_{u,v} = 0$ otherwise. In this setting, $\boldsymbol{S}^{\mathcal{D}}$ denotes the 0–1 adjacency matrix of the skeleton associated with the DAG $\mathcal{D}$, which is obtained by disregarding the orientations of edges. Consequently, the prior probability of $\mathcal{D}$ satisfies $p(\mathcal{D}) \propto p\left(\boldsymbol{S}^{\mathcal{D}}\right)$ for each DAG $\mathcal{D} \in \mathcal{S}_q$, the set of all DAGs on q nodes, where

$$p\left(\boldsymbol{S}^{\mathcal{D}}\right) = \pi^{|\boldsymbol{S}^{\mathcal{D}}|}(1-\pi)^{\frac{q(q-1)}{2} - |\boldsymbol{S}^{\mathcal{D}}|}, \tag{6.2}$$

and $|\boldsymbol{S}^{\mathcal{D}}|$ denotes the number of edges in the skeleton. The quantity $q(q-1)/2$ represents the maximum number of edges in a DAG on q nodes.

6.3.2.1 Normal-Inverse-Gamma (NIG) Prior

In this subsection, following [23], conditional on the assumption that the precision matrix $\boldsymbol{\Omega}$ is Markov with respect to the DAG $\mathcal{D}$, we assign a prior distribution to the Cholesky parameters $(\mathbf{D}, \mathbf{L})$ associated with the nodes, as described below.

$$\begin{aligned} \boldsymbol{D}_{jj} \mid \mathcal{D} &\sim IG\left(\frac{1}{2}\alpha^{\mathcal{D}}_j, \frac{1}{2}\boldsymbol{U}_{jj|pa_{\mathcal{D}}(j)}\right), \\ \boldsymbol{L}_{\prec j]} \mid \boldsymbol{D}_{jj}, \mathcal{D} &\sim \mathcal{N}_{|pa_{\mathcal{D}}(j)|}\left(-\boldsymbol{U}^{-1}_{\prec j \succ}\boldsymbol{U}_{\prec j]}, \boldsymbol{D}_{jj}\boldsymbol{U}^{-1}_{\prec j \succ}\right). \end{aligned} \tag{6.3}$$

Here, $\boldsymbol{U}$ denotes a $q \times q$ symmetric positive definite matrix, and the parameters $\alpha^{\mathcal{D}}_1, \ldots, \alpha^{\mathcal{D}}_q$ represent shape hyperparameters, as defined in [22]. In general, both the hyperparameters $\alpha^{\mathcal{D}}_j$ and the matrix $\boldsymbol{U}$ are tailored to each specific DAG model under consideration. A commonly adopted choice for the shape parameters is given by $\alpha^{\mathcal{D}}_j = \alpha + |pa_{\mathcal{D}}(j)| - q + 1$, $(\alpha > q - 1)$, which ensures that the prior distributions remain compatible across Markov equivalent DAGs [24]. Notably, it can be shown that this specification guarantees that any two Markov equivalent DAGs are assigned identical marginal likelihoods. Moreover, a standard choice for $\boldsymbol{U}$, is $\boldsymbol{U} = gI_q$ with $g > 0$, reflecting a prior belief in the (marginal) independence among variables. The hyperparameter g regulates the informativeness of the prior on the precision matrix $\boldsymbol{\Omega}$; smaller values of g correspond to less informative priors. Under these assumptions, the prior distribution on the Cholesky parameters $(\boldsymbol{D}, \boldsymbol{L})$, given $\mathcal{D}$, factorizes as

$$p(\boldsymbol{D}, \boldsymbol{L} \mid \mathcal{D}) = \prod_{j=1}^{q} p\left(\boldsymbol{L}_{\prec j]} \mid \boldsymbol{D}_{jj}\right) p\left(\boldsymbol{D}_{jj}\right). \tag{6.4}$$

In what follows, we present the posterior distribution of the node-specific parameters, as well as the marginal likelihood associated with each node in the DAG $\mathcal{D}$.

Assuming the likelihood function $\prod_{j=1}^{q} d\mathcal{N}_n\left(\boldsymbol{X}_j \mid -\boldsymbol{X}_{pa_{\mathcal{D}}(j)}\boldsymbol{L}_{\prec j]}, \boldsymbol{D}_{jj}\boldsymbol{I}_n\right)$ and equation (6.3), the node-marginal likelihood will be as follows:

$$m\left(\boldsymbol{X}_j \mid \boldsymbol{X}_{pa_{\mathcal{D}}(j)}, \mathcal{D}\right) = (2\pi)^{-\frac{n}{2}} \times \frac{\left|U_{\prec j \succ}\right|^{\frac{1}{2}}}{\left|\widetilde{U}_{\prec j \succ}\right|^{\frac{1}{2}}} \times \frac{\Gamma\left(\frac{1}{2}\widetilde{\alpha}_j^{\mathcal{D}}\right)}{\Gamma\left(\frac{1}{2}\alpha_j^{\mathcal{D}}\right)} \times \frac{\left(\frac{1}{2}U_{jj|pa_{\mathcal{D}}(j)}\right)^{\frac{1}{2}\alpha_j^{\mathcal{D}}}}{\left(\frac{1}{2}\widetilde{U}_{jj|pa_{\mathcal{D}}(j)}\right)^{\frac{1}{2}\widetilde{\alpha}_j^{\mathcal{D}}}}$$

Moreover, the marginal likelihood of the DAG $\mathcal{D}$ can be factorized as $m(\boldsymbol{X} \mid \mathcal{D}) = \prod_{j=1}^{q} m\left(\boldsymbol{X}_j \mid \boldsymbol{X}_{pa_{\mathcal{D}}(j)}, \mathcal{D}\right)$.

Now, utilizing the assumptions mentioned in the previous sections and by assuming (6.4) as the prior on $(\boldsymbol{D}, \boldsymbol{L})$. Then, the joint posterior density function of $(\boldsymbol{D}, \boldsymbol{L})$ given the data $\boldsymbol{X}$, $p(\boldsymbol{D}, \boldsymbol{L} \mid \mathcal{D}, \boldsymbol{X})$, for $j = 1, \ldots, q$ is characterized by the following properties:

$$\begin{aligned} \boldsymbol{D}_{jj} \mid \mathcal{D}, \boldsymbol{X} &\sim IG\left(\tfrac{1}{2}\widetilde{\alpha}_j^{\mathcal{D}}, \tfrac{1}{2}\widetilde{\boldsymbol{U}}_{jj|pa_{\mathcal{D}}(j)}\right), \\ \boldsymbol{L}_{\prec j]} \mid \boldsymbol{D}_{jj}, \mathcal{D}, \boldsymbol{X} &\sim \mathcal{N}_{|pa_{\mathcal{D}}(j)|}\left(-\widetilde{\boldsymbol{U}}_{\prec j \succ}^{-1}\widetilde{\boldsymbol{U}}_{\prec j]}, \boldsymbol{D}_{jj}\widetilde{\boldsymbol{U}}_{\prec j \succ}^{-1}\right), \end{aligned} \tag{6.5}$$

such that $\widetilde{\boldsymbol{U}} = \boldsymbol{U} + \boldsymbol{X}^T\boldsymbol{X}$, $\widetilde{\alpha}_j^{\mathcal{D}} = \alpha + |pa_{\mathcal{D}}(j)| + n - q + 1$. See also [25] for more details and extensions.

6.3.2.2 Normal-Gamma (NG) Prior

In parallel with the analysis based on the normal-inverse-gamma prior, following [26], we now consider the normal-gamma prior for the Cholesky parameters $(\boldsymbol{D}, \boldsymbol{L})$. Specifically, we adopt this prior formulation to investigate its implications on the posterior distribution and the marginal likelihood of the DAG $\mathcal{D}$. Following the same inferential framework, we first specify the normal-gamma prior distribution on the model parameters, then derive the corresponding posterior distribution, and finally present the expression for the DAG marginal likelihood under this prior setting.

Based on the assumptions outlined in the previous subsection, conditionally on $\mathcal{D}$, we now consider the following non-conjugate prior for node-parameters $\{\left(\boldsymbol{D}_{jj} = \boldsymbol{\Sigma}_{jj|pa_{\mathcal{D}}(j)}, \boldsymbol{L}_{\prec j]}\right), j = 1, \ldots, q\}$

$$\begin{aligned} \boldsymbol{D}_{jj} \mid \mathcal{D} &\sim \Gamma\left(\frac{1}{2}\alpha_j^{\mathcal{D}}, \frac{1}{2}\boldsymbol{U}_{jj|pa_{\mathcal{D}}(j)}\right), \\ \boldsymbol{L}_{\prec j]} \mid \boldsymbol{D}_{jj}, \mathcal{D} &\sim \mathcal{N}_{|pa_{\mathcal{D}}(j)|}\left(-\boldsymbol{U}_{\prec j \succ}^{-1}\boldsymbol{U}_{\prec j]}, \boldsymbol{D}_{jj}\boldsymbol{U}_{\prec j \succ}^{-1}\right). \end{aligned} \tag{6.6}$$

Ultimately, a prior on $(\boldsymbol{D}, \boldsymbol{L})$ is given by

$$p(\boldsymbol{D}, \boldsymbol{L} \mid \mathcal{D}) = \prod_{j=1}^{q} p\left(\boldsymbol{L}_{\prec j]} \mid \boldsymbol{D}_{jj}\right) p\left(\boldsymbol{D}_{jj}\right). \tag{6.7}$$

Therefore, the posterior distribution $(\boldsymbol{D}, \boldsymbol{L})$ given the data $\boldsymbol{X}$, $p(\boldsymbol{D}, \boldsymbol{L} \mid \mathcal{D}, \boldsymbol{X})$, for $j = 1, \ldots, q$ and the node-marginal likelihood will be as follows, respectively.

$$\begin{aligned} p\left(\boldsymbol{L}_{\prec j]} \mid \boldsymbol{D}_{jj}, \mathcal{D}, \boldsymbol{X}\right) &= (2\pi)^{-\frac{|pa_{\mathcal{D}}(j)|}{2}} \boldsymbol{D}_{jj}^{-\frac{|pa_{\mathcal{D}}(j)|}{2}} \left|\widetilde{\boldsymbol{U}}_{\prec j \succ}\right|^{\frac{1}{2}} \\ &\times \exp\left\{-\frac{1}{2\boldsymbol{D}_{jj}}\left(\boldsymbol{L}_{\prec j]} - \boldsymbol{M}^{-1}\boldsymbol{b}\right)^T \boldsymbol{M}\left(\boldsymbol{L}_{\prec j]} - \boldsymbol{M}^{-1}\boldsymbol{b}\right)\right\} \quad , \end{aligned} \tag{6.8}$$

$$p(\boldsymbol{D}_{jj} \mid, \mathcal{D}, \boldsymbol{X}) = \frac{\boldsymbol{D}_{jj}^{\frac{1}{2}\alpha_j^{\mathcal{D}} - \frac{n}{2} - 1}}{2K_a\left(2\sqrt{\alpha\beta}\right)\left(\frac{\alpha}{\beta}\right)^{\frac{a}{2}}} \times \exp\left\{-\frac{1}{2}\boldsymbol{U}_{jj|pa_{\mathcal{D}}(j)}\boldsymbol{D}_{jj} - \frac{\alpha}{2}\boldsymbol{D}_{jj}\right\}, \tag{6.9}$$

where $a = \frac{1}{2}\alpha_j^{\mathcal{D}} - \frac{n}{2}$,

$$\begin{aligned}
\boldsymbol{M} &= \widetilde{\boldsymbol{U}}_{\prec j \succ} = \boldsymbol{U}_{\prec j \succ} + \boldsymbol{X}_{pa_{\mathcal{D}}(j)}^T \boldsymbol{X}_{pa_{\mathcal{D}}(j)}, \\
\boldsymbol{b} &= -\boldsymbol{X}_{pa_{\mathcal{D}}(j)}^T \boldsymbol{X}_j - \boldsymbol{U}_{\prec j]} \\
\alpha &= \frac{1}{2}\left[-\boldsymbol{b}^T \boldsymbol{M}^{-1}\boldsymbol{b} + \boldsymbol{U}_{\prec j]}^T \boldsymbol{U}_{\prec j \succ}^{-1} \boldsymbol{U}_{\prec j]} + \boldsymbol{X}_j^T \boldsymbol{I}_n \boldsymbol{X}_j\right] \\
\beta &= \frac{1}{2}\boldsymbol{U}_{jj|pa_{\mathcal{D}}(j)},
\end{aligned}$$

and $K_\nu(.)$ is the Bessel function of the third kind. Also, we have

$$m\left(\boldsymbol{X}_j \mid \boldsymbol{X}_{pa_{\mathcal{D}}(j)}, \mathcal{D}\right) = (2\pi)^{-\frac{n}{2}} \times \frac{\left|\boldsymbol{U}_{\prec j \succ}\right|^{\frac{1}{2}}}{\left|\widetilde{\boldsymbol{U}}_{\prec j \succ}\right|^{\frac{1}{2}}} \times \frac{\left(\frac{1}{2}\boldsymbol{U}_{jj|pa_{\mathcal{D}}(j)}\right)^{\frac{1}{2}\alpha_j^{\mathcal{D}}}}{\Gamma\left(\frac{1}{2}\alpha_j^{\mathcal{D}}\right)} \times 2K_a\left(2\sqrt{\alpha\beta}\right)\left(\frac{\alpha}{\beta}\right)^{\frac{a}{2}}.$$

Finally, for both the normal-inverse-gamma and the normal-gamma priors, the joint posterior distribution of DAG structures and DAG parameters is proportional to $f(\boldsymbol{X} \mid \boldsymbol{D}, \boldsymbol{L}, \mathcal{D})\, p(\boldsymbol{D}, \boldsymbol{L} \mid \mathcal{D})\, p(\mathcal{D})$, where $\boldsymbol{X}$ is the $n \times q$ data matrix. Since the posterior distribution does not have a closed form, we use a MCMC algorithm to sample from the DAG $\mathcal{D}$ and $(\boldsymbol{D}, \boldsymbol{L})$ using their full conditional distributions.

6.4 South Africa Weather Data Analysis

The dataset utilized in this study comprises 144 hourly weather observations from South Africa, including 10 key meteorological variables, including temperature, humidity, wind speed, precipitation, etc. Notably, the dataset is complete, with no missing values, which eliminates the need for imputation or data cleaning procedures that could introduce additional uncertainty. Furthermore, all variables are continuous, making it particularly suitable for Bayesian modeling based on Gaussian assumptions. To deduce the underlying probabilistic structure among these meteorological variables, we adopt Bayesian structural learning techniques for DAGs. This method permits the discovery of potential dependencies and interactions within the dataset, providing a data-driven perspective on climatic dynamics. We employ two prior distribution frameworks: the conjugate normal-inverse gamma and the non-conjugate normal-gamma priors. By leveraging these Bayesian structures, we want to develop an interpretable probabilistic graphical model that captures the latent dependency structures inherent in the dataset. The findings derived from this analysis contribute to a nuanced understanding of regional climatic interactions in South Africa.

TABLE 6.4
The descriptive statistics of South Africa weather dataset.

Variable	Mean	Std. Dev.	Min	Max
Temperature (°F)	72.91	7.20	60.6	88.2
Feelslike (°F)	72.76	6.92	60.6	85.6
Dew Point (°F)	59.12	2.95	48.8	63.1
Humidity (%)	64.63	16.50	26.69	93.98
Precipitation (inch)	0.0018	0.0048	0	0.031
Wind Speed (mph)	6.87	3.42	0	14.1
Wind Direction (°)	150.49	84.81	0	345.6
Sea Level Pressure (hPa)	1011.85	3.57	1004	1022.1
Cloud Cover (%)	72.03	39.41	0	100
Solar Radiation (W/m^2)	233.24	326.47	0	1041

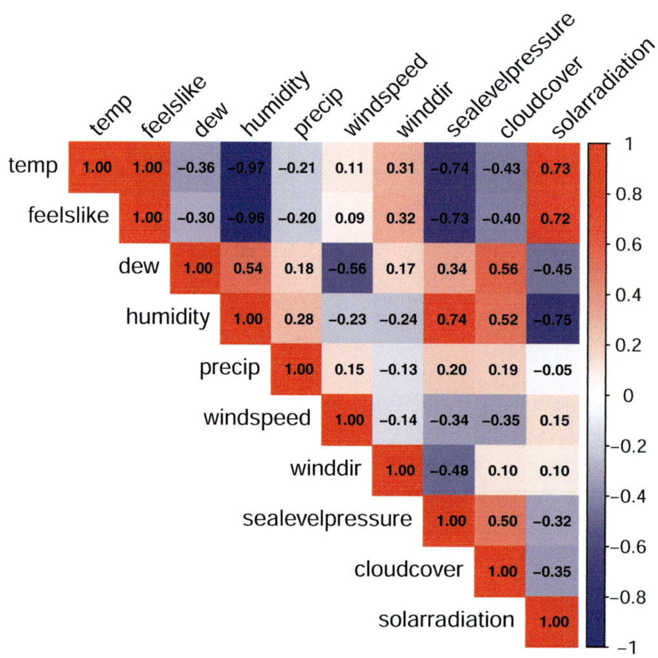

FIGURE 6.2
The correlation heatmap of South Africa weather dataset.

6.4.1 Exploratory Data Analysis (EDA)

The EDA provides valuable insights into the data's structure and interrelationships. Table 6.4 displays the descriptive statistics of the climatic dataset recorded in South Africa. The correlation heatmap in Figure 6.2, reveals significant associations between named variables.

We now proceed with interpreting the relationships among meteorological variables within the South Africa weather dataset using the proposed methods. We implemented the MCMC algorithm (Algorithm 4 in [23]) with parameters $S = 60000$, $\boldsymbol{U} = \frac{1}{n}\boldsymbol{I}_q$, $\alpha = q$, $w = 0.5$, and burning equal to $B = 15000$ on the South Africa weather dataset.

At each MCMC iteration, the process begins by proposing a new DAG, $\mathcal{D}'$, from an appropriate proposal distribution $q(\mathcal{D}' \mid \mathcal{D})$. To this end, we consider three types of operators that locally modify an input DAG $\mathcal{D}$: Insert($i \to j$), Delete($i \to j$) or Reverse($i \to j$) as follows:

1. **Insert** $i \to j$: If the edge $i \to j$ is not present in $\mathcal{D}$, it is added with probability π.
2. **Delete** $i \to j$: If the edge $i \to j$ exists in $\mathcal{D}$, it is removed with probability $1 - \pi$.
3. **Reverse** $i \to j$: If the edge $i \to j$ exists in $\mathcal{D}$, it is reversed by removing $i \to j$ and adding the edge $j \to i$.

In each MCMC iteration, we ensure that the proposed operators preserve the acyclic structure of the DAG. Specifically, any move that results in a cycle is rejected, and the process continues until a valid, acyclic DAG is obtained.

For the proposal distribution in the MCMC (the probability of transition from $\mathcal{D}$ to $\mathcal{D}'$), we take $q(\mathcal{D}' \mid \mathcal{D}) = {}^{1}/_{|\mathcal{O}_{\mathcal{D}}|}$, that determines the transitions between DAGs within the space $\mathcal{S}_q$ (the set of all DAGs on q nodes) and $|\mathcal{O}_{\mathcal{D}}|$ is the number of *valid* operators for the DAG $\mathcal{D}$.

Furthermore, under Insert($i \to j$) and Delete($i \to j$) operators, the acceptance probability for $\mathcal{D}'$ is

$$\alpha_{\mathcal{D}'} = \min\left\{1, \frac{m(\boldsymbol{X}_j \mid \boldsymbol{X}_{pa_{\mathcal{D}'}(j)}, \mathcal{D}')}{m(\boldsymbol{X}_j \mid \boldsymbol{X}_{pa_{\mathcal{D}}(j)}, \mathcal{D})} \cdot \frac{p(\mathcal{D}')}{p(\mathcal{D})} \cdot \frac{q(\mathcal{D} \mid \mathcal{D}')}{q(\mathcal{D}' \mid \mathcal{D})}\right\}$$

while for Reverse($i \to j$) operator, the acceptance probability for $\mathcal{D}'$ is given by

$$\alpha_{\mathcal{D}'} = \min\left\{1, \frac{m(\boldsymbol{X}_i \mid \boldsymbol{X}_{pa_{\mathcal{D}'}(i)}, \mathcal{D}')m(\boldsymbol{X}_j \mid \boldsymbol{X}_{pa_{\mathcal{D}'}(j)}, \mathcal{D}')}{m(\boldsymbol{X}_i \mid \boldsymbol{X}_{pa_{\mathcal{D}}(i)}, \mathcal{D})m(\boldsymbol{X}_j \mid \boldsymbol{X}_{pa_{\mathcal{D}}(j)}, \mathcal{D})} \cdot \frac{p(\mathcal{D}')}{p(\mathcal{D})} \cdot \frac{q(\mathcal{D} \mid \mathcal{D}')}{q(\mathcal{D}' \mid \mathcal{D})}\right\}$$

The term ${p(\mathcal{D}')}/{p(\mathcal{D})}$ in $\alpha_{\mathcal{D}'}$ is equal to ${}^{\pi}/_{1-\pi}$, ${}^{1-\pi}/_{\pi}$, and 1, respectively for Insert, Delete, and Reverse operators with π as the probability of edge inclusion.

Finally, in order to update of DAG-parameters in each iteration, we use their full conditional distribution introduced in 6.5, 6.8, and 6.9. See also [25] for a comprehensive discussion and further developments. The obtained results are reported below, accompanied by a detailed interpretation.

MCMC convergence is evaluated by monitoring variations in graph properties over iterations. A trace plot of DAG edge counts helps evaluate convergence, with an absence of trends indicating satisfactory convergence. Additionally, the average number of edges for the NG and NIG models stabilizes around 16 and 13, respectively (Figure 6.3, right panel), confirming convergence, while the left panel further supports this conclusion by showing no apparent trend. To be even more sure, we look at the posterior probability of edge inclusion, which is the percentage of sampled DAGs that have a certain edge over iterations. Convergence is confirmed when these probabilities remain stable. Figure 6.4 demonstrates these results for selected nodes, reinforcing the reliability of the MCMC process.

Figure 6.5 presents heatmaps illustrating the posterior probability of edge inclusion between meteorological variables. We compute this probability as the average frequency of an edge appearing in DAGs across MCMC iterations. For instance, both models observe a directed edge from the dew point to the humidity. Furthermore, the NG model discovered more dependency compared to the NIG model. Moreover, in the heatmap presented in Figure 6.5, the intensity of the colors represents the probability of an edge (i.e., a dependency) existing between pairs of nodes in the network. High-probability edges indicate strong statistical evidence for a dependency between the corresponding nodes, whereas lower probabilities suggest substantial uncertainty or a lack of meaningful association. This probabilistic representation presents an accurate and principled framework for quantifying

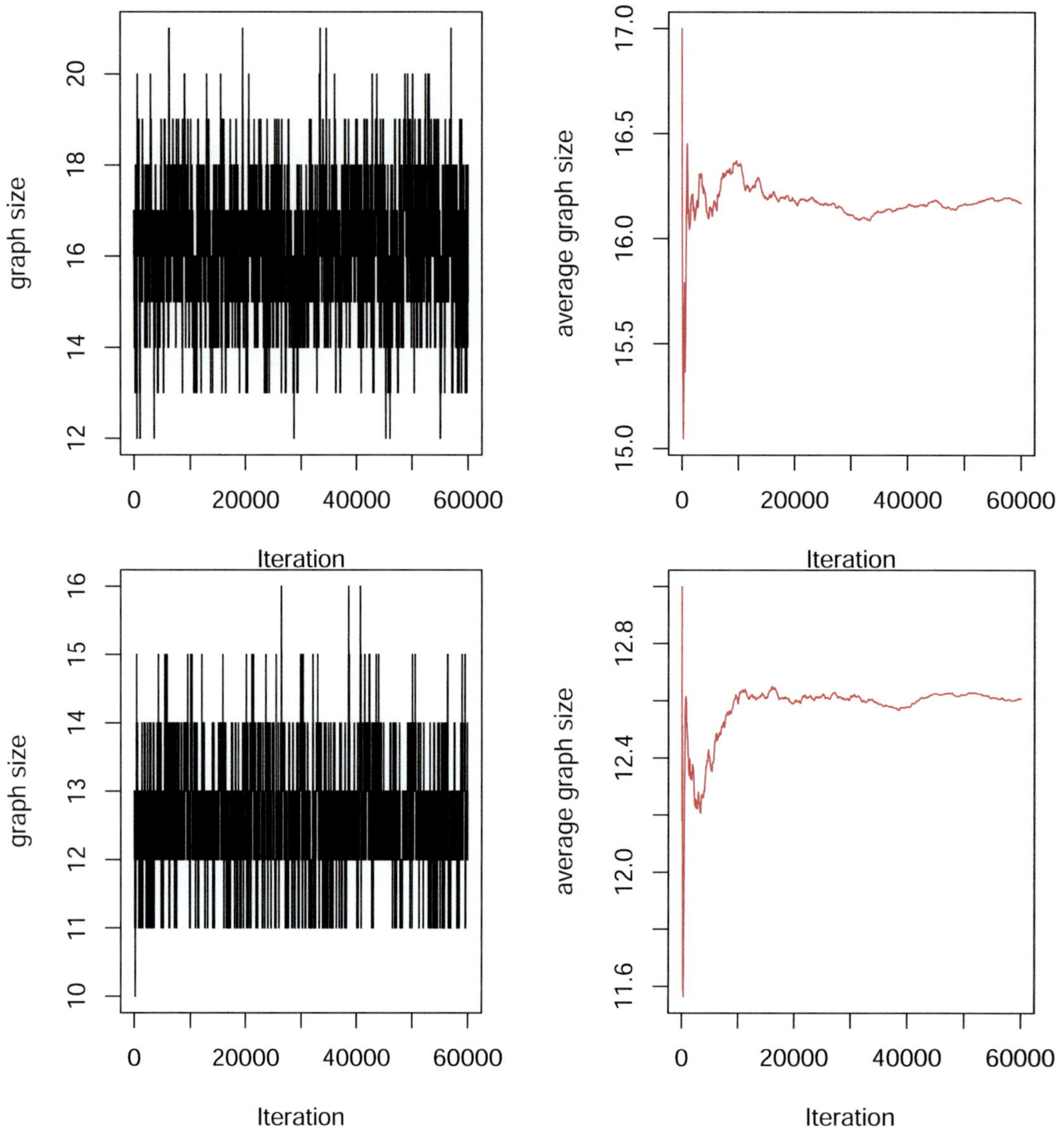

FIGURE 6.3
Trace plot of the number of edges (graph size) in the DAGs visited by the MCMC at each iteration (left panel) and trace plot of the average number of edges in the DAGs computed up to iteration s, $s = 1, \ldots, 60000$ (right panel) for South Africa weather data for the NG model (a) and the NIG model (b).

uncertainty, facilitating a more intelligible understanding of the graph structure. For instance, Figure 6.5(a) illustrates that the relationship between precipitation and humidity is stronger than that between precipitation and wind direction.

Figure 6.6 depicts the maximum a posteriori (MAP) DAG estimates, representing the DAG with the highest posterior probability. In fact, the MAP estimates focus on the most likely structure. In contrast, the median probability model (MPM) DAG estimates are shown in Figure 6.7, constructed by including edges $u \to v$ with a posterior probability

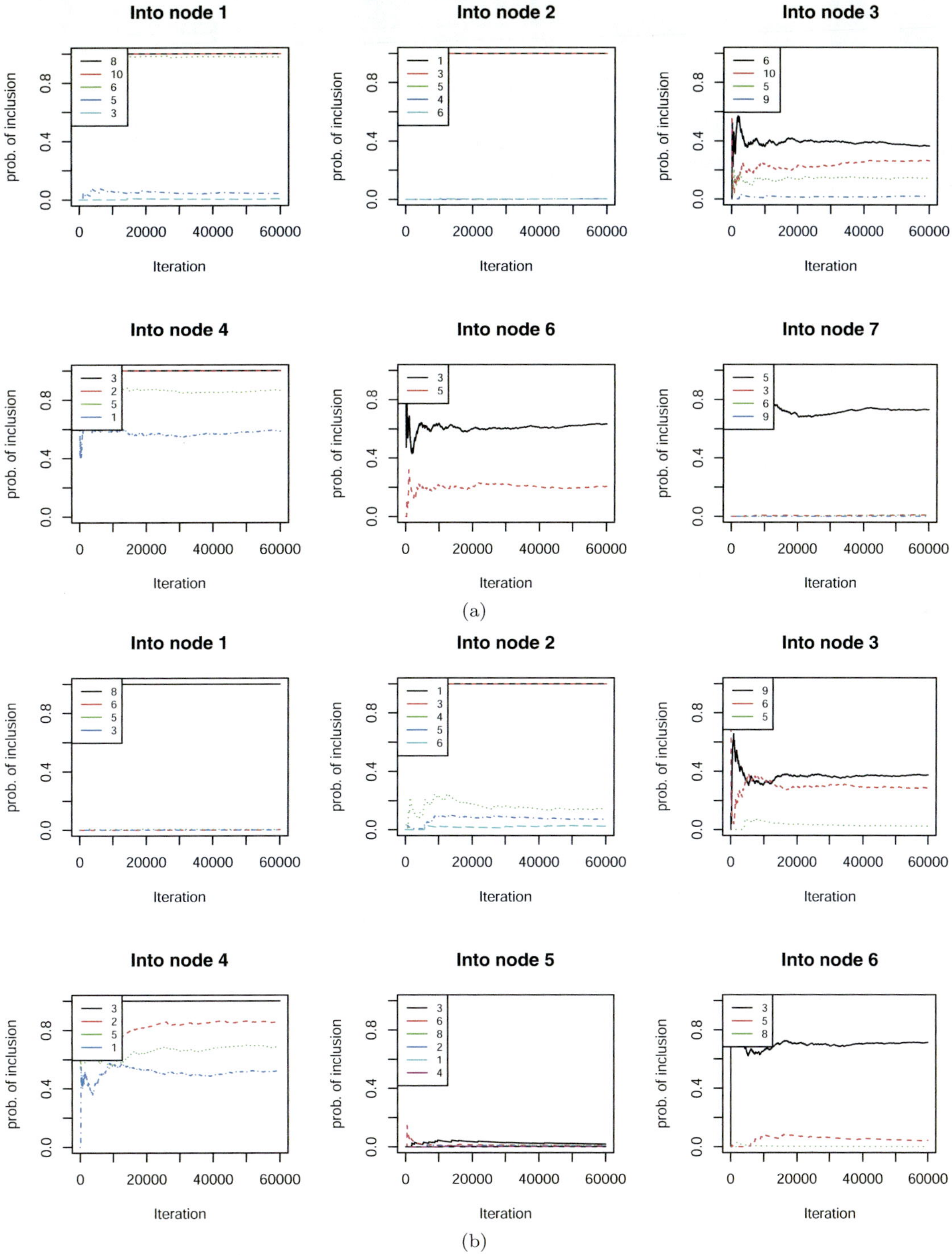

FIGURE 6.4
Trace plot of the posterior probability of inclusion of $u \to v$, computed across iteration s, $s = 1, \ldots, 60000$ for the NG model (a) and the NIG model (b).

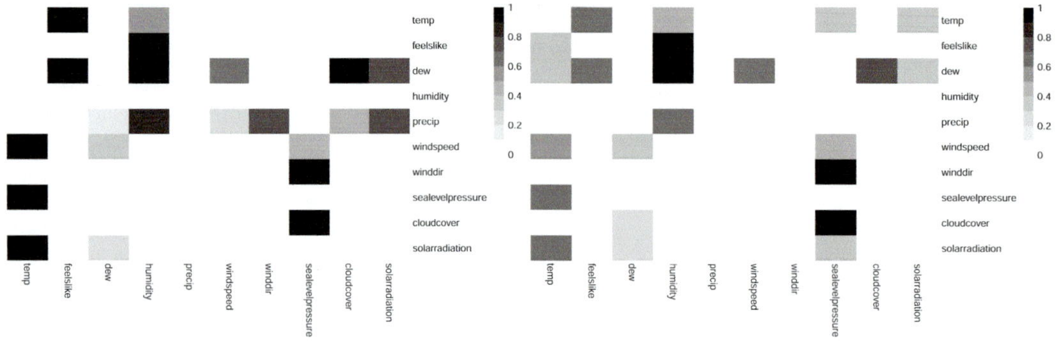

(a) (b)

FIGURE 6.5
Heat map with posterior probabilities of edge inclusion for the NG model (a) and the NIG model (b). Darker colors indicate stronger associations between meteorological variables.

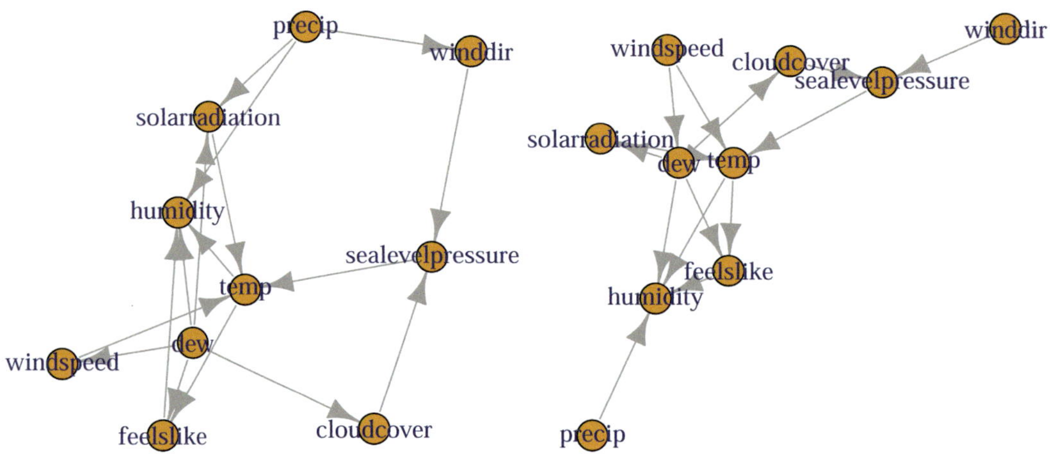

(a) (b)

FIGURE 6.6
The MAP estimates of South Africa weather data for the NG model (a) and the NIG model (b).

greater than 0.5. This model can be effectively visualized through heatmaps, where each cell represents the posterior probability of an edge existing between two nodes in the DAG. In this framework, heatmaps provide a clear visual representation of the strength of connections between variables, with higher probability edges showing more intense colors. Consequently, MPM aligns well with heatmap visualizations.

Climate change is one of the most important environmental and scientific challenges of current time, influenced by key variables such as temperature, humidity, solar radiation, air pressure, and rainfall. As one of the countries exposed to climate change, South Africa faces numerous challenges in this regard. The analysis of the presented graphs (Figures 6.5, 6.6, and 6.7) reveals the intricate relationships among various meteorological variables and their causal interactions within the examined climate system. These results suggest that

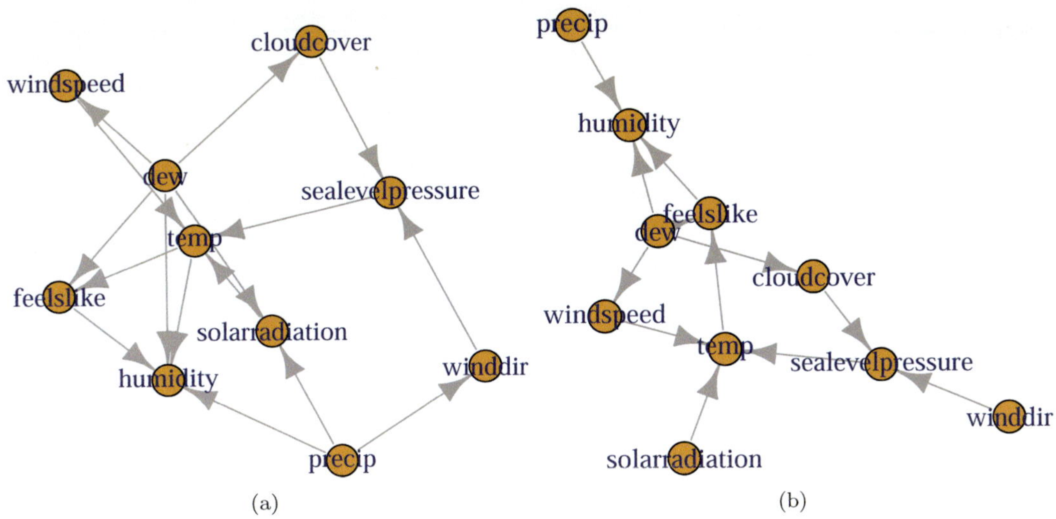

FIGURE 6.7
The MPM estimates of South Africa weather data for the NG model (a) and the NIG model (b).

variations in meteorological variables such as temperature, humidity, solar radiation, and feelslike reflect the potential impacts of climate change on atmospheric systems. The pivotal role of temperature in structuring these relationships underscores its significance as a crucial factor in climate change research. Variations in temperature can induce alterations in other meteorological variables, consequently impacting the forecasting and modeling of atmospheric systems. Therefore, precise monitoring of these variables can contribute to a more profound understanding of climate change and its consequences on weather patterns. The results of the analysis of the aforementioned graphs indicate that temperature (temp) plays a key role in the structure of relationships among meteorological variables and is directly associated with factors such as humidity, solar radiation, and feelslike. These findings align with the basic principles of atmospheric thermodynamics, as rising temperature can accelerate the evaporation rate, ultimately impacting humidity level ([27]). On the other hand, temperature variations are also influenced by solar radiation, which is considered one of the most important energy sources in atmospheric systems. This relationship indicates that solar radiation, as an independent variable, indirectly affects other variables by influencing temperature. In addition, wind speed and direction are recognized as influential variables in the distribution of energy and moisture in the atmosphere. For instance, wind speed and direction are directly related to dew point and sea level pressure, which can be explained by the role of wind in the transport of air masses and the modification of atmospheric pressure. The causal relationship between these variables indicates that changes in wind direction and speed can induce significant alterations in atmospheric system and even lead to variations in precipitation levels. Furthermore, precipitation, as another important variable, influences various variables, including humidity and solar radiation. Increased humidity enhances the probability of cloud formation and precipitation ([28]), whereas solar radiation can impacts humidity levels through the evaporation process, thereby indirectly affecting precipitation. So, the causal relationships clearly demonstrate that the DAG model is capable of identifying the dependency structures among atmospheric variables and determining the causal pathways between them.

6.5 Conclusion

As a nation highly vulnerable to climate change, South Africa confronts a multitude of environmental challenges. Climate change stands as one of the most pressing global issues, driven by key variables such as temperature, humidity, solar radiation, air pressure, and precipitation. These factors not only shape weather patterns but also have profound implications for ecosystems, water resources, and human livelihoods. This study highlights the effectiveness of Bayesian structure learning with DAGs in identifying the causal relationships between weather variables in South Africa. The proposed models illustrate a high level of accuracy in recognizing complex dependencies among atmospheric elements and in delineating the causal pathways between them. Beyond its methodological contributions, the findings of this study offer practical value for climate-related decision-making in South Africa. The identified causal relationships among key meteorological variables provide valuable insights that can be utilized to strengthen early warning systems, guide agricultural and water resource management, and inform the development of adaptive strategies for enhancing climate resilience ([4], [11], and [29]). For instance, understanding how temperature, humidity, and precipitation interact can enable policymakers and practitioners to make more informed decisions regarding planting schedules, drought preparedness, and climate-resilient infrastructure planning ([15]). Furthermore, the probabilistic framework developed in this study can serve as a valuable decision-support tool for regional climate policy and disaster risk management ([14]). We introduce the presented method as an innovative instrument for analyzing weather data and investigating the effects of climate change in South Africa.

While this study successfully applied Bayesian structure learning based on Gaussian DAG models to analyze causal relationships among meteorological variables, extending this approach in future research could yield significant advancements. One promising direction is to extend the model to non-Gaussian variables, offering greater flexibility in modeling the actual distributions of weather data, which are often skewed or heavy-tailed. Moreover, applying this method to larger datasets or with higher temporal and spatial resolution—including more stations and longer time periods—can enhance the generalizability and accuracy of causal findings. In addition, as another future research direction, the utilization of dynamic time series models, such as Dynamic Bayesian Networks (DBNs) or other advanced temporal analysis techniques would enable the investigation of causal relationships among weather variables over time and under changing climatic conditions. This approach will allow us to assess the stability and validity of the current findings over time and gain a more precise understanding of the causal relationships and temporal dependencies between variables; see also [29], [30], and [31]. These extensions would contribute to improved forecasts, a deeper understanding of causal structures, and more informed decision making in the field of climate science.

Funding Statement

Mohammad Arashi's work is based on the research supported in part by the Iran National Science Foundation (INSF) grant No. 4015320.

Bibliography

[1] M. F. Chersich, C. Y. Wright, F. Venter, H. Rees, F. Scorgie, and B. Erasmus. Impacts of climate change on health and wellbeing in South Africa. *International Journal of Environmental Research and Public Health*, 15(9):1884, 2018.

[2] B. M. Ramin and A. J. McMichael. Climate change and health in sub-saharan Africa: A case-based perspective. *EcoHealth*, 6:52–57, 2009.

[3] J. M. Olwoch, B. Reyers, F. A. Engelbrecht, and B. F. N. Erasmus. Climate change and the tick-borne disease, theileriosis (east coast fever) in sub-saharan Africa. *Journal of Arid Environments*, 72(2):108–120, 2008.

[4] B. Erasmus, A. van Jaarsveld, J. van Zyl, and N. Vink. The effects of climate change on the farm sector in the western cape. *Agrekon*, 39(4):559–573, 2000.

[5] B. Mathur. Predicting atmospheric variables in the MERRA-2 database using neural networks. In *2022 International Conference on Emerging Techniques in Computational Intelligence (ICETCI)*, pages 125–131. IEEE, August 2022.

[6] G. De la Torre-Gea, G. M. Soto-Zarazúa, R. Guevara-González, and E. Rico-García. Bayesian networks for defining relationships among climate factors. *International Journal of Physical Sciences (IJPS)*, 6(18):4412, 2011.

[7] R. J. Chase, D. R. Harrison, G. M. Lackmann, and A. McGovern. A machine learning tutorial for operational meteorology. part ii: Neural networks and deep learning. *Weather and Forecasting*, 38(8):1271–1293, 2023.

[8] V. Singh, C. Kumar, and N. Choudhary. Exploring the effectiveness of artificial neural networks and regression models in weather prediction. In *International Conference on Paradigms of Communication, Computing and Data Analytics*, pages 219–233. Springer Nature Singapore, April 2024.

[9] B. A. Smith. *Air Temperature Prediction Using Artificial Neural Networks.* Doctoral dissertation, University of Georgia, USA, 2006.

[10] S. Balluff, J. Bendfeld, and S. Krauter. Meteorological data forecast using rnn. In *Deep Learning and Neural Networks: Concepts, Methodologies, Tools, and Applications*, pages 905–920. IGI Global, 2020.

[11] A. S. Cofíno, R. Cano, C. Sordo, and J. M. Gutierrez. Bayesian networks for probabilistic weather prediction. In *15th European Conference on Artificial Intelligence (ECAI)*, 2002.

[12] T. J. O'Kane, D. Harries, and M. A. Collier. Bayesian structure learning for climate model evaluation. *Journal of Advances in Modeling Earth Systems*, 16(5):e2023MS004034, 2024.

[13] E. R. Hruschka Jr, E. R. Hruschka, and N. F. Ebecken. Applying Bayesian networks for meteorological data mining. In *International Conference on Innovative Techniques and Applications of Artificial Intelligence*, pages 122–133. Springer London, December 2005.

[14] I. Ebert-Uphoff and Y. Deng. Causal discovery for climate research using graphical models. *Journal of Climate*, 25(17):5648–5665, 2012.

[15] C. Vitolo, M. Scutari, M. Ghalaieny, A. Tucker, and A. Russell. Modeling air pollution, climate, and health data using Bayesian networks: A case study of the English regions. *Earth and Space Science*, 5(4):76–88, 2018.

[16] T. Vairoac, M. Leccaa, E. Trovatorea, A. P. Reverberib, and B. Fabianoc. A Bayesian belief network for local air quality forecasting. *Chemical Engineering*, 74, 2019.

[17] A. Sharma and M. K. Goyal. Bayesian network model for monthly rainfall forecast. In *2015 IEEE International Conference on Research in Computational Intelligence and Communication Networks (ICRCICN)*, pages 241–246. IEEE, 2015.

[18] S. L. Lauritzen. *Graphical Models.* Oxford University Press, UK, 1996.

[19] J. Pearl. *Causality.* Cambridge University Press, UK, 2009.

[20] M. Arashi, A. Bekker, and J. van Niekerk. Weighted-type wishart distributions with application. *REVSTAT-Statistical Journal*, 15(2):205–222, 2017.

[21] A. Bekker, J. van Niekerk, and M. Arashi. Wishart distributions: Advances in theory with Bayesian application. *Journal of Multivariate Analysis*, 155:272–283, 2017.

[22] E. Ben-David, T.Li., H Massam, and B. Rajaratnam. High dimensional Bayesian inference for Gaussian directed acyclic graph models. arxiv pre-print. *arXiv preprint arxiv:1109.4371*, 2015.

[23] F. Castelletti and A. Mascaro. Bcdag: An R package for Bayesian structure and causal learning of Gaussian DAGs. *arXiv preprint arXiv:2201.12003*, 2022.

[24] F. Castelletti and G. Consonni. Discovering causal structures in Bayesian Gaussian directed acyclic graph models. *Journal of the Royal Statistical Society Series A: Statistics in Society*, 183(4):1727–1745, 2020.

[25] S. Nazari, M. Arashi, and A. Sadeghkhani. Weighted inverse gamma innovation for the structure learning of DAGs. *Japanese Journal of Statistics and Data Science*, pages 1–17, 2024.

[26] S. Nazari, M. Arashi, and A. Sadeghkhani. Adaptive Bayesian structure learning of DAGs with non-conjugate prior. *arXiv preprint arXiv:2403.17489*, 2024.

[27] J. Hao and Er. Lu. Variation of relative humidity as seen through linking water vapor to air temperature: An assessment of interannual variations in the near-surface atmosphere. *Atmosphere*, 13(8):1171, 2022.

[28] M.R. Needham and D.A. Randall. Linking atmospheric cloud radiative effects and tropical precipitation. *Geophysical Research Letters*, 48(14):e2021GL094004, 2021.

[29] Jingjing Chang, Yongxin Bai, Jie Xue, Lu Gong, Fanjiang Zeng, Huaiwei Sun, Yang Hu, Hao Huang, and Yantao Ma. Dynamic Bayesian networks with application in environmental modeling and management: A review. *Environmental Modelling & Software*, 170:105835, 2023.

[30] Shijun Zhao, Yulong Shan, and Ismail Gultepe. Prediction of visibility in the arctic based on dynamic Bayesian network analysis. *Acta Oceanologica Sinica*, 41(4):57–67, 2022.

[31] Yuzhen Zhang, Yonghua Qu, Jindi Wang, Shunlin Liang, and Yan Liu. Estimating leaf area index from MODIS and surface meteorological data using a dynamic Bayesian network. *Remote Sensing of Environment*, 127:30–43, 2012.

Part II

Directional Statistics

7

Hidden Semi-Markov Models for Directional Time Series

Francesco Lagona

The statistical analysis of time series that include directional components is inherently different from traditional time series analysis, due to the wraparound nature of their domain and the difficulties in modeling dependence structures when directional measurements are involved. Directional hidden semi-Markov models (HSMMs) provide a flexible framework where these issues can be efficiently addressed. This contribution summarizes the theory of these models, discussing methods of parameter estimation and illustrating an application on marine data.

7.1 Introduction

Directional time series are temporal sequences of data that include angular observations. In their simplest form, these series appear as univariate sequences of angles that lie on the circumference of the unit circle $\mathbb{S}^1$ of $\mathbb{R}^2$. In this case, they are usually referred to as circular time series. More generally, directional time series arise as multivariate sequences. Typical examples are either spherical time series, where each observation is an angle on the sphere $\mathbb{S}^{p-1}$ with $p > 2$, or toroidal time series where each observation includes multiple angles, or cylindrical time series, where spherical variables are observed together with traditional linear variables, observed on the real line.

These time series appear often in biometrical and environmental studies. Example of biometrical applications are peak systolic blood pressure times, converted to angles, during two separate time periods [1], protein backbone conformational angles [2], phase angles of circadian-related genes in two tissues [3], and orthologous genes shared by circular prokaryotic genomes [4]. Examples of environmental studies that include directional time series are earthquake data consisting of the pre-earthquake direction of steepest descent and the direction of lateral ground movement [5], time series of wind directions and pollutant concentrations [6], wind directions and speeds [7], wildfire sizes and orientations [8], wave directions and heights [9, 10, 11].

The statistical analysis of directional data is inherently different from the traditional analysis of multivariate data, due to the wraparound nature of their domain and the difficulties in modeling the dependence between angular measurements and mixed angular and linear measurements. The spread of directional data across multiple disciplines has however pushed the literature toward the definition of several distributions on spherical, toroidal,

DOI: 10.1201/9781003564423-7

and cylindrical manifolds [12, 13], and a whole library of parametric distributions is nowadays available. Methods of nonparametric density estimation have been further extended to the directional setting [14]. As a result, when the data are in the form of a sequence of independent and identically distributed observations, they can be efficiently analyzed by taking either a parametric or a nonparametric approach.

In most case studies, however, directional data are heterogeneous and dependent across time, motivating models where directional densities are just a building block among other model components. For example, hidden Markov models (HMMs) [15] provide a general framework to analyze directional time series. Under a HMM, the joint distribution of the data is approximated by a mixture of directional densities, whose parameters evolve according to a latent Markov chain. These models are particularly appropriate when the marginal distribution of the data is multimodal and observations are taken under latent heterogeneous conditions that vary over time. Under this setting, the directional density appropriately accommodates circular and linear-circular dependence structures, a mixture of densities allows for multimodality, and finally, a latent Markov chain accounts for temporal correlation.

HMMs have gained popularity in the analysis of directional time series since the seminal paper by Holzmann *et al.* [16], where the authors propose an HMM approach to analyze cylindrical time series, motivated by a biometrical study of animal movement (speed and change of direction of Drosophila larvae) and by an environmental study of weather conditions (wind speed and direction). In that work, the proposal is constrained by two restrictive assumptions on the latent process: homogeneity and Markovianity. These assumptions facilitate parameter estimation, but they are often hard to motivate.

Homogeneity is seldom fulfilled because state-switching probabilities often depend on time-varying exogenous conditions that make the observed system nonhomogeneous. For this reason, whenever time-varying covariates are available, homogeneous HMMs are often extended to their nonhomogeneous versions [17, 18, 19].

Time-varying covariates however complicates the form taken by the state-specific dwell time hazard in a nonhomogeneous HMM. This hazard is the discrete counterpart of the hazard function, traditionally exploited in survival analysis to characterize the distribution of absolutely continuous, positive random variables. Given the current state of the chain, the hazard indicates the (time-varying) conditional probability of switching state given the current dwell time, where the current dwell time is defined as the current amount of time since the last transition. The dynamics of a nonhomogeneous HMM is described by a hazard that is exogenously updated at each time by the current covariate information, regardless of the dwell time. This is a consequence of the memoryless property of the latent Markov chain in an HMM, a direct result from the Markov property. The Markov assumption is both a strength and a weakness of the model. On one side, it is conceptually appealing to assume that state switches should depend only on current conditions. On the other side, when these conditions do not change, the hazard is time-constant and dwell times are constrained to be geometrically distributed.

Markovianity is therefore a potential limitation of HMMs. This has opened the way to alternative approaches where the latent Markov chain is replaced by a semi-Markov chain, leading to the class of HSMMs. Under a semi-Markov chain, the hazard is endogenously updated by the current dwell time. This dissolves the geometric constraint on the dwell time distribution by relaxing the Markov property to a setting where the evolution of the chain depends on its history since the last transition. Due to their flexibility, HSMMs have gained popularity in environmental research and they have been successfully implemented in studies of regime shifts in ocean density variability [20], GPS tracking data of animal movements [21], multivariate pollutant concentrations [22], and migratory bird count data [23]. Despite all these uses, HSMMs have been only recently introduced in directional statistics for the analysis of toroidal time series [24].

This contribution extends [24]to a more general directional setting, showing that HSMMs are not limited to the analysis of toroidal data series and illustrating a novel application on cylindrical time series.

7.2 Directional Hidden Semi-Markov Models

A directional time series is, in general, a sequence $\boldsymbol{z} = (\boldsymbol{z}_t, t = 1, \ldots, T)$, where $\boldsymbol{z}_t = (\boldsymbol{x}_t, \boldsymbol{y}_t)$ includes a vector of circular (or, more generally, spherical) data, $\boldsymbol{x}_t$, and a vector of linear observations $\boldsymbol{y}_t$. Under a HSMM, the joint distribution of the observed sequence is driven by a sequence of latent multinomial variables $\boldsymbol{u} = (\boldsymbol{u}_t, t = 1, \ldots T)$, $\boldsymbol{u}_t = (u_{t1} \ldots u_{tK})$ with one trial and K classes (or regimes), whose binary components represent class membership at time t. Precisely, the joint distribution of the time series is obtained by integrating a parametric conditional distribution $f(\boldsymbol{z} \mid \boldsymbol{u}; \boldsymbol{\theta})$ of the observed data given the latent classes (the observation process) with respect to a parametric distribution $p(\boldsymbol{u}; \boldsymbol{\eta})$ of the latent chain, namely

$$f(\boldsymbol{z}; \boldsymbol{\theta}, \boldsymbol{\eta}) = \sum_{\boldsymbol{u}} f(\boldsymbol{z} \mid \boldsymbol{u}; \boldsymbol{\theta}) p(\boldsymbol{u}; \boldsymbol{\eta}). \tag{7.1}$$

The observation process is specified by assuming that the observations are conditionally independent, given the latent classes, say

$$f(\boldsymbol{z} \mid \boldsymbol{u}; \boldsymbol{\theta}) = \prod_{t=1}^{T} \prod_{k=1}^{K} f(\boldsymbol{z}_t; \boldsymbol{\theta}_k)^{u_{tk}}, \tag{7.2}$$

where $f(\boldsymbol{z}_t; \boldsymbol{\theta}_k)$ represents the data distribution under each regime.

The directional statistics literature has developed a battery of multivariate distributions with spherical, toroidal, and cylindrical manifolds [12, 13]. As a result, $f(\boldsymbol{z}_t; \boldsymbol{\theta}_k)$ can be chosen among the available proposals. It is however possible that none of the existing densities is appropriate for the data under study, because the associated marginal densities fit poorly the empirical marginal distributions of the data. In this case, a viable solution is provided by copulas. Copula-based densities allow us to specify a distribution with the desired marginals, improving on model flexibility. While several copulas are proposed in the literature, most proposals do not fulfill the periodicity conditions required by directional data. This has led to ad-hoc proposals that are appropriate in directional statistics [6, 25, 26, 27]. While copula-based directional densities have been recently integrated within a HMM framework [28], they have been not yet extended to more general settings, such as that provided by HSMMs.

The specification of a directional HSMM is completed by the distribution of the latent chain $\boldsymbol{u}$, say

$$p(\boldsymbol{u}) = p(\boldsymbol{u}_1) \prod_{t=2}^{T} \prod_{k=1}^{K} \prod_{h=1}^{K} \gamma_{tkh}^{u_{tkh}}, \tag{7.3}$$

where $p(\boldsymbol{u}_1)$ is the initial distribution of the chain and γ_{tkh} indicates the transition probability of moving from state k to state h at time t. We assume that these transition probabilities depend on both a row profile of (possibly time-varying) Q covariates at time t, $\boldsymbol{w}_t$, and the dwell time d_t at time t, i.e., the time that at t has been spent by the chain in state k. Vector $\boldsymbol{w}_t^{\mathsf{T}}$ is a $1 \times Q$ row vector that indicates time-varying exogeneous information that might influence a switching from one state to another. In the application illustrated in Section 7.4, $Q = 1$ and $\boldsymbol{w}_t$ indicates wind speed at time t, as changes in wind speed often prelude changes in weather conditions.

There are several approaches to include time-varying exogenous information in a hidden semi-Markov setting [29, 30, 31]. A very general specification [24] relies on assuming that

$$\gamma_{tkh} = \gamma_{kh}(d_t, \boldsymbol{w}_t^{\mathsf{T}}) = \begin{cases} 1 - q_k(d_t, \boldsymbol{w}_t^{\mathsf{T}}) & h = k \\ q_k(d_t, \boldsymbol{w}_t^{\mathsf{T}})\omega_{kh} & h \neq k. \end{cases} \tag{7.4}$$

In the equation above, $q_k(d_t, \boldsymbol{w}_t^{\mathsf{T}})$ indicates the hazard of a state switch at time t, for a chain that is in state k at time t, while ω_{kh} indicates the conditional probability of a transition to state h, given a state switch, $\sum_h \omega_{kh} = 1$, for any $k = 1, \ldots K$. A parametric form of the hazard $q_k(d_t, \boldsymbol{w}_t^{\mathsf{T}})$ can be obtained by a suitable link function h, namely

$$h(q_k(d_t)) = \beta_{0k} + \beta_{1k} d_t + \boldsymbol{w}_t^{\mathsf{T}} \boldsymbol{\beta}_k \qquad k = 1, \ldots, K. \tag{7.5}$$

Equation (7.5) is a generalized linear model where state-specific regression coefficients $\boldsymbol{\beta}_k$ modulate the influence of endogeneous and exogeneous information on the (link-transformed) hazard function. When $\beta_{1k} = 0$ and $\boldsymbol{\beta}_k = \mathbf{0}$, then $g(q_{tk}) = \beta_{0k}$, and the process reduces to a homogeneous Markov chain, since in this case knowledge of the current state k suffices to predict the next move of the chain. Otherwise, if $\beta_{1k} \neq 0$, such knowledge must be augmented with the information on the current dwell time, d_t, and the process reduces to a homogeneous semi-Markov chain. If also some of the regression coefficients $\boldsymbol{\beta}_k$ are not equal to zero, then transitions of the chain do not only depend on the dwell time but also on exogenous covariates, and the process is a non-homogeneous semi-Markov chain. Typical choices for h are the logit, the probit and the complementary log-log transformation. The complementary log-log is a natural choice for modeling discrete dwell times because it can be seen as the discrete-time counterpart of a continuous-time proportional hazards model [32, 33]. Under this setting, the transition probabilities are known up to a vector of regression coefficients $\boldsymbol{\beta} = \{\beta_{0k}, \beta_{1k}, \boldsymbol{\beta}_k, k = 1, \ldots, K\}$, and an array of conditional transition probabilities $\boldsymbol{\omega} = \{\omega_{kh}, k = 1, \ldots, K, h \neq k\}$, say

$$\gamma_{tkh} = \gamma_{tkh}(d_t, \boldsymbol{w}_t; \boldsymbol{\omega}, \boldsymbol{\beta}).$$

7.3 Maximum Likelihood Estimation

The likelihood function of the model is obtained by integrating the contribution of the observation process (7.2) with the contribution of the latent process (7.3), where the transition probabilities are modeled as in (7.4) namely

$$L(\boldsymbol{\theta}, \boldsymbol{\eta}) = \sum_{\boldsymbol{u}} \prod_{t=1}^{T} \prod_{k=1}^{K} f(\boldsymbol{z}_t; \boldsymbol{\theta}_k)^{u_{tk}} p(\boldsymbol{u}_1; \boldsymbol{\pi}) \prod_{t=2}^{T} \prod_{k=1}^{K} \prod_{h=1}^{H} \gamma_{tkh}(\boldsymbol{\omega}, \boldsymbol{\beta})^{u_{tkh}}, \tag{7.6}$$

where $\boldsymbol{\pi} = (\pi_1, \ldots, \pi_K)$, $\sum_{k=1}^{K} \pi_k = 1$, indicates the probabilities of the initial distribution of the chain. The maximization of (7.6) is however complicated by the summation over all the possible paths of the latent semi-Markov chain. This is a typical difficulty in the literature of hierarchical models, and it is often overcome by an Expectation-Maximization (EM) algorithm that alternates the maximization of a weighted complete-data log-likelihood function (M step) with weights updates (E step). The complete-data log-likelihood function is defined by appropriately augmenting the missing information by pseudo-observations. In our setting, the missing information at each time t is not only the state of the chain, but also the current dwell time, and it is conveniently described by an array of multinomial pseudo-observations $\boldsymbol{u}_t = (u_{tkd}, k = 1, \ldots, K, d = 1, \ldots, T-1)$, where $u_{tkd} = 1$ if at time t the chain is in state k and the current dwell time is d, and 0 otherwise [24]. Under this setting, the complete-data log-likelihood function is given by

$$\log L_{\mathrm{c}}(\boldsymbol{\theta}, \boldsymbol{\eta}) = \sum_{k=1}^{K} u_{1k1} \log \pi_k + \sum_{t=2}^{T} \sum_{k=1}^{K} \sum_{h=1}^{K} \sum_{d,d'=1}^{1,\ldots,T-1} u_{t-1,kd} u_{thd'} \log \gamma_{kh}(d, \boldsymbol{w}_t^{\mathsf{T}}; \boldsymbol{\omega}, \boldsymbol{\beta}) + \sum_{t=1}^{T} \sum_{k=1}^{K} \sum_{d=1}^{T-1} u_{tkd} \log f(\boldsymbol{z}_t; \boldsymbol{\theta}_k). \quad (7.7)$$

The dimension of the pseudo-observation arrays depends on the size of the time series, and potential issues of storage memory may arise. A possible solution [34] relies on truncating these arrays at a value $m < T - 1$, leading to the approximated complete-data log-likelihood function

$$\log L_{\mathrm{c}}^{(m)}(\boldsymbol{\theta}, \boldsymbol{\eta}) = \sum_{k=1}^{K} u_{1k1} \log \pi_k + \sum_{t=2}^{T} \sum_{k=1}^{K} \sum_{h=1}^{K} \sum_{d,d'=1}^{1,\ldots,m} u_{t-1,kd} u_{thd'} \log \gamma_{kh}(d, \boldsymbol{w}_t^{\mathsf{T}}; \boldsymbol{\omega}, \boldsymbol{\beta}) + \sum_{t=1}^{T} \sum_{k=1}^{K} \sum_{d=1}^{m} u_{tkd} \log f(\boldsymbol{z}_t; \boldsymbol{\theta}_k). \quad (7.8)$$

This method does not constrain the maximum dwell time to be m but, for $d > m$, the transition probabilities of the semi-Markov chain are approximated by those of a nonhomogeneous Markov chain. Note that setting $m = 1$ reduces equation (7.8) to the complete-data log-likelihood of a HMM.

The proposed EM algorithm iteratively relies on the approximated complete-data log-likelihood in (7.8). Given the parameter estimates, say $\hat{\boldsymbol{\theta}}$ and $\hat{\boldsymbol{\eta}}$, obtained at the end of the s-th EM iteration, the $(s+1)$-th iteration is initialized by the E-step, which evaluates the expected value of the complete-data log-likelihood in (7.8) with respect to the conditional distribution of the pseudo-observations u_{tkd} and their pairwise products $u_{t-1,kd}u_{thd'}$ given the observed data. The E-step boils down to the computation of the univariate posterior probabilities $\hat{\pi}_{tkd} = P(u_{tkd} = 1 \mid \boldsymbol{z}, \boldsymbol{w}, \hat{\boldsymbol{\theta}}, \hat{\boldsymbol{\eta}})$, and the bivariate posterior probabilities $\hat{\pi}_{tkhdd'} = P(u_{t-1,kd} = 1, u_{thd'} = 1 \mid \boldsymbol{z}, \boldsymbol{w}, \hat{\boldsymbol{\theta}}, \hat{\boldsymbol{\eta}})$. The task of computing these posterior probabilities is generally referred to as the smoothing numerical issue and it is typically solved by specifying the posterior probabilities in terms of suitably normalized functions, which can be computed recursively, avoiding unpractical summations over the state space of latent semi-Markov chain and numerical under- and over-flows.

The M-step of the algorithm updates the parameter estimates by maximizing the expected value of the complete-data log-likelihood in (7.8), obtained from the previous E-step. This expected value is the sum of functions that depend on independent sets of parameters and can therefore be maximized separately, namely

$$Q(\boldsymbol{\theta}, \boldsymbol{\pi}, \boldsymbol{\omega}, \boldsymbol{\beta}) = \underbrace{\sum_{k=1}^{K} \hat{\pi}_{1k} \log \pi_k + \sum_{t=2}^{T} \sum_{k=1}^{K} \sum_{h=1}^{K} \sum_{d,d'=1}^{m} \hat{\pi}_{tkhdd'} \log \gamma_{kh}(d, \boldsymbol{w}_t^{\mathsf{T}}; \boldsymbol{\omega}, \boldsymbol{\beta})}_{Q(\boldsymbol{\pi}, \boldsymbol{\omega}, \boldsymbol{\beta})} + \underbrace{\sum_{t=1}^{T} \sum_{k=1}^{K} \hat{\pi}_{tk} \log f(\boldsymbol{z}_t; \boldsymbol{\theta}_k)}_{Q(\boldsymbol{\theta})},$$

where $Q(\boldsymbol{\pi}, \boldsymbol{\omega}, \boldsymbol{\beta}) = Q(\boldsymbol{\pi}) + Q(\boldsymbol{\omega}) + Q(\boldsymbol{\beta})$, and

$$Q(\boldsymbol{\pi}) = \sum_{k=1}^{K} \hat{\pi}_{1k} \log \pi_k,$$

$$Q(\boldsymbol{\omega}) = \sum_{t=2}^{T} \sum_{k=1}^{K} \sum_{h \neq k} \sum_{d,d'=1}^{m} \hat{\pi}_{tkhdd'} \log \omega_{kh},$$

$$Q(\boldsymbol{\beta}) = \sum_{t=2}^{T} \sum_{k=1}^{K} \sum_{h \neq k} \sum_{d,d'=1}^{m} \hat{\pi}_{tkhdd'} \log q_k(d, \boldsymbol{w}_t^{\mathsf{T}}; \boldsymbol{\beta}) + \sum_{t=2}^{T} \sum_{k=1}^{K} \sum_{d,d'=1}^{m} \hat{\pi}_{tkkdd'} \log(1 - q_k(d, \boldsymbol{w}_t^{\mathsf{T}}; \boldsymbol{\beta})).$$

Functions $Q(\boldsymbol{\pi})$ and $Q(\boldsymbol{\omega})$ are weighted multinomial log-likelihoods, and they are respectively maximized at $\hat{\pi}_k = \hat{\pi}_{1k}$ and

$$\hat{\omega}_{kh} = \frac{\sum_{t=2}^{T} \sum_{h \neq k} \sum_{d,d'=1}^{m} \hat{\pi}_{tkhdd'}}{\sum_{t=2}^{T} \sum_{k=1}^{K} \sum_{h \neq k} \sum_{d,d'=1}^{m} \hat{\pi}_{tkhdd'}}.$$

Function $Q(\boldsymbol{\beta})$ is instead a weighted binomial likelihood. Upon choosing an appropriate link function (e.g., the complementary log-log), it can be therefore maximized by fitting a weighted binomial regression model.

Maximization of the function $Q(\boldsymbol{\theta})$ reduces to K separate optimization problems, one for each latent class k. It can be undertaken by an unconstrained maximization algorithm, after a suitable re-parametrization of the involved parameters [15], exploiting a quasi-Newton procedure as provided by the function `optim` in `R`. Special care is needed when the state-specific densities are specified by a copula. In this case, optimization of $Q(\boldsymbol{\theta})$ conveniently relies on an inference-function-for-margins (IFM) approach [28]. Note that this EM algorithm works in principle with any HSMM. Only the M step is specific of directional statistics, as it involves maximum likelihood estimation of directional densities.

Under this setting, uncertainty quantification for the model parameters is conveniently based on a parametric bootstrap approach. It is obtained by re-fitting the model to B bootstrap samples, which are simulated from the model at the maximum likelihood estimates, hence obtaining $\{\boldsymbol{\eta}\}_{b=1}^{B}$ and $\{\boldsymbol{\theta}\}_{b=1}^{B}$ sets of estimates, exploited to compute equal-tail confidence intervals.

Model selection reduces to finding the optimal number K of components according to a suitable criterion. Like any other mixture model, a HSMM can be either exploited for density estimation to get a good approximation of the data distribution or for clustering, where the goal is to find the *best* data segmentation [35]. When the goal is interpreting the data distribution in terms of a finite number of regimes, segmentation is more important than density estimation and model choice should rely on the minimization of the integrated complete likelihood [ICL; 36], where the goodness of fit is penalized by classification entropy.

7.4 An Illustration on Marine Data

7.4.1 Buoy Data

HSMMs for directional data are here illustrated on a time series of cylindrical data, an example of bivariate times series with mixed linear and circular observations.

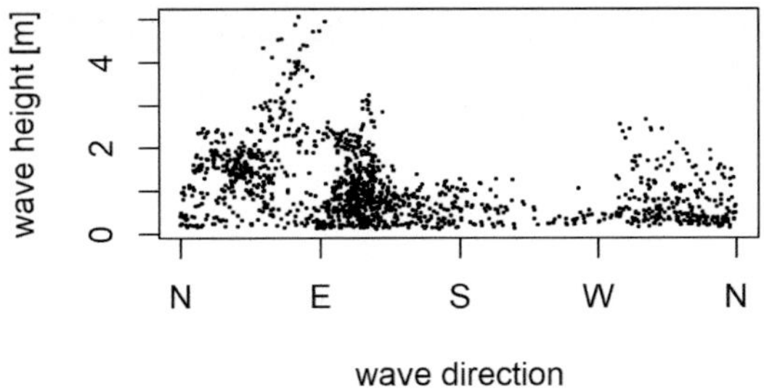

FIGURE 7.1
Wave direction and height, observed at a buoy of the Adriatic Sea in wintertime.

Specifically, we consider a time series of semi-hourly wave directions and heights, recorded in wintertime by the buoy of Ancona, located in the Adriatic Sea at about 30 km from the coast. The Adriatic Sea is a semi-enclosed, long narrow basin, bordered by mountains on three sides, and extending for about 800 km along the major axis from SE to NW, with a width of about 200 km. In wintertime, relevant events in the Adriatic Sea are typically generated by the southeastern Sirocco wind and the northern Bora wind. The complex dependence structure between wave height and direction in the Adriatic Sea is shown by Figure 7.1, which displays the scatter plot of the data. For simplicity, the data are plotted on the plane, although data points are actually on a cylinder. Angles indicate the directions from which waves travel.

Although some distinct patterns appear in this scatter plot, their interpretation is difficult due to the weak correlation between the measurements, the skewness, and the multimodality of the data. Multimodality, weak correlation, and skewness are often held responsible for the inaccuracy of numerical wave models in the Adriatic sea and they are traditionally explained as the result of the complex orography of this basin, combined with time-varying weather conditions. This motivates the development of special methods in order to segment these data into a small number of latent classes, conditionally on which the distribution of the data takes a shape that is easier to interpret than the shape taken by the marginal distribution.

7.4.2 The Abe-Ley Distribution

We model wave data by a cylindrical HSMM (7.1), by plugging an Abe-Ley density [37] in the observation process equation (7.2). It is defined on the cylinder $[0, 2\pi) \times [0, +\infty)$ and takes the form

$$f(x, y) = \frac{\alpha\beta^\alpha}{2\pi\cosh(\kappa)} (1 + \lambda\sin(x - \mu))\, y^{\alpha-1} \exp\left(-(\beta y)^\alpha (1 - \tanh(\kappa)\cos(x - \mu))\right), \quad (7.9)$$

where $x \in [0, 2\pi)$ and $y \in \times[0, +\infty)$.

This density enjoys some desirable properties. First, the meaning of the parameters is intuitively appealing. Precisely, $\alpha > 0$ is a (linear) shape parameter, $\beta > 0$ is a (linear) scale parameter, $\mu \in [0, 2\pi)$ is a (circular) location parameter, $\kappa > 0$ is a (circular) concentration parameter and, finally, $\lambda \in [-1, 1]$ is a (circular) skewness parameter. Second, the normalizing constant $2\pi\cosh(\kappa)$ is numerically tractable, and this facilitates maximum likelihood estimation. Third, under (7.9), the univariate marginal and conditional distributions are available in closed form, and this simplifies simulation routines. The form taken by the conditional distributions is particularly appealing in environmental studies because they reflect typical features of environmental cylindrical data. Specifically, the conditional distribution of the intensity y given the angular direction x is Weibull with scale α and direction-dependent shape β_x, say

$$\begin{aligned} f(y \mid x) =& \alpha(\beta_x)^\alpha y^{\alpha-1} \exp(-(\beta_x y)^\alpha) \\ \beta_x =& \beta(1 - \tanh(\kappa)\cos(x-\mu))^{1/\alpha}. \end{aligned} \tag{7.10}$$

Under model (7.10), the shape parameter β_x is a monotone decreasing function of the circular concentration parameter κ. As a result, $1/\beta_x$ increases as κ increases. More precisely, under (7.10), the regression function of intensity y on direction x is given by

$$\mathbb{E}(Y \mid x) = \frac{1}{\beta_x}\Gamma\left(1+\frac{1}{\alpha}\right), \tag{7.11}$$

while the conditional variance is given by

$$\mathbb{V}(Y \mid x) = \left(\frac{1}{\beta_x}\right)^2 \left(\Gamma\left(1+\frac{2}{\alpha}\right) - \left(\Gamma\left(1+\frac{1}{\alpha}\right)\right)^2\right). \tag{7.12}$$

On the other hand, under (7.9), the conditional distribution of the direction x given the intensity y is a skew von Mises distribution with location parameter μ and concentration parameter κ_y, say

$$\begin{aligned} f(x \mid y) =& \frac{1+\lambda\sin(x-\mu)}{I_0(\kappa_y)} \exp(\kappa_y \cos(x-\mu)), \\ \kappa_y =& (\beta y)^\alpha \tanh(\kappa), \end{aligned} \tag{7.13}$$

where I_0 is the modified Bessel function of the first kind and order zero. Specifically, equation (7.13) allows both the mean direction and the spread around this mean to be functions of the intensity. More precisely, under (7.13), the regression function of direction x is given by

$$\mathbb{E}(X \mid y) = \mu + \tan^{-1}\left(\frac{\lambda}{(\beta y)^\alpha \tanh(\kappa_y)}\right) \tag{7.14}$$

while the conditional circular variance (i.e., the complement of the mean resultant length) is given by

$$\mathbb{V}(X \mid y) = 1 - \frac{I_1(\kappa_y)}{\kappa_y I_0(\kappa_y)}\sqrt{\lambda^2+\kappa_y^2}. \tag{7.15}$$

Equations (7.14) and (7.15) can be easily derived, by exploiting known results on the skew von Mises density [38].

7.4.3 A Cylindrical Hidden Semi-Markov Model

The wave data described above were fitted by a nonhomogeneous, cylindrical HSMM, driven by a hazard function (7.5) that depends on wind speed.

These data have been already analyzed by a directional HMM model with cylindrical densities [39], obtaining a segmentation according to three regimes that can be respectively associated with Sirocco and Bora events, and conditions of calm sea.

With a nonhomogeneous HSMM, we obtain a similar segmentation results, although augmented with additional information on the dwell time distribution of each regime and the influence of wind speed on the hazard of state switching.

Table 7.1 displays the estimates under this model, along with bootstrap percentiles, computed by simulating 300 samples. Figure 7.2 overlaps the state-specific densities on the data points, filled with gray levels according to the estimated class membership probabilities

TABLE 7.1
Parameter estimates and associated bootstrap percentiles of a cylindrical hidden semi-Markov model with three states.

State	Parameter	Estimate	2.5% Percentile	97.5% Percentile
1	linear shape	2.73	2.27	2.74
	linear scale	0.80	0.74	0.82
	concentration	1.33	0.85	1.35
	circular mean	1.77	1.15	1.25
	skewness	−0.90	−1.00	−0.83
	intercept	−7.10	−9.75	−4.66
	dwell time	0.06	0.03	0.20
	wind speed	0.25	−0.17	0.77
2	linear shape	2.36	2.16	2.48
	linear scale	3.08	2.89	3.28
	concentration	0.36	0.19	0.59
	circular mean	5.09	2.98	6.10
	skewness	0.42	−0.42	0.71
	intercept	−2.01	−9.82	−5.69
	dwell time	−0.01	−0.14	0.21
	wind speed	1.53	0.68	3.43
3	linear shape	2.67	2.22	3.23
	linear scale	2.11	1.93	2.46
	concentration	1.99	1.50	2.82
	circular mean	2.21	1.97	2.00
	skewness	0.90	0.71	0.95
	intercept	−2.94	−3.01	5.88
	dwell time	−0.02	−0.14	0.01
	wind speed	−1.07	−2.44	−0.56

Origin	Destination	Estimate	2.5% Percentile	97.5% Percentile
1	2	0.62	0.41	0.69
1	3	0.38	0.29	0.43
2	1	0.45	0.31	0.58
2	3	0.55	0.39	0.64
3	1	0.12	0.05	0.22
3	2	0.88	0.71	0.93

(black indicates probability equal to 1). Finally, Figure 7.3 shows the estimated distribution of the dwell times in each state.

The first component of the model (state 1) is associated with high waves coming from north-northeast. These waves are generated by northern Bora jets that blow along the major axis of the Adriatic basin. These waves are highly concentrated around one modal direction and tend to be higher than the waves clustered under other states. Interestingly, the distribution of wave directions is negatively skewed, because of the orographic conditions of the basin: only the waves that travel along the major axis of the basin are allowed to reach elevated heights.

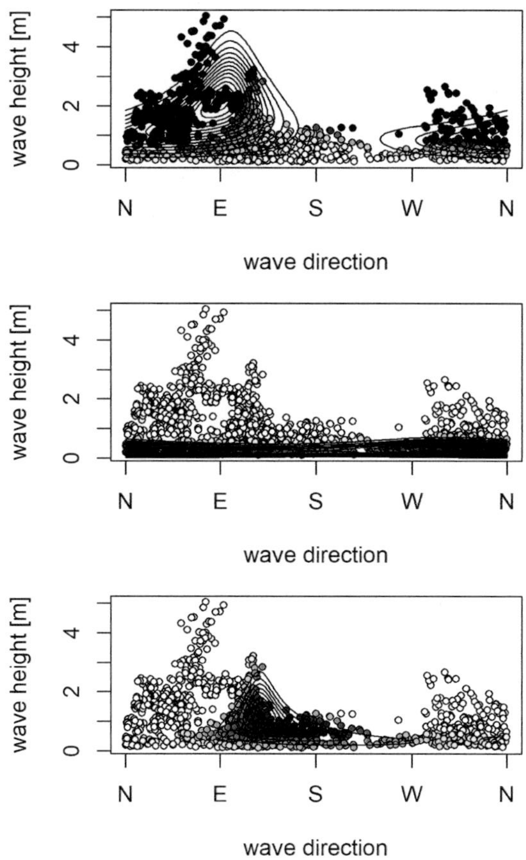

FIGURE 7.2
Wave data, clustered by a hidden semi-Markov model with three states. Dots are colored according to estimated class membership probabilities. Contours indicate the three class-specific cylindrical densities.

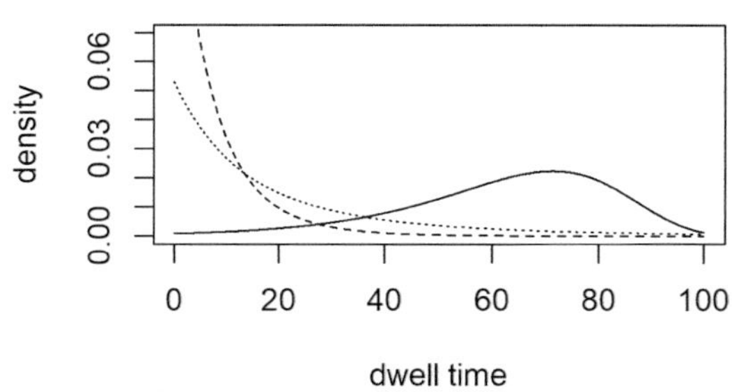

FIGURE 7.3
Distribution of the dwell time of state 1 (solid), state 2 (dashed), and state 3 (dotted).

The second component of the model (state 2) is associated with periods of calm sea. Under this regime, moderate waves are uniformly distributed around the circle of directions and the dependence between height and direction is weak.

The third component (state 3) is associated with the Sirocco episodes. In studies of the Adriatic Sea, detection of the Sirocco regime is very important because it exposes

Venice to the famous flooding tides when occurring in combination with astronomical forces. Compared with the first regime, wave heights and directions appear more synchronized. Specifically, waves travel southeasterly along the major axis of the basin, driven by winds that blow from a similar directional angle. The positive skewness parameter is the result of the orography effect. While this effect was negative for the northern waves under state 1, it is positive under state 3, which clusters waves that travel southeasterly.

The bottom rows of Table 7.1 includes the estimated transition probabilities of the latent semi-Markov chain. These probabilities clearly indicate that state 2 is a transition state: direct transitions between state 1 and 3 are unlikely, while the system favors transitions from state 1 and 3 to state 2. This is a confirmation that the sea surface in the study area tends to alternate relevant marine events (states 1 and 3) with periods of good sea conditions (state 2).

The distinct advantage of a HSMM, compared to a more restrictive HMM, is that the temporal segmentation of the observed process can be interpreted not only in terms of transitions between latent regimes (as it is normally done with a HMM), but also in terms of the duration of each regime. Figure 7.3 shows the distribution of the dwell time of each latent state, estimated at baseline (wind speed equal to 0). The shape of the dwell time distributions under each state is driven by the dwell time coefficient, whose estimate is displayed in Table 7.1. Large values of this parameter indicate departures from a geometric shape in the dwell time distribution. Accordingly, the nonsignificant estimates under regimes 2 and 3 are associated with dwell time distributions that are essentially geometric. Conversely, the estimate under state 1 reflects a significant departure from the geometric shape. According to these findings, the state associated with a Bora episode tends to have longer durations than the states associated with Sirocco episodes and calm see. On the other side, the significantly positive influence of wind speed under state 2 indicates that an increase of wind speed is associated with an episode of state switching, when the system is under good weather conditions. Conversely, such effect becomes negative when the system is under state 3, indicating that strong winds tend to prolong the duration of this state.

7.5 Discussion

Directional HSMMs offer a useful strategy to interpret directional time series by parsimoniously combining directional statistics and survival analysis tools. While directional statistics is exploited at the observation level of the hierarchy using directional densities, survival analysis is employed at the latent level of the hierarchy using hazard regression. Such integration is identifiable (and hence estimable) because the hazard of a transition is modeled separately from the conditional probability of reaching a specific state, given a transition. By working with the hazard function, furthermore, the impact of time-varying covariates on the duration of the system in a given regime can be efficiently tested.

In this way, the model is not only capable of offering a clear-cut description of the dependence structure in time series of directional data in terms of intuitively appealing latent regimes. It is also able to distinguish between transitory and persistent regimes by providing regime-specific dwell time distributions.

In a case study of marine data, it offered a classification that reflects the orography of the study area, a feature often neglected by oceanographic numerical models. It offered valid data segmentation by relaxing the restrictions that affect other models. It finally captured the influence of environmental conditions (wind speed) on the duration of specific regimes of the sea.

Typically, directional HSMMs are fully parametric and therefore exposed to misspecification issues. Non-parametric extensions can in principle be developed either at the observation level of the model hierarchy or at the latent level. At the observation level, an option could be replacing parametric directional densities with a non-parametric density. Methods of non-parametric estimation of toroidal densities are available [40] and can, in principle, be integrated into an HSMM framework. This approach could be pursued when the interest is on the accurate estimation of the marginal data density and appropriate resources for the additional computational burden are available.

At the latent level, a non-parametric approach can be pursued by assuming a non-parametric dwell time distribution and separately estimating the whole set of probabilities masses under each regime [41]. This strategy is computationally tractable only when the study involves a long time series with short dwell times and it does not guarantee against wiggly dwell time distributions with implausible gaps and spikes [42]. Recently [21], penalized likelihood methods have been suggested as a method to estimate the dwell time distribution in a (homogeneous) HSMM. This idea could in principle be integrated into our proposal to estimate hazard functions non-parametrically, at the price of additional computational burden.

Acknowledgments

This work has been supported by MIUR, grant number 2022XRHT8R—The SMILE project: Statistical Modelling and Inference for Living the Environment.

Bibliography

[1] Nick I Fisher and Alan J Lee. A correlation coefficient for circular data. *Biometrika*, 70(2):327–332, 1983.

[2] Kristin P Lennox, David B Dahl, Marina Vannucci, and Jerry W Tsai. Density estimation for protein conformation angles using a bivariate von Mises distribution and Bayesian nonparametrics. *Journal of the American Statistical Association*, 104(486):586–596, 2009.

[3] Delong Liu, Shyamal D Peddada, Leping Li, and Clarice R Weinberg. Phase analysis of circadian-related genes in two tissues. *BMC Bioinformatics*, 7:1–10, 2006.

[4] Grace S Shieh, Shurong Zheng, Richard A Johnson, Yi-Feng Chang, Kunio Shimizu, Chia-Chang Wang, and Sen-Lin Tang. Modeling and comparing the organization of circular genomes. *Bioinformatics*, 27(7):912–918, 2011.

[5] Louis-Paul Rivest. A decentred predictor for circular-circular regression. *Biometrika*, 84(3):717–726, 1997.

[6] E. García-Portugués, R.M. Crujeiras, and W. González-Manteiga. Exploring wind direction and SO2 concentration by circular-linear density estimation. *Stochastic Environmental Research and Risk Assessment*, 27(5):1055–1067, 2013.

[7] Gianluca Mastrantonio, Antonello Maruotti, and Giovanna Jona-Lasinio. Bayesian hidden Markov modelling using circular-linear general projected normal distribution. *Environmetrics*, 26(2):145–158, 2015.

[8] Eduardo García-Portugués, Ana M.G. Barros, Rosa M. Crujeiras, Wenceslao González-Manteiga, and José Pereira. A test for directional-linear independence, with applications to wildfire orientation and size. *Stochastic Environmental Research and Risk Assessment*, 28:1261–1275, 2014.

[9] Jan Bulla, Francesco Lagona, Antonello Maruotti, and Marco Picone. Environmental conditions in semi-enclosed basins: A dynamic latent class approach for mixed-type multivariate variables. *Journal de la Société Française de Statistique*, 156:114–137, 2015.

[10] Francesco Lagona, Marco Picone, Antonello Maruotti, and Simone Cosoli. A hidden Markov approach to the analysis of space-time environmental data with linear and circular components. *Stochastic Environmental Research and Risk Assessment*, 29(2):397–409, 2015.

[11] Fangpo Wang, Alan E. Gelfand, and G. Jona Lasinio. Joint spatio-temporal analysis of a linear and a directional variable: modeling of wave heights and wave directions in the Adriatic Sea. *Statistica Sinica*, 25:25–39, 2015.

[12] Christophe Ley and Thomas Verdebout. *Modern Directional Statistics*. Wiley, 2017.

[13] Arthur Pewsey and Eduardo García-Portugués. Recent advances in directional statistics. *TEST*, 30(1):1–58, March 2021.

[14] Marco Di Marzio, Agnese Panzera, and Charles C. Taylor. Kernel density estimation on the torus. *Journal of Statistical Planning and Inference*, 141(6):2156–2173, 2011.

[15] Walter Zucchini, Iain L MacDonald, and Roland Langrock. *Hidden Markov models for time series: an introduction using R*. Chapman and Hall/CRC, 2016.

[16] Hajo Holzmann, Axel Munk, Max Suster, and Walter Zucchini. Hidden Markov models for circular and linear-circular time series. *Environmental and Ecological Statistics*, 13:325–347, 2006. 10.1007/s10651-006-0015-7.

[17] Pierre Ailliot, Julie Bessac, Valérie Monbet, and Françoise Pène. Non-homogeneous hidden Markov-switching models for wind time series. *Journal of Statistical Planning and Inference*, 160:75–88, 2015.

[18] Antonello Maruotti, Jan Bulla, Francesco Lagona, Marco Picone, and Francesca Martella. Dynamic mixtures of factor analyzers to characterize multivariate air pollutant exposures. *The Annals of Applied Statistics*, 11(3):1617–1648, 2017.

[19] Qin Jiang, Francesco Cioffi, Federico Rosario Conticello, Mario Giannini, Vito Telesca, and Jun Wang. A stacked ensemble learning and non-homogeneous hidden Markov model for daily precipitation downscaling and projection. *Hydrological Processes*, 37(9):e14992, 2023.

[20] Theodoros Economou and Matthew B Menary. A hidden semi-Markov model for characterizing regime shifts in ocean density variability. *Journal of the Royal Statistical Society Series C: Applied Statistics*, 68(5):1529–1553, 2019.

[21] Jennifer Pohle, Timo Adam, and Larissa T. Beumer. Flexible estimation of the state dwell-time distribution in hidden semi-Markov models. *Computational Statistics & Data Analysis*, 172:107479, 2022.

[22] Luca Merlo, Antonello Maruotti, Lea Petrella, and Antonio Punzo. Quantile hidden semi-Markov models for multivariate time series. *Statistics and Computing*, 32(4):61, 2022.

[23] Sam Nicol, Marie-Josée Cros, Nathalie Peyrard, Régis Sabbadin, Ronan Trépos, Richard A Fuller, and Bradley K Woodworth. Flywaynet: A hidden semi-Markov model for inferring the structure of migratory bird networks from count data. *Methods in Ecology and Evolution*, 14(1):265–279, 2023.

[24] Francesco Lagona and Marco Mingione. Nonhomogeneous hidden semi-Markov models for toroidal data. *Journal of the Royal Statistical Society Series C: Applied Statistics*, 74(1):142–166, 2025.

[25] Shogo Kato and Kunio Shimizu. Dependent models for observations which include angular ones. *Journal of Statistical Planning and Inference*, 138(11):3538–3549, 2008. Special Issue in Honor of Junjiro Ogawa (1915–2000): Design of Experiments, Multivariate Analysis and Statistical Inference.

[26] Jose A. Carnicero, M. Concepcón Ausín, and Michael P. Wiper. Non-parametric copulas for circular-linear and circular-circular data: an application to wind directions. *Stochastic Environmental Research and Risk Assessment*, 27(8):1991–2002, 2013.

[27] M. C. Jones, Arthur Pewsey, and Shogo Kato. On a class of circulas: copulas for circular distributions. *Annals of the Institute of Statistical Mathematics*, 67(5):843–862, Oct 2015.

[28] Francesco Lagona. Copula-based segmentation of cylindrical time series. *Statistics & Probability Letters*, 144:16–22, 2019.

[29] Zekun Xu, Eric Laber, Ana-Maria Staicu, and B. Duncan X. Lascelles. Novel approach to modeling high-frequency activity data to assess therapeutic effects of analgesics in chronic pain conditions. *Scientific Reports*, 11:7737, 2021.

[30] Benny Ren and Ian Barnett. Combining mixed effects hidden Markov models with latent alternating recurrent event processes to model diurnal active–rest cycles. *Biometrics*, 79(4):3402–3417, 2023.

[31] Jan-Ole Koslik. Hidden semi-Markov models with inhomogeneous state dwell-time distributions. *Computational Statistics & Data Analysis*, 209:108171, 2025.

[32] John D. Kalbfleisch and Ross L. Prentice. *The Statistical Analysis of Failure Time Data*. Wiley, 1980.

[33] Gerhard Tutz and Matthias Schmid. Basic regression models. In *Modeling Discrete Time-to-Event Data*, pages 35–72. Springer, 2016.

[34] R. Langrock and W. Zucchini. Hidden Markov models with arbitrary state dwell-time distributions. *Computational Statistics & Data Analysis*, 55(1):715–724, 2011.

[35] S. Fruhwirth-Schnatter, G. Celeux, and C.P. Robert. Model selection for mixture models - perspectives and strategies. In *Fruhwirth-Schnatter, S., Celeux, G., Robert, C.P. (Eds.). Handbook of Mixture Analysis*. Chapman and Hall/CRC, 2018.

[36] C Biernacki, G Celeux, and G Govaert. Assessing a mixture model for clustering with the integrated completed likelihood. *IEEE Transactions on Pattern Analysis and Machine Intelligence*, 22:719–725, 2000.

[37] Toshihiro Abe and Christophe Ley. A tractable, parsimonious and flexible model for cylindrical data, with applications. *Econometrics and Statistics*, 4:91–104, 2017.

[38] T. Abe and A. Pewsey. Sine-skewed circular distributions. *Statistical Papers*, 52:683–707, 2011.

[39] Francesco Lagona, Marco Picone, and Antonello Maruotti. A hidden Markov model for the analysis of cylindrical time series. *Environmetrics*, 26(8):534–544, 2015.

[40] Marco Di Marzio, Agnese Panzera, and Charles C. Taylor. Kernel density estimation on the torus. *Journal of Statistical Planning and Inference*, 141(6):2156–2173, 2011.

[41] John Sansom and Peter Thomson. Fitting hidden semi-Markov models to breakpoint rainfall data. *Journal of Applied Probability*, 38(A):142–157, 2001.

[42] Jan Bulla, Ingo Bulla, and Oleg Nenadić. HSMM—an R package for analyzing hidden semi-Markov models. *Computational Statistics & Data Analysis*, 54(3):611–619, 2010.

8

Models for Environmental Cylindrical Time Series

Maria Maddalena Barbieri, Francesco Battaglia, and Domenico Cucina

Environmental studies often produce observations characterized by a bivariate response, where one observation is linear and the other is circular, i.e., an angle or direction. These data are frequently referred to as cylindrical, as they can be represented by points on a cylinder of radius one. Typical examples include wind direction and speed, sea wave direction and height, and changes in direction and speed in animal movement. When such observations are collected at discrete time points, the resulting data are usually referred to as cylindrical time series.

In this chapter, we review various models for bivariate linear-circular time series that are flexible enough to account for possible regime-switching behavior in the underlying process, as well as the within-regime serial dependence through an autoregressive (AR) structure. Regime switching may be random (as in a hidden Markov model, HMM) or driven by past observations (as in a threshold autoregressive model, TAR).

We provide empirical examples to compare these models using information criteria such as AIC and mean squared prediction errors, with applications to forecasting problems.

8.1 Introduction

Cylindrical time series are recordings of a circular and a linear variable at discrete time points [1, 2]. Typical applications include the analysis of wind speed and direction, wave direction and height, turning angle and step length in animal movement, and hour of the day and pollutant concentration. In all these cases, the data are rarely stationary; rather, the dynamics alternate between two or more different regimes. For this reason, most of the literature focuses on hidden Markov models [3], where regime changes are driven by a Markov chain, and observations within each regime are assumed (serially) independent. Hidden Markov models for modeling cylindrical data observed over time were proposed by [4] and have been considered in several other studies (e.g. [5], [6]). For forecasting purposes, however, it is essential to account for within-regime serial dependence. Therefore, we consider multi-regime models in which an explicit dynamic time series model governs the data within each regime, typically an AR model in its simplest form. Transition between regimes may be driven by a (homogeneous) Markov chain or determined by past values of the series, as in self-exciting threshold models [7]. In hidden Markov models, the regime of the next observation is not known and is treated as random at each time point, whereas

DOI: 10.1201/9781003564423-8

in threshold models, the regime is completely determined by past observations, which can have a favorable impact on forecasting.

In this framework, [8] and [9] introduced several Markov-switching AR models for wind speed and direction time series based on non-homogeneous hidden Markov chains in which the evolution depends on past wind conditions. Self-exciting threshold models and Markov-switching AR models for forecasting wind power production were considered in [10].

In this chapter, we introduce different approaches to address the presence of multiple regimes. In particular, we specify several hidden Markov models suitable for this context, corresponding to different joint distributions for the components of the cylindrical data. Various model-fitting strategies, all based on the EM algorithm, are considered. Similarly, in the case of threshold AR models applied to each component of the cylindrical time series, we explore different types of bivariate thresholds that depend on both the linear and the circular components, following the suggestions given by [11] and [12] in the context of general linear multivariate models. To address the challenges associated with fitting this class of models, we make use of a Genetic Algorithm (GA) [13].

8.2 Distinctive Features of Cylindrical Time Series

We denote a cylindrical time series by $(\theta_t, x_t), t = 1, \ldots, n$, where θ_t is the circular component and x_t is the linear component. In our example, θ_t is an angle in the plane, measured in radians and, without loss of generality, taking values in $[0, 2\pi)$, while x_t takes non-negative real values. We also define $y_t = \log x_t$.

To discuss the main characteristics of a cylindrical time series and illustrate our proposals, we use an example based on semi-hourly sea wave directions and heights, collected at 1440 equally spaced time points between February 15 and March 16, 2010, by the Ancona buoy, located in the Adriatic Sea [5]. In the circular component, due north corresponds to zero radians, with the subsequent values increasing clockwise: east corresponds to $\pi/2$, south to π, and west to $3\pi/2$. The four isolated missing values in the original series were imputed by averaging the neighboring observations for the linear component, and using the circular mean for the circular component. The circular mean of a set of observations $\{\theta_t, t = 1, \ldots, n\}$ is defined as:

$$\bar{\theta} = \arg\left(\sum_{t=1}^{n} e^{i\,\theta_t}\right), \tag{8.1}$$

where $\arg(\cdot)$ denotes the argument of the complex number in the parentheses.

In our analysis, the results and plots are based on the first 1000 observations, while the remaining data are retained as a holdout sample to evaluate the forecasting performance of the fitted models.

A key issue in the analysis of such data is that no available graphical representation appears capable of simultaneously capturing all potentially relevant features.

The univariate distribution of the circular variable can be represented using a rose plot, which is the circular analogue of a histogram on the real line (see Figure 8.1)

A natural way to represent this type of data is through a three-dimensional plot, where the bivariate observations are displayed as points on the surface of a unit-radius cylinder, as illustrated in Figure 8.2(a). However, such plots are difficult to use for exploratory analysis, as the features of the joint distribution are not easily visualized. A commonly used alternative is a scatter plot, as shown in Figure 8.2(b), which effectively corresponds to flattening the cylinder to facilitate interpretation. Nevertheless, this representation is not ideal, since

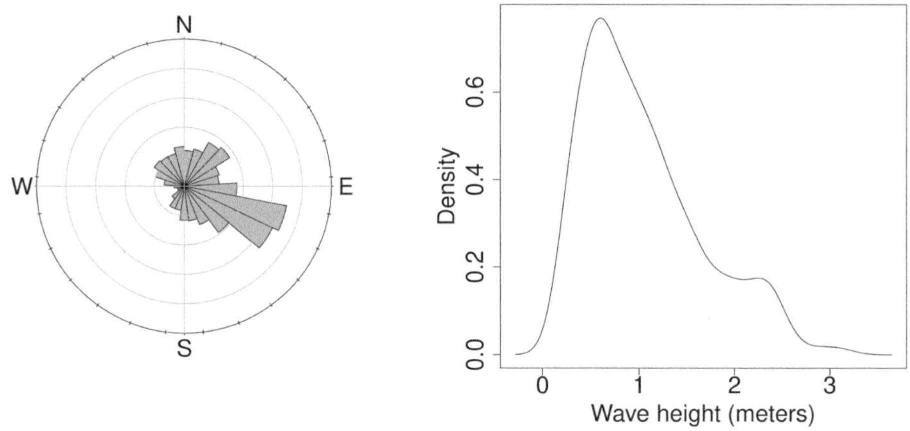

(a) Wave direction: rose diagram (b) Wave height: empirical density

FIGURE 8.1
Marginal distributions of Ancona wave direction and height. (North corresponds to zero radians, east to $\pi/2$, south to π and west to $3\pi/2$.)

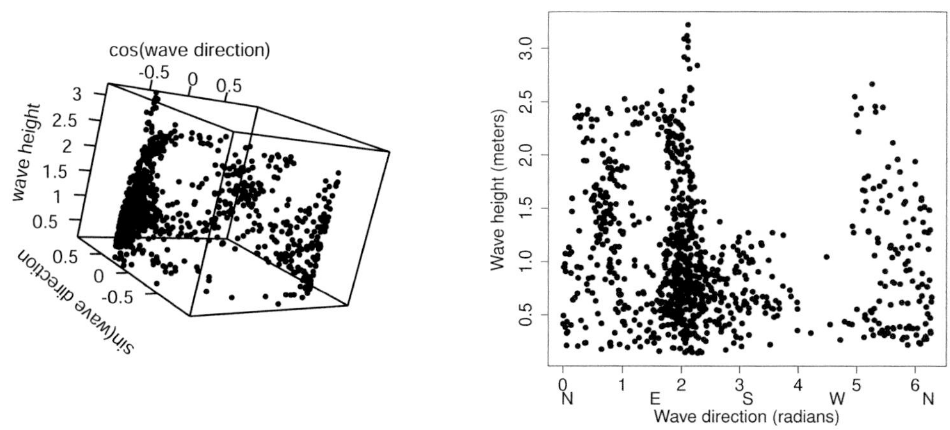

(a) Three-dimensional scatter plot (b) Scatter plot

FIGURE 8.2
Ancona wave direction and height: panel (a) shows a three-dimensional plot of the data; panel (b) displays a scatter plot of the bivariate observations.

the visual appearance of the plot can be highly sensitive to the choice of origin for the angular coordinate, potentially obscuring structural features such as regime changes. Moreover, scatter plots of this kind are generally inadequate for representing temporal dynamics. To better capture the time-dependent structure of the process, it is preferable to plot each component against time, as shown in Figure 8.3. The linear metric used to plot the circular component in Figure 8.3 (middle panel) induces apparent fluctuations at points near the boundaries of the interval $[0, 2\pi)$, which may appear as level shifts or outliers. In reality, these observations are actually close neighbors when considering a metric that respects the circular topology.

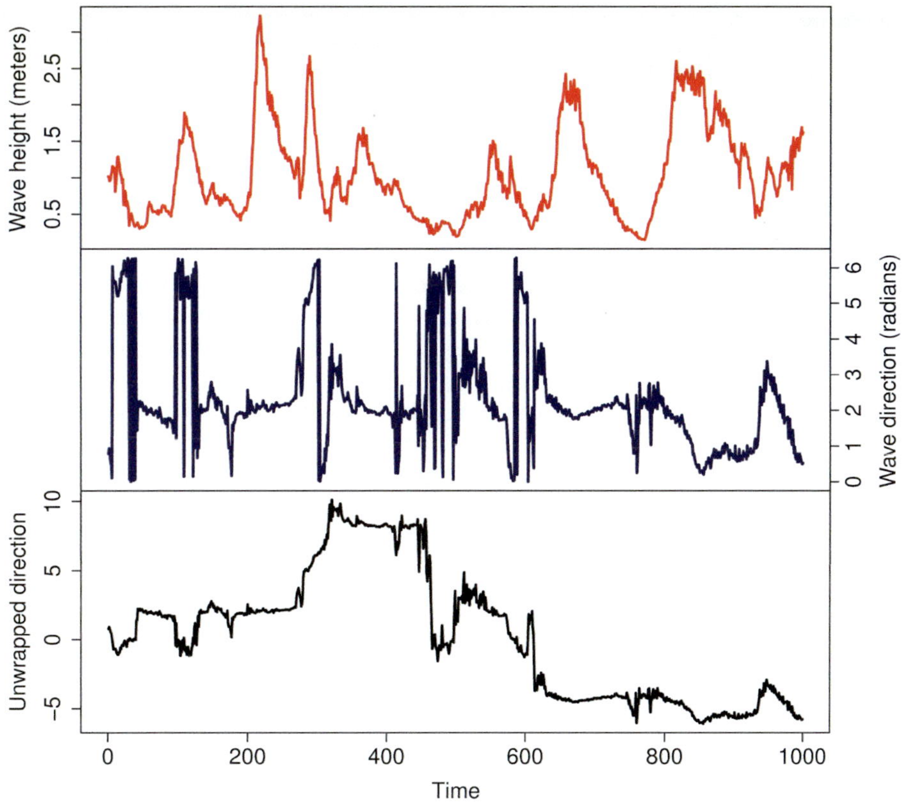

FIGURE 8.3
Upper and middle panel: Wave height (red) and Wave direction (blue) recorded in the buoy of Ancona. Lower panel: unwrapped direction.

A common solution to smooth the time trend and avoid the presence of spikes is to transform the observed circular time series θ_t into a linear one u_t as follows:

$$u_1 = \theta_1, \qquad u_t = u_{t-1} + dist(\theta_t, u_{t-1}), \quad t = 2, \dots, n, \tag{8.2}$$

where $dist(\theta_t, u_{t-1})$ is the angular distance between the two angles θ_t and u_{t-1}:

$$dist(\theta_t, u_{t-1}) = \arg\{\exp[i(\theta_t - u_{t-1})]\} = \theta_t - u_{t-1} - 2\,\pi\,h \tag{8.3}$$

if $(2\,h - 1)\pi < \theta_t - u_{t-1} \leq (2h + 1)\pi$ and h is an appropriate integer.

The transformation, commonly referred to as unwrapping, is applied to avoid discontinuities in a phase signal [14]. It produces the angle displacement of θ_t from u_{t-1} with the minimum absolute value. Thus, it maximizes linear continuity by minimizing the sum of squared increments $u_t - u_{t-1}$. As a result, similar directions on the circle always correspond to similar values after the transformation. Note that the values of the new series, called the *unwrapped* series and shown in Figure 8.3 (lower panel), are not confined to the interval $[0, 2\pi)$, and all trigonometric functions give the same value whether computed on θ_t or u_t. This property also holds for all the densities introduced in the following section.

8.3 Models for Cylindrical Data

Several distributions on the circle have been proposed [1]. The most popular is the von Mises distribution, with probability density function:

$$f(\theta|\mu_\theta, \kappa) = \frac{1}{2\pi\,\mathcal{I}_0(\kappa)} \exp\{\kappa(\cos(\theta - \mu_\theta))\}$$

where $\mathcal{I}_0(\kappa)$ denotes the modified Bessel function of the first kind and of order zero. The parameter $\mu_\theta \in [0, 2\pi)$ is the circular mean, while κ is a non-negative concentration parameter, inversely related to scale. When $\kappa = 0$, the distribution of θ is uniform, whereas for large κ, it is approximately Normal with mean μ_θ and variance κ^{-1} [1].

When also a linear variable X is observed, several joint distributions have been proposed for cylindrical observations (θ, x), depending on whether the range of the linear component is $\mathbb{R}$ or $\mathbb{R}^+$. In the case where the range is $\mathbb{R}$, a very well-known proposal is due to Mardia and Sutton [15], who introduced the density (see also [1]):

$$\begin{aligned} f(\theta, x \mid \mu_\theta, \kappa, \mu_x, \sigma, \rho, \nu) &= f(\theta \mid \mu_\theta, \kappa) f(x \mid \theta, \mu_x, \sigma, \rho, \nu) = \\ &= \frac{1}{2\pi\mathcal{I}_0(\kappa)} \exp\{\kappa\cos(\theta - \mu_\theta)\} \cdot \\ &\cdot \frac{1}{\sqrt{2\pi\sigma^2(1-\rho^2)}} \exp\left\{ -\frac{1}{2\sigma^2(1-\rho^2)} \left(x - \mu_x + \sigma\sqrt{\kappa}\rho\left[cos\nu\cos\theta + \sin\nu\sin\theta_t\right]\right)^2 \right\} \end{aligned} \tag{8.4}$$

where $\mu_\theta \in [0, 2\pi)$ and $\kappa \geq 0$ are, as before, the circular mean and the circular concentration, μ_x and σ are the mean and the standard deviation of the linear variable, respectively and $\nu \in [0, 2\pi)$ and $0 \leq \rho \leq 1$ are the circular-linear parameters. The latter regulates the dependence structure: when $\rho = 0$, formula (8.4) reduces to the product of independent von Mises and Normal densities, whereas when $\rho \neq 0$, the marginal circular density remains a von Mises, but the form of the marginal linear part is complicated. However, the conditional distribution is particularly suitable for regression analysis, in fact, it is Normal with a mean that can be expressed as

$$E[x|\theta] = b_0 + b_1\cos\theta + b_2\sin\theta$$

and variance equal to $\sigma^2(1 - \rho^2)$. Maximum likelihood estimation is straightforward with closed-form expressions for the estimators.

More recently Abe and Ley [16] introduced the joint probability density function:

$$\begin{aligned} f(\theta, x|\mu_\theta, \kappa, \alpha, \beta, \lambda) = &\frac{\alpha\beta^\alpha}{2\pi\cosh(\kappa)}[1 + \lambda\sin(\theta - \mu_\theta)]\, x^{\alpha-1} \cdot \\ &\cdot \exp\{-(\beta x)^\alpha[1 - \tanh(\kappa)\cos(\theta - \mu_\theta)]\} \end{aligned} \tag{8.5}$$

where $x \in [0, +\infty)$, μ_θ and λ are circular location and skewness parameters, β and α are linear scale and shape parameters, while κ plays the role of circular concentration and circular–linear dependence parameter. When $\kappa = 0$, the two components are independent and the density (8.5) results in the product of a linear Weibull and a circular Cardioid distribution. Marginal and conditional densities are available in closed form.

The association between the circular and the linear components may be measured using the sample multiple correlation coefficient between x and $(\cos\theta, \sin\theta)$ [1]:

$$r^2_{x.cs} = \frac{r^2_{xc} + r^2_{xs} - 2r_{xc}r_{xs}r_{cs}}{1 - r^2_{cs}} \tag{8.6}$$

where $r_{xc} = corr(x, \cos\theta)$, $r_{xs} = corr(x, \sin\theta)$, and $r_{cs} = corr(\cos\theta, \sin\theta)$ are the sample correlation coefficients. While to measure the circular-serial dependence, we may resort to the sample circular autocorrelations [17]:

$$\rho_\theta(h) = \frac{\sum_{t=1}^{n-h} \sin(\theta_t - \bar{\theta}) \sin(\theta_{t+h} - \bar{\theta})}{\sum_{t=1}^{n} \sin^2(\theta_t - \bar{\theta})}$$

where $\bar{\theta}$ is the sample circular mean defined in (8.1).

Under stationarity, a popular choice for modeling the serial dependence in the linear component of the cylindrical variable is to use an AR model of appropriate order (say p):

$$y_t = \phi_0 + \phi_1 y_{t-1} + \ldots + \phi_p y_{t-p} + \epsilon_t,$$

where $\{\epsilon_t\}$ is a Gaussian white noise.

Several models have been proposed to model the circular component (see [1], [18], [19], [20]). In this chapter, we focus on wrapped AR models. In this approach, it is assumed that the observed directions θ_t are obtained by wrapping a Gaussian AR process $\mathcal{U}_t$ around the circle, such that $\theta_t = \mathcal{U}_t(\bmod 2\pi)$, where *mod* denotes the modulo operator, i.e., the remainder after dividing $\mathcal{U}_t$ by 2π. The linear process $\mathcal{U}_t$ can be decomposed as $\mathcal{U}_t = \theta_t + 2\pi k_t$, where k_t are unobservable integers. The process $\mathcal{U}_t$ follows the AR model

$$\mathcal{U}_t = \varphi_1 \mathcal{U}_{t-1} + \ldots + \varphi_p \mathcal{U}_{t-p} + \varepsilon_t, \tag{8.7}$$

where $\{\varepsilon_t\}$ is a Gaussian white noise.

As suggested in [18], a suitable solution to the missing data problem when fitting a wrapped AR model is to use the EM algorithm. Details for the low-order AR case, where the algorithm has a reasonably simple form, can be found in [18] and [20]. Note that the AR model in (8.7) does not include a constant term, in order to avoid identifiability issues; that is the process $\mathcal{U}_t$ is assumed to be centered.

An other line of research leads to the class of linked AR models, which are designed for time series with low dispersion. This approach consists of transforming the angles using a link function that projects the circle onto a straight line. Let $g(\cdot)$ be an invertible function defined on $[-\pi, \pi]$ and taking values in $\mathbb{R}$. A direct linked model is then defined as:

$$g(\theta_t - \mu) = \sum_{j=1}^{p} \phi_j g(\theta_{t-j} - \mu) + \varepsilon_t.$$

where μ is the mean direction.

The conditional mean of θ_t given the past can be obtained by mapping the result back to the circular domain using the link function

$$E(\theta_t | \theta_{t-1}, \ldots, \theta_{t-p}) = \mu + g^{-1}\left\{\sum_{j=1}^{p} \phi_j g(\theta_{t-j} - \mu)\right\}.$$

This formulation is often referred to as an inverse linked or circular AR model.

Two main forms of the link function have been proposed: the tangent link $g(\theta) = \tan\{\theta/(2+c)\}$, where the positive constant c is added to avoid excessively large responses to angles near $\pm\pi$, and the inverse Normal $g(\theta) = \Phi^{-1}\left(\frac{\theta}{2\pi} + \frac{1}{2}\right)$, where $\Phi(\cdot)$ denotes the cumulative distribution function of the standard Gaussian distribution. We note that, over a wide range of angles, the values of the inverse normal link function are very well approximated by the tangent link function with $c \approx 0.75$.

An alternative approach that partially avoids circularity is the projected or component model. This method consists of transforming the pair of circular and linear variables, expressed in polar coordinates, into Cartesian coordinates as follows:

$$c_x(t) = x_t \sin(\theta_t - \theta_0) \ , \ c_y(t) = x_t \cos(\theta_t - \theta_0).$$

where θ_0 is the mean direction. Linear AR models are then fitted separately to $c_x(t)$ and $c_y(t)$. Based on the forecasted components $\tilde{c}_x(t)$ and $\tilde{c}_y(t)$, the corresponding forecasts for the circular and the linear components are obtained as: $\tilde{\theta}(t) = \tan^{-1}\{\tilde{c}_x(t)/\tilde{c}_y(t)\} + \theta_0$ and $\tilde{x}(t) = \sqrt{\tilde{c}_x(t)^2 + \tilde{c}_y(t)^2}$.

8.4 Multi-regime Models

We consider AR multi-regime models, in which the order and parameters of the models are allowed to vary across each of a fixed (finite) number m of subsets of observations, referred to as regimes.

Conditional on being in regime $S_t = j$ at time t (where $j = 1, \dots, m$), we assume that:

$$y_t = \phi_0^{(j)} + \phi_1^{(j)} y_{t-1} + \dots + \phi_{p(j)}^{(j)} y_{t-p^{(j)}} + \beta_1^{(j)} \cos\theta_t + \beta_2^{(j)} \sin\theta_t + \epsilon_t \tag{8.8}$$

as in (8.4), and

$$\mathcal{U}_t = \varphi_1^{(j)} \mathcal{U}_{t-1} + \dots + \varphi_{q(j)}^{(j)} \mathcal{U}_{t-q^{(j)}} + \varepsilon_t, \tag{8.9}$$

where $\theta_t = \mathcal{U}_t(\bmod 2\pi)$, $p^{(j)}$ and $q^{(j)}$ are the respective AR orders, and ϵ_t and ε_t are Gaussian white noise processes with variances $\sigma^2_{\epsilon(j)}$ and $\sigma^2_{\varepsilon(j)}$, respectively.

If we denote by $\mathcal{A}_t = \{y_{t-1}, \dots, y_{t-p^{(j)}}, \theta_t\}$ and $\mathcal{B}_t = \{\theta_{t-1}, \dots, \theta_{t-q^{(j)}}\}$ the information available at time t, and by $\boldsymbol{\xi}^{(j)} = \left(\phi_0^{(j)}, \dots, \phi_{p(j)}^{(j)}, \beta_1^{(j)}, \beta_2^{(j)}, \sigma^2_{\epsilon(j)}\right)$ and $\boldsymbol{\zeta}^{(j)} = \left(\varphi_0^{(j)}, \dots, \varphi_{q(j)}^{(j)}, \sigma^2_{\varepsilon(j)}\right)$ the parameter vectors, the conditional densities are

$$f_j(y_t|\mathcal{A}_t, \boldsymbol{\xi}^{(j)}) = (2\pi\sigma^2_{\epsilon(j)})^{-1/2} \cdot \exp\left\{ -\frac{1}{2\sigma^2_{\epsilon(j)}} \left(y_t - \phi_0^{(j)} - \sum_{h=1}^{p^{(j)}} \phi_h^{(j)} y_{t-h} + \beta_1^{(j)} \cos\theta_t + \beta_2^{(j)} \sin\theta_t \right)^2 \right\} \tag{8.10}$$

and

$$f_j(\theta_t|\mathcal{B}_t, \boldsymbol{\zeta}^{(j)}) = (2\pi\sigma^2_{\varepsilon(j)})^{-1/2} \sum_{k-\infty}^{\infty} \exp\left\{ -\frac{1}{2\sigma^2_{\varepsilon(j)}} \left(\theta_t - \sum_{h=1}^{q^{(j)}} \varphi_h^{(j)} \theta_{t-h} - 2\pi k \right)^2 \right\}. \tag{8.11}$$

To select the optimal model in terms of the number of regimes and model orders, we rely on the Akaike Information Criterion (AIC)

$$AIC = -2\log\hat{L} + 2\,np \tag{8.12}$$

where $\hat{L}$ is the maximized likelihood and np is the number of parameters in the model.

In this setting, we are often concerned with forecasting the time series at $T + h$, when we are currently standing at T. This forecast is said to be made at origin T for lead time h.

The minimum mean square error forecasts at lead time h, denoted $\theta_T(h)$ and $y_T(h)$, can be directly generated by models (8.9) and (8.8). These forecasts can be recursively built from previous values as:

$$\mathcal{U}_T(h) = \varphi_1^{(j)}\mathcal{U}_T(h-1) + \ldots + \varphi_{q^{(j)}}^{(j)}\mathcal{U}_T(h-q^{(j)}) \tag{8.13}$$

with $\theta_T(h) = \mathcal{U}_T(h)(\bmod 2\pi)$ and

$$y_T(h) = \phi_0^{(j)} + \phi_1^{(j)} y_T(h-1) + \ldots + \phi_{p^{(j)}}^{(j)} y_T(h-p^{(j)}) + \beta_1^{(j)} \cos\theta_T(h) + \beta_2^{(j)} \sin\theta_T(h) \tag{8.14}$$

where $\theta_T(i) = \theta_{T+i}$ and $y_T(i) = y_{T+i}$ when $T+i < n$; that is, forecasts at times within the observed period are equal to the corresponding observed values.

The forecasts depend directly on the model parameters of the regime corresponding to time $T+h$, but also indirectly on the regimes of previous time points, through the forecasts at lead times smaller than h.

8.4.1 Hidden Markov Models

A hidden Markov model involves an m-state homogeneous Markov chain $\{S_t, t = 1, \ldots, n\}$, defined by its initial state distribution $\{P(S_1 = 1), \ldots, P(S_1 = m)\}$ and a transition probability matrix $P(S_t = j \mid S_{t-1} = i)$, where $i, j = 1, \ldots, m$.

We consider several model specifications that differ both in the assumed joint distribution of the circular and linear components and in the aspects deemed relevant for regime classification.

In all cases, the likelihood function can be written as:

$$L = \sum P(S_1 = s_1) f(\theta_1, x_1 \mid S_1 = s_1) \prod_{t=2}^{n} P(S_t = s_t \mid S_{t-1} = s_{t-1}) f(\theta_t, x_t \mid S_t = s_t) \tag{8.15}$$

where the summation is over all possible sequences of hidden regimes $(s_1, s_2, \ldots, s_n)$, with each s_t taking values in the hidden state space $\{1, 2, \ldots, m\}$. For simplicity of notation, we omit the dependence on the parameter set, which includes the initial state distribution, the elements of the transition probability matrix, and the parameters of the cylindrical variable densities.

In the following, we adopt two different approaches. In the first, which we refer to as the static approach, regime classification is based on a hidden Markov model where observations are assumed to be conditionally independent given the latent regimes. In this case, the model is specified using either the Mardia–Sutton (8.4) or Abe–Ley (8.5) distribution for the joint distribution of the circular and linear variable conditional on the regime. Models (8.8) and (8.9) are then fitted to the observations assigned to each regime, and this fitted models are used for forecasting. We note that when, as in our example, the linear variable x takes positive real values, the model with the Mardia–Sutton distribution is specified on $y = \log x$.

In the second approach, which we refer to as the optimal approach, we assume that each term $f(\theta_t, x_t \mid S_t = s_t)$ in the likelihood function (8.15) is also conditional on the past and equal to the product of expressions (8.10) and (8.11). In this case, model fitting simultaneously provides both regime classification and estimation of an AR model for each component.

Essentially, in the first approach, regime classification is based solely on the magnitude and relative values of the two observed variables, whereas in the second approach, the temporal dependence among observations also plays a key role.

In any case, to fit the model we rely on the EM algorithm, as proposed by [3].

Equations (8.13) and (8.14) provide the forecasts at origin T for lead time h, conditional on the regimes occupied at time $T+h$, and at previous time points involved in the formulas. The final forecast can be obtained as a weighted average of the forecasts across the different regimes, where the weights correspond to the joint probability that the chain occupies those regimes. These probabilities can be derived from the results of the EM algorithm used to fit the model.

8.4.2 Threshold Models

The threshold idea, originally introduced by [21] (see also [12]), essentially explains nonlinear dynamics through a piecewise linear AR structure. If $\{P_1, P_2, \ldots, P_m\}$ denotes a partition of the range of possible values of a (possibly multivariate) time series $\{z_t, t = 1, \ldots, n\}$, a threshold model is defined as an AR structure whose order and parameters vary across regimes. The process is said to be in regime j at time t if $z_{t-d} \in P_j$, where the pre-specified constant d is called the delay and $j = 1, \ldots, m$.

The first type of partition, which we refer to as the orthogonal threshold, is defined as follows, based on the possible values of the pair (θ, y): $P_1 = \{y \leq H\}$ and $P_j = \{y > H, V_{j-1} \leq \theta < V_j\}$ $(j = 2, \ldots, m)$, where H is a threshold in the range of the linear variable, and the sequence of values $V_1, \ldots, V_{m-1}, V_m$ (with $0 < V_1 \leq V_2 \leq \ldots V_{m-1} \leq 2\pi$ and $V_m = V_1$) defines a partition of the unit circle into $(m-1)$ arcs. Thus, the partition is defined by the horizontal boundary H and the $(m-1)$ lower endpoints $V_1, \ldots, V_{m-1}$ of the arcs. This partitioning is particularly suitable for applications involving wind, see waves or animal movement, as it defines one regime corresponding to low speed, regardless of direction, and one or more additional regimes associated with higher values of the linear variable and prevailing directions.

The second specification, referred to as the slanted threshold, defines the partition of the cylindrical surface by means of two intersecting lines. First, a reference value V is fixed on the circle as the origin of directions ($V \in (0, 2\pi]$). Then the surface is partitioned into up to four sub-regions. This partition is fully determined by V, the coordinates of the intersection point (I_1, I_2), and the slopes of the two lines, denoted by l_1 and l_2.

By increasing the vertical coordinate of the intersection point beyond the observed range of the linear component, the two lines become nearly parallel, resulting in a partition that is primarily based on direction. If one or more sub-regions contain no observations, they can be merged with the nearest adjacent region.

Given the partition, the bivariate observation (θ_{t-d}, y_{t-d}) determines the regime at time t.

Conditional on the past and on being in regime j at time t $(j = 1, \ldots, m)$, we assume that y_t has density (8.10), and that θ_t has density (8.11).

To fit a threshold model to the components of a cylindrical time series via maximum likelihood, under both the orthogonal and the slanted threshold specifications, we developed a procedure that performs a search over the space of candidate models. This procedure is based on a genetic algorithm and extends the approach proposed in [22] for the univariate case. Genetic algorithms are optimization methods designed to solve combinatorial problems where the solution space is discrete and large. An introduction to the theory and applications of genetic algorithms as optimization techniques can be found in [23], and a review of their applications in statistics is provided in [24].

We observed that, while the resulting partition is sensitive to changes in the number of regimes m, the delay d, and the model order within each regime, it is quite robust to small variations in the values of the threshold parameters. For this reason, and to simplify the algorithm, we decided to discretize the space of possible values for these parameters, which are, in principle, real-valued.

The algorithm relies on the AIC (8.12) to select the number of regimes, determine the corresponding partition of the two-dimensional space, choose the model order within each regime, and estimate the associated parameters. It is applied separately to the circular and linear components. As a result, the regimes identified for the linear component may differ from those identified for the circular component, although the corresponding partitions depend on both dimensions.

When using the slanted threshold specification, the algorithm also compares the resulting partition with those involving fewer than four regimes, obtained by merging adjacent regions.

The forecast can be computed directly using equations (8.13) and (8.14) for lead times h up to the delay d. When $h > d$, we refer to [7] for further details.

8.5 Application to Cylindrical Data From the Adriatic Sea

The circular and linear components of the series look heavily autocorrelated, even at large lags. They appear to be dependent: formula (8.6) results in a (zero-lag) correlation of 0.38.

The variance of the wave height is 0.39. We also observed that the circular component (wave direction) exhibit substantial variability (Figure 8.1). Its variance is 1.28, computed as:

$$var(\theta) = \frac{1}{n}\sum_{t=1}^{n} dist(\theta_t, \bar{\theta})^2, \tag{8.16}$$

that is, the average of the squared angular distances from their circular mean direction $\bar{\theta}$, where $dist(\theta_t, \bar{\theta})$ is defined in equation (8.3). The unwrapped data also suggest that the observations circulate around the full circle (Figure 8.3), supporting the use of a wrapped normal AR model, as suggested by [20]. Accordingly, we fitted a (fourth-order) wrapped AR model to the wave direction time series. As a byproduct, we obtained, for each time point t, the reconstructed values of the multiplicative constant k_t, which always coincide with those computed via equation (8.2). Furthermore, the wrapped AR model fitted to the original series via the EM algorithm closely approximates the AR model fitted directly to $\{u_t\}$.

Based on these results, in what follows, we define an unwrapped AR model as one fitted to the unwrapped time series $\{u_t\}$. After model fitting, the circular observations θ_t can be recovered by wrapping: $\theta_t = u_t(\bmod 2\pi)$. The same applies to residuals and forecasts.

After first differencing, the variance of wave direction is 0.194 (computed either using formula (8.16) or as the variance of the first-differenced unwrapped series), while the variance of both wave height and the log-height is 0.010.

An exploratory analysis of the scatter plot of the bivariate data in Figure 8.2(b) and the histograms of the observations for each variable in Figure 8.1 suggest the possible presence of at least two regimes, with the indication being more evident for the wave direction than for the wave height. In fact, this suggestion is confirmed by the results of the procedures introduced in the previous Sections, which we will discuss.

Five different classes of models are considered. Three of the classes refer to dynamic hidden Markov models, and they differ in the joint distribution of the circular and linear components and the strategy used for model fitting: the first two adopt the *static* procedure and assume the Abe-Ley (*Abe-Ley, static*) and the Mardia-Sutton (*Mardia-Sutton, static*) distributions; the third refers to the *optimal* approach (*optimal*). The last two classes contain threshold models using orthogonal (*Threshold, orthogonal*) or slanted thresholds (*Threshold, slanted*).

Models from the classes described, with both the number of regimes and the AR order in the set $\{1, 2, 3, 4\}$ were fitted to the data. In each case, we identified the optimal model using the AIC.

When considering threshold models, comparisons also included the choice of the delay d (in the set $\{1, \ldots, 7\}$) and the positions of the lines that delimit the partitions in the space of the two variables.

In order to properly initialize the EM algorithm to estimate the models in the HMM class, we randomly selected several starting values from a set of sensible parameter values and chose those corresponding to the minimum value of AIC.

For each of the classes considered, we report a summary of the results for the model identified by AIC and fitted to the circular component (Table 8.1) and the linear component (Table 8.2) of the cylindrical time series.

As a measure of fit we report the mean squared error (MSE), computed as the lead-1 forecast variance over the entire series, i.e., the variance of the residuals $z_t - \hat{z}_{t|past}$, where z_t is either the log-height or the direction of the waves and $\hat{z}_{t|past}$ is the estimate of the conditional mean (i.e., the one-step-ahead forecast) under the model for each time t. For the direction, the residuals are wrapped back around the circle after fitting the model to the unwrapped series $\{u_t\}$, and the variance is computed accordingly.

TABLE 8.1
Main summaries of models fitted to Ancona wave direction (AL stands for Abe-Ley, MS for Mardia-Sutton, orth and slant for orthogonal and slanted thresholds).

Model	Number of regimes	Observations in each regime	AR orders within regimes	MSE	$\bar{H}$
AL static	4	273, 278, 252, 193	3, 2, 2, 3	0.180	0.03
MS static	4	212, 257, 288, 239	4, 3, 2, 2	0.183	0.03
optimal	3	307, 428, 261	2, 2, 3	0.184	0.17
TAR orth d=6	4	176, 52, 490, 273	3, 4, 3, 2	0.181	
TAR slant d=7	3	185, 97, 711	1, 2, 2	0.186	

TABLE 8.2
Main summaries of models fitted to the log of Ancona wave height (AL stands for Abe-Ley, MS for Mardia-Sutton, orth and slant for orthogonal and slanted thresholds).

Model	Number of regimes	Observations in each regime	AR orders within regimes	MSE	$\bar{H}$
AL static	4	273, 278, 252, 193	4, 2, 4, 1	0.0098	0.03
MS static	4	212, 257, 288, 239	4, 2, 3, 2	0.0097	0.03
optimal	3	307, 428, 261	3, 2, 2	0.0093	0.17
TAR orth d=6	4	696, 58, 48, 191	4, 4, 2, 3	0.0094	
TAR slant d=5	4	264, 384, 124, , 223	3, 2, 2, 4	0.0093	

We note that the number of identified regimes is almost always four. However, it is three in the solution obtained using the hidden Markov AR model fitted via the optimal procedure, as well as in the slanted threshold model for wave direction. In most cases, the fitted AR models have an order greater than one, suggesting that intra-regime autocorrelation should be taken into consideration.

As reported in Tables 8.1 and 8.2, the three HMM classes considered yield MSE values in the range of (0.18, 0.19) when fitted to the wave direction, and slightly below 0.01 when fitted to the log-height. In contrast, the MSE of the HMM model based on the Abe-Ley joint distribution with serial independence is 0.784 for the direction and 0.246 for the log-height. Thus, the switching AR models provide a better fit to the data than the switching models without an AR component. However, they seem to offer little improvement in fit compared to the random walk model.

The solutions obtained through the different strategies proposed appear to be quite similar, although the distinctive features of each can still be highlighted.

The splitting suggested by each of the three HMM strategies appears to be strong, as indicated by the very small values of the average relative entropy:

$$\bar{H} = -\frac{1}{n}\sum_{t=1}^{n}\sum_{j=1}^{m} P(S_t = j)\log\{P(S_t = j)\}/\log(m),$$

a measure of the uncertainty in assigning each observation to a regime, where $P(S_t = j)$ is the probability of the observation at t being classified in regime j. In fact, at most time points t, one of the estimated $P(S_t = j)$ is close to 1, while the others are close to zero. In particular, for the two *static* procedures, more than 92% of the observations have one of the $P(S_t = j)$ values greater than 0.95, while in less than 1% of the observations, the largest value of $P(S_t = j)$ with respect to j is between 0.5 and 0.6.

The partitions induced by the Abe-Ley and Mardia-Sutton static models are largely consistent, with only minor differences in the probabilities of being classified into each component.

If we classify each observation according to the state with the highest probability, we find that each of the four regimes corresponds to a distinct weather condition. One regime corresponds to calm sea conditions: wave height is low, and direction varies widely around the circle. The other three regimes are associated with active wave motion, characterized by higher waves and more concentrated wave directions over specific arcs. Within this second group, one regime is associated with higher waves predominantly coming from the northeast. The remaining two regimes correspond to waves coming from the southeast and are distinguished by wave height: in one regime, the waves are lower; in the other, they are higher and show greater variability. This regime configuration is consistent with the findings of [5], where a hidden Markov model based on the joint Abe-Ley distribution, assuming conditional independence given the latent regime, was applied to the full time series.

The fact that both the Abe-Ley and Mardia-Sutton static models are suitable and largely equivalent for this series is further supported by the Kullback–Leibler divergence between the empirical joint distribution in each regime and the corresponding theoretical distributions, as reported in Table 8.3.

Regarding the hidden Markov model using what we refer to as the *optimal* strategy, we observe a degree of temporal inertia in switching between regimes. As a result, observations near the boundaries of neighboring regimes may exhibit less polarized probabilities, leading to higher entropy values. In this case, only slightly more than 67% of the observations are classified into a specific regime with a probability greater than 0.95. Nonetheless, the highest value of $P(S_t = j)$ is below 0.6 in only 4% of cases. Under this specification, the regimes exhibit a slightly different configuration, although they can still be broadly characterized as periods of sea calm or sea movement in a specific direction. If, as before, we classify the observations into the most likely regime, we obtain the plots shown in

TABLE 8.3
Empirical Kullback-Leibler divergence from the theoretical distribution in each regime identified on the Ancona time series.

	Regime			
Model	**1**	**2**	**3**	**4**
Abe-Ley static	0.368	0.165	0.295	0.366
Mardia-Sutton static	0.257	0.370	0.338	0.288

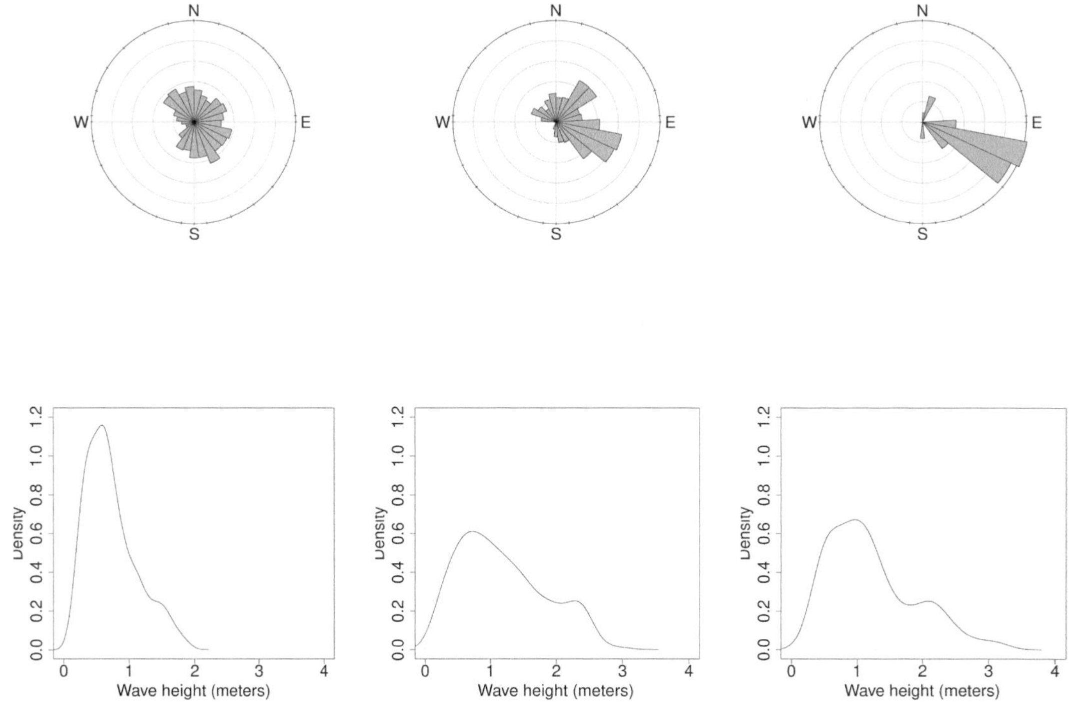

FIGURE 8.4
Rose diagrams for Ancona wave direction and empirical density plots for wave height in the regimes identified using the optimal procedure: the left panels refer to regime 1, the middle panels to regime 2, and the right panels to regime 3. Observations are classified according to the most likely regime. North corresponds to zero radians, east to $\pi/2$, south to π, and west to $3\pi/2$.

Figure 8.4. The first regime corresponds to low wave height and low temporal variability and can occur in any wave direction. The last two regimes are associated with higher wave height and greater variability, predominantly in the northeast and southeast directions. As illustrated in Figure 8.5, in the second regime, wave height tends to increase, while in the third, it generally decreases. Since these patterns are not always sharply defined, the procedure may exhibit some uncertainty when a local change in trend occurs, resulting in a lower maximum classification probability. Likewise, even during calm sea conditions, local increases or decreases in wave height may lead to similarly uncertain classifications, with no single value of $P(S_t = j)$ approaching 1.

The differences compared to the previous results are related to how the procedures operate. In the Abe-Ley and Mardia-Sutton static models, the classification of observations

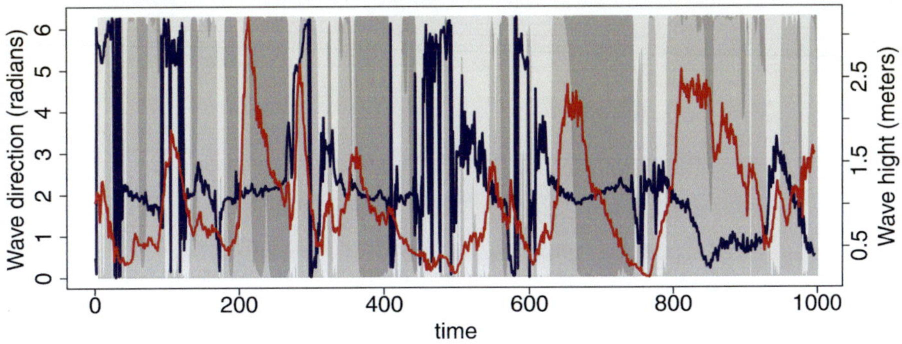

FIGURE 8.5
Ancona wave direction (blue) and height (red). Shades of gray indicate the probability of each regime for the model fitted using the optimal approach. The regimes are ordered according to the average wave height (the darker the shade, the higher the average).

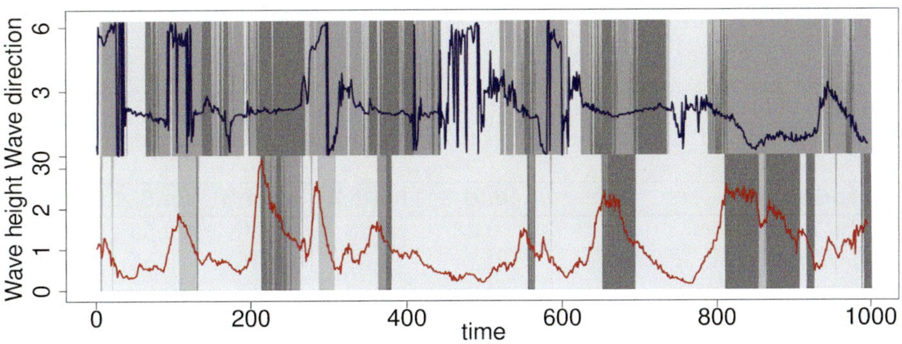

FIGURE 8.6
Ancona wave direction (radians—blue) and height (meters–red). The shades of gray indicate the subdivision of the observations into regimes, as determined by fitting the threshold AR model with slanted thresholds.

occurs prior to accounting for autocorrelation within each regime. As a result, the regimes are primarily characterized by the magnitude and variability of the observations. In contrast, the optimal procedure also incorporates serial correlation when assigning observations to different regimes.

The regimes identified by estimating a threshold AR model, both under the orthogonal and slanted specifications, are largely comparable to those obtained using the optimal approach. The main difference is that, in this case, the classification of the linear and circular components may differ, as illustrated in Figure 8.6. The shades of gray indicate the classification of observations into regimes, as determined by fitting the threshold AR model with slanted thresholds.

The orthogonal threshold model fitted to wave direction defines a partition of the two-dimensional space (θ, y), where y is the logarithm of wave height, using the values $H = 0.47$, $V_1 = 1.89$, $V_2 = 2.26$, and $V_3 = 5.64$. The model fitted to wave height uses $H = 1.25$, $V_1 = 0.37$, $V_2 = 2.11$, and $V_3 = 5.93$. When the slanted threshold model is applied to wave direction, the estimate of V is 5.54, and it corresponds to the origin of the θ-axis when the

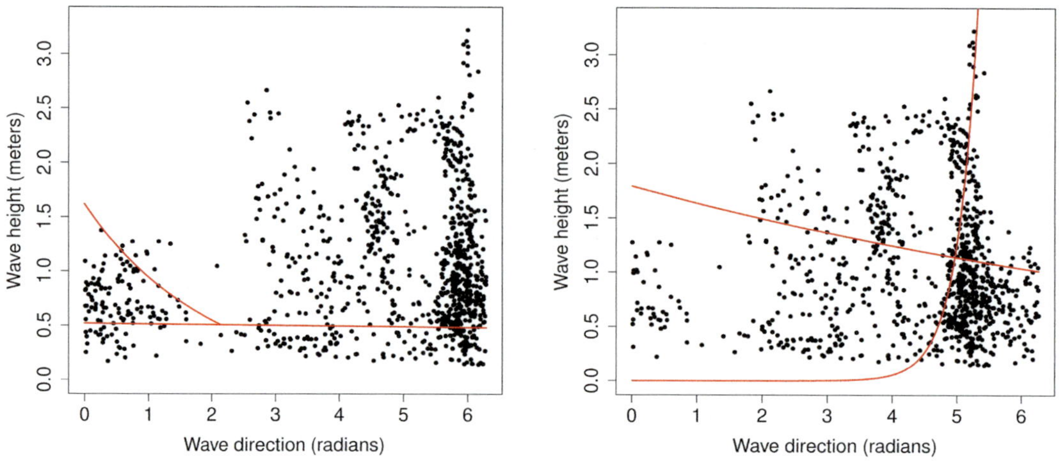

(a) Model fitted to wave direction (b) Model fitted to wave height

FIGURE 8.7
Partition of the (θ, x) space resulting from the slanted threshold AR model fitted to wave direction and wave height (with the circular origin redefined accordingly).

TABLE 8.4
Mean square error of out of sample forecasts of Ancona wave direction.

Model	**lead 1**	**lead 2**	**lead 3**	**lead 4**
Abe-Ley, static	0.23	0.38	0.46	0.55
Mardia-Sutton, static	0.24	0.38	0.46	0.55
optimal	0.24	0.39	0.48	0.58
Threshold, orthogonal $d = 6$	0.23	0.37	0.46	0.54
Threshold, slanted $d = 7$	0.23	0.37	0.44	0.52

model is applied to wave height. This value represents the new origin on the circle. The partition of the (θ, y) space is defined by lines passing through the point $(2.15, -0.68)$, with slopes -0.55 and -0.012 for wave direction, and through the point $(4.98, 0.13)$, with slopes 3.11 and -0.09 for wave height.

In Figure 8.7, we show the partition of the (θ, x) space resulting from the slanted threshold AR model fitted to wave direction and wave height, after adjusting the circular origin. In the plot on the left, the partition consists of only three regions, as the algorithm merged the two adjacent lower regions.

To compare the forecasting performance of the models, we compute the average squared l-step-ahead forecast errors on the holdout sample, for $l = 1, 2, 3, 4$. The results for wave direction are reported in Table 8.4, and those for the log-height in Table 8.5. These results show that the threshold models, especially the slanted threshold specification, produce more accurate forecasts than the HMM models, although the differences are generally not very large.

TABLE 8.5
Mean square error of out of sample forecasts of Ancona wave height.

Model	**lead 1**	**lead 2**	**lead 3**	**lead 4**
Abe-Ley, static	0.011	0.022	0.037	0.053
Mardia-Sutton, static	0.012	0.024	0.041	0.061
optimal	0.011	0.020	0.033	0.046
Threshold, orthogonal $d = 6$	0.011	0.022	0.036	0.053
Threshold, slanted $d = 5$	0.011	0.020	0.032	0.044

8.6 Concluding Remarks

Cylindrical time series usually exhibit considerable autocorrelation, and the simple Markovian mechanism of HMMs may not be sufficient to capture this feature. To account for this aspect, we consider the possibility of using a dynamic model within each regime. For the linear component, a sensible choice is an AR model. Simple and manageable models for the circular component proposed in the literature include wrapped, linked, and projected AR models. These generally yield similar results.

As far as the choice of regimes is concerned, allowing for time dynamics implies that regimes may differ not only in location and variability, but also in their dynamic structure. For this reason, we have considered HMMs where the regimes differ in the parameters of the AR models.

Valid alternatives to dynamic HMMs are threshold AR models. We propose using a bivariate threshold based on the cylindrical delayed observations. In our applications, the threshold models provide a slightly better fit than the HMMs with some advantage in terms of forecast accuracy.

8.7 Acknowledgments

We are grateful to Francesco Lagona for providing the Ancona data used in [5].

The third author acknowledges financial support under MIUR grant number 2022XRHT8R—The SMILE project: Statistical Modelling and Inference for Living the Environment.

Bibliography

[1] K.V. Mardia and P.E. Jupp. *Directional Statistics.* Wiley Series in Probability and Statistics. Wiley, Chichester, 2000.

[2] A. Pewsey and E. García-Portugués. Recent advances in directional statistics. *Test*, 30(1):1–58, 2021.

[3] W. Zucchini, I.L. MacDonald, and R. Langrock. *Hidden Markov models for time series: an introduction using R, 2nd edition.* Chapman and Hall/CRC, 2016.

[4] H. Holzmann, A. Munk, M. Suster, and W. Zucchini. Hidden Markov models for circular and linear-circular time series. *Environmental and Ecological Statistics*, 13:325–347, 2006.

[5] F. Lagona, M. Picone, and A. Maruotti. A hidden Markov model for the analysis of cylindrical time series. *Environmetrics*, 26(8):534–544, 2015.

[6] G. Mastrantonio, A. Maruotti, and G. Jona-Lasinio. Bayesian hidden Markov modelling using circular-linear general projected normal distribution. *Environmetrics*, 26(2):145–158, 2015.

[7] H. Tong. *Non-linear time series: a dynamical system approach.* Oxford university press, 1990.

[8] P. Ailliot and V. Monbet. Markov-switching autoregressive models for wind time series. *Environmental Modelling & Software*, 30:92–101, 2012.

[9] P. Ailliot, J. Bessac, V. Monbet, and F. Pene. Non-homogeneous hidden Markov-switching models for wind time series. *Journal of Statistical Planning and Inference*, 160:75–88, 2015.

[10] P. Pinson, L.E.A. Christensen, H. Madsen, P.E. Sørensen, M.H. Donovan, and L.E. Jensen. Regime-switching modelling of the fluctuations of offshore wind generation. *Journal of Wind Engineering and Industrial Aerodynamics*, 96(12):2327–2347, 2008.

[11] H. Chen, T. Chong, and J. Bai. Theory and applications of TAR model with two threshold variables. *Econometric Reviews*, 31:142–170, 2012.

[12] R.S. Tsay. Testing and modeling multivariate threshold models. *Journal of the American Statistical Association*, 93(443):1188–1202, 1998.

[13] J.H. Holland. *Adaptation in Natural and Artificial Systems.* University of Michigan Press, Ann Arbor, 1975.

[14] K. Itoh. Analysis of the phase unwrapping algorithm. *Applied optics*, 21(14):2470–2470, 1982.

[15] KV Mardia and TW Sutton. A model for cylindrical variables with applications. *Journal of the Royal Statistical Society Series B: Statistical Methodology*, 40(2):229–233, 1978.

[16] T. Abe and C. Ley. A tractable, parsimonious and flexible model for cylindrical data, with applications. *Econometrics and Statistics*, 4:91–104, 2017.

[17] S. R. Jammalamadaka and A. SenGupta. *Topics in Circular Statistics*, volume 5 of *Series on Multivariate Analysis.* World Scientific, Singapore, 2001.

[18] J. Breckling. *The analysis of directional time series: applications to wind speed and direction*, volume 61. Springer-Verlag, New York, 1989.

[19] E. Erdem and J. Shi. ARMA based approaches for forecasting the tuple of wind speed and direction. *Applied Energy*, 88(4):1405–1414, 2011.

[20] N.I. Fisher and A.J. Lee. Time series analysis of circular data. *Journal of the Royal Statistical Society: Series B (Methodological)*, 56(2):327–339, 1994.

[21] H. Tong and K.S. Lim. Threshold autoregression, limit cycles and cyclical data. *Journal of the Royal Statistical Society: Series B (Methodological)*, 42(3):245–268, 1980.

[22] R. Baragona, F. Battaglia, and D. Cucina. Fitting piecewise linear threshold autoregressive models by means of genetic algorithms. *Computational Statistics & Data Analysis*, 47:277–295, 2004.

[23] D. E. Goldberg. *Genetic algorithms in search, optimization and machine learning.* Addison-Wesley, Reading, MA, 1989.

[24] R. Baragona, F. Battaglia, and I. Poli. *Evolutionary Statistical Procedures.* Series Statistics and Computing. Springer-Verlag, Berlin, 2011.

9

A Unified Approach to Optimal Model-Based Detection of a Change-Point in the Mean Direction of Circular Distributions

Ashis SenGupta and Arnab K. Laha

This chapter addresses the retrospective detection of a single change-point in the mean direction parameter of the von Mises (circular normal) distribution, a model widely used for circular data. The focus is on the scenario where at most one change in mean direction is suspected, and some or all nuisance parameters may be unknown. The problem is non-regular due to the parameter space structure. We discuss likelihood-based and Neyman-Rao type tests (NRTT), with focus on the change point problem for the von Mises distribution. Two real-world examples are used to illustrate the approach. A derivation of the NRTT for the Wrapped Cauchy distribution is also provided. The chapter concludes with remarks on possible extensions in change-point detection for circular data.

9.1 Introduction

Detecting abrupt changes in the distribution of temporally ordered independent observations—a problem first formalized by Page [1955]—arises in many real-life applications. In manufacturing, for example, sudden shifts in process parameters can lead to increased defects and costs, necessitating rapid detection of such changes. While industrial settings often require online (real-time) detection, retrospective (off-line) change point analysis is particularly relevant in scientific and environmental contexts, such as identifying shifts in climate regimes [Reeves et al., 2007] or detecting abrupt changes in hydrological time series [Rougé et al., 2013].

Beyond these domains, change point detection plays a crucial role in finance, where it is used to identify structural breaks in financial markets [Chen and Gupta, 2011], and in healthcare, where it enables early detection of critical changes in patient monitoring data [Assareh et al., 2011]. Other important applications include speech segmentation in audio processing, video surveillance, cybersecurity, and human activity recognition using wearable sensors. These diverse applications underscore the broad utility of change point methods for both retrospective and real-time data analysis across scientific, industrial, and technological domains.

DOI: 10.1201/9781003564423-9

Statistical quality control methods such as control charts, cumulative sum (CUSUM) charts, and exponentially weighted moving average (EWMA) charts are commonly employed to detect such abrupt changes. Change point detection has been extensively studied within both parametric and nonparametric statistical frameworks [Chen and Gupta, 2000, Chernoff and Zacks, 1964, Hinkley, 1970, Sen and Srivastava, 1973, Siegmund, 1988, Truong et al., 2020].

The problem of change point detection is especially relevant for *directional* (or *circular*) data, which arise naturally in many scientific and applied contexts. Unlike linear data, directional data take values on a circle (e.g., angles in $[0, 2\pi)$), making standard change point methods inapplicable without appropriate adaptation. The following examples illustrate the importance and application of change point detection in such settings.

In meteorology and oceanography, wind direction represents a classic example of circular data. Detecting abrupt changes in wind direction is crucial for understanding weather fronts, cyclonic events, and transitions between atmospheric regimes. During the passage of a cyclone or frontal system, significant shifts in wind direction can indicate the onset of new flow regimes, which are often associated with changes in sea surface temperature and pressure patterns [Reddem et al., 2023]. Accurate detection of these changes aids in weather forecasting and enhances understanding of air-sea interactions.

Animal movement studies frequently involve tracking the direction of movement over time using GPS or other tracking devices. Change point analysis applied to these angular trajectories enables researchers to identify behavioral transitions, such as the onset of migration, changes in foraging strategy, or responses to environmental cues [Gurarie et al., 2009]. For example, in migratory birds, change points in movement direction can correspond to departure from or arrival at stopover sites, providing insights into migration strategies and habitat utilization patterns.

Many biological processes exhibit inherently periodic or cyclical behavior, including circadian rhythms, seasonal flowering patterns, and daily activity cycles in animals. In such systems, the timing or phase of biological events can be naturally represented as angular data on a circle, where each point corresponds to a specific time within the cycle (e.g., time of day mapped to angles from 0 to 2π). Change point detection in these circular time series enables researchers to identify abrupt shifts in biological timing that may result from environmental perturbations, disease onset, pharmacological interventions, or other external factors. For instance, detecting a sudden change in the phase of the circadian gene expression could indicate disruption of the internal biological clock, while shifts in the timing of flowering events might reveal responses to climate change or seasonal environmental variations.

These examples underscore the broad applicability of change point detection for circular data, which is essential for revealing regime shifts and behavioral transitions across a variety of scientific disciplines. Despite its importance, parametric inference for change-points in circular data has received comparatively less attention. This chapter primarily investigates the change-point problem for the mean direction parameter μ in the von Mises (vM) distribution, defined by the density

$$f(\theta; \mu, \kappa) = \frac{1}{2\pi I_0(\kappa)} \exp\left(\kappa \cos(\theta - \mu)\right),$$

where $0 \leq \theta < 2\pi$, $0 \leq \mu < 2\pi$, and $\kappa > 0$. The hypothesis of interest is as follows.

- H_0: All observations $\theta_1, \ldots, \theta_n$ are $\text{vM}(\mu_0, \kappa)$.
- H_1: There exists a change-point r $(1 \leq r \leq n-1)$ such that $\theta_1, \ldots, \theta_r$ are $\text{vM}(\mu_0, \kappa)$ and $\theta_{r+1}, \ldots, \theta_n$ are $\text{vM}(\mu_1, \kappa)$

This can be re-expressed as $H_0 : \delta = 0$ versus $H_1 : \delta > 0$, where $\mu_1 = \mu_0 + \delta \pmod{2\pi}$, $0 < \delta < 2\pi$. The presence of nuisance parameters μ_0, κ, and r complicates inference.

Early work in the circular data set-up includes generalized likelihood ratio and Bayesian procedures [SenGupta and Laha, 2008,].

The chapter is organized as follows: Section 9.2 reviews likelihood-based methods for the case of known κ, with various scenarios for known and unknown directions. Section 9.3 presents the Neyman-Rao type test (NRTT) for the general case with all nuisance parameters unknown. Section 9.4 illustrates the methods with two real datasets. Section 9.5 discusses the construction of the NRTT for the Wrapped Cauchy distribution. Section 9.6 discusses the key contributions of this paper and concludes.

9.2 Likelihood-Based Tests

This section explores likelihood-based approaches for detecting a change-point in the mean direction of the von Mises distribution, assuming the concentration parameter κ is known. Several cases are considered.

9.2.1 Change Point for the Mean Direction—All Parameters Known

Let $\theta_1, \theta_2, \ldots, \theta_n$ be independent random variables. Here, the initial direction μ_0, the changed direction μ_1, and the concentration parameter κ are all assumed known (without loss of generality, μ_0 is set to zero).

Define

$$\nu = E[\sin(\theta_i - \frac{\mu_1}{2})] = -\sin\frac{\mu_1}{2}A(\kappa),$$

$$\tau^2 = V[\sin(\boldsymbol{\theta}_i - \frac{\mu_1}{2})] = \cos^2\frac{\mu_1}{2} - \frac{I_0''(\kappa)}{I_0(\kappa)}\cos\mu_1 - \sin^2\frac{\mu_1}{2}A^2(\kappa)$$

where $A(\kappa) = I_0'(\kappa)/I_0(\kappa)$. Let $S_l^k(\eta, \alpha) = \sum_{i=l}^k \sin(\eta_i - \alpha)$ and $C_l^k(\eta, \alpha) = \sum_{i=l}^k \cos(\eta_i - \alpha)$. Denote by $B_0^*(t)$ the time-reversed Brownian motion on $[0, 1]$ with drift 0 and diffusion coefficient 1.

Theorem 9.1 *For testing H_0 against H_1, the likelihood ratio test (LRT) statistic is*

$$\tilde{\Lambda} = \max_r S_{r+1}^n(\theta, \frac{\mu_1}{2}).$$

The asymptotic null distribution of

$$\max_r \frac{S_{r+1}^n(\theta, \frac{\mu_1}{2}) - (n-r)\nu}{n^{1/2}\tau}$$

is the same as that of $\sup_{0\leq t\leq 1} B_0^*(t)$.

Proof *Let $\theta_1, \ldots, \theta_n, 0 \leq \theta_i < 2\pi$ be the given observations. We apply the Likelihood ratio test. Fix r, $1 \leq r \leq n-1$. Under H_0, the likelihood is*

$$f_0(\theta_1, \ldots, \theta_n) = (2\pi I_0(\kappa))^{-n}\exp(\kappa C_1^n(\theta, 0))$$

and under the alternative H_{1r}, there is a change point at r and we have

$$f_1(\theta_1, \ldots, \theta_n) = (2\pi I_0(\kappa))^{-n}\exp(\kappa(C_1^r(\theta, 0) + C_{r+1}^n(\theta, \mu_1))$$

Then, the test statistic for fixed r is

$$\Lambda_r = \exp(\kappa[C^n_{r+1}(\theta, 0) \;-\; C^n_{r+1}(\theta, \mu_1)])$$

Since r is unknown, we estimate it by $\hat{r}$ where $\hat{r}$ is that r for which Λ_r is minimum. Define $\Lambda = \Lambda_{\hat{r}}$. Then, the rule is to reject H_0 if $\Lambda < c$ for some constant c depending on the level of significance. Note that the test $\Lambda < c$ is equivalent to

$$S^n_{\hat{r}+1}(\theta, \frac{\mu_1}{2}) > k.$$

Let

$$\widetilde{\Lambda} = \max_r \; S^n_{r+1}(\theta, \frac{\mu_1}{2})$$

Also, note that under the null hypothesis, the random variables $\sin(\theta_i - \frac{\mu_1}{2})$ are i.i.d with finite expectation ν and finite variance τ^2. Then, by an application of the functional central limit theorem [Bhattacharya and Waymire, 1992], we get that

$$\frac{S^n_{r+1}(\theta, \frac{\mu_1}{2}) - (n-r)\nu}{n^{1/2}\tau}$$

converges in distribution (in the Skorohod topology sense on D) to the time-reversed Brownian motion in the interval [0,1] with drift 0 and diffusion coefficient 1 as $n \to \infty$. Since sup is a continuous function on $D[0,1]$ with the Skorohod topology, we see that under the null hypothesis

$$\max_r \frac{S^n_{r+1}(\theta, \frac{\mu_1}{2}) - (n-r)\nu}{n^{1/2}\tau}$$

$$= \sup_{\frac{1}{n} \le t \le 1 - \frac{1}{n}} \frac{S^n_{[nt]+1}(\theta, \frac{\mu_1}{2}) - (n-[nt])\nu}{n^{1/2}\tau}$$

which converges in distribution to $\sup_{0 \le t \le 1} B^*_0(t)$ as $n \to \infty$.

Remark 9.1 *The exact sampling distribution of $\tilde{\Lambda}$ is complex, but cut-off points and p-values can be approximated using the asymptotic result:*

$$P\left(\sup_{0 \le t \le 1} B^*_0(t) \ge \alpha\right) = 2P(Z \ge \alpha),$$

where $Z \sim N(0,1)$ and $\alpha > 0$.

An alternative statistic with a tractable asymptotic distribution is as follows:

$$u(n,t) = \sqrt{2\ln\ln(n-1)} + \frac{\ln\ln\ln(n-1)}{2\sqrt{2\ln\ln(n-1)}} + \frac{t}{\sqrt{2\ln\ln(n-1)}}.$$

Theorem 9.2 *Let*

$$U = \max_{1 \le r \le n-1} \frac{S^n_{r+1}(\theta, \frac{\mu_1}{2}) - (n-r)\nu}{(n-r)^{1/2}\tau}.$$

Then, as $n \to \infty$,

$$P(U < u(n,t)) \to \exp\left(-\frac{e^{-t}}{2\sqrt{\pi}}\right).$$

Proof *The proof follows immediately on application of the results in Darling and Erdos [1956].*

9.2.2 Change Point for the Mean Direction—Changed Direction Unknown

Now, consider the scenario where μ_1 is unknown. Let $\sigma^2 = V(\sin\theta_i) = 1 - \frac{I_0''(\kappa)}{I_0(\kappa)}$ under H_0.

Theorem 9.3 *(a) For testing H_0 against H_{1r}, the locally most powerful (LMP) test rejects H_0 if $S_{r+1}^n(\theta,0) > c$ for some constant c.*

(b) For testing H_0 against H_1 using $\max_r S_{r+1}^n(\theta,0)$, the asymptotic distribution under H_0 of

$$\max_r \frac{S_{r+1}^n(\theta,0)}{\sqrt{n}\sigma}$$

is the same as that of $\sup_{0\le t\le 1} B_0^*(t)$.

Proof *(a) Let us fix $r, 1 \le r \le n-1$. Let $\theta_1,\ldots,\theta_n$, $0 \le \theta_i < 2\pi$ denote the observations. Observe that the log-likelihood is*

$$l(\mu_1;\theta_1,\ldots,\theta_n) = \text{const} + \kappa\{C_1^r(\theta,0) + C_{r+1}^n(\theta,\mu_1)\}.$$

$$\text{Then, } \frac{dl}{d\mu_1} = \kappa S_{r+1}^n(\theta,\mu_1) \text{ and hence } \frac{dl}{d\mu_1}(0) = \kappa S_{r+1}^n(\theta,0).$$

Thus, the LMP test statistic for testing H_0 against H_{1r} is $S_{r+1}^n(\theta,0)$ (since κ is known). We reject H_0 if $S_{r+1}^n(\theta,0) > c$ for some constant c.
(b) Imitating the proof of Theorem 1 above, we get that under the null hypothesis, as $n \to \infty$, $\max_r \frac{S_{r+1}^n(\theta,0)}{\sqrt{n}\,\sigma}$ converges in distribution to $\sup_{0\le t\le 1} B_0^*(t)$.

Remark 9.2 *The LRT is computable but does not have a tractable asymptotic distribution in this case.*

9.2.3 Change Point for the Mean Direction—Both Directions Unknown

Here, both μ_0 and μ_1 are unknown. The LRT statistic is:

$$\Lambda = \min_r \Lambda_r, \quad \Lambda_r = \exp\left\{\kappa\left[C_1^n(\theta,\hat\mu_0) - C_1^r(\theta,\hat\mu_{0r}) - C_{r+1}^n(\theta,\hat\mu_{1r})\right]\right\},$$

where $\hat\mu_0$ is the MLE of μ_0 under H_0, and $\hat\mu_{0r}$ and $\hat\mu_{1r}$ are the MLEs of μ_0 and μ_1 under H_{1r}, respectively.

The exact sampling distribution of Λ is intractable. The large sample distribution of LRT does not work here because of the violation in the assumptions (for e.g., the parameter space here is not an open set) of the relevant theorem. If it can be assumed that there is no change point among the first n_0 observations (i.e., $n_0 < r$) where n_0 is sufficiently large, we can treat $\mu_0 = \hat\mu_0$ and obtain the (approximate) cut-off value using simulation.

9.3 Neyman-Rao Type Test

This section introduces the Neyman-Rao Type Test (NRTT), based on the asymptotically optimal Neyman-Rao test [Hall and Mathiason, 1990, Rao, 2005]. For $1 \le r \le n-1$, define H_{1r} as: $\theta_1, \dots, \theta_r$ are von Mises(μ_0, κ) and $\theta_{r+1}, \dots, \theta_n$ are von Mises(μ_1, κ), with all parameters unknown. Let $\hat{\mu}_2$ and R_2 be the estimated mean direction and resultant for $\theta_{r+1}, \dots, \theta_n$, and R for the full sample.

In what follows, since there is no cause for confusion, we will denote μ_0 simply by μ and μ_1 as $\mu+\delta$. The log-likelihood of the observations will be denoted by $\ell(\delta, \mu, \kappa)$. We note that for the validity of the NR-test it is sufficient to check assumption 4 of Hall and Mathiason [1990]. With some effort, it can be proved that this assumption holds true in this case (see Laha [2001] pp. 57-61).

Theorem 9.4 *Suppose n is large. The NR-test for testing H_0 against H_{1r} is: Reject H_0 if*

$$\frac{n^2 \hat{\kappa} R_2^2 \sin^2(\hat{\mu_2} - \hat{\mu})}{r(n-r)R} > \chi^2_{\alpha,1}$$

at an approximate level of significance α.

Proof *We use the terminology given in Hall and Mathiason [1990]. Since we are constructing a test for $H_0 : \delta = 0$, δ is our parameter of interest and $\eta = (\mu, \kappa)$ are the nuisance parameters. The log-likelihood function $\ell(\mu, \kappa, \delta)$ for the n observations is*

$$\ell(\mu, \kappa, \delta) = -n \log(2\pi) - n \log(I_0(\kappa)) + \kappa \sum_{i=1}^{r} \cos(\theta_i - \mu) + \kappa \sum_{i=r+1}^{n} \cos(\theta_i - \mu - \delta).$$

Note that the score vector $s = (s_\mu, s_\kappa, s_\delta)'$ where

$$\begin{aligned} s_\mu &= \frac{\partial \ell}{\partial \mu} = \kappa \sum_{i=1}^{r} \sin(\theta_i - \mu) + \kappa \sum_{i=r+1}^{n} \sin(\theta_i - (\mu + \delta)), \\ s_\kappa &= \frac{\partial \ell}{\partial \kappa} = -nA(\kappa) + \sum_{i=1}^{r} \cos(\theta_i - \mu) + \sum_{i=r+1}^{n} \cos(\theta_i - (\mu + \delta)), \ \textit{and} \\ s_\delta &= \frac{\partial \ell}{\partial \delta} = \kappa \sum_{i=r+1}^{n} \sin(\theta_i - (\mu + \delta)). \end{aligned}$$

It may be noted that in Hall and Mathiason [1990], the normalized score vector $S = \frac{1}{\sqrt{n}} \begin{pmatrix} s_\delta \\ s_\mu \\ s_\kappa \end{pmatrix}$ is referred to as the score vector. The matrix B defined in Hall and Mathiason [1990] is the per-unit Fisher Information which is obtained by dividing the information matrix I by the sample size n. The expectations are taken under H_0. The elements of I are

$$\begin{aligned} I_{\mu\mu} &= n\kappa A(\kappa), \\ I_{\kappa\kappa} &= nA'(\kappa), \\ I_{\delta\delta} &= (n-r)\kappa A(\kappa), \\ I_{\mu\kappa} &= 0, \\ I_{\mu\delta} &= (n-r)\kappa A(\kappa), \\ I_{\kappa\delta} &= 0. \end{aligned}$$

Thus, we get

$$B = \begin{pmatrix} B_{\delta\delta} & B_{\delta\mu} & B_{\delta\kappa} \\ B_{\mu\delta} & B_{\mu\mu} & B_{\mu\kappa} \\ B_{\kappa\delta} & B_{\kappa\mu} & B_{\kappa\kappa} \end{pmatrix} = \begin{pmatrix} \frac{n-r}{n}\kappa A(\kappa) & \frac{n-r}{n}\kappa A(\kappa) & 0 \\ \frac{n-r}{n}\kappa A(\kappa) & \kappa A(\kappa) & 0 \\ 0 & 0 & A'(\kappa) \end{pmatrix}.$$

We partition B with respect to δ *and* $\eta = (\mu, \kappa)$ *as*

$$\begin{pmatrix} B_{\delta\delta} & B_{\delta\eta} \\ B_{\eta\delta} & B_{\eta\eta} \end{pmatrix},$$

where

$$B_{\delta\eta} = \begin{pmatrix} \frac{n-r}{n}\kappa A(\kappa) & 0 \end{pmatrix}, \qquad B_{\eta\eta} = \begin{pmatrix} \kappa A(\kappa) & 0 \\ 0 & A'(\kappa) \end{pmatrix} \quad \text{and} \qquad B_{\eta\delta} = B'_{\delta\eta}.$$

The NR-test statistic is

$$T_r = (\hat{S}^*_\delta)^2 / \hat{B}^*_{\delta\delta},$$

where $(\hat{S}^*_\delta)^2$ *is the estimated effective normalized score obtained by substituting the MLEs of the nuisance parameters under* H_0 *in the effective normalized score* $S^*_\delta = S_\delta - B_{\delta\eta}B^{-1}_{\eta\eta}S_\eta$. *Similarly,* $\hat{B}^*_{\delta\delta}$, *the estimated effective per-unit information, is obtained from the effective per-unit information* $B^*_{\delta\delta} = B_{\delta\delta} - B_{\delta\eta}B^{-1}_{\eta\eta}B_{\eta\delta}$. *By straightforward computations, we get*

$$S^*_\delta = S_\delta - \frac{n-r}{n}S_\mu,$$
$$B^*_{\delta\delta} = \frac{r(n-r)}{n^2}\kappa A(\kappa).$$

When these are evaluated at the MLEs of (μ, κ) *under* H_0 *say,* $(\hat{\mu}_0, \hat{\kappa}_0)$, *we get*

$$\hat{S}^*_\delta = \hat{S}_\delta = \frac{\hat{\kappa}_0}{\sqrt{n}} \sum_{i=r+1}^{n} \sin(\theta_i - \hat{\mu}_0),$$

since $\hat{S}_\mu = S_\mu|_{(\hat{\mu}_0, \hat{\kappa}_0, 0)} = 0$ *and*

$$\hat{B}^*_{\delta\delta} = \frac{r(n-r)}{n^2}\hat{\kappa}_0 A(\hat{\kappa}_0).$$

Thus, the NR-test statistic is

$$T_r = \frac{\left(\hat{\kappa}_0 \sum_{i=r+1}^{n} \sin(\theta_i - \hat{\mu}_0)\right)^2}{\frac{r(n-r)}{n}\hat{\kappa}_0 A(\hat{\kappa}_0)} = \frac{n^2 \hat{\kappa} R_2^2 \sin^2(\hat{\mu_2} - \hat{\mu})}{r(n-r)R}$$

As shown in Hall and Mathiason [1990], under H_0, *the statistic* T_r *is asymptotically distributed as* χ^2_1. *Hence, for large* n, *we Reject* H_0 *if*

$$T_r > \chi^2_{\alpha,1}$$

at an approximate level of significance α.

Since the change point r is unknown, we use $T = \max_{1 \le r \le n-1} T_r$ as the test statistic for testing H_0 against H_1. We call T the NR-type test (NRTT) statistic. The cutoff points of this test can be approximately obtained using simulations.

9.4 Examples

In this section, we look at two examples of change point problems with circular data. The first problem is the one discussed in Lombard [1986], while the second looks at a dataset of

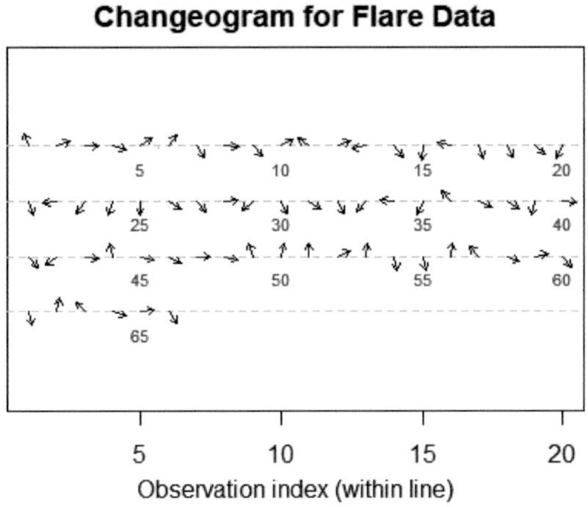

FIGURE 9.1
Changeogram for flare data.

wind directions given by Weijers et al. [1995]. For both these examples, it is assumed that the observations are independent and come from von Mises distribution with constant κ.

Example 1 : As a first example, we treat the flare data analyzed by Lombard [1986]. The raw data is reproduced in Fisher [1993]. The Changeogram [Laha, 2001, SenGupta, 1996] is given in Figure 9.1. The dataset consists of 66 observations. However, Lombard [1986] considered the first 60 observations and detected two change points at 12 and 42, respectively using his change point statistic for single change iteratively. Based on that, he used a test statistic for detecting the presence of two change points and detected 12 and 37 as change points. Since our test statistics are also designed to detect at most one change point, we decided to apply the iterative technique. In keeping with Lombard [1986], we also work with the first 60 observations. Based on the information provided by Lombard [1986], we break up the data set into two subsets, one subset consisting of the first 42 observations and the other subset consisting of the remaining 48 observations after deletion of the first 12. Since the initial direction, the changed direction (if a change point is present), and the concentration parameter are all unknown, the NRTT is applied to the dataset consisting of the observations 1 to 42. The test gave a very significant result, indicating the presence of change and the change point to be at 17. Now considering the observation nos. 13–60 of the flare data, we again apply the NRTT and it again gave a very significant result, indicating the presence of change and identified the change point to be at 42.

Example 2: As a second example, we analyze the data set given by Weijers et al. [1995]. They investigated the horizontal perturbation wind field within thermal structures encountered in the atmospheric surface layer boundary. We retrieved the actual data from the graphical representation presented in the paper. From the Changeogram [Laha, 2001, SenGupta, 1996] one notices that there are possibly two change points one around 17 and the other around 23 (see Figure 9.2).

We decided to consider only the observation numbers 1 to 22 for our analysis to avoid the confounding impact of multiple change points in a small dataset. The NRTT was applied and it gave a very significant result indicating the presence of change and identified the change point to be at 17.

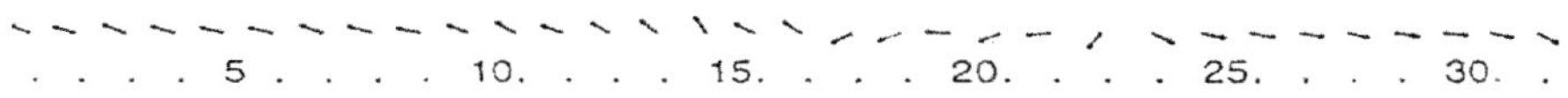

FIGURE 9.2
Changeogram for wind direction data.

9.5 NRTT for Wrapped Cauchy Distribution

The two most popular circular densities in applied sciences, though unimodal and symmetric, are von Mises and Wrapped Cauchy distributions. For developing statistical inference procedures, these are quite appealing, as the former is and the latter is not a member of the multiparameter regular exponential family. In this section, we take up the change-point problem for the Wrapped Cauchy distribution. The probability density function (PDF) of the Wrapped Cauchy $WC(\mu, \rho)$ distribution for an angular random variable θ is given by:

$$f(\theta|\mu,\rho) = \frac{1-\rho^2}{2\pi[1-2\rho\cos(\theta-\mu)+\rho^2]}, \tag{9.1}$$

where $\mu \in [0, 2\pi)$ is the mean direction and $\rho \in [0, 1)$ is the concentration parameter. Fix $1 \le r \le n-1$. We consider a sequence of n independent observations, $\theta_1, \dots, \theta_n$ of which $\theta_1, \dots, \theta_r$ are distributed as $WC(\mu, \rho)$ and $\theta_{r+1}, \dots, \theta_n$ are distributed as $WC(\mu+\delta, \rho)$. We want to test $H_0 : \delta = 0$ against the alternative $H_{1r} : \delta \ne 0$. The log-likelihood function of the data is

$$\begin{aligned}\ell(\mu,\rho,\delta) = n\log\left(1-\rho^2\right) - n\log 2\pi - \sum_{i=1}^{r}\log(1-2\rho\cos(\theta_i-\mu)+\rho^2)\\ - \sum_{i=r+1}^{n}\log(1-2\rho\cos(\theta_i-(\mu+\delta))+\rho^2).\end{aligned}$$

Hence,

$$s_\mu = \frac{\partial\ell}{\partial\mu} = \sum_{i=1}^{r}\frac{2\rho\sin(\theta_i-\mu)}{1-2\rho\cos(\theta_i-\mu)+\rho^2} + \sum_{i=r+1}^{n}\frac{2\rho\sin(\theta_i-(\mu+\delta))}{1-2\rho\cos(\theta_i-(\mu+\delta))+\rho^2},$$

$$s_\rho = \frac{\partial\ell}{\partial\rho} = \frac{-2\rho n}{1-\rho^2} + \sum_{i=1}^{r}\frac{2(\cos(\theta_i-\mu_i)-\rho)}{1-2\rho\cos(\theta_i-\mu_i)+\rho^2} + \sum_{i=r+1}^{n}\frac{2(\cos(\theta_i-(\mu_i+\delta))-\rho)}{1-2\rho\cos(\theta_i-(\mu_i+\delta))+\rho^2},$$

and

$$s_\delta = \frac{\partial\ell}{\partial\delta} = \sum_{i=r+1}^{n}\frac{2\rho\sin(\theta_i-(\mu+\delta))}{1-2\rho\cos(\theta_i-(\mu+\delta))+\rho^2}.$$

Following Hall and Mathiason [1990], the score vector is normalized by $n^{-1/2}$, to get

$$S = \frac{1}{\sqrt{n}}\begin{pmatrix} s_\mu \\ s_\rho \\ s_\delta \end{pmatrix}.$$

The matrix B defined in Hall and Mathiason [1990] is the per-unit Fisher Information which is obtained by dividing the Information matrix I by the sample size n. The elements of B can be computed using the results from Ghosh et al. [2019]. We have

$$\begin{aligned}
B_{\mu\mu} &= \frac{1}{n} I_{\mu\mu} = \frac{2\rho^2}{(1-\rho^2)^2},\\
B_{\rho\rho} &= \frac{1}{n} I_{\rho\rho} = \frac{2}{(1-\rho^2)^2},\\
B_{\delta\delta} &= \frac{1}{n} I_{\delta\delta} = \frac{n-r}{n}\frac{2\rho^2}{(1-\rho^2)^2},\\
B_{\mu\delta} &= \frac{1}{n} I_{\mu\delta} = \frac{n-r}{n}\frac{2\rho^2}{(1-\rho^2)^2},\\
B_{\mu\rho} &= B_{\rho\delta} = 0.
\end{aligned}$$

Since we are interested in testing $H_0 : \delta = 0$, the parameter of interest is δ, and the nuisance parameters are $\eta = (\mu, \rho)$. As in Hall and Mathiason [1990], we write $S = (S_\delta\ S_\eta)'$ where $S_\eta = (S_\mu\ S_\rho)$, and partition B as

$$B = \begin{pmatrix} B_{\delta\delta} & B_{\delta\eta} \\ B_{\eta\delta} & B_{\eta\eta} \end{pmatrix} \quad \text{where} \quad B_{\delta\eta} = \begin{pmatrix} B_{\delta\mu} & B_{\delta\rho} \end{pmatrix} \quad \text{and} \quad B_{\eta\eta} = \begin{pmatrix} B_{\mu\mu} & B_{\mu\rho} \\ B_{\rho\mu} & B_{\rho\rho} \end{pmatrix}.$$

The NR test statistic is

$$T_r = (\hat{S}^*_\delta)^2 / \hat{B}^*_{\delta\delta},$$

where $\hat{S}^*_\delta$ is the estimated effective normalized score for δ and $\hat{B}^*_{\delta\delta}$ is the estimated effective per-unit information for δ. These are obtained by evaluating S^*_δ and $B^*_{\delta\delta}$ at the MLEs of (μ, ρ) under H_0 say, $(\hat{\mu}_0, \hat{\rho}_0)$.

We get through straightforward computations

$$S^*_\delta = S_\delta - \frac{n-r}{n} S_\mu$$

and

$$B^*_{\delta\delta} = \frac{2r(n-r)\rho^2}{n^2(1-\rho^2)^2}.$$

Since the estimated normalized scores of the parameters μ, ρ, and δ under H_0 obtained by substituting the MLE's of μ and ρ under H_0, we get $\hat{S}_\mu = 0$ and $\hat{S}_\rho = 0$. Thus, $\hat{S}^*_\delta = \hat{S}_\delta$.

Hence, the NR-Test statistic is

$$T_r = \frac{\left(\sum_{i=r+1}^{n} \frac{2\hat{\rho}_0 \sin(\theta_i - \hat{\mu}_0)}{1 - 2\hat{\rho}_0 \cos(\theta_i - \hat{\mu}_0) + \hat{\rho}_0^2} \right)^2}{\frac{2r(n-r)\hat{\rho}_0^2}{n(1-\hat{\rho}_0^2)^2}}.$$

The NRTT-statistic based on the NR-Test is

$$T = \max_{1 \le r \le n-1} T_r.$$

We reject H_0 at level of significance α, if $T > k_\alpha$. The cut-off point k_α can be obtained using simulation.

9.6 Discussion and Conclusion

In this chapter, we have considered the problem of change point detection in the parameters of the von Mises distribution. We assume that at most one change point may be present in the given dataset. This problem is non-regular since the parameter space is not open, and one of the parameters (the changed direction) is not present under the null hypothesis of no change and is present only under the alternative hypothesis. We have considered several cases, namely (a) when all the nuisance parameter values are known, (b) when only the changed direction is unknown, (c) when the initial and changed direction are both unknown, and finally (d) when all the nuisance parameters are unknown. For the case (a) we derive the LRT and obtain its asymptotic null distribution, for (b) we derive a test based on the LMP test when the possible location of the change point is known and obtain its asymptotic distribution, for (c) we derive the LRT whose cut-off values can be obtained using simulations if the change point is not located near the extremities of the dataset and, for (d) we derive the NRTT based on the Neyman-Rao test when the possible location of change point is known, whose cut-off values can be obtained using simulation. We consider two real-life examples where it is seen that the NRTT performs well in detecting the presence of change points. The NRTT is a general method that can be derived for studying the change point problem with other circular distributions for which $\sqrt{n}$-consistent estimates of the nuisance parameters can be obtained. This is illustrated by deriving the NRTT for the wrapped Cauchy distribution. We also tried to show, following Lombard(1986), how our method for detecting at most one change-point may be iteratively used for detecting subsequent (multiple) change points in exploratory data analysis. However, since the data sets may overlap in this iterative procedure, one will face the challenging multiple hypotheses testing problem to achieve the desired level of significance. In conclusion, we hope this work will enhance in the future theoretically non-trivial and practically impactful research in the arena of circular statistics.

Bibliography

H. Assareh, I. Smith, and K. Mengersen. Bayesian change point detection in monitoring cardiac surgery outcomes. *Quality Management in Health Care*, 20(3):207–222, 2011.

R. N. Bhattacharya and E. C. Waymire. *Stochastic Processes with Applications.* John Wiley, 1992.

J. Chen and A. K. Gupta. *Parametric Statistical Change Point Analysis.* Birkhauser, Boston, 2000.

J. Chen and A. K. Gupta. *Parametric Statistical Change Point Analysis: with applications to Genetics, Medicine, and Finance.* Birkhäuser, 2nd edition, 2011.

H. Chernoff and S. Zacks. Estimating the current mean of a normal distribution which is subject to changes in time. *Annals of Mathematical Statistics*, 35:999–1018, 1964.

D. A. Darling and P. Erdos. A limit theorem for the maximum of normalized sums of independent random variables. *Duke Mathematical Journal*, 23:143–155, 1956.

N. I. Fisher. *Statistics of Circular Data.* Cambridge University Press, Cambridge, 1993.

M. Ghosh, X. Zhong, A. SenGupta, and R. Zhang. Non-subjective priors for wrapped Cauchy distributions. *Statistics & Probability Letters*, 153:90–97, 2019.

E. Gurarie, R. D. Andrews, and K. L. Laidre. A novel method for identifying behavioural changes in animal movement data. *Ecology Letters*, 12(5):395–408, 2009.

W. J. Hall and D. J. Mathiason. On large-sample estimation and testing in parametric models. *International Statistical Review*, 58:77–97, 1990.

D. V. Hinkley. Inference about a change point in a sequence of random variables. *Biometrika*, 57:1–17, 1970.

A. K. Laha. *Slippage and change point problems with directional data.* PhD thesis, Indian Statistical Institute, 2001.

F. Lombard. The change point problem for angular data: A nonparametric approach. *Technometrics*, 28:391–397, 1986.

E. S. Page. A test for a change in a parameter occurring at an unknown time point. *Biometrika*, 42:523–526, 1955.

C. R. Rao. Score test: Historical review and recent developments. In N. Kannan N. Balakrishnan, H. N. Nagaraja, editor, *Advances in ranking and selection, multiple comparisons, and reliability*, pages 3–20. Birkhäuser, Boston, 2005.

V. S. Reddem, V. Jampana, R. Muthalagu, V. R. Bekkam, P. R. R. Eluri, and S. K. Tummala. On the non-parametric changepoint detection of flow regimes in cyclone Amphan. *Oceanologia*, 65(2):310–317, 2023.

J. F. Reeves, J. Chen, X. L. Wang, R. Lund, and Q. Lu. A review and comparison of changepoint detection techniques for climate data. *Journal of Applied Meteorology and Climatology*, 46(6):900–915, 2007.

C. Rougé, Y. Ge, and X. Cai. Detecting gradual and abrupt changes in hydrological records. *Advances in Water Resources*, 53:33–44, 2013.

A. K. Sen and M. S. Srivastava. On multivariate tests for detecting change in mean. *Sankhya, A*, 35:173–185, 1973.

A. SenGupta. *DDSTAP - Directional Data Statistical Analysis Package.* Indian Statistical Institute, 1996.

A. SenGupta and A. K. Laha. A Bayesian analysis of the change-point problem for directional data. *Journal of Applied Statistics*, 37:693–700.1783–1791, 2008a.

A. SenGupta and A. K. Laha. A likelihood integrated method for exploratory graphical analysis of change point problem with directional data. *Communications in Statistics - Theory and Methods*, 35:693–700, 2008b.

D. Siegmund. Confidence sets in change point problem. *International Statistical Review*, 56:31–48, 1988.

C. Truong, L. Oudre, and N. Vayatis. Selective review of offline change point detection methods. *Signal Processing*, 167:107299, 2020.

E. P. Weijers, A. van Delden, H. F. Vugts, and A. G. C. A. Meesters. The composite horizontal wind field within convective structures of the atmospheric surface layer. *Journal of the Atmospheric Sciences*, 52:3866–3878, 1995.

10

Analysis of Maritime Conditions via Nonparametric Directional Methods

María Alonso-Pena

Understanding maritime conditions, particularly wave direction and height, is vital for informed decision-making in activities such as shipping, navigation, and seafood gathering. This study applies nonparametric directional methods to analyze these critical variables, employing methods such as kernel density estimation (KDE) and flexible regression models to uncover complex relationships. By addressing the circular nature of the data and the cylindrical relationships between wave direction, wave height, and covariates such as wind direction and speed, the analysis provides a solid framework for understanding these interactions.

10.1 Introduction

A key variable in understanding maritime conditions is wave direction, which is inherently circular in nature (see, e.g., [1, 2, 3, 4]), as is wind direction, another fundamental factor influencing sea behavior. Due to their circular support (defined on the unit circle rather than the real line), it is crucial to apply appropriate statistical methods that respect the geometry of these variables.

Beyond individual variables, it is important to consider interactions between wave direction and wave height, since the origin and strength of waves are often jointly influenced by meteorological forces. Likewise, wind direction and wind speed interact in ways that impact not only sea state but also navigational safety, coastal erosion, and energy extraction. These pairings, direction with magnitude, constitute examples of cylindrical variables, where one component is circular and the other is linear (see, e.g., [5, 6]).

Although models for cylindrical data do exist, many impose assumptions that are overly restrictive, limiting their capacity to capture complex real-world behaviors. In the context of real-valued variables, data-driven methods for nonparametric density estimation have been well developed ([7, 8, 9]), and several works have extended these ideas to circular ([10, 11, 12]), toroidal ([13]), and cylindrical domains ([14]). Moreover, [15] proposed a goodness-of-fit test for parametric families of directional-linear densities.

In addition to examining joint distributions, it is also crucial to model how wave direction varies in response to wind behavior. Several parametric regression models have been proposed for circular responses with real-valued covariates ([16, 17, 18, 19]) and with

DOI: 10.1201/9781003564423-10

circular covariates ([20, 21]). However, in practice, the regression functions implied by these models are often too simplistic to capture the true underlying structure of the data.

To address this, data-driven regression models based on kernel smoothing have been extensively studied in the real-valued setting ([22, 23]) and adapted to the case of circular responses by, among others, [24]. Moreover, such estimators can be used to develop significance tests and group comparisons ([25]) as well as goodness-of-fit tests ([26]). More advanced regression models have also been developed to go beyond the conditional mean, such as quantile regression for circular responses ([27]) and modal regression ([28]), offering a more nuanced understanding of directional data under covariate effects.

In this work, we study the behavior of wave direction at a key location on the Spanish coast, focusing on its interaction with both wave height and wind conditions. We adopt a nonparametric, data-driven approach that allows for flexible modeling of circular and cylindrical variables, capturing complex dependencies that parametric models may miss. The paper begins with a description of the dataset, which includes maritime conditions (wave direction and height) and meteorological variables (wind direction and speed) recorded at a strategic point along the Spanish coast. Next, we introduce the methodology used to analyze wave behavior, including kernel-based density estimation techniques for circular and cylindrical variables, regression models for circular responses, nonparametric significance tests, and modal regression. This is followed by a detailed analysis of wave dynamics on the northwestern Spanish coast, with interpretations of the observed patterns. We conclude with final remarks and suggestions for future research.

10.1.1 Wave Data in Galicia, Spain

We analyze maritime and meteorological data collected from a key coastal location in northwestern Spain. The variables of interest include wave origin direction, wave height, wind origin direction, and wind speed. These data were obtained from the Spanish Ministry of Transport and Sustainable Mobility (https://www.puertos.es/es-es/oceanografia/paginas/portus.aspx) and are part of their ongoing oceanographic monitoring efforts.

The measurements were taken at the Villano-Sisargas Buoy, positioned at latitude 43°29.4' N and longitude 9°12.6' W, just off the coast of the Costa da Morte, literally *Coast of Death*, in the region of Galicia. This area is well known for its dangerous maritime conditions, where frequent storms meet a rugged, rocky shoreline. The buoy plays a critical role in monitoring this high-risk zone, contributing to maritime safety and environmental awareness. Beyond its environmental significance, this region has strong economic ties to the sea. Local livelihoods depend heavily on marine activities such as fishing and shellfish harvesting, both of which are highly sensitive to changes in oceanographic and meteorological conditions.

Observations were recorded every 12 hours over a one-year period, yielding a total of 712 measurements. Wave and wind directions are expressed in degrees with 0°indicating North and 90°indicating East.

10.2 Statistical Methodology

To analyze the relationships between maritime variables such as wave direction, wave height, and wind characteristics, we use statistical tools that allow for flexible exploration of the data without imposing rigid assumptions about its structure. In particular, because some of these variables, such as wave and wind direction, are directional in nature, we require methods that can properly account for their circular properties. This section introduces the nonparametric techniques used in our analysis, with a focus on kernel-based methods that are well-suited for handling circular and cylindrical data. These tools make it possible to estimate distributions and model relationships in a data-driven way, capturing key features without relying on predefined parametric forms. We begin by discussing kernel density estimation for real-valued, circular, and cylindrical variables, followed by regression models for circular responses, techniques for assessing the influence of covariates, and a kernel-based approach to circular modal regression.

10.2.1 Kernel Density Estimation for Cylindrical Variables

Let X be a real-valued random variable and $X_1, \ldots, X_n$ a sample of n independent observations from X, which has density function f. The kernel density estimator of f ([29, 30]) is given by

$$\widehat{f}(x) = \frac{1}{n}\sum_{i=1}^{n} L_h(X_i - x), \tag{10.1}$$

with the kernel function L_h being a symmetric density, with support on the real line (or on an interval of the real line) and centered at zero, with h, called bandwidth, controlling the dispersion of the density. Large values of h lead to an estimator $f(x)$ with large bias and small variance, while smaller values of h reduce the bias at the cost of increasing the variance. Therefore, in practice, a data-driven selection of this parameter is of crucial importance.

Kernel density estimation of a circular density f was introduced in the spherical context by [10] and [11], and considered specifically for circular data by [12]. Let Θ be a circular random variable, and $\Theta_1, \ldots, \Theta_n$ a sample of n independent observations from Θ. We consider the estimator

$$\widehat{f}(\theta) = \frac{1}{n}\sum_{i=1}^{n} K_\rho(\Theta_i - \theta), \tag{10.2}$$

where K_ρ is a circular kernel function, i.e., a symmetric circular density function centered at zero with ρ controlling the concentration of the density. The parameter ρ plays a role opposite to the bandwidth in linear kernel density estimation: it governs the trade-off between bias and variance. A large value of ρ results in a more concentrated kernel, producing a rougher estimate with lower bias but higher variance. Conversely, a lower ρ leads to a smoother density estimate with increased bias and reduced variance.

Bivariate kernel density estimation in the case where one of the variables is spherical was studied by [14], which includes cylindrical variables as a particular case. Let (Θ, X) be a cylindrical random variable with density function f. The estimator of f is given by a product kernel of the form

$$\widehat{f}(\theta, x) = \frac{1}{n}\sum_{i=1}^{n} K_\rho(\Theta_i - \theta) L_h(X_i - x). \tag{10.3}$$

Thus, this estimator depends on two smoothing parameters: ρ and h, which should also be chosen in a data-driven way.

10.2.2 Kernel Regression for a Circular Response

Now, we consider the problem of estimating the regression function when the response variable, denoted by Φ, has a circular support. Let Δ be a general covariate, which can have support on the circumference or on the real line. We consider the model

$$\Phi = [m(\Delta) + \varepsilon](\text{mod } 2\pi), \tag{10.4}$$

where ε is a circular random error such that $\mathbb{E}[\sin \varepsilon | \Delta = \delta] = 0$. The regression function m is the mean direction of Φ conditioned to the value of Δ, that is

$$m(\delta) = \text{atan2}[\mathbb{E}(\sin \Phi | \Delta = \delta), \mathbb{E}(\cos \Phi | \Delta = \delta)].$$

Given a bivariate random sample of (Δ, Φ), i.e., $\{(\Delta_i, \Phi_i)\}_{i=1,\dots,n}$, [24] proposed estimating m as

$$\widehat{m}_\nu(\delta) = \text{atan2}[\widehat{m}_{1,\nu}(\delta), \widehat{m}_{2,\nu}(\delta)], \tag{10.5}$$

with

$$\widehat{m}_{1,\nu}(\delta) = \sum_{i=1}^{n} R_\nu(\Delta_i - \delta) \sin \Phi_i,$$

$$\widehat{m}_{2,\nu}(\delta) = \sum_{i=1}^{n} R_\nu(\Delta_i - \delta) \cos \Phi_i,$$

where the atan2 operator is the two-argument arctangent, which returns the angle between the x-axis and the vector from the origin to (x, y). Additionally, R_ν is either a circular kernel function ($R_\nu(\Delta_i - \delta) = K_\rho(\Theta_i - \theta)$) if the covariate is circular or a regular kernel function for a real-valued covariate ($R_\nu(\Delta_i - \delta) = L_h(X_i - x)$).

As in the density estimation problem, the selection of the smoothing parameter ν (equivalently ρ when $\Delta = \Theta$ and h if $\Delta = X$) is of great importance, as it will control the bias-variance trade-off.

Next, we consider the problem of a circular response variable Φ and two covariates of different nature, Θ and X. Therefore, we assume the model

$$\Phi = [m(X, \Theta) + \varepsilon](\text{mod } 2\pi), \tag{10.6}$$

where the regression m depends on the two covariates and is analogously defined as the expected direction of Φ conditioned to the values of the covariates. [31] extended the previous kernel estimator to the case of multiple real-valued covariates by using multivariate kernels. In our case, where there is one circular and one real-valued covariate, we employ a product kernel as in (10.3). Thus, given a trivariate iid sample $\{(\Theta_i, X_i, \Phi_i)\}_{i=1,\dots,n}$ from (Θ, X, Φ), the kernel estimator takes the form

$$\widehat{m}_\nu(x, \theta) = \text{atan2}[\widehat{m}_{1,\nu}(x, \theta), \widehat{m}_{2,\nu}(x, \theta)], \tag{10.7}$$

with

$$\widehat{m}_{1,\nu}(x, \theta) = \sum_{i=1}^{n} L_h(X_i - x) K_\rho(\Theta_i - \theta) \sin \Phi_i,$$

$$\widehat{m}_{2,\nu}(x, \theta) = \sum_{i=1}^{n} L_h(X_i - x) K_\rho(\Theta_i - \theta) \cos \Phi_i.$$

The bivariate regression estimator thus depends on two different smoothing parameters, h and ρ.

10.2.3 Testing the Effect of a Covariate

Before estimating the regression functions, it is useful to test if there is an actual effect of the covariate on the mean direction of the response when considering model (10.4). A methodology to test the significance of the covariate in this context was proposed by [25]. The hypotheses of the test are given by

$$H_0: \quad \Phi = [\gamma + \varepsilon](\text{mod } 2\pi) \tag{10.8}$$
$$H_1: \quad \Phi = [m(\Delta) + \varepsilon](\text{mod } 2\pi), \tag{10.9}$$

with $\mathbb{E}[\sin \varepsilon | \Delta = \delta] = 0$ and where, under the alternative hypothesis, there exists a set of values of Δ of non-zero measure for which the regression function, $m(\Delta)$, is not equal to γ.

Under the null hypothesis, γ is the mean direction of Φ, and can be estimated as

$$\widehat{\gamma} = \text{atan2}\left(\sum_{i=1}^{n} \sin \Phi_i, \sum_{i=1}^{n} \cos \Phi_i\right).$$

Under the alternative, the regression function is estimated as in (10.5). The test statistic is then given by

$$T = \frac{\sum_{i=1}^{n}[1 - \cos(\Phi_i - \widehat{\gamma})] - \sum_{i=1}^{n}[1 - \cos(\Phi_i - \widehat{m}(\Delta_i))]}{\sum_{i=1}^{n}[1 - \cos(\Phi_i - \widehat{m}(\Delta_i))]}.$$

This statistic measures the relative reduction in average angular distance to the conditional mean when moving from the null to the alternative model, normalized by the average distance under the alternative. The distribution of T under H_0 is approximated by a bootstrap strategy described in [25].

When applying this test, one must account for the fact that nonparametric procedures are inherently sensitive to the choice of smoothing parameter, a well-known issue discussed by [23]. Since $\widehat{m}$ suffers from some bias, a smoothing level that works well for estimation might not yield reliable results for hypothesis testing, and [25] recommend using values of the smoothing parameter which undersmooth the regression function. In addition, it's a good practice to evaluate the test over multiple smoothing parameters within a sensible range, rather than relying on a single value.

10.2.4 Circular Modal Regression Based on Kernel Smoothing

In some settings, the conditional distribution of the response may exhibit more than one mode for a given covariate value. To account for this, we also consider a multimodal regression approach ([32]), which focuses on estimating the local modes of the conditional distribution rather than its mean. This method complements the classical kernel regression described earlier and is particularly useful when the response displays clustered or heterogeneous behavior across the range of the covariate. The presence of multiple local modes may suggest the influence of an unobserved covariate or an interaction effect with the covariate under study.

Multimodal regression in the context of circular variables was studied by [28]. For a circular response Φ and a general covariate Δ, the regression multifunction is defined as the set of modes of the conditional density, that is

$$M(\delta) = \left\{\phi \in [0, 2\pi) : \frac{\partial}{\partial \delta} f(\phi|\delta) = 0, \frac{\partial^2}{\partial \delta^2} f(\phi|\delta) < 0\right\}. \tag{10.10}$$

Note that in (10.10) we could replace the conditional density function $f(\phi|\delta)$ by the joint density function $f(\delta, \phi)$. Thus, [28] estimates (10.10) in a nonparametric way by first estimating the joint density function and then computing its local maxima with respect to the response variable:

$$\widehat{M}(\delta) = \left\{ \phi \in [0, 2\pi) : \frac{\partial}{\partial \delta} \widehat{f}(\phi, \delta) = 0, \frac{\partial^2}{\partial \delta^2} \widehat{f}(\phi, \delta) < 0 \right\}. \tag{10.11}$$

The estimator $\widehat{f}(\phi, \delta)$ is given by (10.3) in the case where Δ is a real-valued variable. If Δ has circular support, the estimator is given by a product of circular kernels. To compute the local maxima, a conditional version of the directional mean shift algorithm (see [33, 34]) is used.

Again, two smoothing parameters must be selected because the estimation of the joint density involves two smoothing parameters. However, their role is somewhat different than their roles the density estimation scenario: the parameter ρ associated with the response variable controls the number of local modes for a given value of δ, with higher values leading to multiple local modes and smaller values leading to a unimodal estimator. On the other hand, the parameter associated with the covariate Δ controls the smoothness of the regression multifunction.

10.3 Data Analysis and Discussion

This section presents an exploratory and inferential analysis of the available data, with the goal of understanding the relationship between wave direction and various wind-related variables. We begin by examining the distributional characteristics of each variable individually and in pairs, using both univariate and bivariate kernel density estimates. This is followed by formal testing to assess whether wind speed and direction have a significant effect on the mean wave direction. Finally, we approach the regression problem from both classical and multimodal perspectives, first considering each covariate separately and then jointly, to better capture the structure and possible complexities in the data. For most methods, the *R* package *NPCirc* (see [35]), an updated version of [36], is used.

We first estimate the density function of wave height, wind speed, wave direction, and wind direction individually. For the real-valued variables, we use the kernel estimator in (10.1), where the kernel L_h is a Gaussian density. Many different methods have been proposed in the literature to select the smoothing parameter h in a data-driven way, from a simple rule of thumb ([7]), to more elaborated plug-in approaches or other methods such as cross validation ([37]). Here, we employ the two-step plug-in approach of ([38]). Figures 10.1(a) and 10.1(b) illustrate the resulting kernel density estimates for wave height and wind speed, respectively. Wave height shows a markedly skewed distribution, with low-height waves occurring frequently and waves exceeding 6 meters being relatively rare. Wind speed also exhibits a slight right skewness, with speeds above 18 m/s being uncommon.

Next, we estimate the density functions of the two circular functions with the estimator in (10.2). We employ the von Mises kernel, given by

$$K_\rho(\theta) = \frac{1}{2\pi I_0(\rho)} \exp\{\rho \cos \theta\},$$

with I_0 being the Bessel function of the first kind and order zero. Although the problem of the data-driven selection of ρ has not been as studied as in the real-valued case, there are several proposals ([39, 40, 41, 42, 43]). We follow the approach of [42], selecting ρ as a

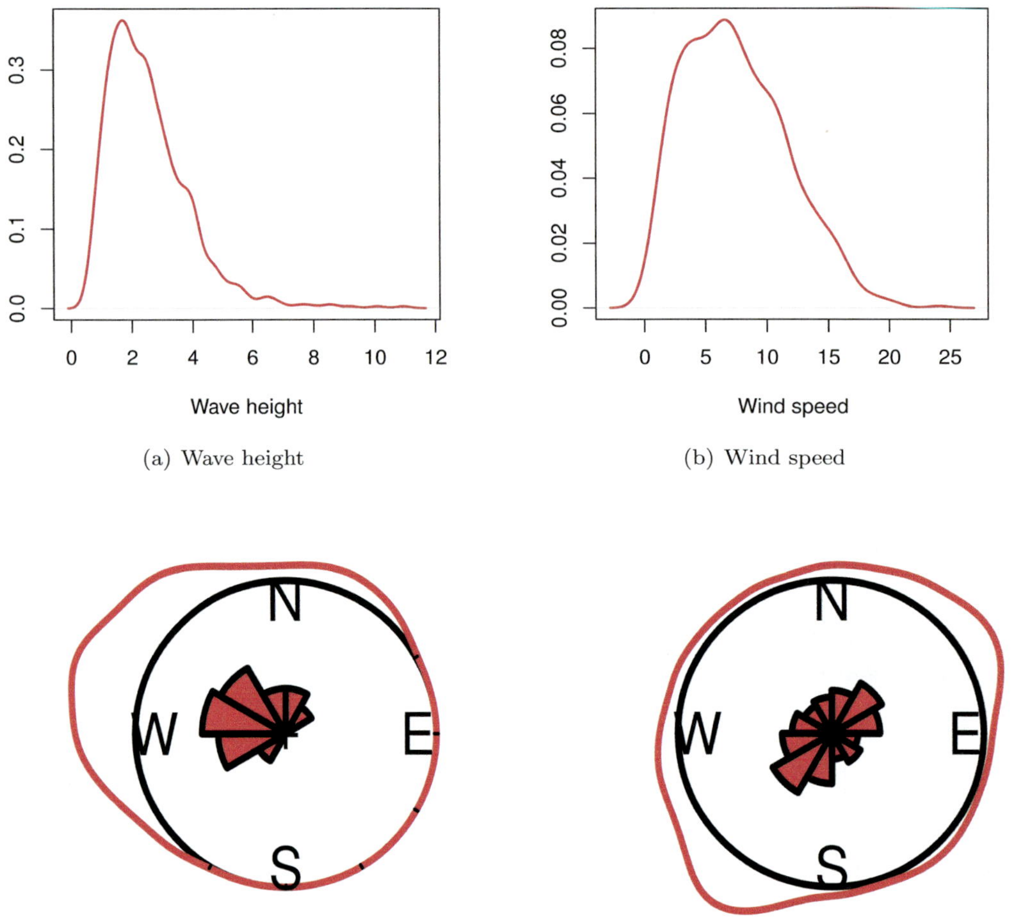

(a) Wave height (b) Wind speed

(c) Wave origin direction (d) Wind origin direction

FIGURE 10.1
Kernel density estimators of wave height (a), wind speed (b), wave direction (c), and wind direction (d).

two-stage plug-in estimate of the optimal concentration minimizing the asymptotic mean integrated squared error of the density estimator. The resulting estimates of the density functions of wave and wind origin directions are displayed in Figures 10.1(c) and 10.1(d), respectively. We see that most waves originate from the North-West, although there seems to be a small mode in the North-East direction. Wind directions exhibit a bimodal pattern, predominantly from the North-East and South-West.

We now examine the joint density of the two cylindrical variables associated with wave and wind behavior. We employ the product kernel density estimate in (10.3), with L_h being a Gaussian kernel and K_ρ being the von Mises kernel. Although the problem of selecting the two smoothing parameters in practice has not received much attention, [14] derived an explicit expressions for the theoretical optimal smoothing parameters in terms of minimizing the asymptotic mean integrated squared errors under a specific parametrization and [44] derived an explicit formula for the least squares cross-validation loss. In this work, we use a plug-in approach were we minimize the mean integrated squared error where the unknown quantities are replaced by parametric estimators based on a mixture of von Mises and Gaussian densities, extending the ideas of [40] and [41] to the bivariate case.

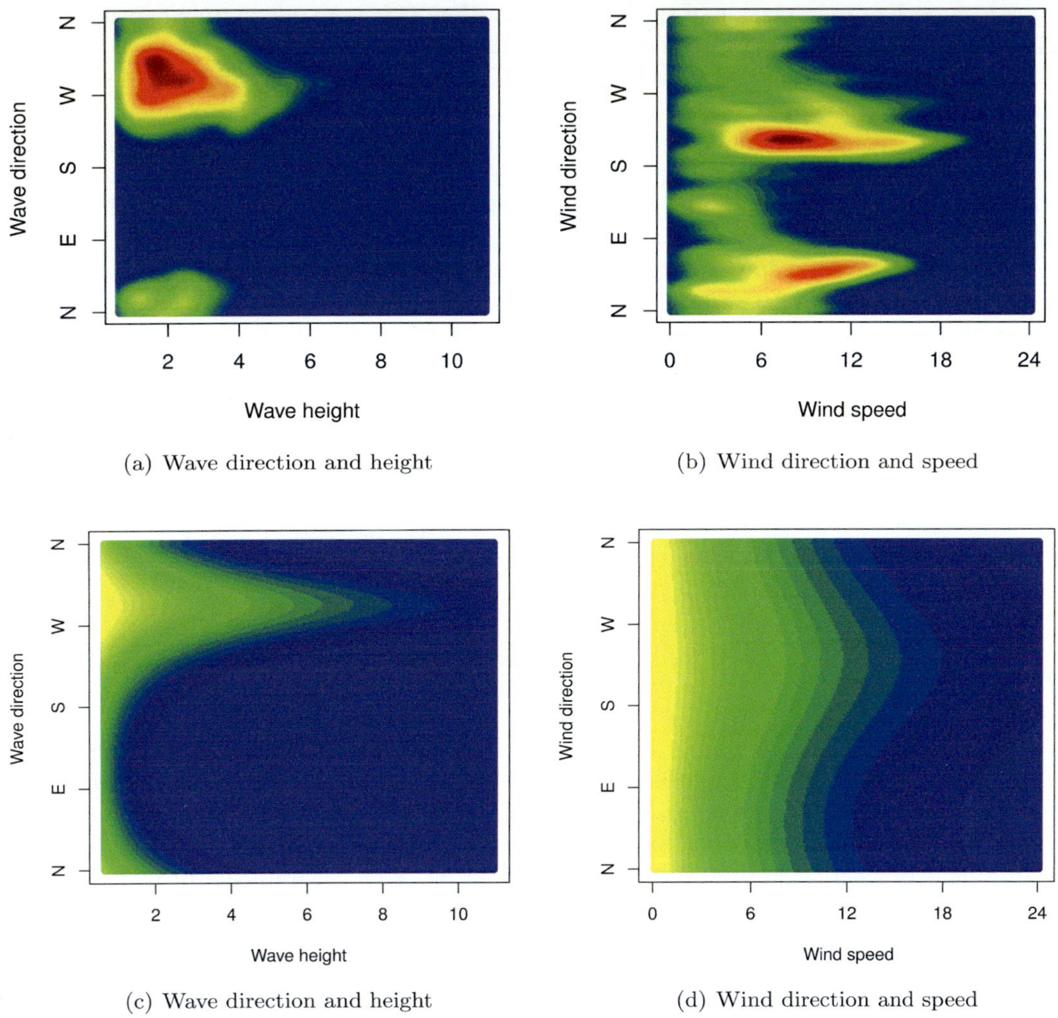

(a) Wave direction and height

(b) Wind direction and speed

(c) Wave direction and height

(d) Wind direction and speed

FIGURE 10.2
Cylindrical density estimators of wave direction and wave height (a) and wind direction and speed (b). Top row shows kernel density estimators and bottom row shows parametric estimators of the Johnson-Wehrly model.

The top panels of Figure 10.2(a) show the cylindrical kernel density estimate for wave direction and height. A distinct mode is observed for low wave heights originating from the North-West. For these lower waves, directions predominantly span the West and North-West, with additional contributions from the North-East and South-West. In contrast, higher waves tend to originate mainly from the West and North-West. Figure 10.2(b) shows the corresponding estimate for wind direction and speed. The primary mode occurs at moderate wind speeds (around 8 m/s) from the South-West, with a secondary local mode for stronger winds from the North-East. Additionally, wind direction is relatively uniform at lower speeds, while higher wind speeds are more concentrated around the South-West and North-East directions. For comparison purposes, the bottom panels of Figure 10.2(a) present the density estimates based on the parametric model of [17], where the parameters are estimated by maximum likelihood. Compared to the nonparametric cylindrical kernel density estimates, the parametric model of Johnson and Wehrly proves too restrictive. While it provides a smooth approximation of the joint distribution, it fails to adequately capture the multimodal structure observed in the data. In particular, the kernel estimates reveal clear bimodality in both the wave and wind direction distributions—most notably for low wave heights and moderate wind speeds—whereas the parametric model tends to smooth over these features, resulting in unimodal or overly simplistic patterns. This limitation highlights the challenges of relying solely on parametric assumptions when modeling complex environmental data, where directional and magnitude interactions often exhibit rich, multimodal behavior.

To assess whether wind speed and wind direction significantly influence the conditional mean direction of the waves, we applied the significance test based on hypotheses (10.9). For each covariate, the test was conducted across 15 different values of the smoothing parameters, including those selected via cross-validation. When considering wind speed as the covariate, the p-value was consistently below 0.001 for all parameter values, indicating a statistically significant effect of wind speed on the mean wave direction. Similar results were obtained for wind direction, supporting the conclusion that it also has a significant influence on the mean wave direction.

We next employ the nonparametric regression estimator in (10.5) to investigate how the mean wave direction varies with wind speed. We employ a Gaussian kernel and select ν by cross-validation, as the minimizer of the function

$$\sum_{i=1}^{n}\{1-\cos[Y_i-\widehat{m}_\nu^{-i}(\Delta_i)]\}. \tag{10.12}$$

The estimated regression function is depicted as a dotted line in Figure 10.3(a). Results indicate that the mean wave origin is primarily from the North-West, gradually shifting toward the West as wind speed increases. For comparison purposes, we also compare with the parametric regression model of [18], which assumes that the conditional distribution is a von Mises density with mean direction given by $\beta_1 + \text{atan}(\beta_2 x)$. Such estimator is depicted as a dashed line, exhibiting an almost constant mean direction centered around the North-West, regardless of wind speed. This contrast highlights a key limitation of the parametric model: its reduced flexibility in capturing nonlinear trends in the directional response.

Notably, a subset of data points shows wave directions from the North-East at moderately high wind speeds. To account for this, we apply the multimodal regression estimator in (10.11), illustrated by the continuous line in Figure 10.3(a). For this, two smoothing parameters must be selected. One might be tempted to use the same strategy as in multiple kernel density estimation; however, it is widely known that smoothing parameters that are optimal for the estimation of the density function are not optimal for the estimation of its local modes (see [45]). Here, we use the modal cross-validation approach proposed by [28]. This estimator reveals a single local mode at low wind speeds and a second mode emerging when wind speed exceeds 10 m/s. This pattern may reflect interactions with wind direction or the influence of additional variables.

We also examine the marginal effect of wind direction on wave direction. We also apply estimator (10.5), where the kernel is given by a von Mises density and the smoothing parameter is again selected by cross-validation. We also compare to the parametric estimator of [46]. In Figure 10.3(b), both the parametric (dashed line) and nonparametric (dotted line) od the conditional mean are shown, in addition to the conditional local mode estimator (solid line). The three estimates largely coincide across most wind directions, indicating good agreement between methods. However, for wind directions between North and East, a noticeable discrepancy arises: the modal estimate shifts toward the North-East, the parametric mean remains closer to the North-West, and the nonparametric mean lies in between. This behavior suggests a skewed conditional distribution in this range of wind

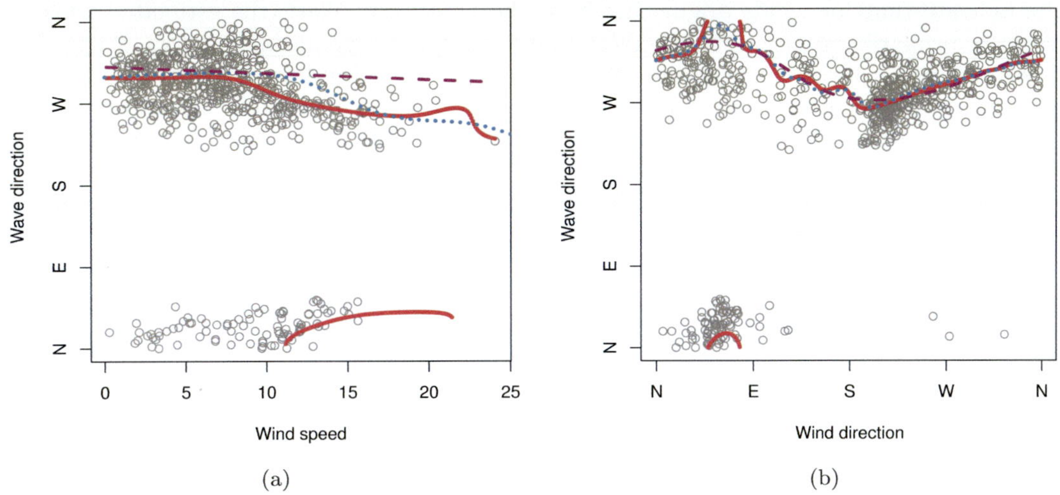

FIGURE 10.3
Scatter plots of wave direction vs. wind speed (a) and wave direction vs. wind direction (b) with parametric estimators of the conditional mean (dashed trace), kernel regression estimators of the conditional mean (dotted trace), and of the conditional modes (continuous trace).

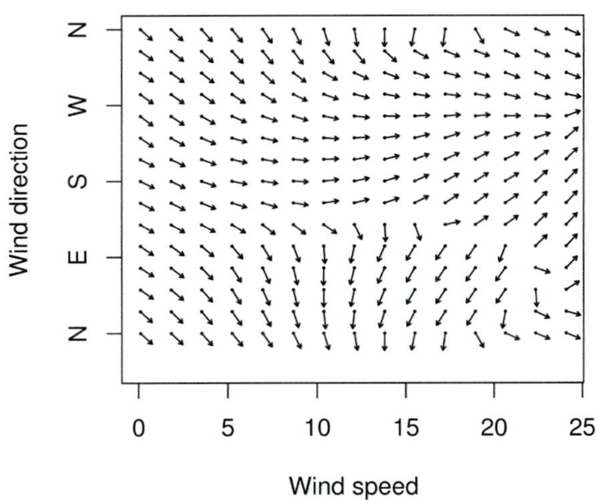

FIGURE 10.4
Representation of the estimated regression function of wave direction with wind direction and speed as covariates. Arrows represent the estimated direction of waves.

directions. While the parametric model appears appropriate in general, this localized deviation highlights its limited ability to accommodate asymmetry, which can be better captured by nonparametric and modal-based approaches.

Finally, we analyze the full regression model (10.6), incorporating both wind speed and wind direction as covariates. We use the nonparametric estimator defined in (10.7), where the two smoothing parameters are selected by cross-validation in an analogue way to (10.12). We visualize the estimated wave direction over a grid of covariate values in Figure 10.4. The results indicate that for winds originating from the South and West, the estimated wave direction is primarily from the North-West under weak wind conditions. As wind speed increases, the wave direction gradually shifts toward the West, and for strong winds from the South, it further shifts toward the South-West. Conversely, for winds coming from the North-East, the estimated wave direction transitions from the North-West to the North, and eventually to the North-East.

These findings suggest that both wind speed and direction jointly influence wave propagation patterns, with stronger winds exerting a more pronounced directional effect. This highlights the importance of considering multivariate interactions when modeling wave behavior.

10.4 Conclusions

In this study, we have examined the behavior of wave direction under the influence of key maritime and meteorological factors, using flexible nonparametric methods designed for circular and cylindrical data. The analysis, based on observations from a coastal monitoring station in northwestern Spain, reveals a rich structure in the relationships between wave direction and covariates such as wave height, wind direction, and wind speed.

Although wave direction prediction is traditionally approached through deterministic models based on differential equations, these models often require statistical fine-tuning to adapt to local or short-term variability. In this context, nonparametric, data-driven methods offer a valuable complement. By avoiding rigid model assumptions, these approaches are particularly well suited to capturing the complex, nonlinear, and possibly multimodal dependencies that characterize real-world oceanographic processes.

Kernel-based estimators enabled us to uncover patterns not easily captured by parametric models, including shifts in dominant wave direction with varying wind speed or direction. Modal regression further highlighted situations where the conditional distribution of wave direction departs significantly from unimodality or symmetry, underscoring the limitations of mean-based analysis in such contexts.

Our significance testing procedures confirmed the relevance of wind speed and direction as individual covariates. However, a key limitation remains: tests were performed separately for each covariate, leaving open the need for multivariate significance testing frameworks that can jointly assess the influence of several covariates. Likewise, modal regression, while useful, was restricted to models with a single covariate. Extending this to handle multiple covariates simultaneously, especially in the cylindrical context, is a promising direction for future work.

Another important avenue is the incorporation of temporal dependence. The sea state is inherently dynamic and successive observations are not independent. In this sense, [47] and [48] proposed parametric methods to account for spatio-temporal processes. Extending these nonparametric methods to account for temporal correlation could be tackled by employing another version of cross-validation for selecting the smoothing parameter, such as the leave-$(2l+1)$-out instead of classical leave-one-out selectors. It would also be convenient to account for temporal dependence in inferential tasks, for example, by adapting the bootstrap strategy in Section 10.2.3. In addition, time-varying density estimation, autoregressive structures, or functional data approaches could be explored to enhance both the descriptive and predictive capabilities of the models.

Altogether, our findings illustrate the strengths of flexible, geometry-aware statistical tools in environmental data analysis. These methods serve not only to refine predictions from physical models but also to offer new insights into the structure of directional phenomena. Future research should continue to bridge the gap between physical and statistical modeling, develop more comprehensive inference tools, and explore dynamic extensions suited to evolving environmental conditions.

Acknowledgement

The author greatfully acknowledges financial support from ID2024-158017NB-I00 funded by MCIN/AEI/10.13039/501100011033 / ERDF, EU and grant ED431C 2025/03 from Xunta de Galicia.

Bibliography

[1] K. V. Mardia and P. E. Jupp. *Directional Statistics.* Wiley, 2000.

[2] S. R. Jammalamadaka and A. SenGupta. *Topics in Circular Statistics.* World Scientific, 2001.

[3] N. I. Fisher. *Statistical Analysis of Circular Data.* Cambridge University Press, 1993.

[4] C. Ley and T. Verdebout. *Modern Directional Statistics.* CRC Press, 2017.

[5] K. V. Mardia and T. W. Sutton. A model for cylindrical variables with applications. *Journal of the Royal Statistical Society: Series B (Statistical Methodology)*, 40:229–233, 1978.

[6] T. Abe and C. Ley. A tractable, parsimonious and flexible model for cylindrical data, with applications. *Econometrics and Statistics*, 4:91–104, 2017.

[7] B. W. Silverman. Using kernel density estimates to investigate multimodality. *Journal of the Royal Statistical Society, Series B (Statistical Methodology).*, 43:97–99, 1981.

[8] D. W. Scott. *Multivariate Density Estimation.* Wiley, 1992.

[9] M. P. Wand and M. C. Jones. *Kernel Smoothing.* Chapman & Hall, 1995.

[10] P. Hall, G. S. Watson, and J. Cabrera. Kernel density estimation with spherical data. *Biometrika*, 74:751–762, 1987.

[11] Z. D. Bai, C. R. Rao, and L. C. Zhao. Kernel estimators of density function of directional data. *Journal of Multivariate Analysis*, 27:24–39, 1988.

[12] N. I. Fisher. Smoothing a sample of circular data. *Journal of Structural Geology*, 11:775–778, 1989.

[13] M. Di Marzio, A. Panzera, and C. C. Taylor. Kernel density estimation on the torus. *Journal of Statistical Planning and Inference*, 141:2156–2173, 2011.

[14] E. García-Portugués, R. M. Crujeiras, and W. González-Manteiga. Kernel density estimation for directional-linear data. *Journal of Multivariate Analysis*, 121:152–275, 2013.

[15] E. García-Portugués, R. M. Crujeiras, and W. González-Manteiga. Central limit theorems for directional and linear random variables with applications. *Statistica Sinica*, 25:1207–1229, 2015.

[16] A. L. Gould. A regression technique for angular variates. *Biometrics*, 25:683–700, 1969.

[17] R. A. Johnson and T. E. Wehrly. Some angular-linear distributions and related regression models. *Journal of the American Statistical Association*, 73:602–606, 1978.

[18] N. I. Fisher and A. J. Lee. Regression models for angular responses. *Biometrics*, 48:645–677, 1992.

[19] B. Presnell, S. P. Morrison, and R. C. Littel. Projected multivariate linear models for directional data. *Journal of the American Statistical Association*, 93:1068–1077, 1998.

[20] T. D. Downs and K. V. Mardia. Circular regression. *Biometrika*, 89(3):683–698, 2002.

[21] S. Kim and A. SenGupta. Multivariate-multiple circular regression. *Journal of Statistical Computation and Simulation*, 87:1277–1291, 2017.

[22] J. Fan and I. Gijbels. *Local Polynomial Modelling and its Applications*. Chapman & Hall, 1996.

[23] A. W. Bowman and A. Azzalini. *Applied Smoothing Techniques for Data Analysis: the Kernel Approach with S-Plus Illustrations*. Oxford University Press, 1997.

[24] M. Di Marzio, A. Panzera, and C. C. Taylor. Non-parametric regression for circular responses. *Scandinavian Journal of Statistics*, 40:238–255, 2012.

[25] M. Alonso-Pena, J. Ameijeiras-Alonso, and R. M. Crujeiras. Nonparametric tests for circular regression. *Journal of Statistical Computation and Simulation*, 19:477–500, 2021.

[26] A. Meilán-Vila, M. Francisco-Fernández, and R. M. Crujeiras. Goodness-of-fit tests for multiple regression with circular response. *Journal of Statistical Computation and Simulation*, 92:1941–1963, 2022.

[27] M. Di Marzio, A. Panzera, and C. C. Taylor. Nonparametric circular quantile regression. *Journal of Statistical Planning and Inference*, 170:1–14, 2016.

[28] M. Alonso-Pena and R. M. Crujeiras. Analyzing animal escape data with circular nonparametric multimodal regression. *Annals of Applied Statistics*, 17:130–152, 2023.

[29] M. Rosenblatt. Remarks on Some Nonparametric Estimates of a Density Function. *The Annals of Mathematical Statistics*, 27(3):832–837, 1956.

[30] E. Parzen. On Estimation of a Probability Density Function and Mode. *The Annals of Mathematical Statistics*, 33(3):1065–1076, 1962.

[31] A. Meilán-Vila, M. Francisco-Fernández, R. M. Crujeiras, and A. Panzera. Nonparametric multiple regression estimation for circular response. *TEST*, 30:650–672, 2021.

[32] Y.-C. Chen, C. R. Genovese, R. J. Tibshirani, and L. Wasserman. Nonparametric modal regression. *The Annals of Statistics*, 44:489–514, 2016.

[33] Y. Zhang and Y.-C. Chen. Kernel smoothing, mean shift, and their learning theory with directional data. *Journal of Machine Learning Research*, 22:1–92, 2021.

[34] Y. Zhang and Y.-C. Chen. The EM perspective of directional mean shift algorithm. *arXiv*, arXiv:2101.10058, 2021.

[35] M. Alonso-Pena, M. Oliveira, J. Ameijeiras-Alonso, R. M. Crujeiras, M. Gijbels, and A. Rodríguez-Casal. *R package* ***NPCirc****: Nonparametric Circular Methods (version 3.1.1)*, 2022.

[36] M. Oliveira, R. M. Crujeiras, and A. Rodríguez-Casal. NPCirc : An R package for nonparametric circular methods. *Journal of Statistical Software*, 61:1–26, 2014.

[37] D. W. Scott and G. R. Terrell. Biased and unbiased cross-validation in density estimation. *Journal of the American Statistical Association*, 82(400):1131–1146, 1987.

[38] S. J. Sheather and M. C. Jones. A reliable data-based bandwidth selection method for kernel density estimation. *Journal of the Royal Statistical Society: Series B (Methodological)*, 53(3):683–690, 2018.

[39] C. C. Taylor. Automatic bandwidth selection for circular density estimation. *Computational Statistics and Data Analysis*, 52(7):3493–3500, 2008.

[40] M. Oliveira, R.M. Crujeiras, and A. Rodríguez-Casal. A plug-in rule for bandwidth selection in circular density estimation. *Computational Statistics and Data Analysis*, 56(12):3898–3908, 2012.

[41] E. García-Portugués. Exact risk improvement of bandwidth selectors for kernel density estimation with directional data. *Electronic Journal of Statistics*, 7:1655–1685, 2013.

[42] J. Ameijeiras-Alonso. A reliable data-based smoothing parameter selection method for circular kernel estimation. *Statistics and Computing*, 34, 2024.

[43] S. Zámečník, I. Horová, S. Katina, et al. An adaptive method for bandwidth selection in circular kernel density estimation. *Computational Statistics*, 39:1709–1728, 2024.

[44] Y. Zhang and Y.-C. Chen. Mode and ridge estimation in Euclidean and directional product spaces: A mean shift approach. *Journal of Computational and Graphical Statistics*, 2025.

[45] A. Casa, J. E. Chacón, and G. Menardi. Modal clustering asymptotics with applications to bandwidth selection. *Electronic Journal of Statistics*, 14:835–856, 2020.

[46] Y. Sarma and S. R. Jammalamadaka. Circular regression. In *Statistical Science and Data Analysis: Proceedings of the Third Pacific Area Statistical Conference*, pages 109–128, Utrecht, Netherlands, 1993. VSP.

[47] F. Wang, A. E. Gelfand, and G. Jona-Lasinio. Joint spatio-temporal analysis of a linear and a directional variable: Space-time modeling of wave heights and wave directions in the Adriatic Sea. *Statistica Sinica*, 25(1):25–39, 2015.

[48] G. Jona-Lasinio, A. E. Gelfand, and G. Mastrantonio. Spatial and spatio-temporal circular processes with application to wave directions. In *Applied Directional Statistics*, pages 145–178. Chapman and Hall/CRC, 2018.

11

Multivariate Bayesian Spatio-Temporal Climate Analysis and Change Point Detection

Giovanna Jona Lasinio, Gianluca Mastrantonio, and Alessio Pollice

Motivated by monthly series of minimum and maximum temperatures and precipitation recorded over six decades across 360 monitoring stations in Italy, we introduce two complementary Bayesian models to investigate multivariate spatio-temporal climate data and identify structural shifts in weather regimes. Both models serve to characterize the Italian ecoregions, which are ecologically homogeneous zones defined by common climatic, physiographic, and ecological attributes, including vegetation, wildlife, hydrography, and landforms.

First, we present a hierarchical model that incorporates spatio-temporal correlation via a non-separable covariance function, multivariate dependencies through a linear model of coregionalization, and cyclic annual effects using a circular representation of time. This approach facilitates missing data imputation, climate surface interpolation, and the climatic characterization of Italian ecoregions, offering a quantitative foundation for their climatic definitions.

The second model is designed to detect change-points in multivariate time series by employing a hierarchical Dirichlet process (DP). This method enables station-specific change-points while supporting shared parameter structures and estimating the number of weather regimes as a random variable. Key features include identifying clusters of spatial units based on parameter stability and differentiating between abrupt and gradual changes. By capturing distinctions between Mediterranean and Temperate ecosystems and revealing contrasting trends across continental plains and mountain ranges, the model highlights regional susceptibilities to climate change.

11.1 Introduction

Climate factors, including temperature, precipitation, and annual cycles, are essential in shaping the distribution and diversity of plants, animals, soils, and their interactions within intricate ecosystems [1, 2]. Gaining insight into climate change and its effects on biodiversity is crucial for safeguarding natural capital and fostering sustainable development at various scales [3]. Although the influence of climate change on individual species has been extensively examined, considerable uncertainties persist regarding its broader effects

DOI: 10.1201/9781003564423-11

on ecosystems, which encompass interacting species within shared environments [4, 5, 6]. To achieve this aim, methods for spatial and spatio-temporal interpolation are widely adopted, giving rise to several challenges. Partially motivated by the latter, a growing body of literature focuses on the interpolation of large spatial and spatio-temporal datasets, with several review papers [7, 8, 9] and books [10] offering detailed insights. Methods for modeling purely spatial covariance matrices while tackling computational challenges associated with large datasets include low-rank models, covariance tapering [10, and references therein], Gaussian Markov Random Fields (GMRF) approximations using Laplace transforms and Stochastic Partial Differential Equations [11], products of lower-dimensional conditional densities [see 12, and references therein], and composite likelihood estimation methods [13]. Multivariate extensions of covariance tapering and composite likelihood methods have been introduced in [14] and [15]. Examples of univariate spatio-temporal models include dynamic low-rank processes [16] and GMRF-based approaches [17]. While spatio-temporal models are commonly defined with discrete time parameters, continuous space-time models have also been investigated, as in [18] and [19], which apply composite likelihood methods for parameter estimation. A recent advancement in modeling large continuous space-time datasets is the extension of the Nearest Neighbor Gaussian Process (NNGP) approximation [20], which has demonstrated superior performance compared to state-of-the-art methods like predictive processes [21].

Change Point (CP) detection has a well-established methodological foundation, with approaches often tailored to specific fields of study. In climate science, CP models have gained prominence, employing methods such as change detection statistics [22], piecewise linear trend estimation [23], sequential and CUSUM tests for regime shifts [24, 25], information-theoretic approaches [26, 27], two-phase time series regression models [28, 29], maximum likelihood estimation [30, 31, 32], and high-dimensional regression variable selection [33]. Software implementations for CP detection are also available [34, 35]. Bayesian methods for CP detection, influenced by seminal works such as [36] and [37], have been comprehensively reviewed in [38]. Recent advancements have explored the use of DPs in Bayesian CP modeling [38, 39], underscoring the diversity and ongoing development in this field. Furthermore, [40] recently emphasized the importance of analyzing multivariate climate departures across different regions relative to historical behaviors. Our approach enables the assessment of simultaneous changes across various geographical areas and multiple dimensions of climate phenomena, including variability, mean, and inter-variable correlations.

The methods and results presented in this chapter are primarily derived from two studies [41] and [42], which address challenges related to climate data and vegetation distribution in Italy through collaboration with botanists.

11.2 Thermo-Pluviometric Historical Data, Italian Ecoregions, and Climate Regimes

Long-term instrumental time series, typically spanning at least 30 years, are essential for contextualizing current weather and climate within a historical framework [43]. For this study, we collected long-term thermo-pluviometric data from a network of 360 weather monitoring stations in Italy. The dataset comprises monthly precipitation and minimum/maximum temperature records over a 60-year period (1951–2010). Missing data are present across all time series, and seasonality is a consistent characteristic of thermo-pluviometric data. The data were sourced from national institutions (ISPRA, CRA/CREA, Meteomont, and ENEA) as well as local authorities.

(a) $k = 5$

(b) $k = 2$ (c) $k = 3$ (d) $k = 4$

FIGURE 11.1
Hierarchical tiers of Italian ecoregions: (a) $k = 5$ Subsections and the climate monitoring network, (b) $k = 2$ divisions, (c) $k = 3$ provinces, (d) $k = 4$ sections.

The ecological classification of climate is a fundamental step in defining and mapping ecoregions, which represent broad ecosystems within distinct geographical areas [44, 45]. A hierarchical classification of Italian ecoregions was recently developed by integrating climatic diagnostic features with biodiversity distribution patterns and environmental characteristics [46]. The Italian ecoregions (see Figure 11.1) are structured into four nested tiers: 2 divisions, 7 provinces, 11 Sections, and 35 Subsections. The climatic features used to diagnose and describe these ecoregions are based on thermo-pluviometric data and bioclimatic indices in the period 1955–2000. This study contributes to characterizing Italian ecoregions in terms of both current and historical climatic conditions, summarizing climate variables at the ecoregion level to assess climate impacts on ecosystems and support biodiversity conservation strategies.

The resilience of complex ecological systems compared to individual species raises essential questions about the impacts of climate change, as ecosystems may exhibit long-term adaptive responses through shifts in species composition or spatial distribution [47]. Abrupt climate changes, detectable through CPs, are likely to have a greater impact on ecosystem structure and functioning than gradual changes [48, 49]. Consequently, models capable of identifying sharp temporal and spatial shifts in climate regimes are crucial, particularly at intermediate scales between local and global levels. We hypothesize that CP detection models incorporating spatial similarities and interactions between bioclimatic features can provide valuable insights into the impacts of climate change on ecosystems. Climate change may affect ecological regions unevenly, especially in areas with pronounced orographic and physiographic heterogeneity, influencing the interdependence between climate parameters and biodiversity patterns. Accordingly, we focus on defining CPs that account for correlations between temperature and precipitation, as well as their mean values and variances across meteorological stations within different ecoregions. This approach enables the spatial distribution of CPs to be mapped at a national scale, identifies common climatic trends across stations, and highlights bioclimatic similarities between ecoregions, independent of potential changes.

11.3 Bayesian Multi-response Modeling of Climate Variables

The challenges outlined in the preceding sections are addressed through a fully model-based approach, utilizing stochastic models that capture key characteristics of the multivariate spatio-temporal field underlying the data, including correlations among climate variables and space-time variability. Estimation is conducted within the Bayesian hierarchical modeling framework, which provides a robust mechanism for managing various sources of uncertainty.

Bayesian estimation offers a flexible and computationally feasible framework for multivariate modeling with continuous space-time data. It facilitates direct imputation of missing data and enables precise evaluation of estimation and prediction uncertainties. The models are estimated using Markov Chain Monte Carlo (MCMC) algorithms. Despite the complexity of the proposed models, the MCMC algorithms primarily rely on Gibbs sampling steps, which are implemented through the introduction of appropriate latent variables.

Let $\boldsymbol{s} \in \mathcal{S} \subset \mathbb{R}^d$, with $d = 2$, and $t \in \mathcal{T} \subset \mathbb{R}^+$ be the spatial and temporal coordinates, respectively. Let $Y_1(\boldsymbol{s}, t)$, $Y_2(\boldsymbol{s}, t)$, and $Y_3(\boldsymbol{s}, t)$ represent respectively the precipitation level, minimum temperature, and maximum temperature observed at $(\boldsymbol{s}, t)$. These variables satisfy the following constraints:

$$Y_1(\boldsymbol{s}, t) \geq 0, \quad \text{and} \quad Y_3(\boldsymbol{s}, t) \geq Y_2(\boldsymbol{s}, t).$$

To facilitate model implementation, the three climate variables are rescaled and standardized. To preserve the information about zero values, the monthly precipitation amount was rescaled by its sample standard deviation, while Y_2 and Y_3 were both standardized.

$$\tilde{Y}_1(\boldsymbol{s}, t) = Y_1(\boldsymbol{s}, t) \mathrm{S}_{Y_1}^{-1}$$

$$\tilde{Y}_i(\boldsymbol{s}, t) = \left(Y_i(\boldsymbol{s}, t) - \bar{Y}_i\right) \mathrm{S}_{Y_i}^{-1}, \quad i = 2, 3,$$

where $\bar{Y}_i$ and $\mathrm{S}_{Y_i}^2$ are the sample means and variances.

11.3.1 Generalized NNGP Spatio-temporal Interpolation

It is commonly assumed that spatio-temporal data are generated by a Gaussian process (GP) and that the covariance function captures space-time dependencies [50, and references therein]. While the richness and flexibility of spatio-temporal stochastic process models are indisputable, their computational feasibility and implementation pose some challenges for *large* datasets. A proposal to model large data continuous in both space and time was provided by [20], who extended the definition of the NNGP approximation for spatial data given in [12], the latter outperforming competitive state of the art approaches such as predictive processes [21]. In this work, we are going to introduce some relevant novelties in the application of the approach proposed by [20] to our multivariate spatio-temporal model setting and use a generalized NNGP approximation for the multi-response interpolation of climate variables. To achieve this aim, we combine the NNGP with the linear model of coregionalization [51] and with a circular representation of time [see for instance 52] that allows incorporating the effects of annual cycles.

Three latent variables $Y_1^*(\boldsymbol{s},t)$, $Y_2^*(\boldsymbol{s},t)$, and $Y_3^*(\boldsymbol{s},t)$ ranging over $\mathbb{R}$, and embedding the constraints given in Section 11.3 are defined as follows:

$$\begin{cases} Y_1^*(\boldsymbol{s},t) = \tilde{Y}_1(\boldsymbol{s},t) & \text{if } \tilde{Y}_1(\boldsymbol{s},t) > 0, \\ Y_1^*(\boldsymbol{s},t) \leq 0 & \text{if } \tilde{Y}_1(\boldsymbol{s},t) = 0, \end{cases}$$

$$Y_2^*(\boldsymbol{s},t) = \tilde{Y}_2(\boldsymbol{s},t),$$

$$\begin{cases} Y_3^*(\boldsymbol{s},t) = \tilde{Y}_3(\boldsymbol{s},t) - \tilde{Y}_2(\boldsymbol{s},t) & \text{if } \tilde{Y}_3(\boldsymbol{s},t) - \tilde{Y}_2(\boldsymbol{s},t) > 0, \\ Y_3^*(\boldsymbol{s},t) \leq 0 & \text{if } \tilde{Y}_3(\boldsymbol{s},t) - \tilde{Y}_2(\boldsymbol{s},t) = 0. \end{cases}$$

Each latent response Y_i^*, $i = 1, 2, 3$ is described by a combination of fixed and random terms:

$$Y_i^*(\boldsymbol{s},t) = \mathbf{X}(\boldsymbol{s})\boldsymbol{\beta}_{i,z_k}(\boldsymbol{s}) + \omega_i(\boldsymbol{s},t) + \lambda_i(\boldsymbol{s},t) + \epsilon_i(\boldsymbol{s},t), \tag{11.1}$$

with $\epsilon_i(\boldsymbol{s},t) \overset{iid}{\sim} N(0, \sigma^2_{\epsilon,i})$. Here, $\mathbf{X}(\boldsymbol{s}) = (1, X(\boldsymbol{s}))$ and $X(\boldsymbol{s})$ is the elevation of site $\boldsymbol{s}$. The integer valued indicator $z_k(\boldsymbol{s}) \in \mathbb{Z}^+$ is the ecoregion label for the kth ecoregion tier: with $k = 1$, we have one ecoregion covering the entire country, while $k = 5$ returns the finer classification with 35 ecoregions. Then, $\boldsymbol{\beta}_{i,z_k} = (\beta_{0,i,z_k}, \beta_{1,i,z_k})'$ are regression coefficients, varying with the ecoregion. The terms $\lambda_i(\boldsymbol{s},t)$ describe the monthly effects of the annual cycle and follow a GP:

$$\lambda_i(\boldsymbol{s},t) \sim GP\left(0, \sigma^2_{c,i} \exp(-\phi_{c,i} h_{circle}) \mathbf{1}_0(h_{space})\right), \quad i = 1, 2, 3. \tag{11.2}$$

$h_{space} = \|s - s'\|$ and $h_{time} = |t - t'|$ indicate spatial and temporal distances, $h_{circle} = h_{time} \bmod (12)$ is the circular distance with 12 months period, and $\mathbf{1}_0(h_{space}) = 1$ when $h_{space} = 0$ and 0 otherwise. The random vector $\boldsymbol{\omega}(\boldsymbol{s},t) = (\omega_1(\boldsymbol{s},t), \omega_2(\boldsymbol{s},t), \omega_3(\boldsymbol{s},t))'$ is defined as:

$$\boldsymbol{\omega}(\boldsymbol{s},t) = \mathbf{A}\mathbf{w}(\boldsymbol{s},t) \tag{11.3}$$

where $\mathbf{A}$ is a 3×3 matrix (see [41] for details on how $\mathbf{A}$ was set) and $\mathbf{w}(\boldsymbol{s},t) = (w_1(\boldsymbol{s},t), w_2(\boldsymbol{s},t), w_3(\boldsymbol{s},t))'$ is a vector of zero mean GPs with independent components and spatio-temporal correlation function $C(h_{space}, h_{time}, \boldsymbol{\theta}_i)$ given by the general non-separable space-time correlation structure proposed by [53]. The vector-valued parameter $\boldsymbol{\theta}_i$ contains two non-negative decay parameters $\phi_{t,i}$ and $\phi_{sp,i}$ associated to time and space, respectively and the space-time interaction parameter η_i ranging in $[0, 1]$. Notice that when $\eta_i = 0$, we achieve space-time separability.

To summarize, the regressive part in (11.1) specifies differences between ecoregions, spatial and temporal dependence is ruled by $\omega_i(\boldsymbol{s},t)$, the seasonal component is described in $\lambda_i(\boldsymbol{s},t)$, the correlation between responses is defined by $\mathbf{A}$ and non-structured effects are summarized in $\epsilon_i(\boldsymbol{s},t)$.

For a parsimonious implementation of the model estimation algorithm, we define a new variable $\omega_i^*(\boldsymbol{s},t) = \omega_i(\boldsymbol{s},t) + \lambda_i(\boldsymbol{s},t)$. Letting N be the number of observations in space and time, we denote their locations by $(\boldsymbol{s}_n, t_n)$, $n = 1,\ldots,N$ and let $\boldsymbol{\omega}_n^* = (\omega_1^*(\boldsymbol{s}_n,t_n), \omega_2^*(\boldsymbol{s}_n,t_n), \omega_3^*(\boldsymbol{s}_n,t_n))'$ with $\boldsymbol{\omega}^* = (\boldsymbol{\omega}_1^*,\ldots,\boldsymbol{\omega}_N^*)'$. With the NNGP, the joint finite-dimensional distribution of the GP $\boldsymbol{\omega}^*$ is approximated as follows:

$$f(\boldsymbol{\omega}^*) = \prod_{n=1}^{N} f(\boldsymbol{\omega}_n^*|\boldsymbol{\omega}_{n-1}^*,\ldots,\boldsymbol{\omega}_1^*) \approx \prod_{n=1}^{N} f(\boldsymbol{\omega}_n^*|\boldsymbol{\Omega}_n(m)). \tag{11.4}$$

where $\boldsymbol{\Omega}_n(m) \subseteq \boldsymbol{\Omega}_n$ be a subset that contains at most m elements of $(\boldsymbol{\omega}_{n-1}^*,\ldots,\boldsymbol{\omega}_1^*)'$. As shown by [12], the quality of the approximation increases with m, and the best results are achieved if we choose the m elements of $\boldsymbol{\Omega}_n$ that have the higher correlation with $\boldsymbol{\omega}_n^*$ (see [41] for details on how to order the observations, how to choose the value of m, and how to choose the elements of $\boldsymbol{\Omega}_n(m)$).

11.3.2 Dirichlet Process-based Change Point Detection

As a secondary objective, we seek to detect change points both in time and space across the Italian territory for the climate data presented in Sections 11.2 and 11.3. Our methodology is motivated by techniques that partition a multivariate time series into separate temporal regimes, each composed of contiguous observations [54]. These regimes are defined under the assumption that observations within the same regime originate from a common distribution. The transitions between regimes are determined by *local change points*, which occur periodically due to seasonal or other cyclic patterns, or by *global change points*, which manifest once and lead to a lasting shift in the underlying distribution.

Local and global CPs are incorporated within mixture-based frameworks for model-based clustering, where the mixture components define clusters or temporal regimes. Our methodology extends the Bayesian mixture-based CP model of [37] to a multivariate setting utilizing the Dirichlet process (DP) [55]. While the original model necessitates predefining the maximum number of CPs, the DP-based framework treats this quantity as a random variable, enhancing model flexibility.

In this study we use these three variables $Y_1^{**}(\boldsymbol{s},t)$, $Y_2^{**}(\boldsymbol{s},t)$, and $Y_3^{**}(\boldsymbol{s},t)$, which are related as follows to the climate variables defined in Sections 11.3 and 11.3.1:

$$Y_i^{**}(\boldsymbol{s},t) \equiv Y_i^*(\boldsymbol{s},t), \quad i = 1,2,$$

$$Y_3^{**}(\boldsymbol{s},t) = \tilde{Y}_3(\boldsymbol{s},t)$$

The model proposed for CP detection assumes that the vector $\mathbf{Y}^{**}(\boldsymbol{s},t) = (Y_1^{**}(\boldsymbol{s},t), Y_2^{**}(\boldsymbol{s},t), Y_3^{**}(\boldsymbol{s},t))$ has a trivariate normal density ($\phi_3(\cdot)$) at each monitoring station $\mathbf{s}$ and time t:

$$f(\mathbf{y}^{**}(\boldsymbol{s},t)|\psi(\boldsymbol{s},t),\boldsymbol{\theta}(\boldsymbol{s},t)) = \phi_3\left(\mathbf{y}^{**}(\boldsymbol{s},t)|\psi(\boldsymbol{s},t) + \boldsymbol{\mu}(\boldsymbol{s},t), \boldsymbol{\Sigma}(\boldsymbol{s},t)\right). \tag{11.5}$$

The vector-valued parameter $\boldsymbol{\theta}(\boldsymbol{s},t)$ is composed by the nine elements in the spatio-temporal mean and covariance $\boldsymbol{\mu}(\boldsymbol{s},t)$ and $\boldsymbol{\Sigma}(\boldsymbol{s},t)$, $\psi(\boldsymbol{s},t)$ represents a seasonal component with $\psi(\boldsymbol{s},t) = \psi(\boldsymbol{s},t+12)$ and $\sum_{t=1}^{12} \psi(\boldsymbol{s},t) = 0$. The parameter $\boldsymbol{\theta}(\boldsymbol{s},t)$ is used to define the CP model by a modified version of the hierarchical DP [56]. At each time-point and monitoring station, the nine-variate parameter vector $\boldsymbol{\theta}(\boldsymbol{s},t)$ is assumed to come from a DP mixture-type model. Mixture components correspond to different distributional features of the underlying three-variate Gaussian distribution, possibly connected with time regimes and clusters of spatial units of the climate variables.

First, we define independent DPs for the nine parameters $\theta_\ell \in \Theta_\ell$, with $\ell = 1, \ldots, 9$. Let G_{θ_ℓ} be a draw from $\text{DP}(\gamma, H_{\theta_\ell})$ that depends on a *scale parameter* $\gamma > 0$ and on the *base distribution* H_{θ_ℓ} over Θ_ℓ [55]:

$$G_{\theta_\ell} = \sum_{p \in \mathbb{N}} \nu_{\ell,p} \delta_{\eta_{\ell,p}}, \tag{11.6}$$

where δ_x is the Dirac delta function. Notice that G_{θ_ℓ} is a discrete distribution and a sample from the $\text{DP}(\gamma, H_{\theta_\ell})$ that can be represented using the sets of atoms $\{\eta_{\ell,p}\}_{p \in \mathbb{N}}$ and weights $\boldsymbol{\nu}_\ell = \{\nu_{\ell,p}\}_{p \in \mathbb{N}}$, where $\eta_{\ell,p} \overset{iid}{\sim} H_{\theta_\ell}$ and $\boldsymbol{\nu}_\ell$ is a draw from a Dirichlet distribution that depends on γ and H_{θ_ℓ} [see 56, 57, for more details]. Next, we combine the sets of atoms $\{\eta_{\ell,p}\}_{p \in \mathbb{N}}$ and weights $\boldsymbol{\nu}_\ell$ associated with each parameter θ_ℓ to construct a multivariate discrete distribution. This distribution serves as the base distribution for a further DP. Let $\boldsymbol{\eta}_k$ denote a vector-valued atom defined as $\boldsymbol{\eta}_k = \{\eta_{\ell, w_{\ell,k}}\}_{\ell=1}^{9}$, where $w_{\ell,k} = p$ indicates that the ℓ-th element of $\boldsymbol{\eta}_k$ corresponds to $\eta_{\ell,p}$. We then define the discrete multivariate distribution:

$$G_0 = \sum_{k \in \mathbb{N}} \xi_k \delta_{\boldsymbol{\eta}_k}, \tag{11.7}$$

where the random weights ξ_k are obtained as products of the associated $\nu_{\ell,p}$. Specifically, $\xi_k = \prod_{\ell=1}^{9} \nu_{\ell, w_{\ell,k}}$, ensuring that no two $\boldsymbol{\eta}_k$ are identical. The set $\{\boldsymbol{\eta}_k\}_{k \in \mathbb{N}}$ comprises all possible unique combinations of the nine parameters, with no repetitions. Each nine-variate atom in G_0 contains three means, three variances, and three correlations, ensuring that no two $\boldsymbol{\eta}_k$ share identical values for all nine parameters.

Since two distinct $\boldsymbol{\eta}_k$ may share some components, the occurrence of a CP does not necessitate a change in all nine parameters. Instead, the minimal condition for detecting a CP is that at least one element of the parameter set changes. Additionally, the mixture-type CP model incorporates a *no-return constraint*: if at any time point, t the process transitions from the k-th mixture component to a new one, it cannot revert to component k at any subsequent time. This constraint governs the temporal evolution of the system at each monitoring station (see Section 3.2 of [42]). Finally, a second constraint is induced by the minimum regime length m^* (see [42] equation (6)) that would imply that all regimes are composed of at least m^* time-points.

11.4 Model-based Analyses of Italian Climate Data

11.4.1 Revision of the Climatic Characterization of Italian Ecoregions

By using the model described in Section 11.3.1, we estimated nine distinct models by varying the number of neighbors in the NNGP and the ecoregional hierarchical tier. Specifically, we considered $m = \{10, 15, 20\}$ for the number of neighbors and $k \in \{3, 4, 5\}$ for the ecoregional tier.

Weakly informative priors were employed across all models. Regression parameters were assigned a $\text{N}(0, 100)$ prior, while all variance parameters were given an inverse gamma prior, $\text{IG}(1, 1)$. The covariance matrix $\boldsymbol{\Sigma}$ was assigned an inverse Wishart prior, $\text{IW}(3, \mathbf{I})$, and the separability parameter was given a uniform prior, $\text{U}(0, 1)$. For the spatial range

TABLE 11.1
Proportions of the variance components for each climate variable.

	σ^2_ω	σ^2_c	σ^2_ϵ
Y_1	0.342	0.512	0.146
Y_2	0.007	0.992	0.001
Y_3	0.155	0.827	0.018

parameter, ϕ_{sp}, we used a uniform prior, $\phi_{sp} \sim U(0.05, 6)$, corresponding to practical spatial ranges in the interval $[6, 60]$ km. Similarly, the temporal range parameters, ϕ_t and ϕ_c, were assigned uniform priors, $\phi_t, \phi_c \sim U(3, 36)$, corresponding to practical temporal ranges in the interval $[1, 12]$ months. The MCMC was implemented with 100,000 iterations, a burn-in phase of 70000 and thinning by 12, keeping 2500 samples for posterior inferences. Thanks to the selected priors, the MCMC algorithm samples regression coefficients, nuggets, missing data, spatio-temporal and circular processes using Gibbs steps, i.e., from their full conditionals, while the other parameters are sampled all together with a Metropolis step using the adaptive proposal of [58], algorithm 4.

The choice among alternative specifications of the same model was performed using the Deviance Information Criterion (DIC) [59] and the continuous ranked probability scores (CRPS) [60] (see [41] for further details). Most model selection criteria indicated that the optimal model structure corresponds to $k = 3$ and $m = 20$. The "best" model suggests aggregating ecoregions into seven distinct provinces (see Figure 11.1), representing the most generalized and aggregated ecoregional tier among those evaluated in the model implementation. This finding aligns with a fundamental principle commonly adopted in hierarchical approaches to ecological land classification. Specifically, this principle asserts that climate is a primary environmental driver influencing broad-scale ecosystem variation. In contrast, geomorphology and soil characteristics become equally or more significant than climate at finer spatial aggregation levels [1, 61].

In Table 11.1, we show the proportion of variance due to the space-time ($\sigma^2_{\omega,i}$ is the ith element of the diagonal of $\mathbf{\Sigma}$), the cyclical and the residual component for each response variable, to appreciate the relevance of each of the three components in explaining the total variation. It is worth noticing that for the minimum temperature Y_2 almost the entire variation can be ascribed to the cyclical component, while for the thermal excursion Y_3 only 15.5% is due to the spatio-temporal term, and a negligible contribution comes from the residual part. The precipitation Y_1 has a different behavior: a large portion of the total variation (51%) is due to the cyclical component, the space-time dynamic accounts for 34.2%, while a quite relevant 14.6% of the total variation is left unexplained. Compatible with the physics of the phenomena, precipitation dynamics seem to be more affected by small-scale events compared to Y_2 and Y_3, causing higher residual variability.

Posterior estimates of the GP and annual cyclical component parameters (see equations (11.1) and (11.2)) are reported in Table 11.2 with their 95% credible intervals (CIs). The three climate variables have non-separable space-time dynamics, as CIs for the η parameters are never close to 0. Practical ranges and covariances of the three components in (11.1) provide useful information on the extent of the spatial, temporal, and annual cyclical dependence. The spatial practical ranges of Y_1, Y_2, and Y_3 are respectively 15.95 *km*, 21.676 *km*, and 6.967 *km*, suggesting a less homogeneous spatial behavior of the temperature range compared to the other two variables. In terms of time dependence, we have the following practical ranges: 37.78 days, 113.73 days, and 45.98 days for Y_1, Y_2 and Y_3, respectively. These values highlight a similar extension of the temporal correlation for the precipitation and the thermal excursion. Finally, as expected, the annual cyclical effect ϕ_c has similar

TABLE 11.2
Posterior estimates and 95% CIs of the GP and annual cyclical component parameters.

		ϕ_{sp}	ϕ_t	ϕ_c	η
Y_1	Est.	0.188	28.979	15.210	0.774
	(CI)	(0.184 0.192)	(28.871 29.072)	(14.972 15.394)	(0.774 0.775)
Y_2	Est.	0.138	9.628	10.176	0.943
	(CI)	(0.137 0.140)	(9.476 9.750)	(10.102 10.239)	(0.942 0.943)
Y_3	Est.	0.431	23.814	9.760	0.166
	(CI)	(0.429 0.432)	(23.576 23.995)	(9.690 9.835)	(0.165 0.168)
		σ_c^2	σ_ϵ^2	σ_ω^2	
Y_1	Est.	0.617	0.176	0.413	
	(CI)	(0.612 0.623)	(0.175 0.178)	(0.409 0.416)	
Y_2	Est.	6.968	0.008	0.050	
	(CI)	(6.830 7.092)	(0.008 0.008)	(0.050 0.051)	
Y_3	Est.	2.799	0.062	0.525	
	(CI)	(2.647 2.919)	(0.061 0.062)	(0.519 0.532)	

behavior for the second and third variables, with practical ranges 71.99 days (Y_1), 107.60 days (Y_2) and 112.19 days (Y_3): annual cycles are longer and almost seasonal (4 months long, as expected) for the minimum temperature and the temperature range. In contrast, a shorter cycle is estimated for the precipitation. Measures of the correlation between climate variables are also obtained and are all far from zero. Precipitation and minimum temperature are positively correlated (0.210 with 95% CI (0.208, 0.213)), while precipitation and temperature range are negatively correlated (−0.214 with 95% CI (−0.218,−0.211)). As expected, minimum temperature and temperature range are negatively correlated with a stronger relation (−0.493 with 95% CI (−0.499,−0.497)).

In Tables 11.3 and 11.4, the posterior estimates of the intercepts and regression coefficients for each of the seven ecoregion provinces are reported with their 95% CIs. All estimates for the fourth province (1D) are not reliable and very different from the other values due to the presence of only one monitoring station in the given area. This suggests aggregating the fourth province to one of its neighbors for future investigations. The estimates of the models' intercepts β_0's allow us to analyze the average behavior of each component in the specific province. The only estimate that shows a value close to zero is for the temperature range in ecoregion 1B, the Po Plain province, suggesting a very small temperature range in that area. All provinces are well characterized with occasional overlapping CIs in Table 11.4, suggesting similarities between areas. Similar behavior in terms of precipitation (Y_1) can be found in provinces 1B, 2A, 2B, and 2C ($j = 2, 5, 6, 7$). At the same time, 1A, 1B, and 2C ($j = 1, 2, 7$) show similarities in terms of minimum temperature (Y_2), and only 1A and 2C ($j = 1, 7$) show interval estimates overlapping for the temperature range (Y_3). Notice that province 1B covers the Po Plain, which is a very heterogeneous area where a transition from continental to Mediterranean behavior occurs. The relation with the elevation described by the estimates of the regression coefficients β_1's often admits the zero value in the 95% CI. This is likely linked to a latitudinal gradient in the area. For example, in the Po Plain province (1B), a large area is divided by the Po River into a northern sector with a continental regime and a southern sector with an Apennine regime, as previously highlighted in relation to the effects of cyclical components. The Alpine province (1A) is associated with regression coefficients that are all quite far from zero. This can be linked to the absence of a latitudinal gradient, as the region is only affected by longitudinal

TABLE 11.3
Point estimates of the intercepts at each province for each GP component and relative 95% CIs.

		$\beta_{0,1}$	$\beta_{0,2}$	$\beta_{0,3}$	$\beta_{0,4}$
Y_1	Est.	1.348	0.741	0.965	-0.050
	(CI)	(1.280 1.400)	(0.668 0.794)	(0.924 1.003)	(-8.155 7.777)
Y_2	Est.	0.207	0.225	0.094	-0.005
	(CI)	(0.189 0.234)	(0.207 0.240)	(0.084 0.106)	(-1.746 1.758)
Y_3	Est.	0.155	0.008	0.779	0.083
	(CI)	(0.055 0.201)	(-0.069 0.097)	(0.737 0.844)	(-4.919 4.748)
		$\beta_{0,5}$	$\beta_{0,6}$	$\beta_{0,7}$	
Y_1	Est.	0.633	0.740	0.783	
	(CI)	(0.520 0.787)	(0.710 0.774)	(0.745 0.838)	
Y_2	Est.	0.878	0.457	0.277	
	(CI)	(0.838 0.910)	(0.445 0.464)	(0.263 0.295)	
Y_3	Est.	-1.965	-0.070	0.196	
	(CI)	(-2.038 -1.856)	(-0.125 -0.038)	(0.127 0.264)	

TABLE 11.4
Point estimates of the regression coefficients of the elevation (multiplied by 10^3) at each province for each GP component and relative 95% CI's. provinces are coded as follows $1 = 1A$, $2 = 1B$, $3 = 1C$, $4 = 1D$, $5 = 2A$, $6 = 2B$, and $7 = 2C$.

		$\beta_{1,1}$	$\beta_{1,2}$	$\beta_{1,3}$	$\beta_{1,4}$
Y_1	Est.	0.029	0.498	0.248	63.022
	(CI)	(−0.022 0.068)	(0.189 0.742)	(0.208 0.294)	(−650.237 788.956)
Y_2	Est.	−0.747	−0.322	−0.452	73.423
	(CI)	(−0.761 −0.729)	(−0.414 −0.242)	(−0.470 −0.437)	(−86.057 231.301)
Y_3	Est.	−0.366	−0.094	−1.073	−117.311
	(CI)	(−0.428 −0.305)	(−0.648 0.423)	(−1.177 −1.023)	(−544.667 338.024)
		$\beta_{1,5}$	$\beta_{1,6}$	$\beta_{1,7}$	
Y_1	Est.	−0.532	0.548	0.416	
	(CI)	(−1.250 0.241)	(0.502 0.592)	(0.272 0.546)	
Y_2	Est.	−2.802	−0.842	−0.452	
	(CI)	(−3.038 −2.541)	(−0.856 −0.826)	(−0.498 −0.405)	
Y_3	Est.	8.256	0.040	−0.736	
	(CI)	(6.891 9.607)	(−0.042 0.112)	(−0.985 −0.570)	

variation. Moreover, the area is characterized by significant relief energy (a large elevation gradient).

After model fitting, we used posterior samples to predict the values of the responses, and hence the cyclical component of the model $\lambda_i(\boldsymbol{s}, t)$, on a square lattice of 3305 spatial points with 15 *km* side over 720 time points. In Figure 11.2, examples of two maps of predicted monthly effects of the annual cyclical component are reported. The maps for January and August have been chosen to represent the factors affecting ecosystems' composition and

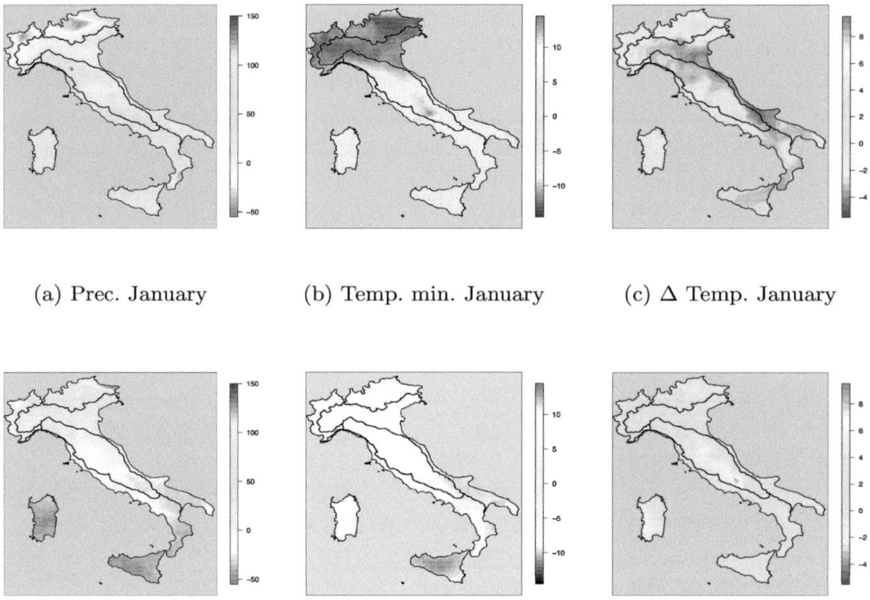

(a) Prec. January (b) Temp. min. January (c) Δ Temp. January

(d) Prec. August (e) Temp. min. August (f) Δ Temp. August

FIGURE 11.2
Maps of the monthly effects of January (a,b,c) and August (d,e,f) on the annual cycles of the three components of the process.

distribution over the Italian territory. Above all, these factors include moisture availability in the different seasons, such as winter cold and summer drought. Firstly, the model was able to show some interesting seasonal patterns of precipitation (Figure 11.2a and 11.2d). These include: (i) the continental regime of the Alpine province, the only region with larger precipitation values in summer than in winter months; (ii) the transitional character of the Po Plain province toward a more Mediterranean regime, with lower summer precipitation; (iii) the apparent latitudinal gradient in both the Apennines and peninsular Tyrrhenian provinces, which mainly reflects the varying distance from the coast of the mountain reliefs. However, more local patterns are also suggested, which need a more profound investigation at the Section and/or Subsection ecoregional levels. These include the longitudinal summer gradient in precipitation between Eastern and Western Alps and the marked summer precipitation decrease in some Southern peninsular and main island sectors. Second, winter cold (Figure 11.2b) characterizes both the Alpine and Po Plain provinces within the Temperate division. On the contrary, the variable behavior within the Apennines province should be further investigated to elaborate on the differences with the Tyrrhenian province of the Mediterranean region. Patterns that need to be characterized at lower ecoregional levels also emerged in this case. These include the latitudinal gradient along the Adriatic province and the differences between the two main Tyrrhenian islands. Among several features, the third component of the process shows the relevance of reduced winter temperatures and their variation in characterizing the thermic continentality of the Po Plain and Adriatic provinces. It also confirms that higher temperatures occur in the Tyrrhenian and the Adriatic Mediterranean provinces.

11.4.2 Change Point Detection of Italian Climate Series

The DP change point model, introduced in Section 11.3.2, is applied to the historical thermo-pluviometric data for Italy described in Sections 11.2 and 11.3. Below, we present the results of multiple estimation runs for alternative model specifications. All estimations were performed using 80,000 MCMC iterations, with 60,000 iterations discarded as burn-in and thinning by a factor of 4. The implementation leveraged parallel computing [62], and priors were specified as follows: $N(0, 100)$ for the means $\mu_{t,\mathbf{s},i}$ and seasonal parameters $\psi_{t,\mathbf{s}}$, InverseGamma$(1, 1)$ for the variances $\sigma_{t,\mathbf{s},i}$, a uniform prior over the space of non-negative definite correlation matrices for $\rho_{t,\mathbf{s},i,i'}$, and Gamma$(1, 1)$ for the DP hyperparameters α and γ. Here, $i, i' = 1, 2, 3$, $t \in \mathcal{T}$, and $\mathbf{s} \in \mathcal{S}$.

A preliminary step in the model implementation involves setting the minimum regime length m^*. This requirement is common in change point detection literature; for example, [24] used a cut-off regime length of 10 years for the Pacific Decadal Oscillation index. Global atmospheric patterns and cycles, such as those related to solar activity or global atmospheric dynamics, can influence regime lengths, particularly in Southern Europe and the Mediterranean region [63]. Inter-annual cycles, driven by seasonal changes, are accounted for by the model term $\psi(\mathbf{s}, t)$. Biennial cycles may be linked to the Quasi-Biennial Oscillation [see 63, and references therein]. Longer cycles, such as 5 to 6-year cycles, are more uncertain in their interpretation and may represent harmonic components of the 11-year sunspot cycle or be influenced by *El Niño* effects. We tested $m^* \in \{1, 24, 60\}$, corresponding to no constraints, and minimum regime lengths of 2 and 5 years, respectively, which are compatible with the underlying physical processes. The $m^* = 1$ case was included to verify that imposing constraints on regime length affects model inferences.

For each time point and variable, 10% of the available stations were randomly selected to form a validation set. This set comprised 22,628 stations and time points for $\mathbf{Y}_1$, 21,321 for $\mathbf{Y}_2$, and 21,272 for $\mathbf{Y}_3$. The values of the three response variables at these stations were treated as missing, and models with $m^* \in \{1, 24, 60\}$ were estimated. Posterior samples of the variables in the validation set were used to compute the *CRPS* [64] for each of the three thermo-pluviometric variables. The CRPS is a widely used metric for evaluating the accuracy of probabilistic forecasting models, representing the mean square error between the predicted and true cumulative distribution functions. In the following, we present results obtained by fitting the model with $m^* = 60$ to the entire dataset.

The automatic CP detection procedure begins by defining the indicator variable $\zeta_{t,\mathbf{s},\ell}$ that assumes value 1 if the 95% HPD intervals of parameter ℓ at times t and $t - 12$ do not overlap, and 0 otherwise. Then, given a sequence of time-adjacent $\zeta_{t,\mathbf{s},\ell} = 1$, the time t corresponding to the first $\zeta_{t,\mathbf{s},\ell} = 1$ is defined to be a CP. The available data contain many time points with missing values for some or all the thermo-pluviometric variables. These time points are not eligible to be CPs for the parameters of the corresponding missing variables. To consider this, we adopt the conservative strategy of discarding CPs corresponding to time points where the corresponding climate variables values are missing. Notice that if, for example, only the minimum temperature is missing at time t, this will not simultaneously affect the detection of a CP for the precipitation mean.

The model output allows a very wide set of inferential usages, here we start by illustrating how to read results at the monitoring station level. At the Monte Cimone-Aeronautica monitoring station a clear increase in temperatures and precipitation occurs around the last two decades of the 20th century. The station is located at 2165 m above the sea level, at the uppermost sector of the Italian Apennine ecoregion (ecoregion 1C1: Northern and northwestern Apennine section), and shows an Alpine cold bioclimatic character. In Figure 11.3 we report the observed time series and the HPD bounds of the three mean parameters for the Monte Cimone station. We find that December 1999 is confirmed as a CP for the

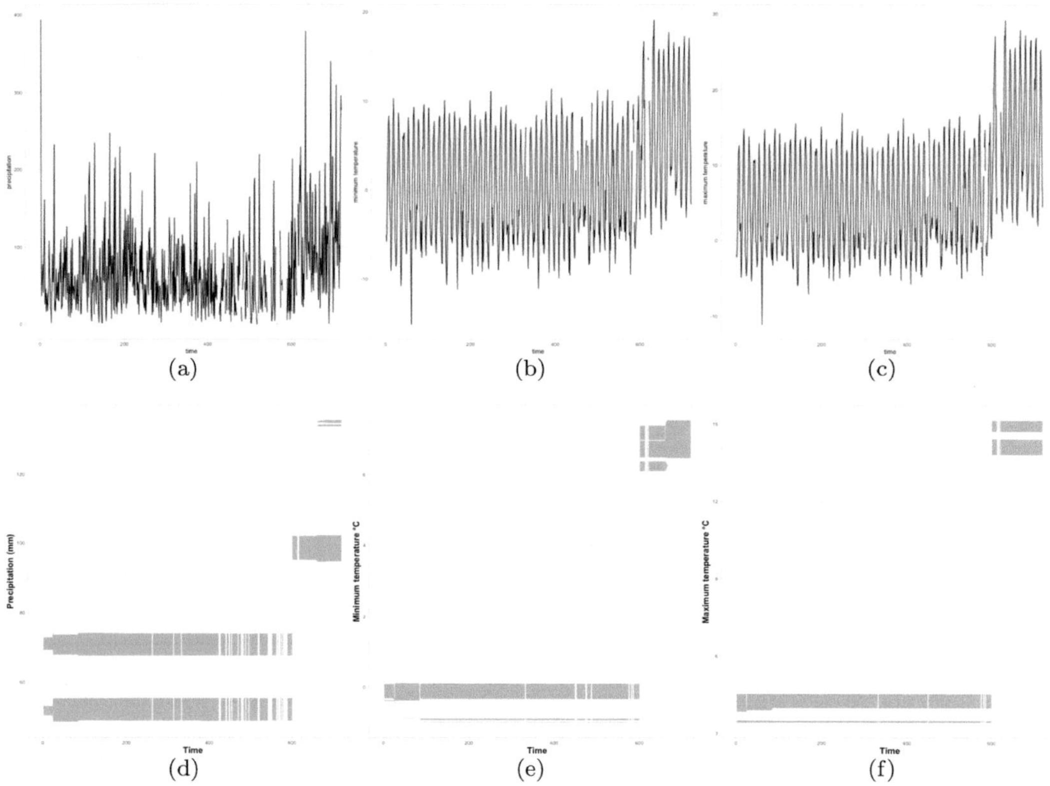

FIGURE 11.3
Time series (first row) and 95% HPD bounds (second row) of the mean parameters for precipitation (first column), minimum temperature (second column) and maximum temperature (third column) at the Monte Cimone monitoring station.

mean of the three variables, for the variance of maximum temperature, for the correlations between temperatures, and for the correlation between precipitation and minimum temperature. The precipitation's variance, instead, shows a CP on January 1952. HPD plots of correlations and variances are not reported for the sake of brevity, but they are available from the authors upon request

As stations can share the atoms of the underlying DP processes, the model output allows to check if one of the nine parameters has the same value at two monitoring stations. The relative frequency of the times that the value of the parameter ℓ is the same in the posterior samples for any pair of stations $\mathbf{s}_1$ and $\mathbf{s}_2$ at time t is hereby denoted by $\pi_{\mathbf{s}_1,\mathbf{s}_2,t,\ell}$ and referred to as *similarity*.

In Figure 11.4, stations with at least one change point over the entire period for the nine parameters are displayed. Many stations exhibit no CPs, particularly for the correlation between precipitation and maximum temperature, where no CPs were detected. Conversely, CPs are more frequently identified for the means of the two temperature variables and the correlation between them.

The intrinsic variability of precipitation in Mediterranean contexts [65] results in sporadic CPs for this variable, primarily occurring in Mediterranean ecoregions and limited to mountainous areas. In contrast, CPs for precipitation are relatively more widespread in Temperate ecoregions, particularly in the Alpine and Apennines sectors. Similarly, for

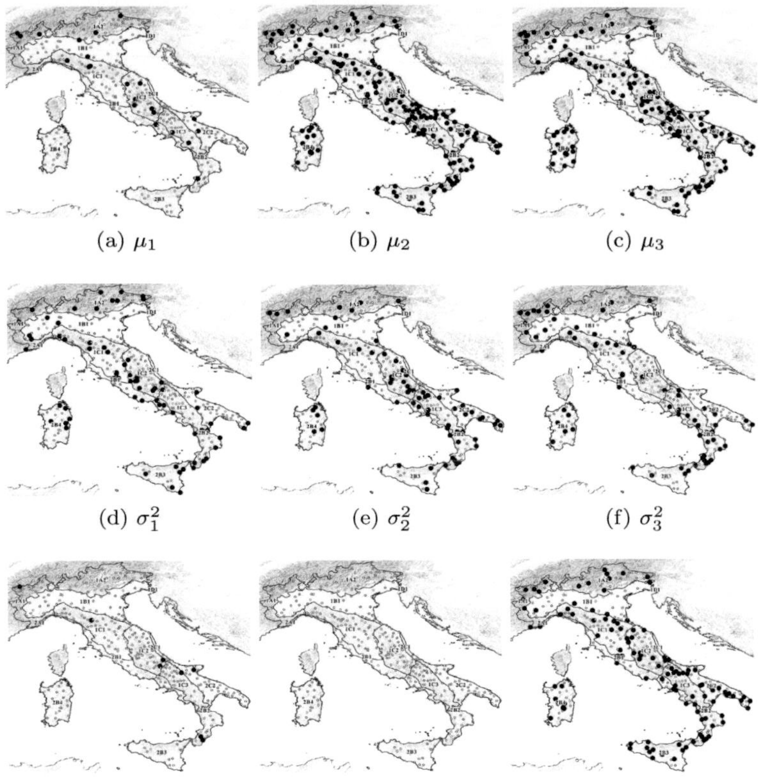

(a) μ_1 (b) μ_2 (c) μ_3

(d) σ_1^2 (e) σ_2^2 (f) σ_3^2

(g) ρ_{12} (h) ρ_{13} (i) ρ_{23}

FIGURE 11.4
Stations with at least one CP, for the nine parameters.

TABLE 11.5
Regressive coefficients referred to similarities of mean and variance parameters (with CIs).

	Δ_{sp}	Δ_{el}
Mean Prec.	−1.313	−0.610
	(−1.323 −1.302)	(−0.610 −0.604)
Mean Min.Temp.	−1.447	−1.125
	(−1.454 −1.439)	(−1.127 −1.123)
Mean Max.Temp.	−1.464	−1.937
	(−1.472 −1.456)	(−1.939 −1.935)
Var. Prec.	−2.244	−0.477
	(−2.255 −2.232)	(−0.481 −0.473)
Var. Min.Temp.	−1.395	−1.058
	(−1.405 −1.386)	(−1.062 −1.056)
Var. Max.Temp.	−1.757	−0.325
	(−1.767 −1.747)	(−0.329 −0.322)

temperature variables, fewer CPs are observed in more continental sectors, such as Sector 1B1 (Po Plain section) and Sector 2C1 (Central Adriatic section). These regions are characterized by significant daily and seasonal temperature variations, which may contribute to the reduced detection of CPs.

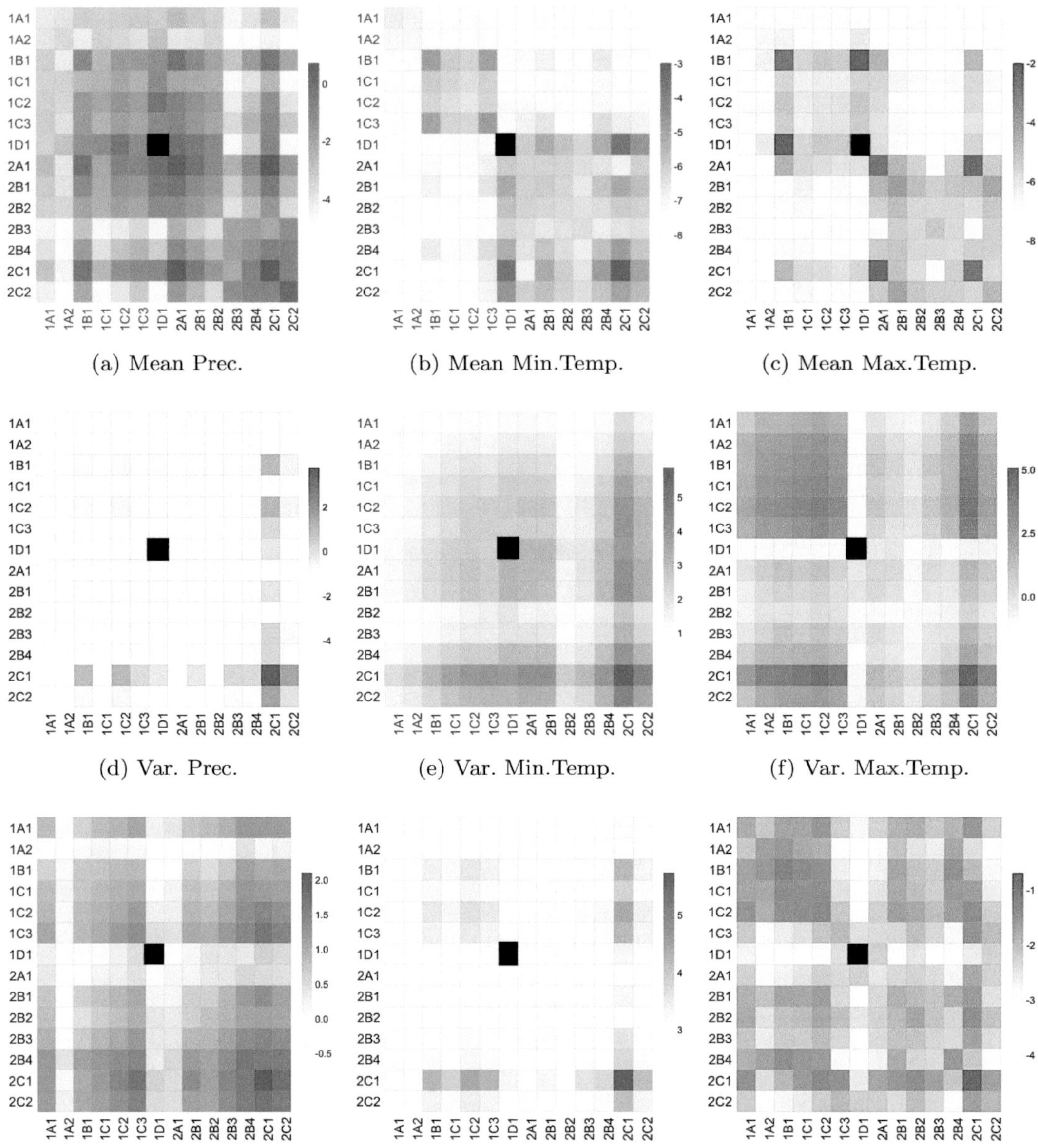

(a) Mean Prec. (b) Mean Min.Temp. (c) Mean Max.Temp.

(d) Var. Prec. (e) Var. Min.Temp. (f) Var. Max.Temp.

(g) Cor. Prec.-Min.Temp. (h) Cor. Prec.-Max.Temp. (i) Cor. Min.Temp.-Max.Temp.

FIGURE 11.5
Predictive values of the logit of the similarities between pairs of ecoregions (sections). As ecoregion 1D1 is composed of one station, we do not compute the self-similarity, leaving the relative square black.

Given the large number of stations and time-points, we summarize the similarities $\pi_{\mathbf{s}_1,\mathbf{s}_2,t,\ell}$ between stations $\mathbf{s}_1$ and $\mathbf{s}_2$ for parameter ℓ at time t by regressing their logistic transformation $\log(\pi_{\mathbf{s}_1,\mathbf{s}_2,t,\ell}/(1-\pi_{\mathbf{s}_1,\mathbf{s}_2,t,\ell}))$ on the spatial distances Δ_{sp} between stations $\mathbf{s}_1$ and $\mathbf{s}_2$, the differences in their elevation Δ_{el} and a grouping factor that identifies the combination of the ecoregions (14 sections) of the two stations. All the regression coefficients of Δ_{sp} and Δ_{el} suggest that the similarity decreases with the spatial distance and elevation difference. In Table 11.5, we report the coefficients referred to the mean and variance parameters. Notice that while the distance between stations seems to have approximately the same effect for the three mean parameters, similarities are more sensitive to distance effects for the precipitation variance than for the temperature ones. Indeed, rainfall and snow events are less linked to geographical variables than temperature [66]. As expected, differences in elevation affect the similarity of the mean temperatures more than they do with the precipitation. For each of the nine parameters, Figure 11.5 describes the relative variation of the logit of the similarities between pairs of ecoregions predicted by the regression model using the mean spatial distance and the mean elevation difference between the ecoregions. Notice that, even if we do not use spatial information in the model, we can see in Figure 11.5 that spatially close regions tend to be more similar.

According to the diagnostic criteria adopted for drawing ecoregion boundaries [46], the similarity in bio-physical features between ecoregional sections is generally expected to increase with their belonging to the same higher tier of the classification (province or division). Results in Figure 11.5 allow this hypothesis to be verified and detailed regarding climatic characteristics. For example, dissimilarities between the Italian Temperate division and the Mediterranean division (set of ecoregions with codes alternatively beginning with "1" or "2") emerged to be more marked in terms of mean temperatures (Figures 11.5b and c) rather than of mean precipitation (Figure 11.5a). At this level, precipitation means become differential just for a subset of the Temperate ecoregions (namely the Alpine sections 1A1 and 1A2, characterized by higher precipitation values and clearly dissimilar from all the Mediterranean sections) or alternatively, for a subset of the Mediterranean ecoregions (namely, the southernmost sections 2B3, 2B4, 2C2, characterized by a very marked summer decrease and clearly dissimilar from most of the Temperate sections). Regarding provinces (identified by codes consisting of the first number and the first letter) within the same division, the Alpine province (codes beginning with "1A") aligns well with the Temperate division in terms of the variance of maximum temperatures (Figure 11.5f) and the correlation between temperatures (Figure 11.5i). However, it exhibits distinct characteristics compared to other Temperate provinces in terms of mean temperatures (Figures 11.5b and c) and the mean and variance of precipitation (Figures 11.5a and d). This distinctiveness can be attributed to two main factors: (1) longer and more intense winter frosts compared to those in the Po Plain province (codes beginning with "1B") at lower altitudes and the Apennine province (codes beginning with "1C") at lower latitudes, and (2) higher precipitation values with a continental regime (characterized by winter minima) compared to lower precipitation values with summer minima observed in peninsular sectors at lower latitudes.

At the section level, some transitional characteristics are evident. For instance, the Central Adriatic section (2C1) fits well within the Adriatic province (codes beginning with "2C") of the Mediterranean division for all parameters. However, it also shows significant similarities with the Po Plain and Apennine provinces of the Temperate division, particularly in terms of the mean and variance of precipitation, the mean and variance of maximum temperatures, and all correlation parameters.

11.5 Discussion and Conclusions

This chapter reviews two model-based approaches for analyzing large-scale climate data. The motivating example is based on a climate dataset comprising monthly values of precipitation, minimum temperature, and maximum temperature recorded across a network of 360 monitoring stations in Italy over a 60-year period. First, we review a multivariate generalization of the NNGP model [20], as proposed in [41]. This approach builds on a series of comparative studies [12, 21, 67], which highlight the computational efficiency of NNGPs. Our proposal integrates the computational advantages of NNGPs with novel strategies for addressing the complex structures inherent in climate data. Specifically, we employ the linear model of coregionalization to account for multivariate spatio-temporal dependencies, introduce a circular representation of the time index to model annual cyclical patterns, and propose an efficient implementation capable of handling large datasets. By explicitly incorporating key features of the underlying processes—such as correlations among climate variables, space-time variability, and cyclical components—our approach achieves improved prediction accuracy compared to alternative methods.

Next, we review the model-based CP detection procedure for large multivariate time series introduced in [42]. This proposal introduces several innovations in CP modeling: (1) not all parameters are required to change at a CP; (2) the identification of changing parameters is part of the model inference rather than being predetermined; (3) monitoring stations can share parameter values; and (4) the number of CPs at each station is treated as a random quantity to be estimated. A key by-product of this model is the identification of clusters of spatio-temporal observations that share values of some or all parameters, enabling the detection of time regimes or groups of stations with similar characteristics over the observed time window. Post-processing of the model output also provides insights into which parameters are more likely to change simultaneously and distinguishes between abrupt and smooth changes. This capability has proven valuable in identifying abrupt changes in individual climate features and their interrelationships, which may threaten the resilience of Italy's natural ecosystems.

The richness of the model output compensates for the high computational complexity of both proposals. In both cases, the output facilitates a detailed analysis of individual time series and their joint behavior across time and space. Local and global features are easily highlighted, providing deep insights into the analyzed phenomena. A unique strength of our approaches is the ability to post-process model output at multiple spatial scales, enabling the evaluation of similarities between stations or regions. Notice that in [41] and [42], the out of sample predictive capability of the proposed models is assessed by cross-validation in comparison with some alternative model settings and competing models.

The modeling results show that climatic variation aligns meaningfully with ecoregional boundaries, with particular differences observed at intermediate tiers of the classification (e.g., sections and provinces). This not only validates the ecological logic of the current tiered classification but also highlights the potential of model-based climate diagnostics to support ecological zoning and conservation planning. The approach also provides a rigorous framework for selecting the optimal ecoregion tier for climate representation, based on predictive performance and uncertainty quantification. Results from this analysis reveal spatial patterns in the timing and nature of climate regime changes that correspond to ecological and topographic gradients. For example, Mediterranean ecoregions often exhibit smoother but earlier changes in precipitation and temperature dynamics compared to Temperate and Alpine zones, which tend to show more abrupt regime shifts. This supports the hypothesis that climate change manifests heterogeneously across ecological regions, influenced by orographic complexity and regional atmospheric patterns. Furthermore, the model uncovers clusters of monitoring stations sharing common climate transition profiles, some of which span across traditional ecoregion boundaries, suggesting that future updates to ecological zoning could benefit from incorporating statistical indicators of temporal coherence.

The inferences drawn from these models significantly enhance the bioclimatic characterization of Italian ecoregions, improving our understanding of complex climate patterns in a country marked by significant physiographic heterogeneity. By leveraging observed rather than interpolated data, our results provide a clearer definition of: (1) similarities and differences within and between ecoregions in terms of climate means and variability; (2) the effects of correlations among climatic variables on vegetation patterns at national and sub-national scales; and (3) the vulnerability of sensitive ecosystems to local climate change trends, in alignment with the IUCN Red List criteria [68]. These outcomes offer valuable benchmarks for refining monitoring programs focused on ecosystem responses (e.g., [69]). For instance, they enable the assessment of how the potential vulnerability to climate change differs between Mediterranean and Temperate ecosystems based on specific parameters. Additionally, detecting CPs across ecoregional sections provides finer spatial resolution for identifying change trends, distinguishing, for example, between continental plains and mountain ranges within the Temperate context.

These features complement auxiliary data sources, such as satellite-based datasets like WorldClim, MODIS, or TRMM. However, such sources often lack the temporal depth required for our analysis (1950–2010) and do not support interpolation at arbitrary resolutions. Our model, in contrast, provides a comprehensive characterization of Italian ecoregions with respect to precipitation, minimum temperature, and maximum temperature, capturing cyclical trends, spatial correlations, and temporal variability.

Together, these models establish a robust framework for climate data analysis, ecosystem monitoring, and long-term climatic impact assessment. They provide crucial insights for conservation and adaptive management while addressing challenges associated with large datasets. The current model settings are well-suited for analyzing relatively smooth phenomena, such as monthly summaries of climatic variables, but would require adaptation for different temporal aggregations. Nevertheless, the model architecture is highly flexible and can be tailored to a wide range of meteorological and climatic studies.

Bibliography

[1] Robert G. Bailey. Identifying ecoregion boundaries. *Environmental Management*, 34(Suppl 1):S14–S26, 2004.

[2] Marc J. Metzger, Robert G. H. Bunce, Rob H. G. Jongman, Roger Sayre, Antonio Trabucco, and Robert Zomer. A high-resolution bioclimate map of the world: a unifying framework for global biodiversity research and monitoring. *Global Ecology and Biogeography*, 22(5):630–638, 2013.

[3] Gretta T. Pecl, Miguel B. Araújo, Johann D. Bell, Julia Blanchard, Timothy C. Bonebrake, I-Ching Chen, Timothy D. Clark, Robert K. Colwell, Finn Danielsen, Birgitta Evengård, Lorena Falconi, Simon Ferrier, Stewart Frusher, Raquel A. Garcia, Roger B. Griffis, Alistair J. Hobday, Charlene Janion-Scheepers, Marta A. Jarzyna, Sarah Jennings, Jonathan Lenoir, Hlif I. Linnetved, Victoria Y. Martin, Phillipa C. McCormack, Jan McDonald, Nicola J. Mitchell, Tero Mustonen, John M. Pandolfi, Nathalie Pettorelli, Ekaterina Popova, Sharon A. Robinson, Brett R. Scheffers, Justine D. Shaw, Cascade J. B. Sorte, Jan M. Strugnell, Jennifer M. Sunday, Mao-Ning Tuanmu, Adriana Vergés, Cecilia Villanueva, Thomas Wernberg, Erik Wapstra, and Stephen E. Williams. Biodiversity redistribution under climate change: Impacts on ecosystems and human well-being. *Science*, 355(6332), 2017.

[4] Gian-Reto Walther. Community and ecosystem responses to recent climate change. *Philosophical Transactions of the Royal Society B: Biological Sciences*, 365(1549):2019–2024, 2010.

[5] Andrew J. Felton and Melinda D. Smith. Integrating plant ecological responses to climate extremes from individual to ecosystem levels. *Philosophical Transactions of the Royal Society B: Biological Sciences*, 372(1723):20160142, 2017.

[6] Deyong Yu, Yupeng Liu, Peijun Shi, and Jianguo Wu. Projecting impacts of climate change on global terrestrial ecoregions. *Ecological Indicators*, 103:114–123, 2019.

[7] M. J. Heaton, A. Datta, A. Finley, R. Furrer, R. Guhaniyogi, F. Gerber, R. B. Gramacy, D. Hammerling, M. Katzfuss, F. Lindgren, D. W. Nychka, F. Sun, and A. Zammit-Mangion. A Case Study Competition Among Methods for Analyzing Large Spatial Data. *ArXiv e-prints*, October 2017.

[8] Songnian Li, Suzana Dragicevic, Francesc Antón Castro, Monika Sester, Stephan Winter, Arzu Coltekin, Christopher Pettit, Bin Jiang, James Haworth, Alfred Stein, and Tao Cheng. Geospatial big data handling theory and methods: A review and research challenges. *ISPRS Journal of Photogrammetry and Remote Sensing*, 115:119–133, 2016. Theme issue 'State-of-the-art in photogrammetry, remote sensing and spatial information science'.

[9] G. Jona Lasinio, G. Mastrantonio, and A. Pollice. Discussing the "big n problem". *Statistical Methods and Applications*, 22(1):97–112, 2013.

[10] Sudipto Banerjee, Alan E. Gelfand, and Bradley P. Carlin. *Hierarchical Modeling and Analysis for Spatial Data*. Chapman and Hall/CRC, New York, second edition, December 2014.

[11] Marta Blangiardo and Michela Cameletti. *Spatial and spatio-temporal Bayesian models with R-INLA*. Wiley, 2015.

[12] Abhirup Datta, Sudipto Banerjee, Andrew O. Finley, and Alan E. Gelfand. Hierarchical nearest-neighbor gaussian process models for large geostatistical datasets. *Journal of the American Statistical Association*, 111(514):800–812, 2016.

[13] Jo Eidsvik, Benjamin A. Shaby, Brian J. Reich, Matthew Wheeler, and Jarad Niemi. Estimation and prediction in spatial models with block composite likelihoods. *Journal of Computational and Graphical Statistics*, 23(2):295–315, 2014.

[14] Moreno Bevilacqua, Alessandro Fassò, Carlo Gaetan, Emilio Porcu, and D. Velandia. Covariance tapering for multivariate Gaussian random fields estimation. *Statistical Methods & Applications*, 25(1):21–37, Mar 2016.

[15] Moreno Bevilacqua, Alfredo Alegria, Daria Velandia, and Emilio Porcu. Composite likelihood inference for multivariate Gaussian random fields. *Journal of Agricultural, Biological, and Environmental Statistics*, 21(3):448–469, 2016.

[16] Andrew O. Finley, Sudipto Banerjee, and Alan E. Gelfand. Bayesian dynamic modeling for large space-time datasets using Gaussian predictive processes. *Journal of Geographical Systems*, 14(1):29–47, 2012.

[17] Ganggang Xu, Faming Liang, and Marc G. Genton. Bayesian spatio-temporal geostatistical model with an auxiliary lattice for large datasets. *Statistica Sinica*, 25:61–79, 2015.

[18] Yun Bai, Peter X.-K. Song, and T. E. Raghunathan. Joint composite estimating functions in spatiotemporal models. *Journal of the Royal Statistical Society: Series B (Statistical Methodology)*, 74(5):799–824, 2012.

[19] Moreno Bevilacqua, Carlo Gaetan, Jorge Mateu, and Emilio Porcu. Estimating space and space-time covariance functions for large data sets: A weighted composite likelihood approach. *Journal of the American Statistical Association*, 107(497):268–280, 2012.

[20] Abhirup Datta, Sudipto Banerjee, Andrew O. Finley, Nicholas A. S. Hamm, and Martijn Schaap. Nonseparable dynamic nearest neighbor Gaussian process models for large spatio-temporal data with an application to particulate matter analysis. *Annals of Applied Statistics*, 10(3):1286–1316, 09 2016.

[21] Sudipto Banerjee, Alan E. Gelfand, Andrew O. Finley, and Huiyan Sang. Gaussian predictive process models for large spatial data sets. *Journal of the Royal Statistical Society: Series B (Statistical Methodology)*, 70(4):825–848, 2008.

[22] Venkata K. Jandhyala, Stergios B. Fotopoulos, and Jing You. Change-point analysis of mean annual rainfall data from Tucumán, Argentina. *Environmetrics*, 21(7-8):687–697, 2010.

[23] A. R. Tomè and P. M. A. Miranda. Piecewise linear fitting and trend changing points of climate parameters. *Geophysical Research Letters*, 31(2), 2004.

[24] Sergei N. Rodionov. A sequential algorithm for testing climate regime shifts. *Geophysical Research Letters*, 31(9), 2004.

[25] Michael Robbins, Colin Gallagher, Robert Lund, and Alexander Aue. Mean shift testing in correlated data. *Journal of Time Series Analysis*, 32(5):498–511, 2011.

[26] QiQi Lu, Robert Lund, and Thomas C. M. Lee. An MDL approach to the climate segmentation problem. *Annals of Applied Statistics*, 4(1):299–319, 03 2010.

[27] Claudie Beaulieu, Jie Chen, and Jorge L. Sarmiento. Change-point analysis as a tool to detect abrupt climate variations. *Philosophical Transactions of the Royal Society A: Mathematical, Physical and Engineering Sciences*, 370(1962):1228–1249, 2012.

[28] Robert Lund and Jaxk Reeves. Detection of undocumented changepoints: A revision of the two-phase regression model. *Journal of Climate*, 15(17):2547–2554, 01 Sep. 2002.

[29] Robert Lund, Xiaolan L. Wang, Qi Qi Lu, Jaxk Reeves, Colin Gallagher, and Yang Feng. Changepoint detection in periodic and autocorrelated time series. *Journal of Climate*, 20(20):5178–5190, 15 Oct. 2007.

[30] P.K. Bhattacharya. Maximum likelihood estimation of a change-point in the distribution of independent random variables: General multiparameter case. *Journal of Multivariate Analysis*, 23(2):183–208, 1987.

[31] Douglas M. Hawkins. Fitting multiple change-point models to data. *Computational Statistics and Data Analysis*, 37(3):323–341, 2001.

[32] Rebecca Killick, Idris A. Eckley, Kevin Ewans, and Philip Jonathan. Detection of changes in variance of oceanographic time-series using changepoint analysis. *Ocean Engineering*, 37(13):1120 – 1126, 2010.

[33] Yuanbo Li, Ngai Hang Chan, Chun Yip Yau, and Rongmao Zhang. Group orthogonal greedy algorithm for change-point estimation of multivariate time series. *Journal of Statistical Planning and Inference*, 212:14–33, 2021.

[34] Nicholas A. James and David S. Matteson. ECP: An R package for nonparametric multiple change point analysis of multivariate data. *Journal of Statistical Software*, 62(7), 2013.

[35] Jonas Kristoffer Lindeløv. MCP: An R package for regression with multiple change points. *OSF Preprints*, 2020.

[36] Bradley P. Carlin, Alan E. Gelfand, and Adrian F. M. Smith. Hierarchical Bayesian analysis of changepoint problems. *Journal of the Royal Statistical Society. Series C (Applied Statistics)*, 41(2):389–405, 1992.

[37] Siddhartha Chib. Estimation and comparison of multiple change-point models. *Journal of Econometrics*, 86(2):221 – 241, 1998.

[38] Stefano Peluso, Siddhartha Chib, and Antonietta Mira. Semiparametric multivariate and multiple change-point modeling. *Bayesian Analysis*, 14(3):727–751, 09 2019.

[39] Stanley I. M. Ko, Terence T. L. Chong, and Pulak Ghosh. Dirichlet process hidden Markov multiple change-point model. *Bayesian Analysis*, 10(2):275–296, 06 2015.

[40] John T. Abatzoglou, Solomon Z. Dobrowski, and Sean A. Parks. Multivariate climate departures have outpaced univariate changes across global lands. *Scientific Reports*, 10(1):3891, 2020.

[41] Gianluca Mastrantonio, Giovanna Jona Lasinio, Alessio Pollice, Giulia Capotorti, Lorenzo Teodonio, Giulio Genova, and Carlo Blasi. A hierarchical multivariate spatio-temporal model for clustered climate data with annual cycles. *The Annals of Applied Statistics*, 13(2):797–823, 2019.

[42] Gianluca Mastrantonio, Giovanna Jona Lasinio, Alessio Pollice, Lorenzo Teodonio, and Giulia Capotorti. A Dirichlet process model for change-point detection withÂ multivariate bioclimatic data. *Environmetrics*, 33(1):e2699, 2022.

[43] WMO. *Guidelines on Best Practices for Climate Data Rescue*. Number 1182 in WMO - Guidelines. World Meteorological Organization, 2016.

[44] Robert G. Bailey. Delineation of ecosystem regions. *Environmental Management*, 7(4):365–373, 1983.

[45] Thomas R. Loveland and James M. Merchant. Ecoregions and ecoregionalization: geographical and ecological perspectives. *Environmental Management*, 34((Suppl 1): S1), 2004.

[46] C. Blasi, G. Capotorti, R. Copiz, D. Guida, B. Mollo, D. Smiraglia, and L. Zavattero. Classification and mapping of the ecoregions of Italy. *Plant biosystems - An International Journal Dealing with all Aspects of Plant Biology*, 148(6):1255–1345, 2014.

[47] Therese C. Frauendorf, Richard A. MacKenzie, Ralph W. Tingley III, Abby G. Frazier, Michael H. Riney, and Rana W. El-Sabaawi. Evaluating ecosystem effects of climate change on tropical island streams using high spatial and temporal resolution sampling regimes. *Global Change Biology*, 25(4):1344–1357, 2019.

[48] R. B. Alley, J. Marotzke, W. D. Nordhaus, J. T. Overpeck, D. M. Peteet, R. A. Pielke, R. T. Pierrehumbert, P. B. Rhines, T. F. Stocker, L. D. Talley, and J. M. Wallace. Abrupt climate change. *Science*, 299(5615):2005–2010, 2003.

[49] John W. Williams, Jessica L. Blois, and Bryan N. Shuman. Extrinsic and intrinsic forcing of abrupt ecological change: case studies from the late quaternary. *Journal of Ecology*, 99(3):664–677, 2011.

[50] A. Gelfand, P. Diggle, M. Fuentes, and P. Guttorp. *Handbook of Spatial Statistics*. Chapman and Hall, 2010.

[51] Alan E. Gelfand, Alexandra M. Schmidt, Sudipto Banerjee, and C. F. Sirmans. Nonstationary multivariate process modeling through spatially varying coregionalization. *TEST*, 13(2):263–312, 2004.

[52] Shinichiro Shirota and Alan E. Gelfand. Space and circular time log Gaussian Cox processes with application to crime event data. *The Annals of Applied Statistics*, 11(2):481–503, 06 2017.

[53] Tilmann Gneiting. Nonseparable, Stationary Covariance Functions for Space-Time Data. *Journal of the American Statistical Association*, 97(458):590–600, 2002.

[54] Allou Samé, Faicel Chamroukhi, Gérard Govaert, and Patrice Aknin. Model-based clustering and segmentation of time series with changes in regime. *Advances in Data Analysis and Classification*, 5(4):301–321, 2011.

[55] Thomas S. Ferguson. A Bayesian analysis of some nonparametric problems. *The Annals of Statistics*, 1(2):209–230, 1973.

[56] Yee Whye Teh, Michael I Jordan, Matthew J Beal, and David M Blei. Hierarchical Dirichlet processes. *Journal of the American Statistical Association*, 101(476):1566–1581, 2006.

[57] Emily B. Fox, Erik B. Sudderth, Michael I. Jordan, and Alan S. Willsky. A sticky HDP-HMM with application to speaker diarization. *The Annals of Applied Statistics*, 5(2A):1020–1056, 06 2011.

[58] Christophe Andrieu and Johannes Thoms. A tutorial on adaptive MCMC. *Statistics and Computing*, 18(4):343–373, 2008.

[59] David J. Spiegelhalter, Nicola G. Best, Bradley P. Carlin, and Angelika Van Der Linde. Bayesian measures of model complexity and fit. *Journal of the Royal Statistical Society: Series B (Statistical Methodology)*, 64(4):583–639, 2002.

[60] E. P. Grimit, T. Gneiting, V. J. Berrocal, and N. A. Johnson. The continuous ranked probability score for circular variables and its application to mesoscale forecast ensemble verification. *Quarterly Journal of the Royal Meteorological Society*, 132(621C):2925–2942, 2006.

[61] Caspar A. Mücher, Jan A. Klijn, Dirk M. Wascher, and Joop H.J. Schaminée. A new European landscape classification (Lanmap): A transparent, flexible and user-oriented methodology to distinguish landscapes. *Ecological Indicators*, 10(1):87–103, 2010. Landscape Assessment for Sustainable Planning.

[62] OpenMP Architecture Review Board. OpenMP application program interface version 3.0, May 2008.

[63] Juan Antonio Luque-Espinar, Rosa María Mateos, Inmaculada García-Moreno, Eulogio Pardo-Igúzquiza, and Gerardo Herrera. Spectral analysis of climate cycles to predict rainfall induced landslides in the western mediterranean (majorca, spain). *Natural Hazards*, 89(3):985–1007, 2017.

[64] Tilmann Gneiting and Adrian E Raftery. Strictly proper scoring rules, prediction, and estimation. *Journal of the American Statistical Association*, 102(477):359–378, 2007.

[65] A. Dükeloh and J. Jacobeit. Circulation dynamics of Mediterranean precipitation variability 1948-98. *International Journal of Climatology*, 23(15):1843–1866, 2003.

[66] Miquel Ninyerola, Xavier Pons, and Joan M. Roure. A methodological approach of climatological modelling of air temperature and precipitation through gis techniques. *International Journal of Climatology*, 20(14):1823–1841, 2000.

[67] Robert B. Gramacy and Daniel W. Apley. Local Gaussian process approximation for large computer experiments. *Journal of Computational and Graphical Statistics*, 24(2):561–578, 2015.

[68] David A. Keith, Jon Paul Rodríguez, Thomas M. Brooks, Mark A. Burgman, Edmund G. Barrow, Lucie Bland, Patrick J. Comer, Janet Franklin, Jason Link, Michael A. McCarthy, Rebecca M. Miller, Nicholas J. Murray, Jeanne Nel, Emily Nicholson, María A. Oliveira-Miranda, Tracey J. Regan, Kathryn M. Rodríguez-Clark, Mathieu Rouget, and Mark D. Spalding. The IUCN red list of ecosystems: Motivations, challenges, and applications. *Conservation Letters*, 8(3):214–226, 2015.

[69] S. Chelli, C. Wellstein, G. Campetella, R. Canullo, R. Tonin, S. Zerbe, and R. Gerdol. Climate change response of vegetation across climatic zones in Italy. *Climate Research*, 71(3):249–262, 2017.

Part III

Spatial and temporal modelling

12

Robust Change Point Detection in Air Pollution

Ana Borges, M. Rosário Ramos, Clara Cordeiro, and Mariana Carvalho

12.1 Introduction

The search for change points in time series is essential in numerous contexts, as these points often reflect shifts in the underlying distribution and abrupt changes in model parameters. Change points may be known or anticipated at specific times in history, such as during a pandemic, natural catastrophe, or war, and may be completely unknown in advance. These changes can be particularly important in studying environmental time series, such as air pollution levels within a monitoring process.

Several investigations have been developed in this context. In [1], a Cumulative Sum (CUSUM) technique was applied to hourly pollutant readings in central London and detected change points in NO2 and PMcoarse levels that aligned with the introduction of congestion charging (95% confidence interval). In [2], residuals from ARMA models using a CUSUM test were analyzed and identified changes in PM10 on 15 and 19 January 2012, attributed to de-icer application in Klagenfurt, Austria.

The Bai and Perron structural break were used in [3] to test on CO2 data from 41 countries and reported abrupt changes that some studies linked to climate protest events. [4] applied Bai's change point test to U.S. emission data, finding breakpoints concurrent with new federal legislation.

A Bayesian likelihood ratio test [5] has been employed for modeling breakpoints in PM2.5 and PM10. In contrast, [6] adopted a Random Forest (RF) model that reduced estimation errors by 30–42% when incorporating local and regional features into trend estimation. Geographic scopes ranged from city-level (e.g., Central London, Klagenfurt) to multinational studies, and temporal resolutions varied from hourly to daily. These articles demonstrate that statistical and machine learning methods can isolate significant shifts in pollutant time series, linking identified breakpoints to specific events or policy interventions.

Recently, this issue was addressed by developing a methodological framework that integrates meteorological normalization, change point detection, and trend analysis to characterize shifts in air pollution levels [7]. Their approach involves applying a machine learning-based RF model to normalize pollutant concentrations by removing the confounding effects of weather variables such as temperature, pressure, humidity, and wind patterns. The authors then detect structural changes using Wild Binary Segmentation (WBS) and analyze long-term trends with Theil-Sen regression. Finally, they compare detected breakpoints with known external events, such as industrial shutdowns, to assess the impact of

DOI: 10.1201/9781003564423-12

emission-related changes. The methodology was applied to a dataset of NOx, SO_2, CO, and H_2S concentrations from 2013 to 2019 in a rural-industrial region of Southern Italy. Results showed that meteorologically normalized data provided a clearer view of emission-driven variations, revealing distinct breakpoints corresponding to operational changes at an oil processing facility. Their findings confirmed a significant downward trend in NOx levels, likely attributed to regulatory and operational factors.

With the increase of computing power and development of new techniques that have taken place in recent years, statistical and machine learning methodologies have significantly improved the ability to analyze changes in air pollution patterns, considering both meteorological variability and anthropogenic activities. The literature has shown that unexpected events, such as the COVID-19 pandemic, can cause abrupt changes in air quality, and advanced methods are needed to assess the real influence of these changes. For example, [8] employed a combination of machine learning, including ensemble learning, competitive learning, and unsupervised clustering techniques, to analyze changes in PM2.5, NO2, and O3, minimizing meteorological effects in the assessment of pollution variations.

Furthermore, new change point detection strategies have been proposed to evaluate the effectiveness of environmental interventions and unexpected air quality events. For example, some studies introduced an interval forecasting model that incorporates change point detection to deal with the high volatility of pollutant data, optimizing initial parameters and improving the stability of forecasts [9]. This approach complements traditional statistical approaches such as meteorological normalization and multiple regression analysis, which were explored by [6] to estimate trends in air pollutants, albeit with limitations in accurately correcting for meteorological variability.

Integrating hybrid and deep learning models has shown promise for forecasting and analyzing air pollution. It was also demonstrated that recurrent neural networks, such as Long Short-Term Memory and Gated Recurrent Unit, outperform conventional statistical models in predicting pollutant concentrations, offering greater accuracy in detecting trends and anomalies ([10],[11]).

Given this challenge, the goal is to develop and apply a robust statistical methodology to a real dataset to identify changes in air pollutant emission patterns over time. Detecting these turning points enables a better understanding of emission trends, assessment of environmental policy effectiveness, and identification of potential anomalies that could impact air quality. By focusing on monthly emissions data for CO, NH3, NMVOC, NOx, OC, PM2.5, PM10, and SO2 between 2000 and 2022, the case study provides a detailed analysis of the trends and variations of these pollutants in Portugal and Italy, allowing for more accurate inferences about the impacts of environmental regulations.

The rest of the chapter is structured as follows: Section 12.2 describes the structure of the methodology and its steps, starting with the decomposition of the time series via Seasonal-Trend Decomposition using Loess (STL), followed by the detection of change points using the Strucchange, PELT (Pruned Exact Linear Time), and Binary Segmentation (BS) methods, available in `R`. Section 12.3 presents the data and the results of the methodology for the emissions time series, highlighting the main change points identified and their possible causes. This section discusses the validity and robustness of the methods used to identify structural breaks in the series. Finally, Section 12.4 discusses achievements, compares the trends observed and explores the implications for future environmental regulations, assessing the effects of the policies adopted in Portugal and Italy and suggesting directions for future research.

12.2 Methodology

The proposed procedure is based on the rationale of [12], where a methodology was suggested to detect and quantify trends in water consumption. Unlike this early work, which focused on changes in the slope of a linear model, the present approach expanded the scope to capture breaks in the mean and variance of the data, such as jumps, adding other change point methods to address the problem. It was intended to provide a robust choice of change points, especially for cases where ground truth is not verifiable and when one unique method may not guarantee the best performance. Additionally, to prevent seasonal effects from interfering, change point analysis is applied to the seasonally adjusted time series. The aim is to find the solution that shows highest agreement rate between different change point methods, as well as for raw and transformed data. All the statistical analysis was performed using the R version 4.4.2 [13].

This procedure is used to analyze and compare time-series data on eight air pollutants from Portugal and Italy (CO, NH3, NMVOC, NOx, OC, PM25, PM10, and SO2). This methodology allows to identify, characterize, and evaluate the significance of structural breakpoints. Each step of the proposed method is explained in the following subsections.

12.2.1 Estimating Time Series Components

A time series can be viewed as the result of the combination of the following components: T, a trend component, which represents an upward or downward movement over the time horizon; S, a seasonal component, which captures repetitive patterns over time; and I, an irregular component, which remains after the other elements have been removed. In an additive model, a time series Y_t can be decomposed as:

$$Y_t = T_t + S_t + I_t,$$

where $t = 1, \ldots, n$ represents the time period, and n is its length.

Classical methods for decomposing time series allow us to identify the trend, seasonal, and irregular components. However, these methodologies typically lack flexibility in specifying the seasonal component, and the trend component is usually represented by a deterministic time function, which can be sensitive to outliers. The non-parametric Seasonal-Trend decomposition using Loess (STL) [14] addresses these limitations by allowing the identification of a time-varying seasonal component, a non-linear trend, and providing robustness in the presence of outliers.

Like many non-parametric regression methods, STL requires the subjective selection of smoothing parameters. The two main parameters are the seasonal (s.window) and trend (t.window) window widths. Therefore, to overcome this limitation, the stl.fit() procedure [15] [16] was used since it allows an "objective" choice of STL smoothing parameters. This algorithm selects the best STL model for each combination of the s.window and t.window based on an error measure such as mean absolute error and others. This procedure has been developed to obtain "automatically" the seasonal and trend smoothing parameters by minimizing an error measure.

In the following sections, we denote the deseasonalized series by $\tilde{Y}_t = Y_t - \hat{S}_t$, and only $\tilde{Y}_t$ is used for changepoint analysis.

12.2.2 Seasonality Adjustment

Prior to change point analysis, a test for seasonality is conducted. Each time series was tested for seasonality using the combined seasonality test from the seastests package, which includes the QS-test and the KW-test for seasonality available in `seastest` [17] package in the `R` software to detect significant seasonal variations. This function employs multiple statistical tests, including the Welch Overlap Test (WO), Kruskal-Wallis, and Friedman. The WO test measures the degree of statistical overlap between consecutive seasonal periods, such as months or quarters. The Kruskal-Wallis Test is useful when the data does not follow a normal distribution. The Friedman test examines differences between seasonal periods within paired blocks, such as consecutive years.

The decision to include this initial step is based on the importance of testing for the existence of seasonality before removing it, especially in analyses to detect change points in time series. Removing a non-existent seasonal component can introduce distortions in the data, jeopardizing accurately identifying change points. As discussed by [18], the correct identification and treatment of seasonality is essential to ensure the robustness of structural change detection methods. Therefore, the deseasonalized series $\tilde{Y}_t = Y_t - \hat{S}_t$ was retained for the changepoint detection methods described in the following section.

12.2.3 Breakpoint Detection and Validation

In time series analysis, evaluating and identifying periods that exhibit different statistical properties is a way of understanding and predicting the evolution of phenomena more accurately. Change points can take various forms, such as a change in mean or level in a time series, a change in trend measured by the slope of a linear model with time as predictor or abrupt change in variance of a sequence of observations. Generally speaking, a change point splits a time series into two or multiple segments where each segment can have its own mean, trend or variance, different from its neighbor segments. The change point typologies considered are then the following:

Mean: there is shift in the mean of a time series $\tilde{Y}_t$ at time k where $1 < k < n$ if

$$\mu_t = \begin{cases} \mu_1 & : t \leq k \\ \mu_2 & : t > k \end{cases} \quad \mu_1 \neq \mu_2.$$

Trend: the change point in trend means that the slope of a linear trend in time series changes at time k, independently of a change in the intercept.

$$\tilde{Y}_t = \begin{cases} \alpha_1 + \beta_1 t + \varepsilon_t & \text{for } t \leq k \\ \alpha_2 + \beta_2 t + \varepsilon_t & \text{for } t > k \end{cases} \quad \text{with } (\alpha_1, \beta_1) \neq (\alpha_2, \beta_2)$$

where ε_t is the random noise assumed to have zero mean.

Variance: finally, the change point in variance at time k is described by

$$\sigma_t^2 = \begin{cases} \sigma_1^2 & : t \leq k \\ \sigma_t^2 & : t > k \end{cases} \quad \sigma_1^2 \neq \sigma_2^2.$$

There are many change point methods available, but they all have weaknesses such as the estimation at the ends of the time series, consistency in break dates in multiple change points, depend on the length of the series or are not suitable for a particular purpose. Therefore a selection of three methods is applied simultaneously to detect and estimate break dates. Based on the percentage of overlapping change points with the benchmark method, the final breaks set is defined.

The tested methods included PELT and BS to analyze changes in either the mean or variance. These methods were compared with the breakpoints technique, implemented in R package *strucchange* [19].

Like in [12], the R package *strucchange* [19] is used to detect t structural breaks, which tests for structural changes in linear regression models, estimating the number of segments (m) and the set of the breakpoints $bp = \{ t_1^*, t_2^*, \cdots, t_{m-1}^* \}$, minimizing the Bayesian information criterion (BIC) and the residual sum of squares (RSS) [20]. This procedure has a bandwidth parameter (h), which determines the minimal segment size ($h \times n$ observations). In function `breakpoints()` [19] the default bandwidth parameter is $h = 0.15$. However, it is up to the user to change h to a more suitable value based on the knowledge of the data. According to [21], statistical tests used to detect multiple structural breaks have less power when sample and/or segment size is reduced, so it is important to establish a minimum size. In this case, a two-year segment was the minimum length considered.

BS to multiple change points (BS): BS is a fast algorithm and begins by applying a single change point method to the entire data. When the point that yields the best loss decrease in the cost function is found, the time series data are split at the change point to create two new sub-series. The method extends this rule to multiple change points by repeating iteratively in the subsets of the time series until no new sub-sequences(segments) are created. PELT is one of the segmentation algorithms for detecting multiple change points in sequential data. It performs by minimizing a cost function of a segment with a penalty, in this case the BIC, a penalty linear in the number of change points. Although some literature notes that it does not perform well at the ends of the series, the PELT method performs efficiently compared to other existing methods, such as BS. BS and PELT allow the detection of change points in variance and are performed through R package *changepoint* [22] [23].

12.2.3.1 Data Transformation and Robust Breakpoint Selection

At the same time, each method was applied to the raw time series and to a Box-Cox transformed version to ensure robustness against skewness.
The Box-Cox transformation is defined as

$$y^{(\lambda)} = \begin{cases} \frac{y^\lambda - 1}{\lambda}, & \text{if } \lambda \neq 0 \\ \log(y), & \text{if } \lambda = 0 \end{cases}$$

where y denotes the original data and λ is a transformation parameter chosen to approximate normality. For further analysis, only breakpoints standard to both methods and to both transformed and untransformed versions of the series were considered, ensuring that identified structural changes were robust across methods and transformations.

12.2.3.2 Analysis of Changes in Mean and Variance

Differences in means before and after each breakpoint were tested using a Newey–West [24] adjusted t-test to account for autocorrelation within segments. The test statistic is given by:

$$t = \frac{\bar{y}_B - \bar{y}_A}{\sqrt{\widehat{\text{Var}}(\bar{y}_B - \bar{y}_A)}},$$

where $\bar{y}_B$ and $\bar{y}_A$ denote the segment means before and after the breakpoint, respectively, and $\widehat{\text{Var}}$ is the heteroskedasticity- and autocorrelation-consistent (HAC) variance estimator using Newey–West adjustments.

This test provided p-values for mean differences, indicating the significance of shifts at each breakpoint.

The Newey–West adjusted t-test relies on the asymptotic normality of the segment means. While the HAC estimator is robust to heteroskedasticity and autocorrelation, approximate normality is still assumed for the inference. Normality of residuals within each segment was visually inspected to validate this assumption.

Furthermore, Levene's test [25] was applied to the segments before and after each breakpoint to assess changes in variance, which is robust to non-normality. The Levene test statistic is defined as

$$W = \frac{(N-k)}{(k-1)} \cdot \frac{\sum_{i=1}^{k} n_i (Z_{i\cdot} - Z_{\cdot\cdot})^2}{\sum_{i=1}^{k} \sum_{j=1}^{n_i} (Z_{ij} - Z_{i\cdot})^2},$$

where $Z_{ij} = |y_{ij} - \bar{y}_i|$, y_{ij} is the j-th observation in group i, $\bar{y}_i$ is the group mean, $Z_{i\cdot}$ is the group mean of the Z_{ij}s, and $Z_{\cdot\cdot}$ is the overall mean of the Z_{ij}s.

This makes it possible to determine whether the variance of each segment differed significantly at the breakpoints.

12.2.3.3 Segment Trend Analysis

After finding the method that presented the highest percentage of coincident breakpoints with the benchmark method, a set of robust breakpoints is created (i.e., coincident in both methods and data with Box-Cox transformation). For each specified breakpoint, data segments before and after were extracted. The trend (slope) within each segment was estimated using Sen's slope estimate [26], a robust non-parametric method.

Let Y_t denote the value of the time series at time t. For a segment of n observations, Sen's slope is computed as the median of all pairwise slopes:

$$\hat{\beta}_{\text{Sen}} = \text{median}\left(\frac{Y_j - Y_i}{j - i}\right), \quad \forall i, j : 1 \leq i < j \leq n \tag{12.1}$$

This estimator is robust to outliers and does not assume a specific distribution for the errors.

The general form of the linear model fitted to each segment is

$$Y_t = \alpha + \beta t + \varepsilon_t, \tag{12.2}$$

where

- α is the intercept,
- β is the trend (estimated using Sen's slope), and
- ε_t is the error term at time t.

The statistical significance of each slope was evaluated with associated p-values, providing information on whether the trend in each segment was statistically significant.

12.2.4 Quantifying Changes

For both mean and slope, the Relative Magnitude of Change (RMC) is calculated as a measure of the proportional change between segments across breakpoints, as follows:

$$RMC_{\text{mean}} = \left|\frac{\mu_{\text{after}} - \mu_{\text{before}}}{\mu_{\text{before}}}\right| \times 100 \tag{12.3}$$

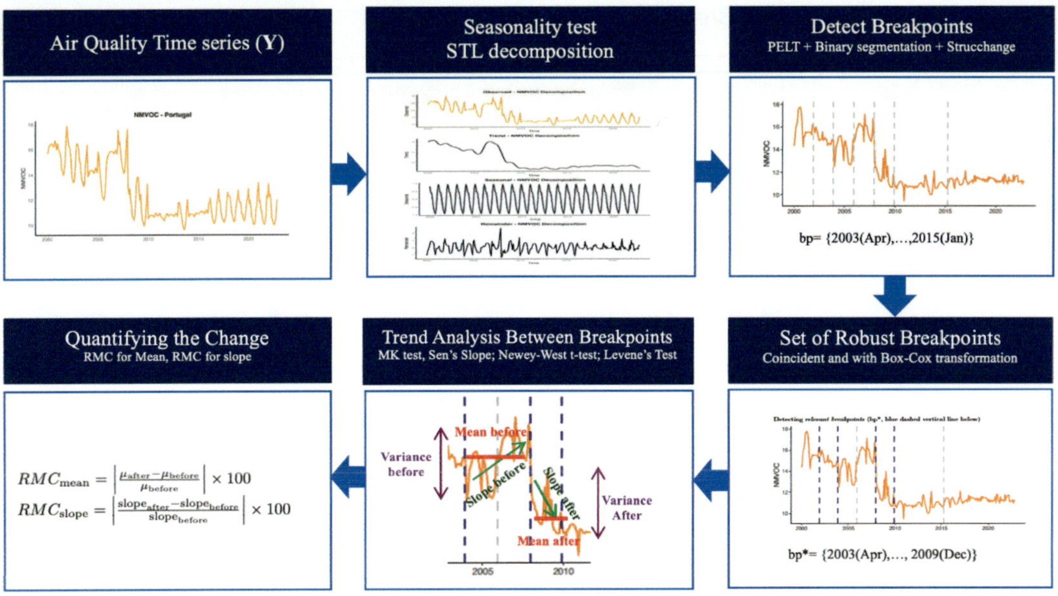

FIGURE 12.1
Flowchart showing the steps of the methodology applied to the NMVOC time series.

$$RMC_{\text{slope}} = \left| \frac{\text{slope}_{\text{after}} - \text{slope}_{\text{before}}}{\text{slope}_{\text{before}}} \right| \times 100 \tag{12.4}$$

RMC for the mean (12.3) was calculated as the absolute percentage change in the mean values before and after each breakpoint, while the RMC for the slope (12.4) measured the proportional change in trend between segments, as in [12]. Figure 12.1 presents the scheme that shows the steps of the methodology.

12.3 Air Pollution Case Study

12.3.1 Selection and Description of Air Pollutants

Time series on emissions from 2000 up to 2022 were extracted from EDGAR (Emissions Database for Global Atmospheric Research)[1]. To illustrate how significant breakpoints are detected after decomposition, three of the major outdoor air pollutants were selected from the original data set of eight air pollutants: nitrogen oxides (NOx), carbon monoxide (CO), particle pollution, and particulate matter (PM10). The selection criteria considered each pollutant's environmental impact, the observed breakpoint patterns within the data, and the trend differences between the two countries. The following is a detailed rationale for the inclusion of each pollutant.

Nitrogen Oxides (NOx): Emissions of NOx, derived mainly from combustion in the transportation and industrial sectors, play a significant role in air pollution, acid deposition, and respiratory health risks. Given their relevance in environmental policy, NOx emissions are frequently targeted through regulatory measures to reduce vehicular and industrial

[1] https://edgar.jrc.ec.europa.eu/

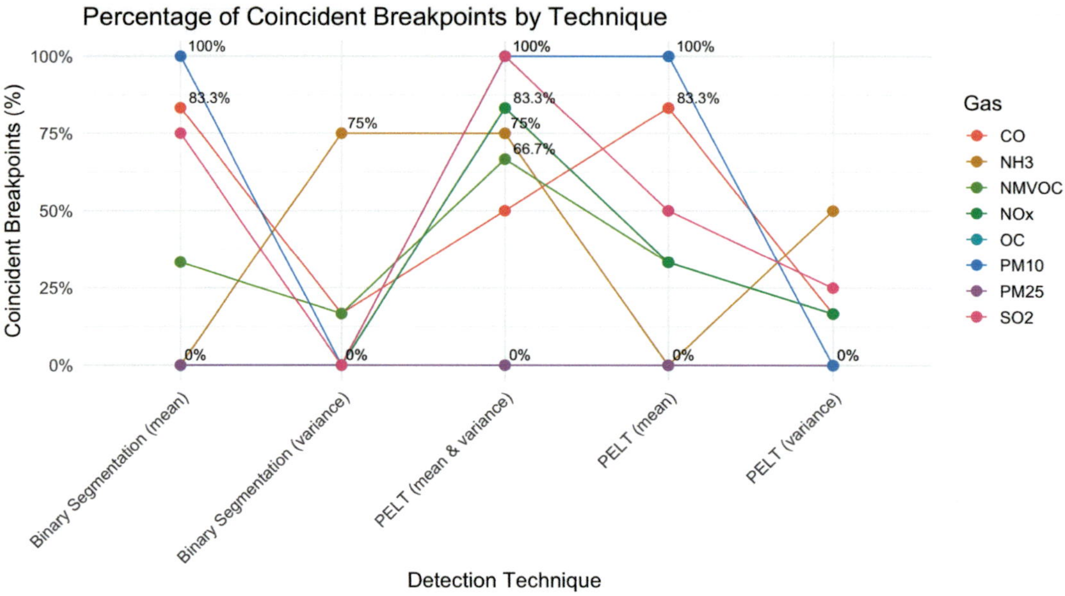

FIGURE 12.2
Coincident breakpoints for each pollutant and detection technique.

emissions. Analysis of the NOx time series for both countries reveals multiple breakpoints, especially around 2002–2005, 2010, and recent years, suggesting links to significant policy actions or economic shifts within each country. This pattern makes NOx emissions an ideal candidate for evaluating the comparative effectiveness of policies in Portugal and Italy.

Carbon Monoxide (CO): Emissions of CO, primarily generated from vehicle exhaust and industrial activities, pose serious public health risks. CO is closely monitored and regulated due to its adverse effects on air quality and human health. The CO time series for Portugal and Italy exhibits several breakpoints that may reflect policy efforts to control emissions from transportation and industry. Variations in the timing and magnitude of these breakpoints across countries offer insights into how each nation has managed or mitigated CO emissions in response to policy changes or economic developments.

Particulate Matter (PM10): PM10 includes particles such as dust, soot, and smoke, which can penetrate deeply into the respiratory tract, posing risks to respiratory and cardiovascular health. PM10 levels are influenced by various sources, including transportation, industrial activity, and natural events, making it a key indicator of air quality. The PM10 time series highlights notable differences between Italy and Portugal and significant breakpoints that may correspond to specific air quality regulations or industrial controls implemented in each country. These variations in PM10 trends and breakpoints serve as a valuable basis for assessing how effectively each country has managed sources of particulate matter over time.

12.3.2 Breakpoint Analysis

To analyze and compare trends in eight air pollutant time series for Portugal and Italy (CO, NH3, NMVOC, NOx, OC, PM25, PM10, and SO2), Figure 12.1, a sequence of statistical methods to identify, characterize, and evaluate the significance of structural breakpoints were applied.

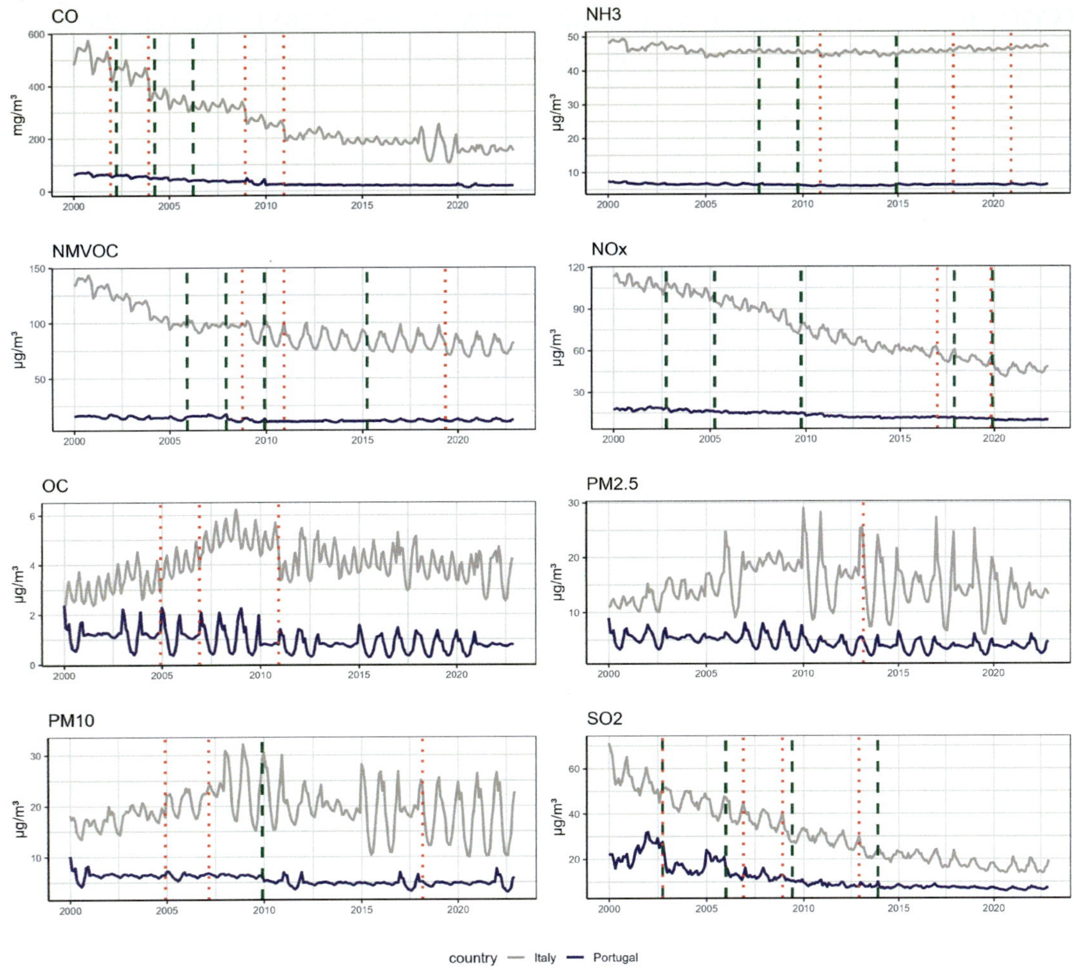

FIGURE 12.3
Air pollutant time series with breakpoints for Portugal (**green**) and Italy (**red**).

The plot presented in Figure 12.2 compares the ability of different methods (BS and PELT) to detect breakpoints coinciding with the validated baseline technique. Each line represents a pollutant, and the y-axis shows the percentage of coincident breakpoints. For further robust selection, we retained the consistent breakpoints with and without Box-Cox transformation. Results for Portugal and Italy time-series are presented in Figure 12.3.

A comparison of air quality time series for Portugal and Italy shows differences in breakpoint timings and pollutant concentrations. Overall, pollutant levels in Italy are generally higher than in Portugal.

CO: Italy's concentrations reached over 600 mg/m^3 in the early 2000s, while Portugal's levels remained below 100 mg/m^3 throughout the period. Breakpoints for CO in Portugal are detected in April 2002, April 2004, and April 2006, whereas in Italy they occur in December 2001, December 2003, December 2008, and December 2010.

NH3: Italy's concentrations exceed 40 µg/m^3 compared to Portugal's levels, which remain below 20 µg/m^3. Breakpoints in Portugal are detected in October 2007, October 2009, and December 2014, while in Italy, they are found in December 2010, December 2017, and December 2020.

NMVOC: Italy exhibits breakpoints in October 2008, December 2010, and May 2019, while in Portugal, breakpoints are identified in December 2005, December 2007, December 2009, and April 2015.

NOx: Concentrations in Italy are nearly three times higher than in Portugal during the early 2000s, though both countries show a declining trend over time. Breakpoints for NOx in Portugal are found in October 2002, April 2005, October 2009, December 2017, and December 2019, while in Italy, they occur in January 2017 and November 2019.

OC: No breakpoints are detected in Portugal, while Italy shows breakpoints in December 2004, December 2006, and December 2010.

PM2.5: Breakpoints are only detected in Italy occurring in March 2013.

PM10: Portugal shows one breakpoint in December 2009, while Italy exhibits breakpoints in December 2004, March 2007, and March 2018.

SO2: Portugal's breakpoints are detected in October 2002, January 2006, June 2009, and December 2013, while Italy's are observed in October 2002, December 2006, December 2008, and December 2012.

In terms of simplicity, only the results associated with the three time series of each country, Tables 12.1, 12.2 and 12.3, are presented.

Regarding CO, Portugal showed breakpoints in 2002, 2004, and 2006, significantly reducing slope and mean levels after each breakpoint. The RMC for the slope ranged from 61.94% to 82.49%, indicating a steady slowdown in the rate of decline. Mean levels decreased significantly, with RMCs around 54.4% to 56.8%, while variance exhibited statistically significant changes at the first and third breakpoints. Similarly, breakpoints were observed in Italy in 2001, 2003, 2008, and 2010. CO levels showed a marked decline, with slope RMCs between 35% and 78% and mean RMCs between 51.2% and 52.4%, accompanied by significant variance shifts.

Analyzing NOx emissions, Portugal exhibited five breakpoints (2002, 2005, 2009, 2017, and 2019), with the slope becoming progressively less negative, reflecting a slowdown in the reduction rate. The slope RMC varied widely, from 6% to 218%, suggesting significant trend intensity shifts. Mean NOx levels decreased with RMCs between 29.7% and 32.5%, while variance reductions were also significant. In Italy, NOx breakpoints in 2017 and 2019 indicated a stabilization of the downward trend, with slope RMC at 43.3% and mean RMC between 40.8% and 42.2%. Variance decreased notably following these breakpoints.

For PM10, Portugal displayed a single breakpoint in 2009, showing a modest slope reduction (RMC 10.9%) and a mean decrease with an RMC of 24.5%. Variance remained statistically unchanged. In Italy, three breakpoints (2004, 2007, and 2018) were identified,

TABLE 12.1
Breakpoints and Metrics for CO.

Country	BP	Slope B	Slope A	Mean B	Mean A	Var B	Var A	RMC Mean	RMC Slope
Portugal	2002–04	−0.381*	−0.145*	64.41	29.37*	25.56	168.43*	0.544	0.619
	2004–04	−0.345*	−0.110*	60.62	26.50*	39.60	98.68	0.563	0.681
	2006–04	−0.382*	−0.067*	55.96	24.17*	79.53	58.20*	0.568	0.825
Italy	2001–12	−1.773	−1.150*	514.32	250.88*	969.05	8939.80*	0.512	0.351
	2003–12	−2.396*	−0.994*	482.80	229.79*	1758.27	5136.22*	0.524	0.585
	2008–12	−2.343*	−0.640*	399.95	192.69*	6559.76	1564.15*	0.518	0.727
	2010–12	−2.218*	−0.468*	374.51	181.47*	8326.47	906.61*	0.516	0.789

Note. Statistically Significant: * ($p < 0.05$), reported for slopes, mean differences, and variance differences.

TABLE 12.2
Breakpoints and metrics for NO_x.

Country	BP	Slope B	Slope A	Mean B	Mean A	Var B	Var A	RMC Mean	RMC Slope
Portugal	2002–10	0.028*	−0.033*	17.95	12.62*	0.54	5.85*	0.703	0.002
	2005–04	−0.033*	−0.031*	17.25	12.07*	1.10	4.20*	0.700	0.002
	2009–10	−0.034*	−0.023*	16.23	11.07*	1.94	1.56*	0.682	0.002
	2017–12	−0.038*	−0.029*	14.22	9.88*	6.34	0.48*	0.695	0.003
	2019–12	−0.035*	0.007	13.86	9.35*	6.84	0.05*	0.675	0.003
Italy	2017–01	−0.301*	−0.170*	83.97	49.63*	336.80	22.49*	0.591	0.0036
	2019–11	−0.283*	0.023	79.65	46.01*	403.03	8.38*	0.578	0.004

Note. Statistically Significant: * ($p < 0.05$), reported for slopes, mean differences, and variance differences.

TABLE 12.3
Breakpoints and metrics for PM_{10}.

Country	BP	Slope B	Slope A	Mean B	Mean A	Var B	Var A	RMC Mean	RMC Slope
Portugal	2009–12	−0.0019*	−0.0017*	6.40	4.83*	0.42	0.46	0.246	0.109
Italy	2004–12	0.052*	−0.028*	17.46	19.87	3.20	22.23*	0.139	1.538
	2007–03	0.071*	−0.038*	18.53	19.72	6.09	24.65*	0.064	1.543

Note. Statistically Significant: * ($p < 0.05$), reported for slopes, mean differences, and variance differences.

with substantial slope changes (RMCs between 138% and 154%) and significant shifts in both mean and variance, reflecting variations in average PM10 levels and their dispersion.

12.4 Discussion and Conclusion

The results of this study show that the proposed methodology effectively identified significant change points in the time series of atmospheric emissions in Portugal and Italy between 2000 and 2022. Applying STL decomposition, combined with the Strucchange, PELT and BS methods, made it possible to detect structural breaks associated with possible changes in environmental policies, economic crises, and variations in emission sources. As observed, the main periods of change in the CO, NOx, and PM10 series coincide with the implementation of European regulations, such as the National Emission Ceilings Directive (2001/81/EC) and the successive Euro standards for vehicle emissions, which have significantly impacted the reduction of pollutants over time.

Previous studies, such as [1], showed that regulatory changes can be detected by analyzing structural breaks in environmental time series. Structural break tests were applied to CO2 series in 41 countries and identified similar patterns of changes associated with environmental policies and climate protests [3]. Comparative analysis of the data between Portugal and Italy revealed that although both countries showed downward trends in NOx and CO, the magnitudes of the changes were different, possibly due to differences in the implementation of policies and the pace of vehicle fleet renewal. These findings reinforce the argument of [4], who suggest that the effectiveness of environmental regulations depends not only on legislation but also on each country's enforcement capacity and economic structure.

The methods used demonstrated high robustness, especially when comparing raw and Box-Cox transformed time series results. Moreover, segments detected by the structural change techniques were consistent between different approaches, strengthening the reliability of the change points identified.

However, there are limitations to be considered. First, the method captures statistical changes but does not establish direct causality between the breaks identified and specific events. In addition, the change points detected may not have fully reflected the possible delayed effects of environmental policies. As pointed out by [6], integrating machine learning techniques can improve analysis by considering spatial and temporal factors in trend detection. Thus, future research could expand this study by incorporating socioeconomic and meteorological variables for a more comprehensive diagnosis of the determinants of changes in air quality.

In conclusion, this work has shown that statistical analysis of environmental time series is a valuable tool for monitoring the evolution of atmospheric emissions and evaluating the effectiveness of environmental policies. The application of the proposed approach revealed essential patterns of change in pollutant emissions in Portugal and Italy, highlighting the impact of implemented regulations.

For future studies, the recommendation is to include more countries and include statistical approaches that normalize meteorological effects and advanced machine learning techniques to better model trends and anomalies.

12.5 Glossary

Binary Segmentation: A recursive algorithm to detect multiple change points by splitting the data at the most significant change point.

Box-Cox Transformation: A family of power transformations used to stabilize variance and make the data more normally distributed.

Breakpoint: A point in a time series where there is a statistically significant change in trend, mean, or variance.

Change Point Detection: The process of identifying times at which the statistical properties of a time series change.

EDGAR (Emissions Database for Global Atmospheric Research): A global database that provides consistent and independent emission data for atmospheric research.

Levene's Test: A statistical test used to assess the equality of variances between groups, robust to non-normal data.

Newey-West Adjustment: A method for adjusting standard errors in regression models to account for autocorrelation and heteroskedasticity.

PELT (Pruned Exact Linear Time): An algorithm for detecting multiple change points in time series data efficiently by minimizing a penalized cost function.

Relative Magnitude of Change (RMC): A measure of the proportional difference in mean or slope before and after a change point.

Seasonality Test (`combined_test`): A function from the `seastests` R package that combines multiple statistical tests (e.g., Kruskal-Wallis, Welch, Friedman) to assess the presence of seasonality.

Sen's Slope Estimate: A non-parametric method to estimate the trend in a time series, robust to outliers.

STL Decomposition (Seasonal-Trend Decomposition using Loess): A method that decomposes a time series into seasonal, trend, and residual components using locally weighted regression.

Acknowledgments

AB and MC work was supported by FCT - Fundação para a Ciência e a Tecnologia, through project UIDB//4728/2025, DOI: https://doi.org/10.54499/UID/04728/2025. CC work is supported by CIDMA (https://ror.org/05pm2mw36) under the Portuguese Foundation for Science and Technology (FCT, https://ror.org/00snfqn58), Grants UID/04106/2025 (https://doi.org/10.54499/UID/04106/2025) and UID/PRR/04106/2025 (https://doi.org/10.54499/UID/PRR/04106/2025). MRR is is funded by national funds through FCT – Fundação para a Ciência e a Tecnologia, I.P., under CEAUL Research Unit, UID/00006/2025, DOI: https://doi.org/10.54499/UID/00006/2025.

Bibliography

[1] David C. Carslaw, Karl Ropkins, and Margaret C. Bell. Change-point detection of gaseous and particulate traffic-related pollutants at a roadside location. *Environmental Science; Technology*, 40(22):6912–6918, Oct 13 2006.

[2] Christopher Dienes and Alexander Aue. Online Monitoring of pollution concentrations with autoregressive moving average time series. *Journal of Time Series Analysis*, 35(3):239–261, Jan 30 2014.

[3] Festus Adedoyin, Ilhan Ozturk, Isah Abubakar, Terver Kumeka, Oludele Folarin, and Festus Victor Bekun. Structural breaks in CO2 emissions: Are they caused by climate change protests or other factors? *Journal of Environmental Management*, 266:110628, 7 2020.

[4] Thomas B. Fomby and Limin Lin. A change point analysis of the impact of "environmental federalism" on air quality in the United states: 1940–98. *Economic Inquiry*, 44(1):109–120, 1 2006.

[5] Muhammad Rizwan Khan and Biswajit Sarkar. Change point detection for airborne particulate matter (PM2.5, PM10) by using the Bayesian approach. *Mathematics*, 7(5):474, May 24 2019.

[6] Minghao Qiu, Corwin Zigler, and Noelle E. Selin. Statistical and machine learning methods for evaluating trends in air quality under changing meteorological conditions. *Atmospheric Chemistry and Physics*, 22(16):10551–10566, Aug 19 2022.

[7] Roberta Valentina Gagliardi and Claudio Andenna. Change points detection and trend analysis to characterize changes in meteorologically normalized air pollutant concentrations. *Atmosphere*, 13(1):64, Dec 30 2021.

[8] M. Hulkkonen, A. Lipponen, T. Mielonen, H. Kokkola, and N. Prisle. Changes in urban air pollution after a shift in anthropogenic activity analyzed with ensemble learning, competitive learning and unsupervised clustering. *Atmospheric Pollution Research*, 2022.

[9] F. Jiang, Q. Zhu, and T. Tian. An ensemble interval prediction model with change point detection and interval perturbation-based adjustment strategy: A case study of air quality. *Expert Systems with Applications*, 222:119823, 2023.

[10] F. Naz, C. McCann, M. Fahim, T. Cao, R. Hunter, N. Viet, L. Nguyen, and T. Duong. Comparative analysis of deep learning and statistical models for air pollutants prediction in urban areas. *IEEE Access*, 11:64016–64025, 2023.

[11] Z. Han. Application of statistical models in air quality monitoring. *Theoretical and Natural Science*, 2025.

[12] Clara Cordeiro, Ana Borges, and M. Rosário Ramos. A strategy to assess water meter performance. *Journal of Water Resources Planning and Management*, 148(2):05021027, 2022.

[13] R Core Team. *R: A Language and Environment for Statistical Computing.* R Foundation for Statistical Computing, Vienna, Austria, 2024.

[14] Robert B. Cleveland, William S. Cleveland, Jean E. McRae, and Irma Terpenning. STL: A seasonal-trend decomposition procedure based on Loess. *Journal of Official Statistics*, 6(1):3–73, 1990.

[15] S. Cristina, C. Cordeiro, S. Lavender, P. C. Goela, J. Icely, and A. Newton. Meris phytoplankton time series products from the SW Iberian peninsula (Sagres) using seasonal-trend decomposition based on Loess. *Remote Sensing*, 8(6):449, 2016.

[16] Clara Cordeiro. stl.fit(): Function developed in Cristina et al., 2016.

[17] Daniel Ollech. *seastests: Seasonality Tests*, 2021. R package version 0.15.4.

[18] S. Sharma, D. A. Swayne, and C. Obimbo. Trend analysis and change point techniques: A survey. *Energy, Ecology and Environment*, 1(3):123–130, 2016.

[19] Achim Zeileis, Friedrich Leisch, Kurt Hornik, and Christian Kleiber. strucchange: An R package for testing for structural change in linear regression models. *Journal of Statistical Software*, 7(2):1–38, 2002.

[20] Achim Zeileis, Christian Kleiber, Walter Krämer, and Kurt Hornik. Testing and dating of structural changes in practice. *Computational Statistics & Data Analysis*, 44:109–123, 2003.

[21] Jushan Bai and Pierre Perron. Multiple structural change models: A simulation analysis. *Econometric Theory and Practice: Frontiers of Analysis and Applied Research*, 1:212–237, 2006.

[22] Rebecca Killick and Idris A. Eckley. Changepoint: An R package for changepoint analysis. *Journal of Statistical Software*, 58(3):1–19, 2014.

[23] Rebecca Killick, Kaylea Haynes, and Idris A. Eckley. *Changepoint: An R package for changepoint analysis*, 2024. R package version 2.3.

[24] Whitney K. Newey and Kenneth D. West. A simple, positive semi-definite, heteroskedasticity and autocorrelation consistent covariance matrix. *Econometrica*, 55(3):703–708, 1987.

[25] Howard Levene. Robust tests for equality of variances. *Contributions to Probability and Statistics: Essays in Honor of Harold Hotelling*, pages 278–292, 1960.

[26] Pranab Kumar Sen. Estimates of the regression coefficient based on kendall's tau. *Journal of the American Statistical Association*, 63(324):1379–1389, 1968.

13

Testing of Long-Term Granger Causality in Environmental Time Series

Vyacheslav Lyubchich, K. Halimeda Kilbourne, and Geneviève Nesslage

Long-term relationships in environmental time series require specialized methods that can handle large lags. We introduce an approach to testing long-term Granger causality that addresses this challenge. By comparing out-of-sample prediction errors of nested models, our method provides a robust framework for identifying causal relationships with a variety of models. To focus on long-term dynamics, we impose restrictions on the models by excluding near-contemporaneous dependencies, mitigating the influence of short-term fluctuations on model selection based on traditional information criteria such as the Akaike information criterion (AIC) and Bayesian information criterion (BIC). Through extensive simulations, we demonstrate the effectiveness of these restrictions in uncovering genuine long-term relationships. We apply our approach to real-world environmental data, investigating the causal connections between the North Atlantic Oscillation (NAO) and both golden tilefish (*Lopholatilus chamaeleonticeps*) landings and sea surface temperatures. The results underscore the potential of our method to reveal intricate causal relationships in environmental time series, offering valuable insights to researchers and policymakers.

13.1 Introduction

A broad range of modern environmental challenges—such as evolving climate, pollution, and biodiversity loss—can be described by complex spatiotemporal processes linked at different scales. Our understanding of the complexity and interconnectedness of environmental systems improves with the increasing availability of environmental time series data from diverse sources (satellites, sensors, historical records) and with more advanced methods providing us with data-driven insights. Those insights may uncover causal relationships fundamental for understanding environmental dynamics. Causal understanding is crucial for designing targeted interventions and policies, developing more accurate and reliable predictive models, and gaining a deeper understanding of the intricate mechanisms and feedback within environmental systems.

Granger causality is a widely used and valuable framework for investigating directional relationships in time series data [1, 2]. It is based on identifying predictability improvements due to added information from another lagged time series, and can be applied in the analysis

DOI: 10.1201/9781003564423-13

of observational data where controlled experiments are often impossible or unethical. This makes Granger causality a practical solution in large-scale applications such as identification of pollution sources [3, 4] or climate interactions [5, 6].

Because many environmental processes operate over long timescales, the analysis needs to focus on long-term causal relationships rather than just short-term correlations [7]. Traditional information criteria, AIC and BIC, are commonly used for specifying models in Granger causality testing. However, AIC and BIC can be overly sensitive to short-term fluctuations and may prioritize models that fit short-term variations well but fail to capture underlying long-term dynamics. Existing Granger causality methods often fall short in addressing these challenges inherent in environmental time series analysis, particularly when focusing on long-term relationships.

Recent advancements in time series analysis have highlighted the need to move beyond standard stationarity assumptions to capture the full complexity of environmental phenomena. Many environmental processes exhibit cyclostationary or periodically correlated behavior, where statistical properties like mean and covariance vary periodically with time, reflecting natural cycles such as diurnal, seasonal, or annual rhythms [8, 9, 10]. Furthermore, almost-cyclostationary processes generalize this by allowing for slow modulations of these periodicities, offering a more nuanced representation of complex environmental systems that blend strict cycles with evolving trends [10, 11, 12]. While significant strides have been made in modeling and analyzing these periodic and almost-periodic processes (e.g., using periodically autoregressive moving average, PARMA, models [13, 14]), a critical challenge often overlooked is the presence of heavy-tailed distributions. Environmental data frequently exhibit heavy tails, characterized by a higher probability of extreme events (e.g., floods, droughts, and extreme temperatures) than predicted by Gaussian models [15, 16]. Standard Granger causality tests and model selection criteria, which often assume finite second-order moments or normality, can yield misleading results or lack robustness when applied to such data, making accurate long-term causal inference particularly challenging in these contexts.

This chapter introduces a novel framework for assessing long-term Granger causality in environmental time series, designed to handle potentially complex models of underlying relationships. It builds on bootstrapped out-of-sample prediction errors as a robust and reliable framework for model comparison, especially when dealing with noisy data and complex models, and data distributions deviating from normal. Our primary contribution is the development of a framework that explicitly mitigates the influence of short-term noise and prevents information criteria from being misled by short-term correlations, thereby improving our ability to detect genuine long-term causal relationships that is crucial for environmental research and policy. Although we demonstrate the approach using linear autoregressive models, the underlying modeling engine can be switched to more complex model classes capable of producing out-of-sample forecasts, including machine learning models and PARMA models, ensuring its applicability to the cyclostationary, almost cyclostationary, and heavy-tailed processes common in environmental science.

The subsequent section of the chapter introduces the proposed method of restricted autoregressive modeling and out-of-sample Granger causality testing using bootstrap distribution of the test statistics. Next, a simulation study assesses the performance of the proposed tests in finite samples under different scenarios. We then apply our method to real-world environmental datasets, specifically investigating the causal connections between the NAO and both golden tilefish (*Lopholatilus chamaeleonticeps*) landings and sea surface temperatures. Finally, we discuss implications of our findings and conclude with potential avenues for future research.

13.2 Methods

13.2.1 Restricted Autoregressive Model for Testing Granger Causality

Consider a stationary time series Y_t (t is the time index), for which a univariate autoregressive model of order p, AR(p), can be written as

$$Y_t = \phi_0 + \sum_{i=1}^{p} \phi_i Y_{t-i} + w_t, \tag{13.1}$$

where ϕ_0 is the intercept, ϕ_i's are autoregressive coefficients, p is the autoregressive order, and w_t is a time series of white-noise errors, $w_t \sim WN(0, \sigma_w^2)$. Model 13.1 predicts Y_t based on its lagged values $Y_{t-1}, \ldots, Y_{t-p}$. It can be augmented to include up to k lagged values of another stationary time series, X_t:

$$Y_t = \phi_0 + \sum_{i=1}^{p} \phi_i Y_{t-i} + \sum_{j=r+1}^{k} \beta_j X_{t-j} + e_t, \tag{13.2}$$

where β_j's are regression coefficients and $e_t \sim WN(0, \sigma_e^2)$. Model 13.1 is nested in Model 13.2, since the latter can be transformed into the former by setting all its coefficients β_j to zero. When $r = 0$, Models 13.1 and 13.2 represent a classic setup for testing whether the past values of X_t are useful in predicting Y_t [1, 17]. If so, the variable X is termed a Granger cause of Y. For jointly multivariate Gaussian variables, the Granger causality is equivalent to transfer entropy, which is a popular model-agnostic method for causality testing [18].

To focus on long-term dynamics, we impose restrictions by excluding near-contemporaneous dependencies between X and Y, up to lag r ($0 < r < k$). By setting $r > 0$ in Model 13.2, we exclude the terms $\beta_1, \ldots, \beta_r$ from the model and thus mitigate the influence of short-term fluctuations of X_t. This allows us to specify a model for long-term relationships (with large k) while not increasing the number of parameters dramatically. As a result, Model 13.2 can remain parsimonious, and the optimal—potentially, large—k can be selected using traditional information criteria such as the AIC and BIC. Without implementing such restrictions (i.e., if $r = 0$), AIC and BIC-based model selection procedures strongly penalize large lags k, thus limiting further investigations of long-term relationships.

13.2.2 Assumptions of Granger Causality

Granger causality, while a powerful framework for inferring directional relationships in time series, relies on several key assumptions that users should be aware of when interpreting results.

Stationarity: A fundamental assumption for Granger causality testing is that the time series involved are stationary (or can be made stationary through appropriate transformations, such as differencing) since non-stationary series can lead to spurious causal inference [19, 20]. While our examples largely deal with series that are either stationary or processed to approximate stationarity, users should always apply appropriate tests (e.g., augmented Dickey–Fuller test) and transformations as needed.

Linearity: The standard Granger causality framework, as implemented with autoregressive models like Models 13.1 and 13.2, assumes linear relationships between variables [20]. If the underlying relationships are nonlinear, a linear model might provide misleading results. While our proposed methodology is extendable to other, more complex, models, the current implementation focuses on linear AR models.

No Confounding Factors: Granger causality assesses predictability within the observed dataset. A critical assumption is that there are no significant confounding factors or unobserved variables outside the analyzed dataset that are driving both the X_t and Y_t variables. If such common drivers exist but are not included in the model, a spurious causal relationship might be inferred. This requires a thorough theoretical understanding and careful selection of variables [21, 22].

Sufficiency of Lags: It is assumed that enough past lags of both X_t and Y_t are included in the models to capture all relevant dependencies and render the error terms white noise. If the chosen lags are insufficient, the residuals may not be white noise, which can invalidate the test results. There also exists a trade-off between the number of lags and power of the direct Granger procedure [17]. While AIC helps us choose the right lag, these issues could still be a concern, especially in complex systems.

No Instantaneous Causality: While not strictly an assumption that invalidates the test, the standard Granger causality test typically assesses lagged relationships. Instantaneous causality is detected by nonzero correlation of X_t and Y_t at the same time step, but the modeling framework is incapable of deriving the direction of this relationship without additional background knowledge [17, 20, 23]. Our method's exclusion of near-contemporaneous dependencies ($r > 0$) specifically aims to focus on lagged effects.

13.2.3 Direct Testing of Granger Causality

As a baseline approach for this study, we consider direct testing of Granger causality using the F-test in the vector autoregression framework, VAR(p) [20, 24]. We select the order p using AIC. This test uses all T values of each time series and permits near-contemporaneous dependencies between X_t and Y_t. The test was implemented using the R package *vars* [25].

13.2.4 Out-of-sample Testing of Granger Causality

We implement a robust way to test superiority of the full Model 13.2 relative to Model 13.1 by comparing errors of one-step-ahead recursive forecasts (denoted as "$t+1$"). In this setup, the last P observations of Y_t ($t = 1, \dots, T$) are reserved for out-of-sample testing; the lags p and k in Model 13.2 are selected based on AIC or BIC using the first $T - P$ observations; then, the models are estimated and the forecasts are produced with the inclusion of each new data point (total P times). The out-of-sample prediction errors are used to test the null hypothesis:

$$H_0 : \mathrm{E}(e_{t+1}^2) = \mathrm{E}(w_{t+1}^2) \tag{13.3}$$

against the one-sided alternative of X being a Granger cause of Y:

$$H_a : \mathrm{E}(e_{t+1}^2) < \mathrm{E}(w_{t+1}^2), \tag{13.4}$$

where E is the expectation operator. The alternative hypothesis is one-sided because there should be no reason why the forecasts from the nested Model 13.1 would be superior to those from Model 13.2 [26]. We use the following four test statistics based on the out-of-sample errors and approximate their distributions using the bootstrap.

MSEcor statistic based on a correlation test (Chapter 9.3 in [27]), which is analogous to the MSE-reg approach described by [28] and [29] if it is using the regression coefficient itself, not its t-statistic. First, create a pair of random variables $e_{t+1} - w_{t+1}$ and $e_{t+1} + w_{t+1}$. Then, assuming the individual forecasts are unbiased and the errors not autocorrelated, covariance of the new variables is

$$\mathrm{E}\left[(e_{t+1} - w_{t+1})(e_{t+1} + w_{t+1})\right] = \mathrm{E}(e_{t+1}^2 - w_{t+1}^2) = \mathrm{E}(e_{t+1}^2) - \mathrm{E}(w_{t+1}^2)$$

implying that negative correlation of the new variables provides evidence against the H_0 in equation 13.3. Hence, the MSEcor statistic is the Pearson correlation coefficient for the pair of new random variables or the ordinary least squares estimate of coefficient c in the regression

$$(e_{t+1} - w_{t+1}) = c\,(e_{t+1} + w_{t+1}) + u_t,$$

where u_t is white noise, thus giving us

$$\text{MSEcor} = \frac{\hat{c}}{se(\hat{c})}, \tag{13.5}$$

where $se(\hat{c})$ is the standard error of the regression coefficient. Negative c means that the squared forecast errors of the full Model 13.2 are lower than those of the nested Model 13.1, and hence X is a Granger cause of Y.

EN statistic for the new encompassing test of [26]. Similar to the MSEcor statistic, the EN statistic is based on a pair of new variables, now w_{t+1} and $w_{t+1} - e_{t+1}$, and their scaled covariance

$$EN = P\frac{\sum(w_{t+1}^2 - w_{t+1}e_{t+1})}{\sum e_{t+1}^2}. \tag{13.6}$$

Because the difference defining one of the new variables is switched here, compared to the MSEcor definitions, the interpretation of EN is also switched. Positive EN corresponds to the case when errors of the full Model 13.2 are smaller than those of the nested Model 13.1, suggesting that X is a Granger cause of Y.

MSEt statistic based on the average difference of squared errors (e.g., see [28, 29]). Define $d_t = e_{t+1}^2 - w_{t+1}^2$, then estimate an intercept-only model for d_t using ordinary least squares:

$$d_t = a + u_t,$$

where $u_t \sim N(0, \sigma^2)$. Negative intercept a is observed when the prediction errors e_{t+1}^2 are smaller than w_{t+1}^2, implying that the full Model 13.2 is better, and lagged values of X_t are needed to improve forecasts of Y_t (i.e., X is a Granger cause of Y). Then, the MSEt statistic for Granger causality is the t-statistic for the intercept a:

$$\text{MSEt} = \frac{\hat{a}}{se(\hat{a})}, \tag{13.7}$$

where $se(\hat{a})$ is the standard error. Note that the intercept represents $\mathrm{E}(e_{t+1}^2 - w_{t+1}^2)$ and thus the MSEt statistic is asymptotically equivalent to MSEcor [28].

OOSF statistic represents the F-test adapted to the out-of-sample context [28]:

$$OOSF = \frac{-P\sum d_t}{\sum e_{t+1}^2},$$

where d_t is defined as for the MSEt. In this formulation, large OOSF corresponds to the case of the full Model 13.2 errors being smaller, implying X is a Granger cause of Y.

Bootstrap algorithm of [29] was adapted to empirically derive distributions of the test statistics. Our bootstrap testing procedure is available from the R package *funtimes* [30] and includes the following steps:

1. Reserve the last P observations for evaluating one-step-ahead forecasts. Define the proportion of the sample size to be used in this out-of-sample verification as $\pi = P/T$.
2. Use the first $T - P$ observations to specify the full Model 13.2 and the corresponding nested Model 13.1, then estimate both models.
3. From each model, obtain P recursive one-step-ahead forecasts, then estimate the errors $\hat{e}_{t+1}$ and $\hat{w}_{t+1}$.
4. Use $\hat{e}_{t+1}$ and $\hat{w}_{t+1}$ to calculate observed test statistics.
5. Bootstrap the nested model estimated on the full sample:
   (a) Under the null hypothesis of no Granger causality, estimate the nested Model 13.1 on the full sample and extract its fitted values $\hat{Y}_t$ and residuals $\hat{w}_t$.
   (b) Sample with replacement the residuals to obtain a bootstrapped series $\hat{w}_t^*$.
   (c) Generate new time series $\hat{Y}_t^* = \hat{Y}_t + \hat{w}_t^*$.
   (d) Use $\hat{Y}_t^*$ in Steps 1–4 above to obtain bootstrapped versions of the one-step-ahead forecast errors, $\hat{e}_{t+1}^*$ and $\hat{w}_{t+1}^*$, and bootstrapped versions of the test statistics.
   (e) Repeat the steps (b)–(d) a large number of times to obtain bootstrapped distributions of the test statistics. Let B be the number of bootstrap replications.
6. Calculate bootstrapped p-values as the proportion of bootstrapped statistics that are smaller or equal to the observed statistics (MSEcor and MSEt) or greater or equal to the observed statistics (EN and OOSF).

13.3 Simulation Study

13.3.1 The Simulation Design

We ran an extensive simulation study to evaluate the performance of the proposed methodology for testing long-term Granger causality under different scenarios based on the structures of data generating processes (DGPs) and imposing restrictions in Model 13.2.

DGP 0 comprises two independent AR(1) processes with autoregression coefficients of 0.3. In the notations of bivariate VAR, DGP 0 can be written as

$$\begin{bmatrix} Y_t \\ X_t \end{bmatrix} = \begin{bmatrix} 0.3 & 0 \\ 0 & 0.3 \end{bmatrix} \begin{bmatrix} Y_{t-1} \\ X_{t-1} \end{bmatrix} + \begin{bmatrix} \epsilon_{yt} \\ \epsilon_{xt} \end{bmatrix}.$$

The null hypothesis of no Granger causality is true in DGP 0, hence this DGP is used to assess the empirical size (i.e., probability of Type I error) of the tests. DGPs 1–3 represent deviations from the H_0 and are used to assess the empirical power of the tests (i.e., $1 - \beta$, where β is the probability of Type II error).

DGP 1, in addition to autoregressive dependence, incorporates near-contemporaneous dependence between X_{t-1} and Y_t:

$$\begin{bmatrix} Y_t \\ X_t \end{bmatrix} = \begin{bmatrix} 0.3 & 0.3 \\ 0 & 0.3 \end{bmatrix} \begin{bmatrix} Y_{t-1} \\ X_{t-1} \end{bmatrix} + \begin{bmatrix} \epsilon_{yt} \\ \epsilon_{xt} \end{bmatrix}.$$

When testing if X is a Granger cause of Y under this DGP, the null hypothesis should be rejected, hence DGP 1 is used to assess the power of the out-of-sample tests. However, if restrictions of $r > 0$ are imposed, the power is expected to be low, because the only dependence between the time series is not evaluated.

DGP 2, in addition to autoregressive dependence, incorporates long-term dependence between X_{t-15} and Y_t, represented with the coefficient β_{15}:

$$\begin{bmatrix} Y_t \\ X_t \end{bmatrix} = \begin{bmatrix} 0.3 & 0 \\ 0 & 0.3 \end{bmatrix} \begin{bmatrix} Y_{t-1} \\ X_{t-1} \end{bmatrix} + \begin{bmatrix} 0 & \beta_{15} \\ 0 & 0 \end{bmatrix} \begin{bmatrix} Y_{t-15} \\ X_{t-15} \end{bmatrix} + \begin{bmatrix} \epsilon_{yt} \\ \epsilon_{xt} \end{bmatrix}.$$

This DGP is used to check the ability of the tests to detect long-term dependence between time series when short-term dependence does not exist. We evaluate the tests' power under DGP 2 by varying the values $\beta_{15} = \{0.1, 0.3, 0.5, 0.7, 0.9\}$. If not stated otherwise, the reported results are for $\beta_{15} = 0.3$.

DGP 3 combines the effects represented in DGP 1 and DGP 2 by comprising both short- and long-term dependence:

$$\begin{bmatrix} Y_t \\ X_t \end{bmatrix} = \begin{bmatrix} 0.3 & 0.3 \\ 0 & 0.3 \end{bmatrix} \begin{bmatrix} Y_{t-1} \\ X_{t-1} \end{bmatrix} + \begin{bmatrix} 0 & 0.3 \\ 0 & 0 \end{bmatrix} \begin{bmatrix} Y_{t-15} \\ X_{t-15} \end{bmatrix} + \begin{bmatrix} \epsilon_{yt} \\ \epsilon_{xt} \end{bmatrix}.$$

For each DGP and time series length $T = 100$, 300, and 500, we ran 1000 Monte Carlo simulations and implemented each test with $B = 1000$. The simulations were performed using the R package *mlVAR* [31]. The proportion of the data left for out-of-sample testing was $\pi = 0.1$, 0.2, and 0.3. All tests were implemented with $r = 0$ and $r = 10$, AIC-based selection of the orders p and k in Model 13.2, and varying significance level $\alpha = 0.01, 0.02, \ldots, 0.20$. The baseline F-test for direct testing of Granger causality used all data and did not depend on the settings of π and r. For testing the robustness of the proposed approach to heavy-tailed distributions, the simulations were repeated using t_5 distribution for the errors ϵ_{yt} and ϵ_{xt}.

13.3.2 Finite Sample Performance

Based on the Monte Carlo results for DGP 0, size of the tests is maintained well for commonly used significance levels, up to $\alpha = 0.05$ or 0.10, but deviations from the nominal level grow as α increases (Figure 13.1). The statistics MSEcor and MSEt are slightly better at maintaining the test size than the EN and OOSF. The modest proportion $\pi = 0.2$ of data used for out-of-sample testing provides the best size performance across all statistics (Figure 13.1; see Appendix for complete results under all combinations of the simulation parameters).

The size performance of the tests improves with the increase in sample size T. For $T = 500$, the empirical rejection proportions most closely match the nominal levels (Figure 13.2).

Given the adequate performance of all the out-of-sample tests for $\alpha = 0.05$, we consider the corresponding test powers when applied to DGP 1 (Figure 13.3). The highest power is delivered when the EN statistic is used, followed by the OOSF, while the MSEcor and MSEt deliver similar lower performance. In all the cases, test power grows fast with the increase of sample size T; the EN-based inference closely approaches 100% power for larger samples. For shorter time series ($T = 100$), the best performance is achieved when 20% of the data are reserved for out-of-sample testing. For longer time series, on the contrary, the highest power is achieved with higher π (Figure 13.3).

Out-of-sample tests generally maintain test size more effectively than the F-test, which shows an elevated Type I error rate even with large samples ($T = 500$) (Figure 13.4, DGP 0). This implies that the F-test's power results are likely inflated (e.g., Figure 13.4, DGP 1

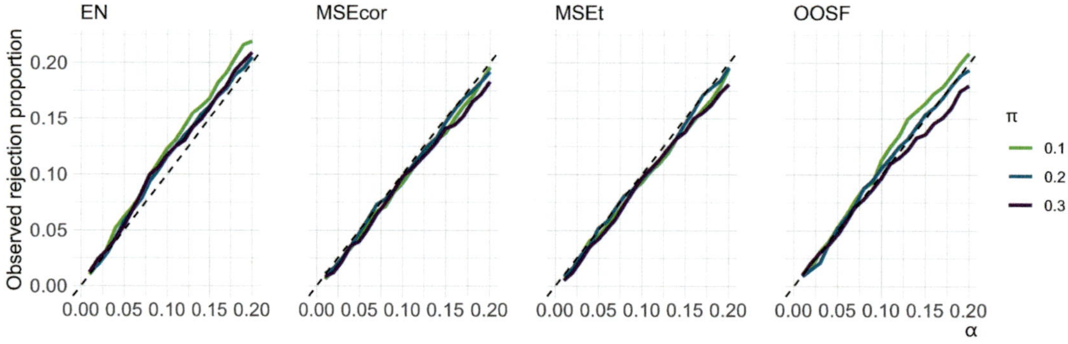

FIGURE 13.1
Size of the tests for the nominal levels α, sample size $T = 500$, and proportions of data used for out-of-sample testing π, based on DGP 0. The dashed diagonal line represents the ideal case of empirical probabilities matching the nominal levels.

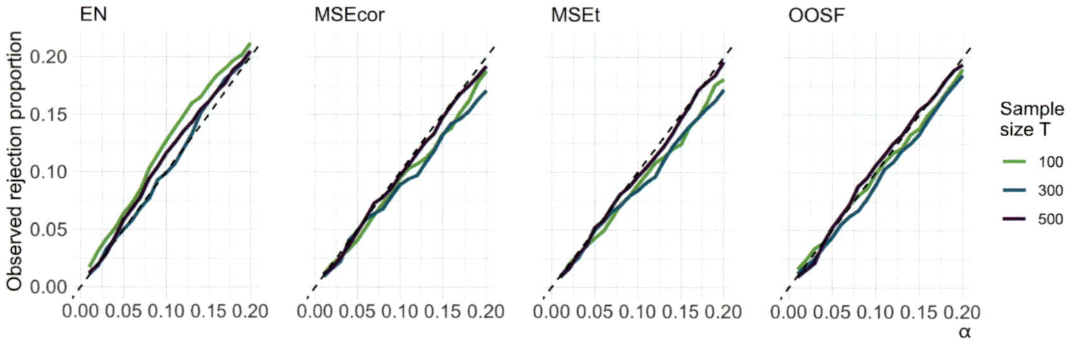

FIGURE 13.2
Size of the tests for the nominal levels α, sample sizes T, and proportion of data used for out-of-sample testing $\pi = 0.2$, based on DGP 0. The dashed diagonal line represents the ideal case of empirical probabilities matching the nominal levels.

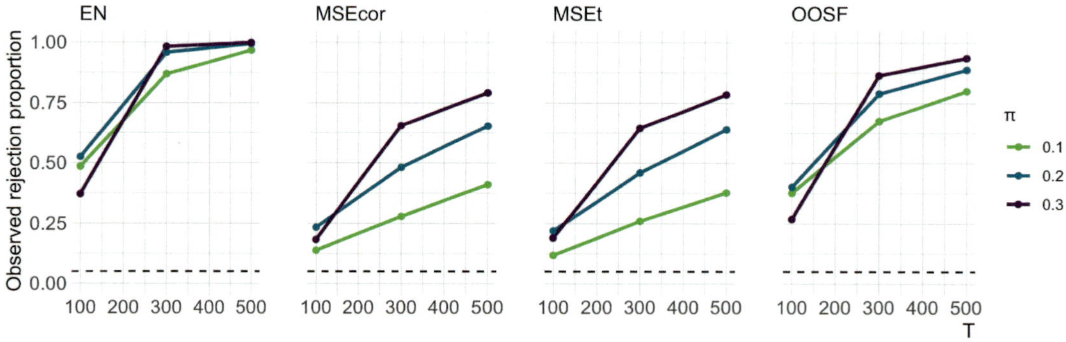

FIGURE 13.3
Power of the tests for the significance level $\alpha = 0.05$ (dashed line), sample sizes T, and proportions of data used for out-of-sample testing π based on DGP 1.

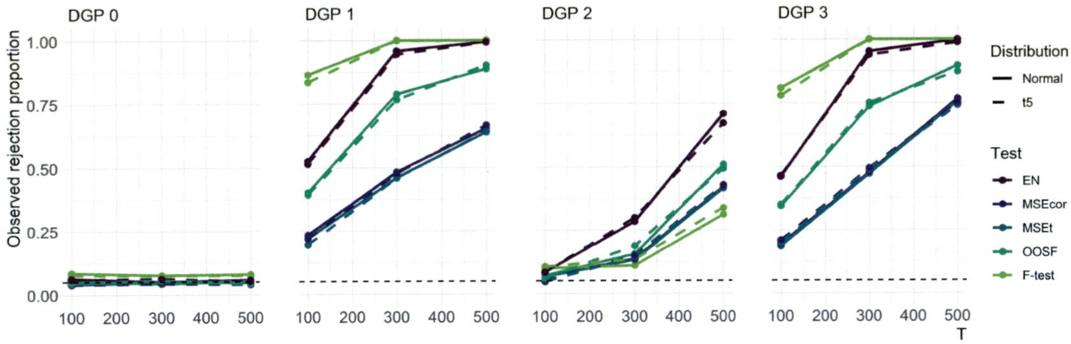

FIGURE 13.4
Observed rejection proportions of the tests for different data generating processes (DGPs) and sample sizes T. The significance level $\alpha = 0.05$ (dashed line) and, for all tests except the F-test, the out-of-sample set proportion $\pi = 0.2$, and $r = 0$.

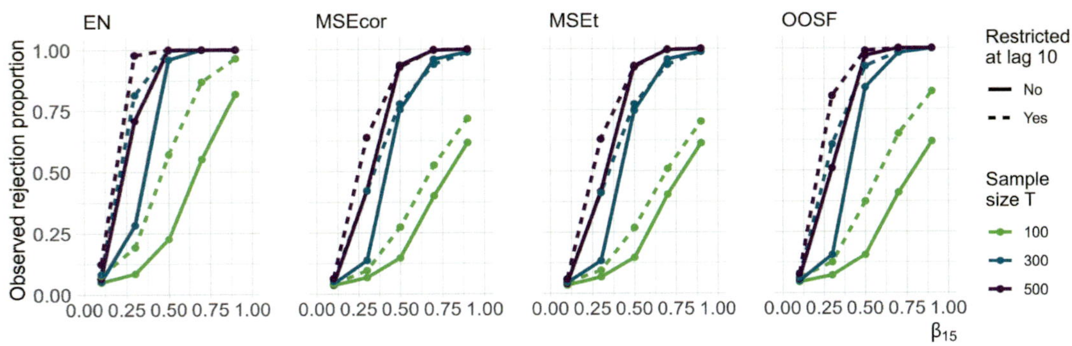

FIGURE 13.5
Empirical power of detecting Granger causality in DGP 2 with different values for the coefficient β_{15}. The plots also demonstrate the effect of lag restrictions and sample size T. The significance level $\alpha = 0.05$, and the out-of-sample set proportion $\pi = 0.2$.

and 3), while out-of-sample tests may have lower power because they use only a subsample for model construction and a smaller subsample for testing.

However, the F-test's power significantly diminishes under DGP 2 (Figure 13.4). In this case, with true lag of 1 for autocorrelation and 15 for cross-correlation, AIC-driven model selection often prioritizes models with fewer lags, preventing the model from fully capturing all relationships and thus reducing the test power. Importantly, the out-of-sample test statistics, particularly EN, demonstrate competitive power across DGPs 1–3 (Figure 13.4).

We found only trivial differences between the results under the normal and heavy-tailed t_5 distributions, suggesting robustness of the presented methodology (compare the solid and dashed lines in Figure 13.4).

Next, we evaluate the effect of using restricted models for detecting long-term causality like in DGP 2. When the regression coefficient β_{15} responsible for the long-term cross-correlation is small to moderate, we obtain substantial gains in the test power by implementing the restricted model with $r = 10$ (Figure 13.5). The effect is most consistent in shorter time series ($T = 100$) where it persists even for large β_{15}. In long time series, AIC is more capable of selecting a correct model even with the classic specification of $r = 0$,

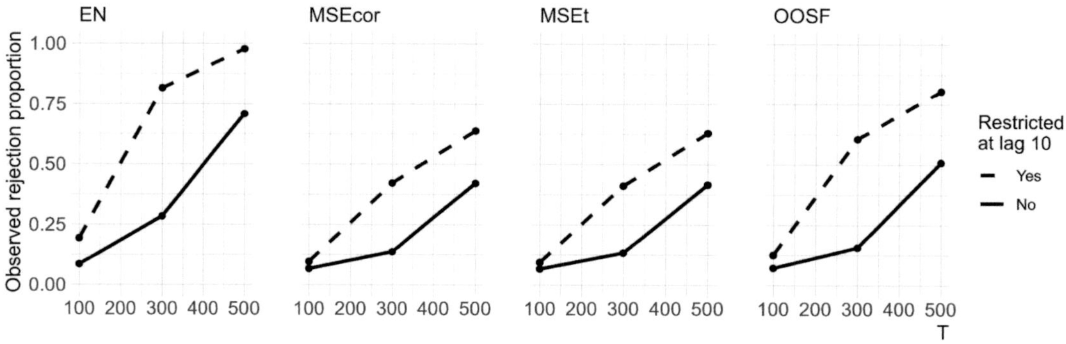

FIGURE 13.6
Empirical power of detecting Granger causality in DGP 2 with $\beta_{15} = 0.3$, under different sample sizes T and model restrictions imposed. The significance level $\alpha = 0.05$, and the out-of-sample set proportion $\pi = 0.2$.

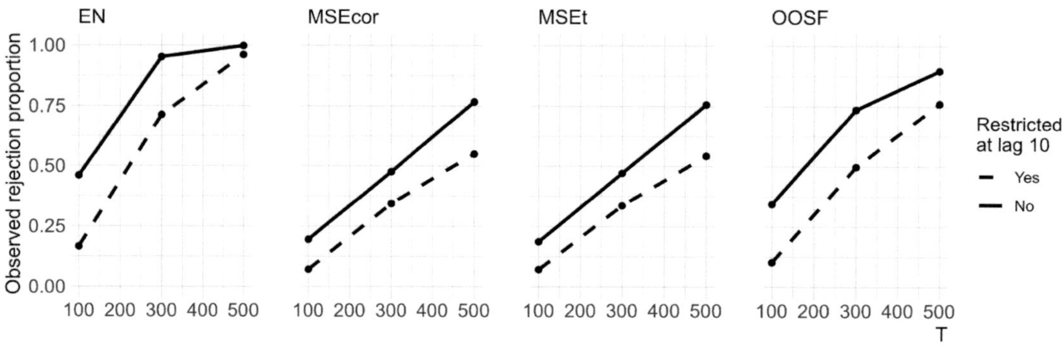

FIGURE 13.7
Empirical power of detecting Granger causality in DGP 3 under different sample sizes T and model restrictions imposed. The significance level $\alpha = 0.05$, and the out-of-sample set proportion $\pi = 0.2$.

thus the need for imposing the restriction is reduced. However, the test power grows more slowly when $r = 0$ than when reasonable restrictions are imposed (Figure 13.6).

When time series exhibit both short- and long-term causality like in DGP 3, unrestricted model specification ($r = 0$) corresponds to higher power than when the short-term dependencies are disregarded by the imposed restrictions (Figure 13.7). However, users focusing on studying long-term relationships might be willing to trade off some of the test power for the long-term inference. Note that the trade-off is necessary only when short-term dependencies are present (compare Figures 13.6 and 13.7).

We supply complete simulation results for each test in the Appendix (Tables 13.3–13.11). We also report empirical probabilities of the H_0 being rejected by all the tests, which replicates the situation when researchers "make sure" the results are statistically significant by requiring all tests to reject the null hypothesis. Such a testing procedure results in poor maintenance of test size under DGP 0 and low test power under other DGPs (Table 13.12).

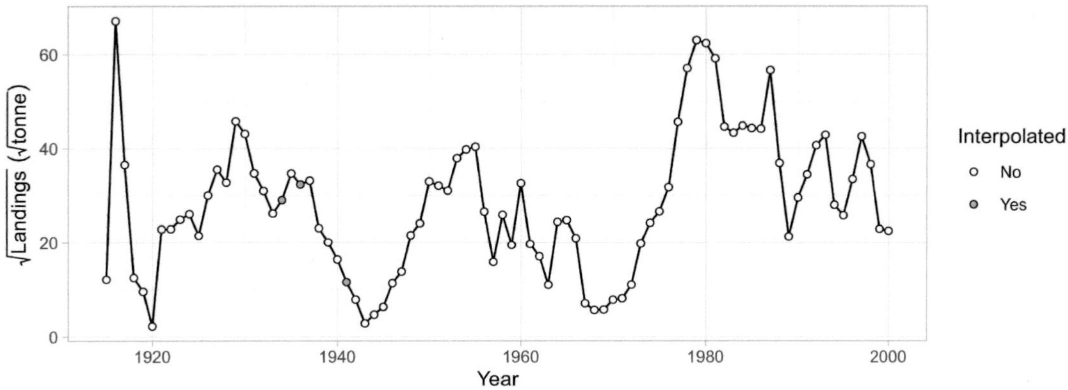

FIGURE 13.8
Time series of golden tilefish landings, 1915–2000. Interpolated values are shown in gray.

13.4 Case Studies

13.4.1 Golden Tilefish Landings

Golden tilefish population dynamics and associated fishery landings are thought to be driven by long-term trends in oceanographic conditions. A time series of historical landings for the US northern stock (off the coast of Virginia and states to the north) of golden tilefish was constructed by summing commercial catches landed annually across all gears from 1915 to 2017 [32]. We removed data from 2001 to 2017 to exclude the period in which landings were quota-limited, then applied a square root transformation to stabilize the variance [33]. Several missing values were interpolated using an autoregressive model (Figure 13.8) in the R package *imputeTS* [34].

We build on previous studies of the relationships between tilefish landings and climate indices such as the Atlantic Multidecadal Variability index (AMV) calculated for the complete year ("annual") and for specific months (December–April, "DJFMA"), and a station-based index of the NAO calculated for December–February ("DJF") [33, 35] (Figure 13.9). The NAO is a measure of the atmospheric pressure gradient between the subtropic (Azores) and subpolar (Iceland) regions of the North Atlantic. The AMV is the linearly detrended average sea surface temperature anomaly in the North Atlantic between 0° and 70° N latitude; it is commonly referred to as the Atlantic Multidecadal Oscillation in biological applications, [36] but it is not oscillatory.

The cross-correlation analysis identified relationships between lagged versions of the climate indices and golden tilefish landings. Most of the contemporaneous correlations, with lag ±2 years, were not statistically significant (Figure 13.10). With such cross-correlation pattern and small sample size ($T = 86$), this case corresponds to the short time series simulated from DGP 2. Based on the empirical test powers for DGP 2 (Figure 13.5), imposing the lag restrictions should boost the test power for detecting Granger causality.

The Granger causality test results for annual AMV generally agree with the previous study [33], where this climate index was not included in a regression model for golden tilefish landings. Using the significance level $\alpha = 0.05$, only MSEcor test rejected the null hypothesis of no Granger causality, while the other three tests did not reject it (Table 13.1). On the contrary, both the AMV for December–April and NAO for December–February were identified by all four tests as Granger causes of tilefish landings (Table 13.1). These

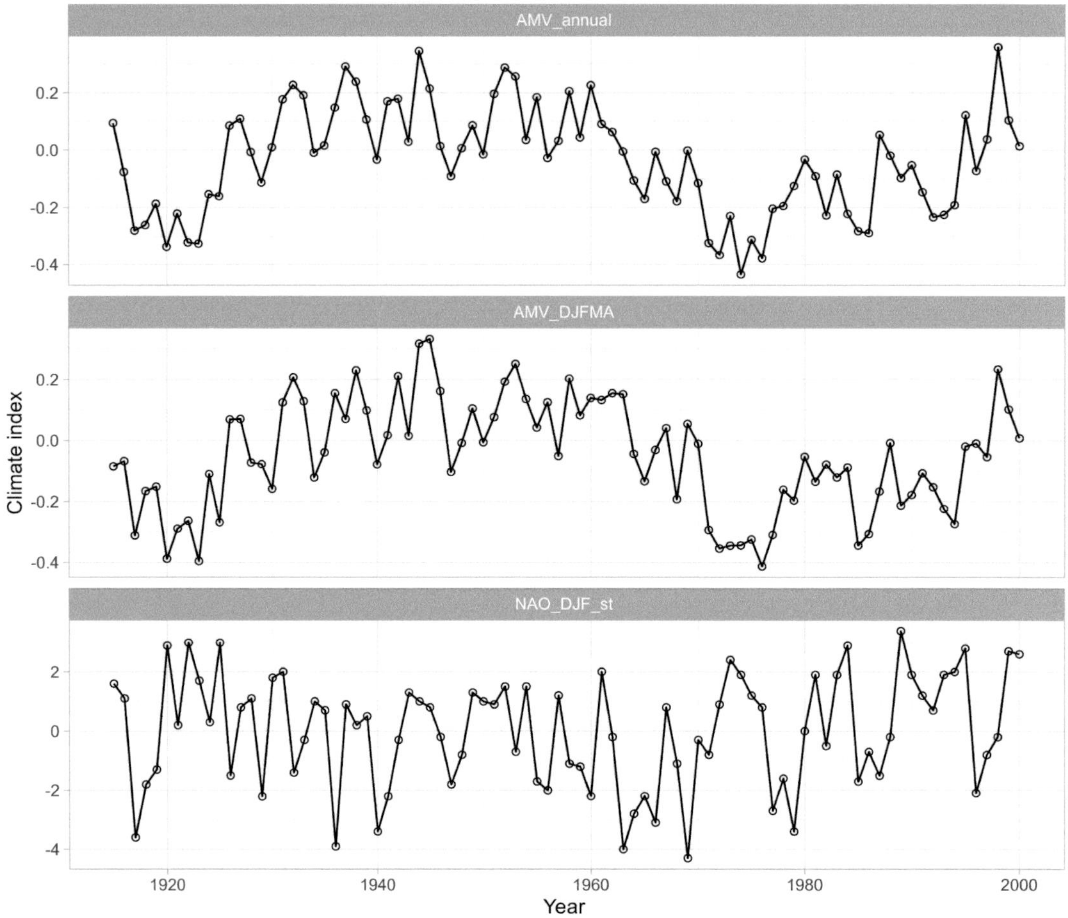

FIGURE 13.9
Time series of climate indices related to golden tilefish landings.

conclusions also match the previous more elaborate analyses, where both of these winter indices were identified as predictors in a random forest and a generalized additive model for tilefish landings [33].

13.4.2 Atlantic Multidecadal Climate Variability

Two fundamentally different ideas have developed about the causes of AMV and the sea surface temperature (SST) anomalies that are used to define it. One idea is that the ocean simply integrates the atmospheric noise of the NAO and causes a low-frequency persistence in the system [37, 38]. Although it is the most parsimonious explanation on the surface, some think it does not explain all of the relevant observational evidence [39]. The alternative hypothesis, proposed by [40], is that a coupled ocean-atmosphere process drives multiyear temperature anomalies in the North Atlantic through changes in heat transport by Atlantic Meridional Overturning Circulation (AMOC). Models show AMOC and NAO coupling on decadal to multidecadal timescales such that an intensification of winds driven by a positive NAO phase will cause initial cooling in the high-latitude ocean that causes a loss of buoyancy and increased deep ocean overturning [41]. The increased overturning ramps up AMOC, which then transports more heat to the high-latitude surface ocean, thereby

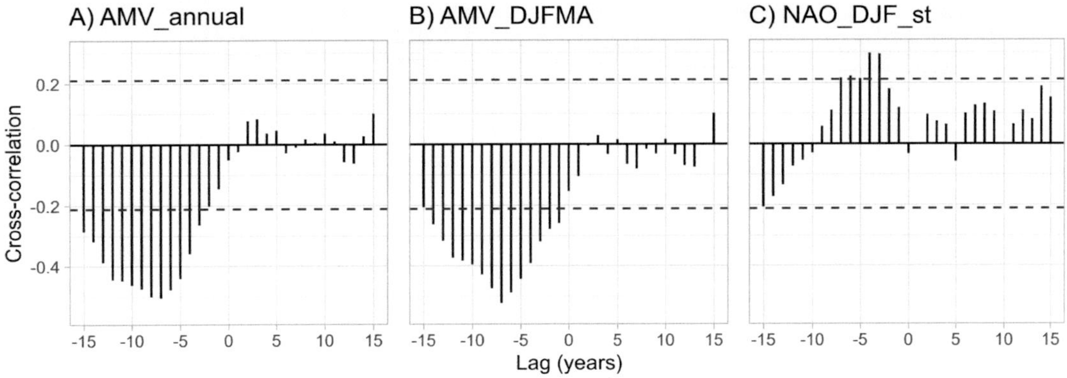

FIGURE 13.10
Cross-correlation of golden tilefish landings, Y_t, with climatic time series, X_{t+Lag}: A) annual AMV, B) AMV for the months December–April, and C) station-based NAO for the months December–February. The dashed lines denote 95% confidence interval under the null hypothesis of no correlation.

TABLE 13.1
Bootstrapped p-values from out-of-sample testing of Granger causality (bold font denotes p-values < 0.05). Each test for the climate time series being a Granger cause of the golden tilefish landings used $B = 10^4$, $r = 2$, and $\pi = 0.2$.

Climate index	EN	MSEcor	MSEt	OOSF
AMV annual	0.094	**0.038**	0.061	0.062
AMV Dec–Apr	**0.037**	**0.018**	**0.027**	**0.026**
NAO Dec–Feb	**0.007**	**0.005**	**0.008**	**0.005**

causing warm SST anomalies in the following decade(s). Warmer SSTs at high latitudes lead to a weakening of North Atlantic atmospheric pressure gradients (associated with a weakly negative NAO phase), which reduces trade winds and thus warms the tropical ocean. Warmer conditions at high latitudes eventually cause a decrease in overturning and decreased heat convergence.

Measurements of SST and sea level pressure have not been collected for long enough to verify that such a multidecadal-scale process occurs outside of models. Paleoclimate proxy data can provide multi-century-long records for testing such a model-based hypothesis, but paleoclimate data are sparsely distributed. We develop an Observing System Simulation Experiment, in which we try to identify known signals from limited data in preparation for using paleoclimate data in similar ways. In this experiment, we artificially limit the analysis to only locations with proxy SST data [42], thus generating a limited observation scenario.

We used NAO for December–March of 1855–2014 ($T = 160$ years) and nine time series of SST from locations corresponding to existing paleoclimate reconstructions [42]. From each time series, we removed the long-term trend by fitting a local weighted regression (lowess) with a span of 200 years. Thus, the filtered time series retained only mid- and high-frequency variability.

Cross-correlation analysis of the filtered NAO and SST time series revealed a complex pattern of correlations between the two processes at different latitudes (Figure 13.11). At zero lag, the classic ocean response to NAO [40] appears as a tripole pattern with negative correlations at high and low latitudes but positive correlations at mid-latitudes. A decadal-scale signal also appears in the correlation data, where SSTs are warmer 20–30 years after

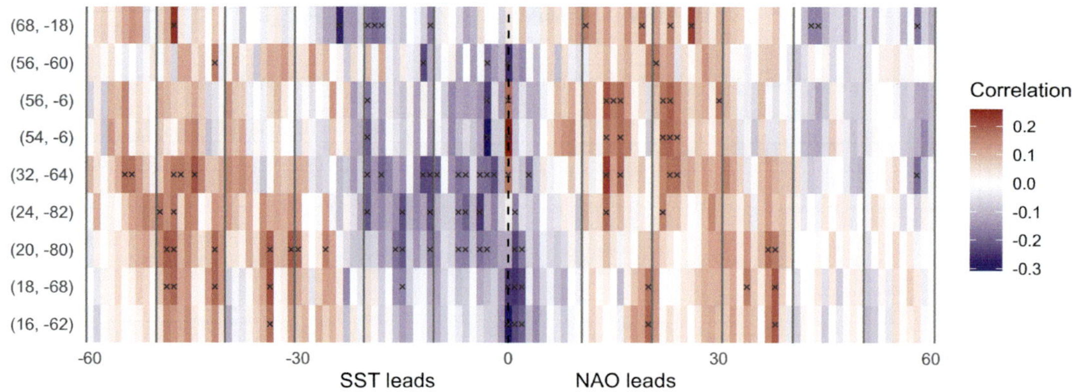

FIGURE 13.11
Cross-correlation of the NAO for the months December–March of 1855–2014 with sea surface temperature (SST) at different locations. The SST coordinates are labeled on the y-axis (latitude, longitude), ordered by latitude from north to south. The x-axis shows lags (years). The crosses denote individually significant correlations at the 95% confidence level.

TABLE 13.2
Bootstrapped p-values from out-of-sample testing of Granger causality (bold font denotes p-values < 0.05) for North Atlantic Oscillation (NAO) with individual time series of sea surface temperature (SST) and averaged across upper, middle, and lower latitudes. Each test used $B = 10^4$, $r = 10$, and $\pi = 0.2$.

Coordinates	**SST leads**				**NAO leads**			
(lat, lon)	**EN**	**MSEcor**	**MSEt**	**OOSF**	**EN**	**MSEcor**	**MSEt**	**OOSF**
(68, −18)	0.059	0.163	0.166	0.074	0.656	0.683	0.895	0.706
(56, −60)	0.307	0.327	0.514	0.396	0.739	0.614	0.897	0.513
(56, −6)	0.601	0.510	0.483	0.458	**0.017**	**0.010**	**0.015**	**0.011**
(54, −6)	0.920	0.889	0.518	0.793	**0.014**	**0.030**	0.052	**0.039**
Upper lats	**0.039**	0.229	0.228	0.123	**0.012**	**0.017**	**0.034**	**0.009**
(32, −64)	0.606	0.337	0.713	0.218	0.188	0.277	0.403	0.388
(24, −82)	0.363	0.403	0.827	0.475	0.518	0.723	0.726	0.837
Mid lats	0.284	0.290	0.689	0.341	0.244	0.385	0.069	0.527
(20, −80)	0.068	0.093	0.153	0.136	0.403	0.857	0.957	0.969
(18, −68)	0.159	0.208	0.308	0.292	0.064	0.406	0.986	0.696
(16, −62)	0.352	0.622	0.315	0.773	0.146	0.601	0.541	0.842
Low lats	0.097	0.163	0.199	0.249	0.137	0.376	0.760	0.605

a positive NAO period. This is consistent with similar correlation analyses [43] and is consistent with the mechanistic relationship between NAO and SST described in the preceding paragraph, including the delay in the tropical SST response compared to the high-latitude SST response. However, correlation is not causation and although SST has been shown to be a Granger cause of NAO on daily [5] and annual [44] timescales, to our knowledge, Granger causality has not yet been tested between NAO and SST over decadal timescales.

Testing of long-term Granger causality (above the lag of $r = 10$ years) revealed significant effects of NAO at two of the locations (Table 13.2). The effect was also detected by all four tests when an aggregated time series for upper latitudes was used. In other tests for individual SST time series, there was not enough evidence of Granger causality. The aggregated time series for SST in upper latitudes was considered a Granger cause of NAO based on the EN statistic but not other statistics. Overall, p-values for aggregated time series were generally lower than for individual time series, suggesting that aggregation might have decreased noise (Table 13.2). However, the small sample size T and the small number of time series within each region might have been responsible for low power of the tests.

13.5 Conclusion

In this chapter, we introduced a novel approach to testing long-term Granger causality in environmental time series, addressing the challenges posed by large lags in autoregressive models. By imposing restrictions to exclude near-contemporaneous dependencies, our method effectively isolates long-term dynamics, ensuring that traditional information criteria such as AIC and BIC can be applied without bias against large lags.

Our extensive simulations demonstrate the robustness of these restrictions in uncovering genuine long-term relationships. The out-of-sample prediction error comparisons provide a reliable framework for identifying causal links, even in the presence of short-term noise. The simulation results also strongly suggest that our proposed approach is robust to deviations from normality such as under a heavy-tailed t_5 distribution. This is a highly desirable property for statistical methods, as real-world data frequently exhibit heavier tails or other non-normal characteristics, making methods that rely on Gaussian assumptions less reliable.

Applying our methodology to real-world environmental data, we explored the causal connections between the NAO and both golden tilefish landings and sea surface temperatures. The results highlighted the potential of our approach to reveal intricate causal relationships, offering valuable insights to researchers and policymakers.

Our proposed method effectively identifies long-term Granger causal relationships under the assumption of fixed, pre-defined lag structures (represented by r and k). However, real-world environmental systems can exhibit variable-lag or time-varying causal relationships, which might not be fully captured by a static lag structure. For instance, the delayed impact of climate phenomena on ecological systems might not always be constant; an event that causes an effect after a certain delay in one period might have a different delay in another, due to changes in the system or its stability. Our current approach, which relies on a single optimal lag selection for the entire time series, may not fully capture such dynamic, variable-lag events. Future research could explore extensions to this methodology, such as time-varying parameter models or adaptive lag selection techniques, to account for these more complex causal patterns.

In summary, our method provides a powerful tool for investigating long-term causal relationships in environmental time series, enhancing our understanding of complex ecological and climatic interactions. The R code for reproducing the results of this study is available from [45]. Future research can build on these findings by possible extensions to nonlinear statistical and machine learning models by applying the approach to other environmental variables and by exploring its potential in different contexts.

Acknowledgment

This work was supported in part by grants #NA20OAR4310481 (Kilbourne, Lyubchich), #A101316 NA14OAR4320158 effort 37045812 (Nesslage, Lyubchich) and #NA23NMF4570429 (Nesslage, Lyubchich) from the US National Oceanic and Atmospheric Administration.

13.6 Appendix: Simulation Results

TABLE 13.3
Empirical probabilities of rejecting the null hypothesis using the direct Granger causality testing under different data generating processes (DGP) with normal (N) or t_5 noise distributions.

	DGP 0		**DGP 1**		**DGP 2**		**DGP 3**	
T	N	t_5	N	t_5	N	t_5	N	t_5
				$\alpha = 0.01$				
100	0.029	0.025	0.689	0.656	0.032	0.045	0.581	0.597
300	0.028	0.018	0.998	0.993	0.064	0.086	0.992	0.989
500	0.016	0.020	1.000	1.000	0.270	0.310	1.000	1.000
				$\alpha = 0.05$				
100	0.086	0.078	0.865	0.836	0.100	0.107	0.809	0.780
300	0.077	0.076	1.000	0.999	0.111	0.135	1.000	0.999
500	0.080	0.076	1.000	1.000	0.311	0.338	1.000	1.000
				$\alpha = 0.10$				
100	0.136	0.129	0.920	0.890	0.152	0.161	0.889	0.849
300	0.144	0.132	1.000	0.999	0.162	0.192	1.000	0.999
500	0.131	0.130	1.000	1.000	0.358	0.379	1.000	1.000

TABLE 13.4
Empirical probabilities of rejecting the null hypothesis using the out-of-sample Granger causality testing with the EN statistic under different data generating processes (DGP) with normal noise distribution.

	DGP 0			DGP 1			DGP 2			DGP 3		
$T\backslash\pi$	0.1	0.2	0.3	0.1	0.2	0.3	0.1	0.2	0.3	0.1	0.2	0.3
$\alpha = 0.01$; without restrictions ($r = 0$)												
100	0.019	0.017	0.011	0.270	0.315	0.226	0.014	0.021	0.031	0.235	0.246	0.171
300	0.011	0.011	0.011	0.772	0.901	0.960	0.147	0.178	0.179	0.680	0.874	0.931
500	0.010	0.012	0.012	0.926	0.978	0.996	0.539	0.625	0.586	0.902	0.990	0.999
$\alpha = 0.01$; with restrictions at lag $r = 10$												
100	0.014	0.016	0.018	0.010	0.010	0.019	0.056	0.066	0.074	0.031	0.056	0.060
300	0.013	0.012	0.015	0.011	0.013	0.011	0.530	0.692	0.736	0.428	0.555	0.592
500	0.017	0.015	0.014	0.013	0.018	0.016	0.817	0.940	0.965	0.750	0.898	0.935
$\alpha = 0.05$; without restrictions ($r = 0$)												
100	0.073	0.064	0.051	0.485	0.525	0.371	0.077	0.085	0.107	0.433	0.460	0.358
300	0.048	0.050	0.051	0.869	0.958	0.982	0.260	0.284	0.254	0.833	0.952	0.974
500	0.062	0.058	0.054	0.967	0.994	0.998	0.674	0.708	0.634	0.948	0.998	0.999
$\alpha = 0.05$; with restrictions at lag $r = 10$												
100	0.069	0.063	0.050	0.058	0.057	0.061	0.171	0.193	0.211	0.122	0.166	0.186
300	0.067	0.065	0.071	0.057	0.045	0.058	0.703	0.816	0.829	0.619	0.713	0.723
500	0.057	0.059	0.061	0.067	0.064	0.067	0.905	0.978	0.976	0.877	0.961	0.967
$\alpha = 0.10$; without restrictions ($r = 0$)												
100	0.118	0.127	0.110	0.577	0.618	0.458	0.130	0.152	0.193	0.504	0.564	0.457
300	0.106	0.099	0.104	0.899	0.975	0.985	0.329	0.340	0.314	0.889	0.968	0.986
500	0.123	0.116	0.117	0.971	0.995	0.999	0.743	0.748	0.659	0.973	0.998	1.000
$\alpha = 0.10$; with restrictions at lag $r = 10$												
100	0.126	0.123	0.103	0.112	0.110	0.114	0.263	0.289	0.312	0.198	0.259	0.268
300	0.119	0.127	0.127	0.107	0.121	0.115	0.776	0.859	0.861	0.700	0.777	0.768
500	0.109	0.103	0.110	0.108	0.123	0.116	0.934	0.982	0.983	0.915	0.973	0.976

TABLE 13.5
Empirical probabilities of rejecting the null hypothesis using the out-of-sample Granger causality testing with the EN statistic under different data generating processes (DGP) with t_5 noise distribution.

	DGP 0			DGP 1			DGP 2			DGP 3		
$T\backslash\pi$	0.1	0.2	0.3	0.1	0.2	0.3	0.1	0.2	0.3	0.1	0.2	0.3
$\alpha = 0.01$; without restrictions ($r = 0$)												
100	0.013	0.007	0.015	0.280	0.323	0.223	0.016	0.015	0.029	0.222	0.275	0.210
300	0.014	0.015	0.018	0.769	0.884	0.949	0.161	0.190	0.176	0.758	0.864	0.931
500	0.017	0.014	0.012	0.923	0.978	0.990	0.485	0.595	0.569	0.893	0.978	0.996
$\alpha = 0.01$; with restrictions at lag $r = 10$												
100	0.016	0.018	0.012	0.016	0.014	0.012	0.051	0.076	0.089	0.062	0.070	0.061
300	0.009	0.015	0.021	0.015	0.013	0.014	0.511	0.677	0.754	0.409	0.549	0.609
500	0.014	0.009	0.013	0.017	0.012	0.016	0.822	0.944	0.970	0.737	0.881	0.928
$\alpha = 0.05$; without restrictions ($r = 0$)												
100	0.063	0.060	0.051	0.473	0.514	0.374	0.080	0.084	0.116	0.419	0.463	0.390
300	0.064	0.063	0.059	0.867	0.945	0.978	0.279	0.299	0.264	0.869	0.938	0.971
500	0.057	0.052	0.054	0.961	0.992	0.994	0.648	0.671	0.620	0.948	0.987	0.998
$\alpha = 0.05$; with restrictions at lag $r = 10$												
100	0.068	0.062	0.068	0.067	0.073	0.060	0.150	0.192	0.218	0.168	0.165	0.174
300	0.058	0.064	0.079	0.054	0.054	0.057	0.697	0.810	0.823	0.614	0.708	0.745
500	0.068	0.056	0.049	0.061	0.057	0.050	0.895	0.977	0.986	0.856	0.953	0.956
$\alpha = 0.10$; without restrictions ($r = 0$)												
100	0.115	0.115	0.106	0.560	0.628	0.471	0.140	0.172	0.188	0.515	0.553	0.484
300	0.118	0.120	0.114	0.892	0.956	0.987	0.370	0.361	0.324	0.901	0.962	0.984
500	0.112	0.098	0.111	0.968	0.993	0.997	0.702	0.709	0.644	0.959	0.993	1.000
$\alpha = 0.10$; with restrictions at lag $r = 10$												
100	0.120	0.119	0.109	0.115	0.129	0.108	0.248	0.273	0.320	0.243	0.250	0.271
300	0.119	0.127	0.121	0.103	0.116	0.112	0.754	0.851	0.856	0.691	0.765	0.787
500	0.111	0.106	0.103	0.114	0.104	0.107	0.928	0.985	0.988	0.894	0.969	0.967

TABLE 13.6
Empirical probabilities of rejecting the null hypothesis using the out-of-sample Granger causality testing with the MSEcor statistic under different data generating processes (DGP) with normal noise distribution.

	DGP 0			**DGP 1**			**DGP 2**			**DGP 3**		
$T\backslash\pi$	**0.1**	**0.2**	**0.3**	**0.1**	**0.2**	**0.3**	**0.1**	**0.2**	**0.3**	**0.1**	**0.2**	**0.3**
$\alpha = 0.01$; without restrictions ($r = 0$)												
100	0.008	0.011	0.007	0.040	0.086	0.068	0.011	0.021	0.015	0.053	0.062	0.062
300	0.011	0.009	0.007	0.104	0.228	0.376	0.037	0.056	0.061	0.110	0.214	0.329
500	0.006	0.011	0.008	0.184	0.381	0.533	0.110	0.208	0.278	0.289	0.516	0.700
$\alpha = 0.01$; with restrictions at lag $r = 10$												
100	0.007	0.010	0.006	0.005	0.008	0.011	0.015	0.023	0.021	0.008	0.012	0.017
300	0.006	0.006	0.006	0.007	0.004	0.006	0.089	0.196	0.302	0.081	0.144	0.207
500	0.010	0.005	0.009	0.014	0.011	0.010	0.154	0.349	0.537	0.131	0.282	0.431
$\alpha = 0.05$; without restrictions ($r = 0$)												
100	0.042	0.041	0.033	0.139	0.234	0.183	0.051	0.067	0.086	0.154	0.195	0.184
300	0.042	0.049	0.047	0.279	0.482	0.655	0.117	0.137	0.148	0.300	0.476	0.615
500	0.044	0.048	0.040	0.411	0.653	0.791	0.305	0.421	0.458	0.542	0.767	0.894
$\alpha = 0.05$; with restrictions at lag $r = 10$												
100	0.055	0.044	0.035	0.045	0.043	0.043	0.081	0.096	0.103	0.051	0.071	0.087
300	0.041	0.035	0.049	0.038	0.042	0.042	0.254	0.422	0.517	0.213	0.344	0.415
500	0.036	0.046	0.053	0.033	0.056	0.050	0.384	0.638	0.770	0.332	0.550	0.691
$\alpha = 0.10$; without restrictions ($r = 0$)												
100	0.100	0.095	0.078	0.259	0.345	0.296	0.100	0.118	0.148	0.249	0.301	0.297
300	0.087	0.089	0.096	0.433	0.620	0.785	0.188	0.216	0.219	0.456	0.639	0.754
500	0.092	0.097	0.098	0.562	0.767	0.868	0.434	0.549	0.553	0.682	0.867	0.937
$\alpha = 0.10$; with restrictions at lag $r = 10$												
100	0.097	0.089	0.070	0.089	0.076	0.080	0.144	0.169	0.191	0.111	0.138	0.153
300	0.094	0.088	0.104	0.086	0.093	0.091	0.386	0.561	0.649	0.333	0.466	0.540
500	0.080	0.090	0.095	0.082	0.102	0.103	0.521	0.758	0.865	0.487	0.679	0.803

TABLE 13.7
Empirical probabilities of rejecting the null hypothesis using the out-of-sample Granger causality testing with the MSEcor statistic under different data generating processes (DGP) with t_5 noise distribution.

	DGP 0			**DGP 1**			**DGP 2**			**DGP 3**		
$T\backslash\pi$	**0.1**	**0.2**	**0.3**	**0.1**	**0.2**	**0.3**	**0.1**	**0.2**	**0.3**	**0.1**	**0.2**	**0.3**
$\alpha = 0.01$; without restrictions ($r = 0$)												
100	0.013	0.005	0.009	0.041	0.059	0.064	0.010	0.006	0.013	0.046	0.064	0.082
300	0.007	0.009	0.008	0.105	0.235	0.364	0.043	0.038	0.052	0.118	0.253	0.366
500	0.014	0.010	0.008	0.186	0.384	0.568	0.108	0.213	0.293	0.278	0.507	0.659
$\alpha = 0.01$; with restrictions at lag $r = 10$												
100	0.011	0.011	0.009	0.008	0.005	0.006	0.016	0.022	0.037	0.016	0.020	0.021
300	0.011	0.006	0.013	0.007	0.005	0.007	0.097	0.181	0.282	0.062	0.123	0.207
500	0.014	0.008	0.010	0.007	0.010	0.011	0.151	0.368	0.539	0.127	0.289	0.429
$\alpha = 0.05$; without restrictions ($r = 0$)												
100	0.045	0.040	0.041	0.149	0.218	0.195	0.055	0.051	0.071	0.143	0.209	0.217
300	0.056	0.044	0.053	0.288	0.476	0.623	0.117	0.149	0.146	0.309	0.493	0.656
500	0.058	0.040	0.044	0.444	0.666	0.805	0.276	0.428	0.472	0.524	0.751	0.852
$\alpha = 0.05$; with restrictions at lag $r = 10$												
100	0.060	0.047	0.042	0.039	0.033	0.027	0.065	0.088	0.121	0.080	0.077	0.096
300	0.041	0.039	0.043	0.044	0.035	0.037	0.272	0.415	0.528	0.208	0.341	0.407
500	0.045	0.031	0.041	0.044	0.044	0.040	0.393	0.619	0.800	0.338	0.562	0.716
$\alpha = 0.10$; without restrictions ($r = 0$)												
100	0.108	0.088	0.076	0.256	0.330	0.310	0.110	0.114	0.129	0.251	0.322	0.318
300	0.102	0.099	0.105	0.457	0.615	0.749	0.212	0.227	0.226	0.471	0.651	0.764
500	0.099	0.081	0.106	0.584	0.781	0.892	0.404	0.536	0.548	0.651	0.836	0.912
$\alpha = 0.10$; with restrictions at lag $r = 10$												
100	0.096	0.084	0.093	0.082	0.085	0.077	0.144	0.152	0.205	0.140	0.151	0.158
300	0.093	0.092	0.093	0.101	0.087	0.082	0.394	0.548	0.648	0.338	0.469	0.555
500	0.083	0.078	0.088	0.098	0.076	0.084	0.541	0.759	0.882	0.475	0.697	0.814

TABLE 13.8
Empirical probabilities of rejecting the null hypothesis using the out-of-sample Granger causality testing with the MSEt statistic under different data generating processes (DGP) with normal noise distribution.

	DGP 0			**DGP 1**			**DGP 2**			**DGP 3**		
$T\backslash\pi$	**0.1**	**0.2**	**0.3**	**0.1**	**0.2**	**0.3**	**0.1**	**0.2**	**0.3**	**0.1**	**0.2**	**0.3**
$\alpha = 0.01$; without restrictions ($r = 0$)												
100	0.010	0.008	0.006	0.029	0.065	0.056	0.012	0.018	0.017	0.032	0.041	0.060
300	0.009	0.010	0.007	0.086	0.206	0.338	0.036	0.052	0.066	0.087	0.195	0.316
500	0.008	0.009	0.005	0.157	0.341	0.510	0.110	0.205	0.271	0.276	0.510	0.688
$\alpha = 0.01$; with restrictions at lag $r = 10$												
100	0.012	0.005	0.003	0.009	0.006	0.009	0.012	0.019	0.017	0.008	0.011	0.019
300	0.008	0.007	0.005	0.005	0.005	0.009	0.082	0.183	0.289	0.075	0.129	0.202
500	0.006	0.005	0.009	0.013	0.009	0.013	0.149	0.333	0.535	0.112	0.272	0.428
$\alpha = 0.05$; without restrictions ($r = 0$)												
100	0.049	0.043	0.033	0.119	0.220	0.190	0.056	0.067	0.096	0.139	0.187	0.183
300	0.044	0.050	0.046	0.260	0.460	0.644	0.124	0.133	0.154	0.294	0.471	0.603
500	0.045	0.052	0.042	0.378	0.639	0.784	0.296	0.416	0.453	0.540	0.757	0.893
$\alpha = 0.05$; with restrictions at lag $r = 10$												
100	0.055	0.044	0.039	0.043	0.039	0.046	0.075	0.094	0.108	0.059	0.072	0.094
300	0.042	0.034	0.052	0.036	0.045	0.045	0.244	0.411	0.519	0.214	0.337	0.417
500	0.032	0.041	0.049	0.035	0.055	0.049	0.370	0.629	0.769	0.329	0.543	0.689
$\alpha = 0.10$; without restrictions ($r = 0$)												
100	0.097	0.089	0.065	0.245	0.337	0.292	0.108	0.116	0.156	0.221	0.305	0.302
300	0.093	0.084	0.095	0.409	0.613	0.780	0.196	0.223	0.220	0.443	0.630	0.744
500	0.093	0.096	0.097	0.542	0.761	0.866	0.432	0.543	0.553	0.686	0.861	0.938
$\alpha = 0.10$; with restrictions at lag $r = 10$												
100	0.094	0.086	0.070	0.089	0.072	0.077	0.140	0.170	0.191	0.114	0.138	0.156
300	0.089	0.087	0.105	0.087	0.090	0.092	0.372	0.564	0.648	0.333	0.461	0.535
500	0.079	0.085	0.097	0.077	0.105	0.103	0.519	0.759	0.864	0.485	0.683	0.804

TABLE 13.9
Empirical probabilities of rejecting the null hypothesis using the out-of-sample Granger causality testing with the MSEt statistic under different data generating processes (DGP) with t_5 noise distribution.

	DGP 0			**DGP 1**			**DGP 2**			**DGP 3**		
$T\backslash\pi$	**0.1**	**0.2**	**0.3**	**0.1**	**0.2**	**0.3**	**0.1**	**0.2**	**0.3**	**0.1**	**0.2**	**0.3**
$\alpha = 0.01$; without restrictions ($r = 0$)												
100	0.013	0.008	0.010	0.032	0.037	0.056	0.010	0.009	0.015	0.024	0.044	0.069
300	0.008	0.008	0.013	0.083	0.194	0.332	0.036	0.043	0.051	0.101	0.236	0.351
500	0.012	0.007	0.007	0.150	0.356	0.539	0.103	0.206	0.301	0.256	0.497	0.653
$\alpha = 0.01$; with restrictions at lag $r = 10$												
100	0.012	0.007	0.008	0.009	0.007	0.005	0.013	0.021	0.034	0.008	0.017	0.018
300	0.013	0.009	0.009	0.009	0.007	0.008	0.090	0.168	0.270	0.060	0.124	0.189
500	0.013	0.012	0.007	0.012	0.012	0.011	0.132	0.335	0.525	0.115	0.277	0.421
$\alpha = 0.05$; without restrictions ($r = 0$)												
100	0.046	0.046	0.046	0.122	0.195	0.185	0.055	0.047	0.075	0.122	0.198	0.228
300	0.046	0.053	0.055	0.270	0.459	0.608	0.121	0.146	0.150	0.305	0.488	0.637
500	0.053	0.041	0.048	0.417	0.641	0.794	0.272	0.426	0.469	0.511	0.741	0.846
$\alpha = 0.05$; with restrictions at lag $r = 10$												
100	0.057	0.037	0.039	0.041	0.037	0.040	0.065	0.089	0.122	0.066	0.081	0.090
300	0.039	0.041	0.045	0.048	0.038	0.029	0.275	0.410	0.535	0.209	0.335	0.402
500	0.047	0.034	0.041	0.042	0.042	0.040	0.387	0.626	0.794	0.324	0.563	0.703
$\alpha = 0.10$; without restrictions ($r = 0$)												
100	0.107	0.088	0.080	0.231	0.314	0.295	0.115	0.121	0.137	0.217	0.310	0.318
300	0.108	0.092	0.110	0.421	0.598	0.741	0.217	0.234	0.226	0.473	0.653	0.767
500	0.098	0.086	0.100	0.573	0.783	0.888	0.406	0.536	0.545	0.647	0.834	0.909
$\alpha = 0.10$; with restrictions at lag $r = 10$												
100	0.097	0.086	0.093	0.087	0.092	0.074	0.142	0.151	0.209	0.139	0.151	0.158
300	0.091	0.085	0.097	0.091	0.088	0.078	0.400	0.532	0.643	0.335	0.466	0.553
500	0.085	0.076	0.090	0.094	0.073	0.081	0.523	0.750	0.873	0.486	0.702	0.817

TABLE 13.10
Empirical probabilities of rejecting the null hypothesis using the out-of-sample Granger causality testing with the OOSF statistic under different data generating processes (DGP) with normal noise distribution.

	DGP 0			DGP 1			DGP 2			DGP 3		
$T\backslash\pi$	0.1	0.2	0.3	0.1	0.2	0.3	0.1	0.2	0.3	0.1	0.2	0.3
				$\alpha = 0.01$; without restrictions ($r = 0$)								
100	0.013	0.016	0.007	0.200	0.246	0.159	0.018	0.021	0.017	0.187	0.171	0.121
300	0.011	0.012	0.012	0.560	0.687	0.789	0.084	0.092	0.096	0.456	0.596	0.684
500	0.009	0.009	0.010	0.724	0.844	0.901	0.296	0.367	0.382	0.666	0.830	0.921
				$\alpha = 0.01$; with restrictions at lag $r = 10$								
100	0.013	0.012	0.007	0.007	0.011	0.012	0.038	0.037	0.043	0.025	0.028	0.034
300	0.013	0.010	0.009	0.010	0.008	0.013	0.328	0.460	0.508	0.257	0.345	0.383
500	0.013	0.011	0.013	0.010	0.018	0.014	0.546	0.735	0.801	0.485	0.636	0.713
				$\alpha = 0.05$; without restrictions ($r = 0$)								
100	0.071	0.052	0.041	0.377	0.402	0.268	0.062	0.072	0.080	0.336	0.344	0.228
300	0.044	0.044	0.043	0.674	0.789	0.864	0.161	0.156	0.161	0.586	0.737	0.805
500	0.053	0.052	0.046	0.800	0.888	0.936	0.448	0.509	0.497	0.769	0.897	0.961
				$\alpha = 0.05$; with restrictions at lag $r = 10$								
100	0.063	0.050	0.040	0.048	0.047	0.047	0.124	0.125	0.123	0.084	0.103	0.102
300	0.059	0.054	0.066	0.050	0.045	0.049	0.486	0.606	0.643	0.439	0.498	0.529
500	0.049	0.046	0.052	0.057	0.061	0.053	0.681	0.806	0.881	0.636	0.761	0.824
				$\alpha = 0.10$; without restrictions ($r = 0$)								
100	0.107	0.099	0.075	0.455	0.464	0.333	0.105	0.121	0.124	0.402	0.428	0.321
300	0.096	0.089	0.102	0.701	0.831	0.894	0.229	0.226	0.222	0.654	0.786	0.853
500	0.113	0.106	0.097	0.821	0.905	0.946	0.530	0.573	0.557	0.803	0.928	0.969
				$\alpha = 0.10$; with restrictions at lag $r = 10$								
100	0.118	0.097	0.074	0.102	0.090	0.090	0.194	0.188	0.191	0.137	0.165	0.156
300	0.117	0.109	0.110	0.102	0.102	0.100	0.565	0.661	0.705	0.501	0.578	0.593
500	0.094	0.092	0.100	0.106	0.116	0.100	0.720	0.840	0.913	0.699	0.805	0.857

TABLE 13.11
Empirical probabilities of rejecting the null hypothesis using the out-of-sample Granger causality testing with the OOSF statistic under different data generating processes (DGP) with t_5 noise distribution.

	DGP 0			DGP 1			DGP 2			DGP 3		
$T\backslash\pi$	0.1	0.2	0.3	0.1	0.2	0.3	0.1	0.2	0.3	0.1	0.2	0.3
				$\alpha = 0.01$; without restrictions ($r = 0$)								
100	0.011	0.008	0.009	0.225	0.235	0.165	0.018	0.008	0.017	0.181	0.200	0.140
300	0.014	0.009	0.012	0.576	0.673	0.757	0.097	0.094	0.081	0.502	0.632	0.726
500	0.018	0.009	0.010	0.724	0.865	0.918	0.273	0.376	0.396	0.628	0.811	0.885
				$\alpha = 0.01$; with restrictions at lag $r = 10$								
100	0.013	0.015	0.010	0.012	0.008	0.007	0.048	0.051	0.056	0.037	0.039	0.036
300	0.009	0.012	0.013	0.013	0.010	0.008	0.330	0.434	0.520	0.260	0.345	0.380
500	0.013	0.005	0.008	0.011	0.012	0.014	0.559	0.720	0.820	0.470	0.635	0.740
				$\alpha = 0.05$; without restrictions ($r = 0$)								
100	0.059	0.048	0.040	0.371	0.393	0.280	0.065	0.061	0.068	0.309	0.351	0.280
300	0.061	0.056	0.061	0.688	0.767	0.851	0.191	0.188	0.162	0.654	0.751	0.830
500	0.053	0.044	0.049	0.799	0.903	0.953	0.408	0.493	0.493	0.754	0.872	0.925
				$\alpha = 0.05$; with restrictions at lag $r = 10$								
100	0.057	0.049	0.052	0.057	0.053	0.033	0.111	0.122	0.139	0.120	0.112	0.105
300	0.057	0.048	0.059	0.053	0.042	0.040	0.478	0.599	0.647	0.432	0.491	0.547
500	0.065	0.044	0.044	0.054	0.049	0.048	0.685	0.833	0.894	0.617	0.773	0.831
				$\alpha = 0.10$; without restrictions ($r = 0$)								
100	0.113	0.093	0.068	0.453	0.467	0.341	0.114	0.131	0.128	0.378	0.438	0.350
300	0.113	0.118	0.112	0.725	0.802	0.876	0.262	0.263	0.219	0.712	0.810	0.863
500	0.106	0.090	0.098	0.829	0.923	0.960	0.493	0.565	0.543	0.795	0.898	0.937
				$\alpha = 0.10$; with restrictions at lag $r = 10$								
100	0.106	0.101	0.098	0.099	0.100	0.074	0.181	0.180	0.204	0.183	0.176	0.165
300	0.105	0.105	0.106	0.099	0.089	0.087	0.550	0.661	0.709	0.517	0.575	0.617
500	0.101	0.085	0.097	0.091	0.089	0.082	0.740	0.873	0.915	0.672	0.814	0.860

TABLE 13.12
Empirical probabilities of rejecting the null hypothesis by all four tests, based on the EN, MSEcor, MSEt, and OOSF statistics under different data generating processes (DGP) with normal noise distribution.

	DGP 0			DGP 1			DGP 2			DGP 3		
$T\backslash\pi$	0.1	0.2	0.3	0.1	0.2	0.3	0.1	0.2	0.3	0.1	0.2	0.3
				$\alpha = 0.01$; without restrictions ($r = 0$)								
100	0.002	0.002	0.002	0.011	0.044	0.036	0.004	0.008	0.009	0.016	0.023	0.036
300	0.001	0.002	0.002	0.068	0.182	0.319	0.028	0.044	0.058	0.074	0.177	0.307
500	0.000	0.005	0.002	0.134	0.326	0.497	0.081	0.191	0.265	0.246	0.490	0.674
				$\alpha = 0.01$; with restrictions at lag $r = 10$								
100	0.000	0.001	0.001	0.001	0.001	0.005	0.003	0.008	0.012	0.001	0.006	0.009
300	0.001	0.001	0.002	0.001	0.001	0.002	0.062	0.168	0.275	0.059	0.113	0.189
500	0.002	0.002	0.003	0.005	0.001	0.003	0.124	0.308	0.519	0.099	0.257	0.412
				$\alpha = 0.05$; without restrictions ($r = 0$)								
100	0.015	0.023	0.018	0.081	0.192	0.154	0.028	0.045	0.054	0.117	0.162	0.135
300	0.016	0.021	0.023	0.230	0.447	0.631	0.095	0.119	0.136	0.264	0.454	0.597
500	0.022	0.022	0.021	0.358	0.626	0.782	0.273	0.403	0.447	0.516	0.749	0.890
				$\alpha = 0.05$; with restrictions at lag $r = 10$								
100	0.018	0.021	0.020	0.014	0.019	0.027	0.044	0.070	0.080	0.031	0.053	0.071
300	0.026	0.021	0.031	0.019	0.017	0.026	0.226	0.394	0.505	0.190	0.317	0.406
500	0.021	0.021	0.028	0.020	0.035	0.030	0.352	0.614	0.762	0.300	0.531	0.682
				$\alpha = 0.10$; without restrictions ($r = 0$)								
100	0.053	0.062	0.045	0.216	0.310	0.249	0.067	0.080	0.096	0.205	0.268	0.230
300	0.048	0.050	0.058	0.390	0.605	0.776	0.164	0.193	0.189	0.428	0.618	0.736
500	0.057	0.066	0.065	0.530	0.756	0.862	0.421	0.534	0.535	0.673	0.859	0.935
				$\alpha = 0.10$; with restrictions at lag $r = 10$								
100	0.061	0.056	0.046	0.052	0.050	0.048	0.106	0.143	0.165	0.078	0.116	0.134
300	0.053	0.063	0.074	0.056	0.060	0.063	0.355	0.548	0.639	0.308	0.445	0.521
500	0.044	0.057	0.064	0.049	0.073	0.068	0.503	0.751	0.862	0.474	0.672	0.799

Bibliography

[1] C. W. J. Granger. Investigating causal relations by econometric models and cross-spectral methods. *Econometrica*, 37(3):424–438, 1969.

[2] A. Shojaie and E. B. Fox. Granger causality: A review and recent advances. *Annual Review of Statistics and Its Application*, 9(1):289–319, 2022.

[3] M. Al-Rdaydeh, N. Schneider, and A. Matar. Examining the linkages among electricity consumption, income and environmental pollution in Saudi Arabia: from a spectral wavelet analysis to the Granger causality test. *International Journal of Environmental Studies*, 78(4):634–662, 2021.

[4] C.-H. Chan, J.-Y. Juang, T.-H. Chu, C.-H. Mao, and S.-Y. Huang. A novel evaluation of air pollution impact from stationary emission sources to ambient air quality via time-series Granger causality. In *Earth Data Analytics for Planetary Health*, pages 33–53. Springer, 2023.

[5] T. J. Mosedale, D. B. Stephenson, M. Collins, and T. C. Mills. Granger causality of coupled climate processes: Ocean feedback on the North Atlantic Oscillation. *Journal of Climate*, 19(7):1182–1194, 2006.

[6] S. M. Samarasinghe, M. C. McGraw, E. A. Barnes, and I. Ebert-Uphoff. A study of links between the Arctic and the midlatitude jet stream using Granger and Pearl causality. *Environmetrics*, 30(4):e2540, 2019.

[7] D. A. Smirnov and I. I. Mokhov. From Granger causality to long-term causality: Application to climatic data. *Physical Review E*, 80(1):016208, 2009.

[8] W. A. Gardner, A. Napolitano, and L. Paura. Cyclostationarity: Half a century of research. *Signal Processing*, 86(4):639–697, 2006.

[9] A. Napolitano. *Generalizations of Cyclostationary Signal Processing: Spectral Analysis and Applications.* John Wiley & Sons, Chichester, United Kingdom, 2012.

[10] A. Napolitano. *Cyclostationary Processes and Time Series: Theory, Applications, and Generalizations.* Academic Press, London, United Kingdom, 2020.

[11] L. Izzo and A. Napolitano. Generalized almost-cyclostationary signals. *Advances in Imaging and Electron Physics*, 135:103–223, 2005.

[12] M. R. Mahmoudi, M. Maleki, K. Borodin, K.-H. Pho, and D. Baleanu. On comparing and clustering the spectral densities of several almost cyclostationary processes. *Alexandria Engineering Journal*, 59(4):2555–2565, 2020.

[13] R. Lund and I. V. Basawa. Recursive prediction and likelihood evaluation for periodic ARMA models. *Journal of Time Series Analysis*, 21(1):75–93, 2000.

[14] J. D. Salas, D. C. Boes, and R. A. Smith. Estimation of ARMA models with seasonal parameters. *Water Resources Research*, 18(4):1006–1010, 1982.

[15] E. Macdonald, B. Merz, B. Guse, V. D. Nguyen, X. Guan, and S. Vorogushyn. What controls the tail behaviour of flood series: Rainfall or runoff generation? *Hydrology and Earth System Sciences*, 28(4):833–850, 2024.

[16] A. Zaffar, S. Abbas, and M. R. K. Ansari. An analysis of heavy tail and long-range correlation of sunspot and El Nino-Southern Oscillation (ENSO) cycles. *Solar System Research*, 53(5):399–409, 2019.

[17] G. Kirchgässner and J. Wolters. *Introduction to Modern Time Series Analysis.* Springer-Verlag, Berlin, 2007.

[18] L. Barnett, A. B. Barrett, and A. K. Seth. Granger causality and transfer entropy are equivalent for Gaussian variables. *Physical Review Letters*, 103(23):238701, 2009.

[19] C. W. J. Granger and P. Newbold. Spurious regressions in econometrics. *Journal of Econometrics*, 2:111–120, 1974.

[20] H. Lütkepohl. *New Introduction to Multiple Time Series Analysis.* Springer Science & Business Media, 2005.

[21] H. Lütkepohl. Non-causality due to omitted variables. *Journal of Econometrics*, 19:367–378, 1982.

[22] C. A. Sims. Exogeneity and causal ordering in macroeconomic models. In *New Methods in Business Cycle Research: Proceedings from a Conference*, pages 23–44. Federal Reserve Bank of Minneapolis, Minneapolis, 1976.

[23] J. M. Price. Causality in temporal systems: A correction. *Journal of Econometrics*, 10:253–256, 1979.

[24] A. Attanasio, A. Pasini, and U. Triacca. Granger causality analyses for climatic attribution. *Atmospheric and Climate Sciences*, 3(4):36931, 2013.

[25] B. Pfaff. *vars: VAR Modelling*, 2024. R package version 1.6-1.

[26] T. E. Clark and M. W. McCracken. Tests of equal forecast accuracy and encompassing for nested models. *Journal of Econometrics*, 105(1):85–110, 2001.

[27] C. W. J. Granger and P. Newbold. *Forecasting economic time series.* Academic Press, London, UK, 2nd edition, 1986.

[28] M. W. McCracken. Asymptotics for out of sample tests of Granger causality. *Journal of Econometrics*, 140(2):719–752, 2007.

[29] A. Attanasio, A. Pasini, and U. Triacca. Has natural variability a lagged influence on global temperature? A multi-horizon Granger causality analysis. *Dynamics and Statistics of the Climate System*, 1(1), 2016.

[30] V. Lyubchich, Y. R. Gel, and S. Vishwakarma. *funtimes: Functions for Time Series Analysis*, 2023. R package version 9.1.

[31] S. Epskamp, M. K. Deserno, and L. F. Bringmann. *mlVAR: Multi-Level Vector Autoregression*, 2024. R package version 0.5.2.

[32] P. Nitschke. Golden tilefish, Lopholatilus chamaeleonticeps, stock assessment update through 2016 in the Middle Atlantic-Southern New England Region. Technical report, Northeast Fisheries Science Center, Woods Hole, MA, 2017.

[33] G. Nesslage, V. Lyubchich, P. Nitschke, E. Williams, C. Grimes, and J. Wiedenmann. Environmental drivers of golden tilefish (Lopholatilus chamaeleonticeps) commercial landings and catch-per-unit-effort. *Fisheries Oceanography*, 30(5):608–622, 2021.

[34] S. Moritz and S. Gatscha. *imputeTS: Time Series Missing Value Imputation*, 2022. R package version 3.3.

[35] V. Lyubchich and G. Nesslage. *Environmental drivers of golden tilefish fisheries v1.0.* Zenodo, 2020, DOI 10.5281/zenodo.3732839.

[36] J. A. Nye, M. R. Baker, R. Bell, A. Kenny, K. H. Kilbourne, K. D. Friedland, E. Martino, M. M. Stachura, K. S. Van Houtan, and R. Wood. Ecosystem effects of the Atlantic multidecadal oscillation. *Journal of Marine Systems*, 133:103–116, 2014.

[37] M. A. Cane, A. C. Clement, L. N. Murphy, and K. Bellomo. Low-pass filtering, heat flux, and Atlantic multidecadal variability. *Journal of Climate*, 30(18):7529–7553, 2017.

[38] A. Clement, K. Bellomo, L. N. Murphy, M. A. Cane, T. Mauritsen, G. Rädel, and B. Stevens. The Atlantic Multidecadal Oscillation without a role for ocean circulation. *Science*, 350(6258):320–324, 2015.

[39] R. Zhang, R. Sutton, G. Danabasoglu, Y.-O. Kwon, R. Marsh, S. G. Yeager, D. E. Amrhein, and C. M. Little. A review of the role of the Atlantic meridional overturning circulation in Atlantic multidecadal variability and associated climate impacts. *Reviews of Geophysics*, 57(2):316–375, 2019.

[40] J. Bjerknes. Atlantic air-sea interaction. In H. E. Landsberg and J. Van Mieghem, editors, *Advances in Geophysics*, volume 10, pages 1–82. Elsevier, 1964.

[41] R. C. J. Wills, K. C. Armour, D. S. Battisti, and D. L. Hartmann. Ocean–atmosphere dynamical coupling fundamental to the Atlantic multidecadal oscillation. *Journal of Climate*, 32(1):251–272, 2019.

[42] P. Moffa-Sánchez, E. Moreno-Chamarro, D. J. Reynolds, P. Ortega, L. Cunningham, D. Swingedouw, D. E. Amrhein, J. Halfar, L. Jonkers, J. H. Jungclaus, K. Perner, A. Wanamaker, and S. Yeager. Variability in the northern North Atlantic and Arctic oceans across the last two millennia: A review. *Paleoceanography and Paleoclimatology*, 34(8):1399–1436, 2019.

[43] T. L. Delworth, F. Zeng, L. Zhang, R. Zhang, G. A. Vecchi, and X. Yang. The central role of ocean dynamics in connecting the North Atlantic Oscillation to the extratropical component of the Atlantic multidecadal oscillation. *Journal of Climate*, 30(10):3789–3805, 2017.

[44] W. Wang, B. T. Anderson, R. K. Kaufmann, and R. B. Myneni. The relation between the North Atlantic Oscillation and SSTs in the North Atlantic basin. *Journal of Climate*, 17(24):4752–4759, 2004.

[45] V. Lyubchich. *Long-Term Granger Causality for Environmental Time Series (R Code) v1.0.* Zenodo, 2025, DOI 10.5281/zenodo.15926393.

14

Efficient Spatio-Temporal Bayesian Modeling with INLA

Janet van Niekerk, Elias Teixeira Krainski, Denis Rustand, and Håvard Rue

14.1 Summary

Spatio-temporal models are essential in explaining various natural phenomena observed from real data. Such insights are necessary for intervention planning, resource allocation and interpolation of missing data. In environmental studies, often the goal is not just about predicting future situations but also about intervening to alter the predictions in a positive way. Spatio-temporal models are often developed as hierarchical Bayesian models with specific structured priors over space, time, and the interaction of space and time. Increasing the domain size or resolution poses a computational complexity, and the task of efficient and accurate inference of such complex models can deter practitioners from employing them on huge datasets, resulting in limited insights. The statistical models (we will call it here just models) are the stochastic models whose parameters are estimated from data and other information. Mechanistic or physical models usually explain physical processes more accurately, although a statistical model can learn from real data, and we can quantify the uncertainty of the model. In this work we show how the integrated nested Laplace approximations (INLA) framework, and recent developments thereof, provide efficient and accurate posterior inference in terms of predictions with accompanying uncertainty quantification. We can also jointly model various processes, such as temperature and air quality, and as such capture an underlying driving mechanism emulating the environment as a whole. A case study based on climate data for South Africa is presented, to illustrate the methodology as well as the near real-time prediction ability of this approximate Bayesian inference framework.

14.2 Introduction to Bayesian Hierarchical Modeling

Statistical modeling provides a method to possibly find drivers for certain phenomena. Hierarchical models are known to provide flexibility while maintaining an interpretable structure of the model. Multiparameter models like regression models are often posed as hierarchical models since the hierarchy follows naturally from the data level to the unobserved latent level. Prior information can be incorporated in this model by adding additional levels in the

DOI: 10.1201/9781003564423-14

hierarchical model and models with structured random effects fit flawlessly into this framework. We provide a concise description of hierarchical models next and refer the reader to [1, 2] for further details.

Bayesian hierarchical models combine the data generating model, with the model for the regression parameters, and the priors for the unobservable latent effects and hyperparameters. Consider datapoints $y_1, y_2, ..., y_n$ as a sample of size n from a population generated by a statistical model based on parameters denoted as vectors by $\boldsymbol{\omega}$ and $\boldsymbol{\theta}$, so we write that the data is generated as follows:

$$\boldsymbol{y} \sim L(\boldsymbol{\omega}, \boldsymbol{\theta}).$$

The vector $\boldsymbol{\omega}$ contains the parameters that are explicitly used in the regression model, and $\boldsymbol{\theta}$ contains the parameters that are not explicitly part of the regression model, but used to explain other features of the data generating model, like noise or shape parameters (often referred to as hyperparameters or nuisance parameters). Suppose we observe some other variables together with the response data $y_1, y_2, ..., y_n$ termed the covariates and denoted by $\boldsymbol{x}_1, \boldsymbol{x}_2, ..., \boldsymbol{x}_n$ so that each vector $\boldsymbol{x}_i$ contains the values of the p covariates that accompany datapoint y_i. The regression model that relates the datapoint to the covariates is defined for some summary statistics of $\boldsymbol{y}$ such as the mean or median as follows:

$$E(y_i|\boldsymbol{\omega}) = g(h(\boldsymbol{\omega}, \boldsymbol{x}_i)),$$

where the function $g(.)$ is called a link function [3] that projects the predictor $h(.)$ to the domain of $\boldsymbol{y}$. A common model is a generalized additive mixed model (GAMM) which implies that the predictor $h(\boldsymbol{\omega}, \boldsymbol{x}_i)$ is a linear function of the parameters $\boldsymbol{\omega}$ such that the regression model can be written as

$$E(y_i|\boldsymbol{\omega}) = g(\boldsymbol{\omega}^\top \boldsymbol{x}),$$

where $\boldsymbol{x}$ is called the design matrix and it contains the values of the corresponding covariates for the specific elements of $\boldsymbol{\omega}$. Usually, the first element in $\boldsymbol{\omega}$ is an intercept so that the first column of $\boldsymbol{x}$ is a vector of 1s. The subsequent elements in $\boldsymbol{\omega}$ correspond to the different fixed effects, followed by the random effects.

An important assumption that is often made is that the data is conditionally independent given the model parameters $\boldsymbol{\omega}$ and $\boldsymbol{\theta}$, which implies that if we know the values of $\boldsymbol{\omega}$ and $\boldsymbol{\theta}$, then the datapoints are independent of each other. Thus,

$$L(\boldsymbol{y}|\boldsymbol{\omega}, \boldsymbol{\theta}) = \prod_{i=1}^{n} L_i(y_i|\boldsymbol{\omega}, \boldsymbol{\theta}),$$

which implies that the joint conditional distribution of the data can be defined by the product of the individual density/probability mass functions L_i. Note that the general notation here implies that we can account for multiple sources of data with different characteristics like measurements, counts, indicator variables etc. in one framework.

A *Bayesian hierarchical model* is then constructed by assigning prior beliefs (expert opinions, historical insights, etc.) to all the unknown parameters. In our formulation, we thus assign some prior beliefs in the form of probability density functions to the elements of $\boldsymbol{\omega}$ and $\boldsymbol{\theta}$. Each parameter can be assigned its own independent prior since it is plausible to consider all parameters as independent of each other, before the data is incorporated.

The question arises of how we can combine these prior beliefs with the data to form an updated belief of our parameters? A popular way to combine these two sources of information is using Bayes' theorem, resulting in Bayesian modeling. Bayes' theorem was proposed by Thomas Bayes in 1763 to provide a way to reverse conditional probabilities. Bayesian modeling uses Bayes theorem to calculate posterior densities (the probabilities of the parameters learned from the data and prior beliefs) as inference answering appropriate research question on the data generating model of the observed data. The joint conditional distribution of the data can be viewed as the likelihood that this data is the result of a certain set of $\boldsymbol{\omega}$ and $\boldsymbol{\theta}$, while what we generally want is to learn what the probabilities for different values of $\boldsymbol{\omega}$ and $\boldsymbol{\theta}$ are based on the observed data, hence the inverse problem arises to which Bayes theorem provides an answer.

Bayes theorem can be viewed as an information processing rule that is 100% efficient which implies that by using Bayes rule no information is lost or added [4].

Define the joint prior density (belief) for $\boldsymbol{\omega}$ and $\boldsymbol{\theta}$ as $\pi(\boldsymbol{\omega}, \boldsymbol{\theta})$, then the joint posterior density of $\boldsymbol{\omega}$ and $\boldsymbol{\theta}$ is calculated using Bayes rule as

$$p(\boldsymbol{\omega}, \boldsymbol{\theta}|\boldsymbol{y}) \propto \left[\prod_{i=1}^{n} L_i(y_i|\boldsymbol{\omega}, \boldsymbol{\theta})\right] \pi(\boldsymbol{\omega}, \boldsymbol{\theta}),$$

where $p(\boldsymbol{y})$ is called the marginal likelihood or the evidence.

One specific type of Bayesian GAMM is called a Latent Gaussian Model (LGM), where some assumptions are made on a subset of the parameters. Many commonly used statistical models are LGMs like spatial models, temporal models, count data models, logistic regression, spline models, etc., and as such this class of models is very useful in practice. An LGM is a hierarchical Bayesian model defined by the following three levels:

$$\begin{array}{lll} \text{Level 1}: & \begin{cases} y_i|\boldsymbol{\omega}, \boldsymbol{\theta} \overset{\text{iid}}{\sim} L_i(y_i|\boldsymbol{\omega}, \boldsymbol{\theta}) \\ E(y) = g(\boldsymbol{\omega}^\top \boldsymbol{x}) \end{cases} & \begin{array}{l}\text{Data generating model} \\ \text{Regression model}\end{array} \\ \text{Level 2}: & \boldsymbol{\omega}|\boldsymbol{\theta} \sim \pi_N(\mathbf{0}, \boldsymbol{Q}_{\boldsymbol{\theta}}^{-1}) & \text{Latent prior} \\ \text{Level 3}: & \boldsymbol{\theta} \sim \pi(\boldsymbol{\theta}) & \text{Hyperparameter prior,} \end{array} \tag{14.1}$$

so that the data is assumed conditionally independent based on the model, the regression parameters $\boldsymbol{\omega}$ are assigned a multivariate Gaussian prior with zero mean vector and covariance matrix (inverse precision matrix) $\boldsymbol{Q}_{\boldsymbol{\theta}}^{-1}$, denoted by $\pi_N(\mathbf{0}, \boldsymbol{Q}_{\boldsymbol{\theta}}^{-1})$ and the other parameters $\boldsymbol{\theta}$, termed the hyperparameters are assigned any sensible prior distribution $\pi(\boldsymbol{\theta})$. The joint prior for the unknown parameters can then be specified as

$$\pi(\boldsymbol{\omega}, \boldsymbol{\theta}) = \pi_N(\mathbf{0}, \boldsymbol{Q}_{\boldsymbol{\theta}}^{-1})\pi(\boldsymbol{\theta}).$$

Note that for structured random effects, like spatial or temporal components the precision matrix for the latent prior is a function of $\boldsymbol{\theta}$ and thus denoted as $\boldsymbol{Q}_{\boldsymbol{\theta}}$. Additionally, $\boldsymbol{\omega}$ is assumed to be a Gaussian Markov Random Field (GMRF) [5] which implies that the precision matrix of $\boldsymbol{\omega}$, $\boldsymbol{Q}_{\boldsymbol{\theta}}$ is a sparse matrix. This assumption is preserved when continuing to add model terms, as most of the data models are built based on a priori independent parameters. Consider the following simple example for illustration of the concepts.

Example

Consider a general linear regression model for the mean parameter, $\boldsymbol{\mu}$. Then the data generating model can be written as

$$y_i|\mu_i, \sigma \sim \pi_N(\mu_i, \sigma^2)$$

where the regression model for the mean is based on covariates $\boldsymbol{x}_1 = \{x_{11}, x_{12}, ..., x_{1n}\}$ and $\boldsymbol{x}_2 = \{x_{21}, x_{22}, ..., x_{2n}\}$, i.e., $\boldsymbol{x}_{n\times 3} = [\mathbf{1}, \boldsymbol{x}_1, \boldsymbol{x}_2]$ and written as

$$\mu_i = \boldsymbol{\beta}^\top \boldsymbol{x}_{i\cdot},$$

where $\boldsymbol{\beta} = [\beta_0, \beta_1, \beta_2]$, $\boldsymbol{x}_{i\cdot}$ is the ith row of the design matrix $\boldsymbol{x}$, β_0 is the overall intercept, β_1 is the slope for covariate $\boldsymbol{x}_1$, and β_2 is the slope for covariate $\boldsymbol{x}_2$.

In this example, the latent field is $\boldsymbol{\omega} = \boldsymbol{\beta}$ and the hyperparameter is σ. For numerical stability and computational efficiency, it is often preferred to transform the estimable parameter from σ to $\log(1/\sigma^2) = \log\tau$. The priors for $\boldsymbol{\beta}$ are weakly informative Gaussian priors (with zero mean and low precision, i.e., large variance), and the prior for $\log\tau$ can be any sensible density function.

Suppose a random intercept, v_i is added to accommodate heterogeneity, then the regression model for datapoint i is changed to

$$\mu_i = \boldsymbol{\beta}^\top \boldsymbol{x}_{i\cdot} + v_i,$$

where the prior for v_i is defined to be a centered Gaussian with common precision τ_v, i.e., $v_i|\tau_v \overset{i.i.d.}{\sim} \pi_N(0, \tau_v^{-1})$. In this case, the latent field is now defined as $\boldsymbol{\omega} = \{\boldsymbol{\beta}, v_1, v_2, ..., v_n\}$ and the hyperparameters are $\boldsymbol{\theta} = \{\log\tau, \log\tau_v\}$. The resulting prior precision matrix of $\boldsymbol{\omega}$ is a diagonal matrix. For the identifiability of the overall intercept β_0, a sum-to-zero constraint on $\boldsymbol{v}$ is necessary so that the model is inferred based on the constraint $\sum_{i=1}^{n} v_i = 0$.

Now suppose we want to model the time series temperature at one location. A correlated random effect can be employed to deal with the fact that consecutive observations in time are correlated. In that regard, suppose that v_i can be modeled as a first order spline (the details are presented in Section 14.4.2). Thus the prior model for v_i can be formulated as

$$v_i - v_{i-1}|\tau_v \sim \pi_N(0, \tau_v^{-1}),$$

such that the prior for $\boldsymbol{v}$ is $\pi(\boldsymbol{v}|\tau_v) \propto \tau^{(n-1)/2} \exp\left(-\frac{1}{2}\boldsymbol{v}^\top \boldsymbol{Q}\boldsymbol{v}\right)$. Thus, the latent field $\boldsymbol{\omega} = \{\boldsymbol{\beta}, \boldsymbol{v}\}$ follows a multivariate Gaussian distribution with precision matrix $\boldsymbol{Q}_{\boldsymbol{\theta}}$. For illustration, consider the intercept-only model for $n = 5$ with $\boldsymbol{\beta} = \beta_0$, then

$$\boldsymbol{Q}_{\boldsymbol{\theta}} = \begin{bmatrix} \checkmark & \cdot & \cdot & \cdot & \cdot & \cdot \\ \cdot & \checkmark & \checkmark & \cdot & \cdot & \cdot \\ \cdot & \checkmark & \checkmark & \checkmark & \cdot & \cdot \\ \cdot & \cdot & \checkmark & \checkmark & \checkmark & \cdot \\ \cdot & \cdot & \cdot & \checkmark & \checkmark & \checkmark \\ \cdot & \cdot & \cdot & \cdot & \checkmark & \checkmark \end{bmatrix} \text{ and } \boldsymbol{\Sigma}_{\boldsymbol{\theta}} = \boldsymbol{Q}_{\boldsymbol{\theta}}^{-1} = \begin{bmatrix} \checkmark & \cdot & \cdot & \cdot & \cdot & \cdot \\ \cdot & \checkmark & \checkmark & \checkmark & \checkmark & \checkmark \\ \cdot & \checkmark & \checkmark & \checkmark & \checkmark & \checkmark \\ \cdot & \checkmark & \checkmark & \checkmark & \checkmark & \checkmark \\ \cdot & \checkmark & \checkmark & \checkmark & \checkmark & \checkmark \\ \cdot & \checkmark & \checkmark & \checkmark & \checkmark & \checkmark \end{bmatrix},$$

where a dot indicates a zero value and a checkmark indicates a non-zero value.

From the resulting covariance matrix, we note that even though the random effects are all correlated since they depend on the same precision parameter and each other, they are *conditionally independent* which results in a very sparse precision matrix for the latent field that results in only $2n$ non-zero values. Note that the prior covariance matrix, however, is dense with $0.5(n^2+n)+1$ non-zero entries. The sparse structure of the precision matrix can be exploited to ensure efficient inference of the Bayesian model while this is not achievable when the covariance matrix is used, especially as n grows.

14.3 INLA as a Computational Framework for Bayesian Inference

The inference of Bayesian models can be achieved in multiple ways that are largely classed into two groups, first, asymptotically exact inference like the sampling-based Markov Chain Monte Carlo and Hamiltonian Monte Carlo approaches and second, approximate inference that targets a sufficient approximation of the true posteriors like integrated nested Laplace approximations (INLA), Variational inference (VI), and Expectation propagation (EP), among others. Sampling-based methods are known to be computationally expensive when compared with approximate methods as the model complexity or data size increases. To this end, approximate methods are very popular for complex models and/or large data. INLA was proposed by Rue et al. [6] as an efficient Bayesian inference methodology for latent Gaussian models. Important methodological advances for INLA were proposed by Van Niekerk et al. [7], which enables the use of INLA for even more complex models and/or huge data.

INLA uses mathematical expressions to calculate an approximation to the posteriors of the parameters, instead of using samples that converges in distribution to the true posterior, as the sample size increases. It is thus much more efficient for LGMs than other approaches and considered as the state-of-the-art for Bayesian inference of LGMs. The approximate posteriors calculated from INLA are in many cases equal to the exact posteriors, without having to wait for sampling or test for convergence. Understandably, such a powerful framework contains many tailored and innovative details, and we present only the key ideas next.

Consider an LGM defined as in (14.2). The parameters to be estimated are the latent field, $\boldsymbol{\omega}$ and the hyperparameters, $\boldsymbol{\theta}$. The INLA methodology can be summarized in three main computation targets.

1. Target 1: The joint posterior of the hyperparameters is approximated using the full joint density of $\{\boldsymbol{y}, \boldsymbol{\omega}, \boldsymbol{\theta}\}$ and a Gaussian approximation for the *conditional* posterior of the latent field as described in [7]. Thus,

$$\tilde{p}(\boldsymbol{\theta}|\boldsymbol{y}) \propto \left. \frac{p(\boldsymbol{y}, \boldsymbol{\omega}, \boldsymbol{\theta})}{p_N(\boldsymbol{\omega}|\boldsymbol{y}, \boldsymbol{\theta})} \right|_{\boldsymbol{\omega}=\boldsymbol{\mu}_\omega}.$$

   Note that the Gaussian approximation in the denominator is based on the Laplace method [8] (the mean is placed at the mode, and the precision matrix is the negative Hessian matrix of $p(\boldsymbol{\omega}|\boldsymbol{y}, \boldsymbol{\theta})$ evaluated at the mode $\boldsymbol{\mu}_\omega$).

2. Target 2: Based on the joint posterior approximation in Target 1, $\tilde{p}(\boldsymbol{\theta}|\boldsymbol{y})$ is explored and integration points with their corresponding integration weights are calculated. Next, the marginal posteriors of the hyperparameters are approximated using numerical integration as follows:

$$\tilde{p}(\theta_j|\boldsymbol{y}) = \int \tilde{p}(\boldsymbol{\theta}|\boldsymbol{y}) d\boldsymbol{\theta}_{-j}$$

3. Target 3: From Target 1, we extract the univariate approximate *conditional* posteriors of the latent field from the low-rank Variational Bayes corrected Gaussian approximation $p_N(\boldsymbol{\omega}|\boldsymbol{y}, \boldsymbol{\theta})$ (see [9] and [10]), and numerically integrate over the hyperparameters, $\boldsymbol{\theta}$, using the integration points and weights from Target 2, to obtain the marginal posteriors of the latent field.

$$\tilde{p}(\omega_l|\boldsymbol{y}) = \int p_N(\omega_l|\boldsymbol{y}, \boldsymbol{\theta}) \tilde{p}(\boldsymbol{\theta}|\boldsymbol{y}) d\boldsymbol{\theta}. \tag{14.2}$$

Various technical details are necessary to make INLA as efficient and accurate as it is, interested readers are referred to [6, 7, 11] for some insights into these details.

The INLA methodology is implemented in an R package, *INLA*, referred to as R-INLA in the rest of the chapter. Installation instructions and various complementary books and tutorials are available on the website `https://www.r-inla.org`.

Note that for simplicity, the case of Gaussian data is presented further, but most kinds of data can be handled with R-INLA so that we can consider count data, binary data, strictly positive data, censored (survival) data among others, with an underlying Gaussian field (GF) with Matérn covariance as part of the linear predictor.

14.4 Spatio-Temporal Models

Spatio-temporal models are models that include random processes that are indexed by space, time, and possibly the interaction of space and time. In the framework of LGMs, a spatially structured random effect is considered simply as a specific structured random effects; similarly for temporal effects and interaction effects. This enables us to use the general methodology that is developed for the entire class of LGMs to efficiently infer spatio-temporal models. By viewing the spatio-temporal components as particular random effects, we can incorporate other covariates in the process and uncover the latent spatio-temporal processes, which is different from smoothing data in space and/or time—the latter approaches would correspond closely to intercept only regression models with spatio-temporal random effects. This approach also allows for multivariate spatio-temporal modeling since we can correlate the latent spatio-temporal processes for different types of endpoints like count data, binary data, or censored data, among others.

In this chapter, we present some common continuously-indexed spatial and discrete-time temporal Gaussian processes and show that they all form GMRFs, and thus fits in the class of LGMs. This is useful since then we can use the INLA methodology for Bayesian inference of these models. Moreover, models for lattice data and discrete time points exist and are implemented in R-INLA, but fall outside the scope of this chapter.

14.4.1 Spatial Effect

Suppose we have an observation, $y(s_i)$ at a point $s_i \in \mathbb{R}^2$. Now, we can observe many points in a study area $\boldsymbol{D}$, such that $\boldsymbol{s} \in \boldsymbol{D} \subset \mathbb{R}^2$. In this chapter, $\boldsymbol{D}$ corresponds to the country of South Africa. We assume that a stochastic process $U(\boldsymbol{s})$ is the latent spatial effect that resulted in the obtained sample. The observations usually contain some noise because of the sampling procedure and we thus use a model to decompose the signal $U(\boldsymbol{s})$ from the noise $\boldsymbol{\epsilon}$ so that if $\boldsymbol{y}$ is a measurement on the real line, we can formulate a model as

$$y(s_i) = u(s_i) + \epsilon_i,$$

where $u(\boldsymbol{s})$ is the value of the latent process at the samples points and $\boldsymbol{\epsilon} \sim \pi_N(\boldsymbol{0}, \tau_\epsilon^{-1}\boldsymbol{I})$. The latent process is usually assumed to be multivariate Gaussian with zero mean and specific precision matrix that captures the dependence over space. A common covariance matrix for a GF over space is the isotropic and stationary Whittle-Matérn (or simply Matérn) covariance defined as

$$\boldsymbol{\Sigma}_{ij} = \sigma_u^2 \rho(U(s_i), U(s_j)),$$

where σ_u is the marginal standard deviation of the latent process, and the correlation function is defined as

$$\rho(U(s_i), U(s_j)) = \frac{2^{1-\nu}}{\Gamma(\nu)}(\kappa||s_i - s_j||)^\nu K_\nu(\kappa||s_i - s_j||),$$

where $||s_i - s_j||$ is the Euclidean distance between points s_i and s_j, and $K_\nu(\cdot)$ is the modified Bessel function of the second kind. Note that the Matérn covariance model is defined based on two parameters, a smoothness parameter $\nu > 0$ and a scale parameter $\kappa > 0$. To understand the effect of κ, we interpret κ in terms of the practical range r where

$$r = \sqrt{\frac{8\nu}{\kappa}},$$

which implies that r will be the distance between two points such that the spatial correlation between these points are approximately 0.14. The practical range thus provides a way to characterize the neighborhood of practical spatial dependence for each point. The smoothness parameter governs the smoothness of the process and can be estimated by the approach proposed by [12], but is commonly fixed to $\nu = 1$ (this is the case in this chapter).

14.4.1.1 Priors for the Matérn Model

The marginal standard deviation, σ_u, and the practical range, r, (or scale parameter κ) are the two hyperparameters of the Matérn model and thus need to be estimated from the data. Traditionally, gamma priors for the precision parameters or inverse-gamma priors for the variance have been proposed, but these are guaranteed to result in overfitting as shown by [13]. Overfitting in our case would imply that a spurious spatial effect can be obtained, and thus care should be taken with the specification of the priors of the hyperparameters. To this end, penalizing complexity (PC) priors were proposed by [13] to prevent overfitting while contracting to a null effect, instead of a spurious spatial effect. A null effect would be when the spatial process does not vary over space by either having a very small marginal standard deviation ($\sigma_u \to 0$) or a very large range ($r \to \infty$). In this chapter, we consider the joint PC prior [14] for σ_u and κ over $\mathbb{R}^2$ defined as follows:

$$\pi(\sigma_u, \kappa) = \frac{1}{2}\lambda_r \lambda_\sigma(\kappa)\kappa^{-1/2} \exp\left(-\lambda_r \kappa^{1/2} - \lambda_\sigma(\kappa)\sigma_u^{-2}\right),$$

where the two user-defined parameters λ_r and λ_σ determine the rate of contraction to the null effect. These values are derived [14] from the statements

$$P(r < r_0) = \alpha_r \quad \text{and} \quad P(\sigma_u > \sigma_0 | \kappa) = \alpha_\sigma,$$

where r_0, σ_0, α_r, and α_σ are given by the user, thus constructing a sensible and usable prior.

14.4.1.2 SPDE Formulation of the Matérn Model

A direct application of the Matérn covariance results in a dense matrix, since under the Matérn model the correlation between any two spatial locations decays with distance but only becomes exactly zero in the limit, as the distance tends to infinity. This results in a high computational burden since the computational complexity grows with n as $O(n^2)$. However, [15] formulated an explicit link between a Gaussian process with a Matérn covariance function, and the field $U(\boldsymbol{s})$ that is the weak solution to a specific stochastic partial differential equation (SPDE) defined as

$$(\kappa^2 - \Delta)^{\frac{\alpha}{2}} U(\boldsymbol{s}) = W(\boldsymbol{s}),$$

where $W(s)$ denotes a white noise Gaussian process with unit variance, Δ is the Laplacian operator and $\boldsymbol{s} \in \mathbb{R}^d, \alpha = \nu + d/2, \kappa > 0, \nu > 0$. Solving this SPDE using the finite element method (FEM) for $d = 2$, results in a solution $U(\boldsymbol{s})$ that follows a Gaussian distribution with a sparse precision matrix, which mimics a GF with the Matérn covariance matrix on $\mathbb{R}^2$. The FEM is based on a discretization of the domain and in this case we consider a mesh of triangles to be able to follow curves in the boundary. The mesh is then extended beyond the study area to ensure a negligible effect of the boundary conditions. This result forms the foundation of the INLA-SPDE approach that encompasses the Matérn model, as well as other diffusion-based spatial processes for geo-statistical analyses. Note that the latent process $U(\boldsymbol{s})$ forms part of the model through a matrix of weights, $w(\boldsymbol{s})$ since the resolution of the mesh can be different (usually higher) than what is observed, i.e.,

$$y(s_i) = \boldsymbol{A}_i w(\boldsymbol{s}) + \epsilon,$$

where $\boldsymbol{A}$ is termed the projector matrix that projects the latent process onto the observational domain, and $\boldsymbol{A}_i$ is the ith row of $\boldsymbol{A}$. This formulation enables the prediction of the latent process value at unobserved points, at a high resolution. Interested readers can find more details about the INLA-SPDE approach from [16, 17].

This section presents the details for a stationary isotropic Matérn covariance, however other complex models can be formulated. One kind of non-stationary Matérn model is the barrier model which defines a GF with Matérn covariance for a study area containing barriers in terms of correlation strength. In such cases, the covariance should not be based on Euclidean distance since the barriers cannot be crossed. Other definitions of distance (like the shortest path around the barrier) are problematic since it is very sensitive to the level of resolution, and moreover computationally inefficient, while not being an automated framework. [18] proposed the non-stationary Matérn model for impermeable barriers which does not depend on an arbitrary definition of distance, and this model is implemented in R-INLA. Several extensions of the SPDE approach were developed including non-stationary, non-Gaussian, general smoothness, as well to different domains such as metric spaces and spatio-temporal domains, see [19, 20] for more details. Some of these extensions are available in the R-package *INLAspacetime*.

14.4.2 Temporal Effect

Suppose we have an observation $y(t_i)$ that is measured at a timepoint $t_i \in \mathbb{R}^+$. Now we can measure this observation at different timepoints in our temporal domain defined as $\Omega \subset \mathbb{R}^+$. These timepoints can be discrete, for example at monthly intervals or continuous, for example times of the day. When considering the class of LGMs and INLA, this distinction is not of utmost mathematical importance since all latent processes can be defined on resolutions that are different from the resolution of the data. Common temporal Gaussian processes are autoregressive models, random walk models, or models formulated on seasonality. Similar as in Section 14.4.1, we assume that a stochastic process $U(\boldsymbol{t})$ is the latent temporal effect that resulted in the obtained sample. The observations usually contain some noise because of the sampling procedure and we thus use a model to decompose the signal $U(\boldsymbol{t})$ from the noise $\boldsymbol{\epsilon}$ so that if $\boldsymbol{y}$ is a measurement on the real line, we can formulate the model as

$$y(t_i) = u(t_i) + \epsilon,$$

where $u(\boldsymbol{t})$ is the value of the latent process at the sampled timepoints. The latent process is usually assumed to be multivariate Gaussian with zero mean and specific precision matrix that captures the temporal dependence. We present only the details for an autoregressive process of order one, and random walk processes of order one and two, respectively, since these are sufficient for the application.

14.4.2.1 Autoregressive Model

Suppose the latent process $U(\boldsymbol{t})$ is assumed to follow an autoregressive process of order one (AR1), then the joint density of $U(\boldsymbol{t})$ is defined by

$$\begin{aligned} u(t_1)|\tau_u, \rho_u &\sim \pi_N\left(0, \frac{1}{\tau_u(1-\rho_u)^2}\right) \\ u(t_i)|u(t_{i-1}) &= \rho_u u(t_{i-1}) + \epsilon_i, \quad \epsilon_i \sim \pi_N(0, \tau^{-1}), \end{aligned}$$

with precision parameter τ_u, and correlation parameter ρ_u. Note that if $\rho_u = 1$, then the temporal effect implies a random walk order one effect in time while $\rho_u = 0$ implies a white noise process in time. Again, we assume PC priors for the precision and correlation parameters [21] defined as follows:

$$\begin{aligned} \pi(\tau_u) &= \frac{\lambda_\tau}{2}\tau_u^{-3/2}\exp(\lambda_\tau \tau^{-1/2}) \qquad (14.3) \\ \pi_{\rho_u \to 0}(\rho_u) &= \frac{\lambda_\rho^0}{2}\exp\left(-\lambda_\rho^0\sqrt{-\ln(1-\rho_u^2)}\right)\frac{|\rho_u|}{(1-\rho_u^2)\sqrt{-\ln(1-\rho_u^2)}}, \\ \pi_{\rho_u \to 1}(\rho_u) &= \frac{\lambda_\rho^1 \exp\left(-\lambda_\rho^1\sqrt{1-\rho_u}\right)}{2\sqrt{1-\rho_u}\left(1-\exp(-2\sqrt{2}\lambda_\rho^1)\right)} \end{aligned}$$

where $\lambda_\tau = -\ln(\alpha_\tau)/a$ for $P(1/\sqrt{\tau_u} > a) = \alpha_\tau$, $\lambda_\rho^0 = -\ln(\alpha_\rho)/\sqrt{-\ln(1-a^2)}$ for $P(|\rho_u| > a) = \alpha_\rho$ and for λ_ρ^1, we calculate it implicitly from $\frac{1-\exp(-\lambda_\rho^1\sqrt{1-a})}{1-\exp(-\sqrt{2}\lambda_\rho^1)} = \alpha_\rho$ for $P(|\rho_u| > a) = \alpha_\rho$.

14.4.2.2 Random Walk Models

Suppose the latent process $U(\boldsymbol{t})$ is assumed to follow a random walk of order k (RWk), then the joint density of $U(\boldsymbol{t})$ is defined by

$$\Delta^k u(t_i)|\tau_u \sim \pi_N(0, \tau_u^{-1})$$

with precision parameter τ_u for which the PC prior (14.3) is assumed. Higher values of k corresponds to a smoother process $u(\boldsymbol{t})$, while high values of τ_u indicate a contraction of the random walk model to its null space model. Thus, a PC prior for the precision parameter is used to specify the rate of contraction to the null space in the absence of evidence from the data for the more flexible random walk model. In practice, order one or two models are considered. The null space for the RW1 model is a horizontal line thus indicating no effect of the covariate, while for RW2 the null space is a linear effect. The random walk models can be viewed as stochastic spline models that converge to the true underlying process as the number of evaluation points (also known as "knots") increase, contrary to deterministic splines like B-splines, P-splines, etc. This is particularly advantageous from a computational viewpoint since we do not need to preform iterative model fitting based on varying the number and placement of knots, but rather we fit the model once for a large number of knots due to the theoretical convergence guarantee for random walk models. In effect, a random walk model cannot overfit the data like deterministic splines [22] and for this reason, we often employ them not only as temporal structure models, but also to model non-linear effects of general covariates [23].

14.4.3 Spatio-temporal Interaction Effect

Separable spatio-temporal models can be constructed as a convolution of the spatial and temporal processes. Suppose we have an observation $y(s_i, t_j)$ that is measured at a location

$s_i \in \mathbb{R}^2$ and timepoint $t_j \in \mathbb{R}^+$. We can measure this observation at different locations and timepoints in our study domain defined as $\boldsymbol{D} \subset \mathbb{R}^2 \times \mathbb{R}^+$. Without loss of generality, assume that $y(s_i, t_j)$ is real-valued, then the data-generating model can be proposed as

$$y(s_i, t_j) = u(s_i, t_j) + \epsilon,$$

where $u(s_i, t_j)$ is the value of the latent process $U(s, t)$, at the location s_i and timepoint t_j, and again $\boldsymbol{\epsilon}$ is assumed to be independent and unstructured Gaussian noise. The latent process $U(s, t)$ should have a valid correlation function defined as $\rho(U(s_i, t_j), U(s_{i^*}, t_{j^*}))$. If the spatio-temporal process is defined to have a separable correlation function, then the correlation function can be constructed as

$$\rho(U(s_i, t_j), U(s_{i^*}, t_{j^*})) = \rho_1(U(s_i), U(s_{i^*}))\rho_2(U(t_j), U(t_{j^*})),$$

where $\rho_1(.)$ denotes the correlation function over space and $\rho_2(.)$ denotes the correlation function over space. It is well-known that the precision matrix of a this spatio-temporal process is the Kronecker product of the precision matrices of the spatial GF and the temporal GF, respectively [24]. Note that the spatio-temporal process itself is not separable. Moreover, if the precision matrices of the spatial and temporal components are sparse, so that the individual processes are GMRFs, then the spatio-temporal process is also a GMRF by construction. Hence, INLA can be used for efficient inference of spatio-temporal models, as well as prediction at unobserved locations in space or unobserved timepoints.

In this chapter, we consider the Matérn model as the spatial latent process and the autoregressive order one, random walk order one or order two models as the temporal latent processes. Each of these components can be defined as a GMRF and the resulting spatio-temporal process constructed with a kronecker product of precision matrices, will thus also be a GMRF. Various other formulations of spatio-temporal models are possible including those with non-separable precision matrices [19], and those considering models for lattice data like the Besag model [24].

14.5 Joint Temperature Models for South Africa 2021–2023—A Geostatistical Spatio-temporal Model

High-resolution climate data for South Africa is available from the Climatic Research Unit, University of East Anglia (`https://crudata.uea.ac.uk/cru/data/hrg/`) for the period 1901–2024, using the WGS84 coordinate reference system. In this chapter, we select data from 2021 to 2023 for various climatological variables such as the monthly minimum and maximum temperatures, the monthly precipitation measurements, and the number of wet days per month, among others. These variables are collected at 536 specific points in South Africa at various timepoints. For illustration, we consider the minimum and maximum temperature in this section, although the other variables can be analyzed similarly. The dependence between temperature and altitude is well-known, but for illustration we omit altitude and all other covariates in this example since they are not available in our data. Our goal is to see if there is a significant change in the minimum and/or maximum temperatures over space and time and identify areas that have experienced unexplained changes.

The observation locations and maximum and minimum temperatures for January and July 2022, are presented in Figure 14.1. Note that the area without any points is Lesotho, which is not part of South Africa. However, climate does not follow administrative boundaries and it is thus important that our model incorporates Lesotho as part of the study

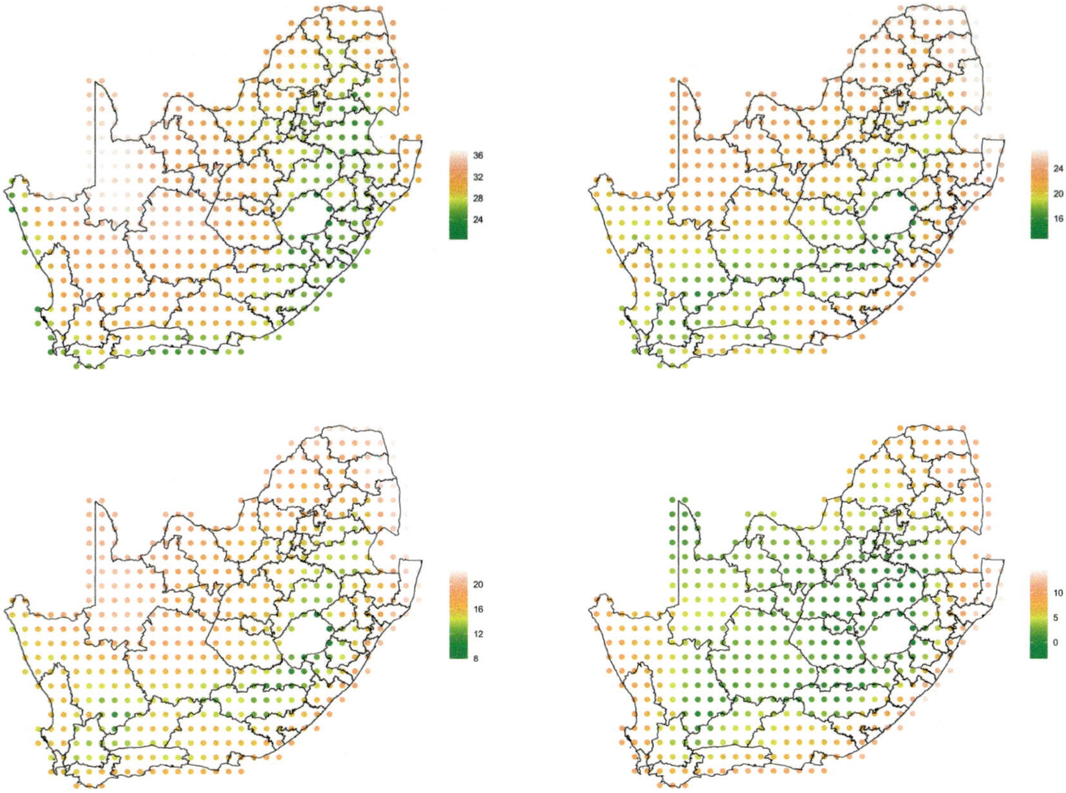

FIGURE 14.1
Observed minimum (bottom) and maximum (top) temperatures for January (left) and July (right) 2022. The different municipalities are illustrated.

area. This consideration results in missing values for the climate variables, although we can use the spatio-temporal model based on the data to estimate these values naturally using the Bayesian approach, if that is desired.

As discussed in Section 14.4.1, a mesh of triangles should be constructed on which the latent spatial process is defined. Since we want to predict at unobserved locations as well we design a mesh with higher resolution than that of the observed locations. In practice, multiple meshes can be constructed and the model inferred for each one. Due to the construction of the Matérn model with the solution of the SPDE, the model will converge stochastically to the true posterior as the resolution of the mesh increases (as we add more triangles in Figure 14.2). Thus, a practitioner can increase the resolution until the posterior inference no longer exhibit large changes. Further increase of resolution will not lead to a gain in accuracy, but will increase the computational cost. Note that the mesh is extended beyond the study area as mentioned in Section 14.4.1. For the high resolution mesh in Figure 14.2 (right), we have 3458 nodes where the latent model will be constructed. Since we have monthly data for three years, the time domain has 36 points and in total we have 19,296 datapoints. Based on the mesh and timepoints, one spatio-temporal latent process will have dimension 124488 with < 5 hyperparameters (depending on the temporal structure chosen). The latent field dimension necessitates efficient inferential approaches since sampling-based inference like MCMC is not feasible for a model of this size. For more complex models, the issue of the computational burden for inference is exacerbated, for instance the simple joint

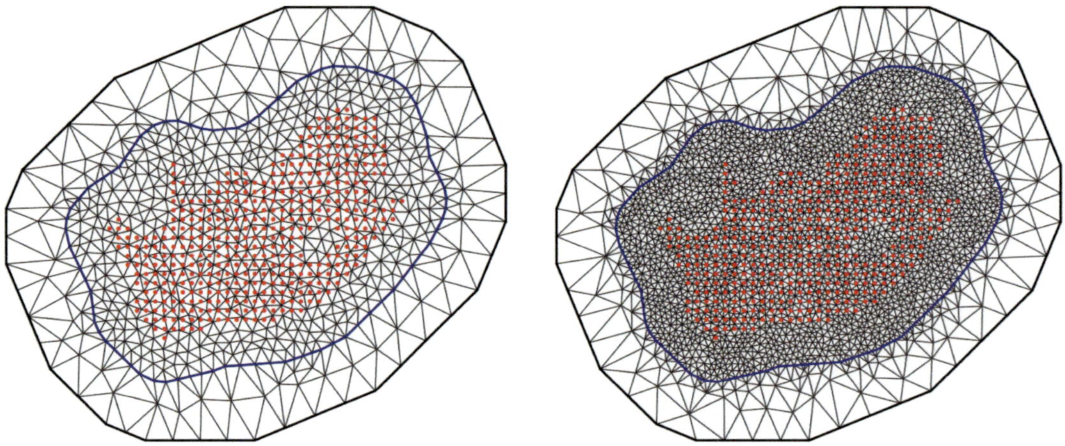

FIGURE 14.2
Discretization of South Africa using a mesh of triangles: lower resolution (left) and high resolution (right) latent spatial processes. The red dots are the observation locations. The model area is indicated by the mesh and the extensions away from the study area is clear.

model in Section 14.5.1.2 has a latent field dimension of 2×62604. In this Section, we show how to use INLA to fit these models within a reasonable computational budget. All code is available on GitHub[1].

14.5.1 Model Formulation

14.5.1.1 Maximum and Minimum Temperature

The maximum temperature can be modeled as a realization from a Gaussian process defined as follows:

$$\begin{aligned} \text{max\_t}(s_i, t_j) &\sim p_N(\mu_{max}(s_i, t_j), \sigma^2_{max}) \\ \mu_{max}(s_i, t_j) &= \beta_0^{max} + \beta_1^{max} I(2022) + \beta_2^{max} I(2023) + \boldsymbol{A}_{ij}\boldsymbol{u}_{max}(\boldsymbol{s}, \boldsymbol{t}) \\ u_{max}(\boldsymbol{s}, \boldsymbol{t}) &\sim p_N(\boldsymbol{0}, \sigma^2_{u,max}\boldsymbol{\rho}_1(\boldsymbol{s}) \otimes \boldsymbol{\rho}_2(\boldsymbol{t})), \end{aligned}$$

where the spatial process is a Gaussian process with a Matérn correlation and the temporal process evolves in time as a Gaussian process with autoregressive order one structure. Hence, $\boldsymbol{\rho}_1(\boldsymbol{s})$ is defined as in Section 14.4.1, and $\boldsymbol{\rho}_2(\boldsymbol{t})$ is defined in Section 14.4.2.1. All the hyper-parameters follow PC priors are discussed before, and the fixed effects $\beta_0^{max}, \beta_1^{max}, \beta_2^{max}$ follows a weakly informative Gaussian prior. $\beta_1^{max}, \beta_2^{max}$ captures a consistent drift over time.

Note that for the data under consideration, the maximum and minimum temperatures appear to exhibit Gaussianity from exploratory data analysis, but in general extreme value distributions might be more appropriate. In INLA, generalized extreme value distributions can be fitted using the family argument *gev*.

14.5.1.2 Joint Model for Temperature

We can construct a joint spatio-temporal model to investigate an underlying spatio-temporal latent process that is common to multiple variables, such as minimum and maximum tem-

[1]`https://github.com/JanetVN1201/Code_for_papers/tree/main/Efficientspatio-temporalBayesian modelingusingINLA`

peratures. This can be a first step to finding a proxy process that represents the *climate state*. The model is defined as:

$$\begin{aligned}
\text{max_t}(s_i,t_j) &\sim p_N(\mu_{max}(s_i,t_j),\sigma^2_{max})\\
\mu_{max}(s_i,t_j) &= \beta_0^{max}+\beta_1^{max}I(2022)+\beta_2^{max}I(2023)+\boldsymbol{A}_{ij}\boldsymbol{u}_{max}(\boldsymbol{s},\boldsymbol{t})\\
u_{max}(\boldsymbol{s},\boldsymbol{t}) &\sim p_N(\boldsymbol{0},\sigma^2_{u,max}\boldsymbol{\rho}_1(\boldsymbol{s})\otimes\boldsymbol{\rho}_2(\boldsymbol{t}))\\
\text{min_t}(s_i,t_j) &\sim p_N(\mu_{min}(s_i,t_j),\sigma^2_{min})\\
\mu_{min}(s_i,t_j) &= \beta_0^{min}+\beta_1^{min}I(2022)+\beta_2^{min}I(2023)+\gamma\boldsymbol{A}_{ij}\boldsymbol{u}_{max}(\boldsymbol{s},\boldsymbol{t})+\boldsymbol{A}_{ij}\boldsymbol{u}_{min}(\boldsymbol{s},\boldsymbol{t})\\
u_{min}(\boldsymbol{s},\boldsymbol{t}) &\sim p_N(\boldsymbol{0},\sigma^2_{u,min}\boldsymbol{\rho}_1^*(\boldsymbol{s})\otimes\boldsymbol{\rho}_2^*(\boldsymbol{t}))
\end{aligned}$$

where the spatial process is a Gaussian process with a Matérn correlation and the temporal process evolves in time as a Gaussian process with autoregressive order one structure. Hence, $\boldsymbol{\rho}_1(\boldsymbol{s})$ and $\boldsymbol{\rho}_1^*(\boldsymbol{s})$ are defined as in Section 14.4.1 with their respective range and marginal standard deviation parameters, and $\boldsymbol{\rho}_2(\boldsymbol{t})$ and $\boldsymbol{\rho}_2^*(\boldsymbol{t})$ are defined in Section 14.4.2.1 with their respective correlation and precision parameters. All the hyperparameters follow PC priors are discussed before, and the fixed effects β_0^{max} and β_0^{max} follow weakly informative Gaussian priors.

The parameter γ follows a weakly informative Gaussian prior with a prior mean of zero. This parameter quantifies the dependence between the two variables across space and time, since a zero value of γ indicates two independent latent processes. This model can be further expanded to include other climatological variables and this is left as future work. This model can be inferred efficiently through INLA.

14.5.2 Results

The posterior inference of the models proposed in the preceding sections are presented in this section. The posterior summaries are presented in Table 14.1. As a point estimator we can use the posterior mean, mode, or median and we present the mode in this work. Note that the interpretation of the spatio-temporal processes slightly differ between the individual and joint models. For the individual models, each spatio-temporal process can be interpreted as the underlying spatio-temporal dynamics of the respective temperatures. For the joint model, $\boldsymbol{u}_{max}(\boldsymbol{s},\boldsymbol{t})$ is the underlying spatio-temporal process that is common to both minimum and maximum temperatures and can thus be viewed as a possible proxy to represent the climate process, while $\boldsymbol{u}_{min}(\boldsymbol{s},\boldsymbol{t})$ is now the spatio-temporal process that explains additional variability in the minimum temperatures that is not explained by the climate process, hence having a smaller standard deviation than in the separate models. Since the posterior mode (MAP) of the dependence parameter γ is estimated to be 0.644, with 95% credible interval (0.18; 0.681), we have evidence of a common underlying driver that acts on maximum and minimum temperatures in the same direction. For illustration, we present the posterior mode of the spatio-temporal processes for the year 2021 for each month in Figures 14.3 and 14.4 where the spatio-temporal dynamics are clear since the spatial field does not merely shift with a constant value between the months. Similar figures can be produced using the code in the GitHub repository for other years, months, and the joint model.

14.5.3 Warning System Based on Exceedance Probability of WHO Standard

The World Health Organization WHO advises that when temperature passes 40 degrees Celsius, the use of a fan leads to an increases in the temperature, as opposed to below 40 degrees

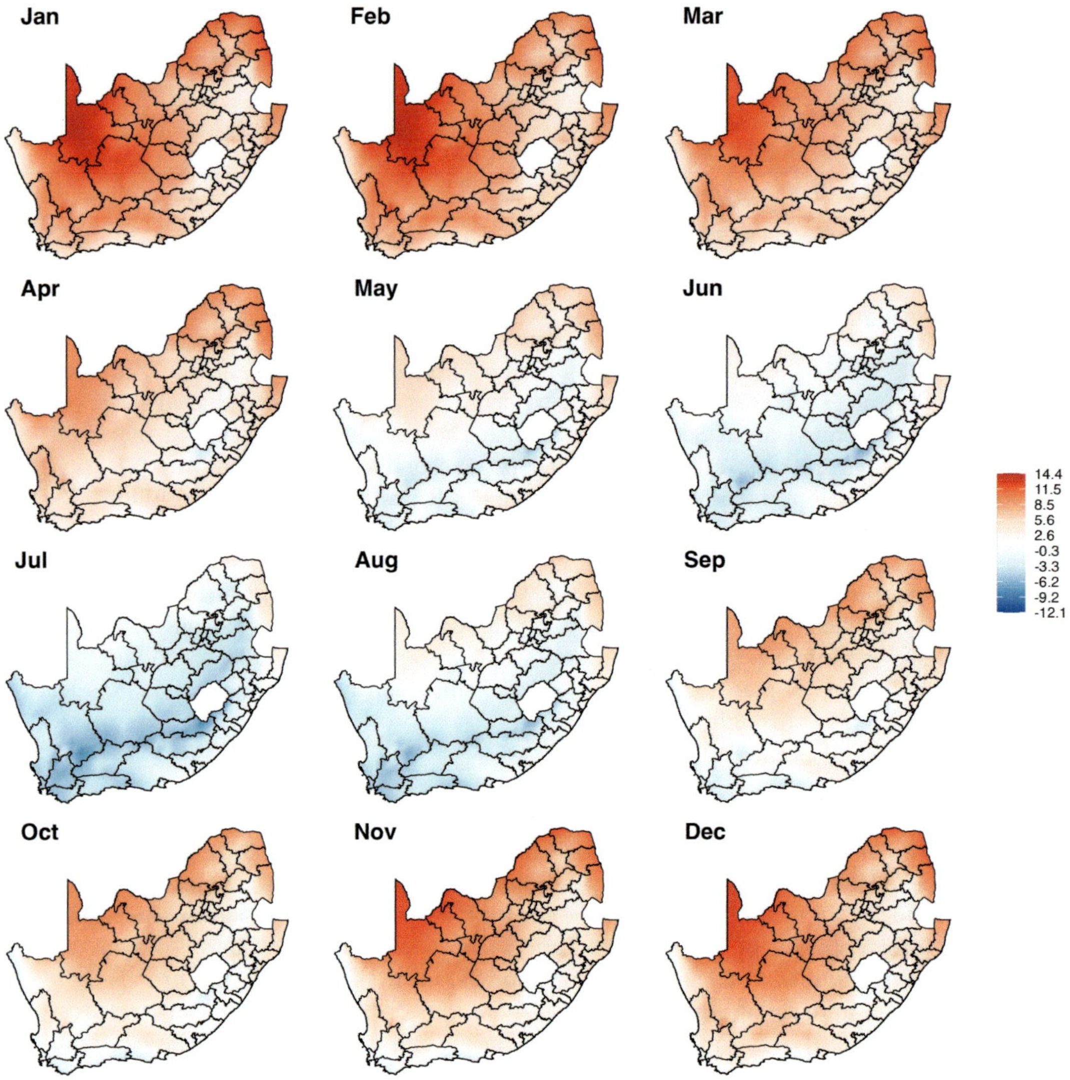

FIGURE 14.3
Spatio-temporal latent process $\boldsymbol{u}_{max}(\boldsymbol{s}, \boldsymbol{t})$ for the maximum temperatures for 2021 from Section 14.5.1.1

Celsius (`https://www.who.int/news-room/fact-sheets/detail/climate-change-heat-and-health`). Since South Africa is a third world country, most houses do not have an air conditioning system and as such the majority of the population use electric fans to cool themselves down. It is thus important to send targeted warning about the use of fans when the expected maximum temperature regularly exceeds 40 degrees Celsius. Based on our inferred joint spatio-temporal model, we can calculate the posterior probability of the maximum temperature exceeding the threshold of 40 over space and time. Figure 14.5 shows some of these posterior probabilities for certain timepoints. Note that the contour plots indicate areas where $P(\text{max\_t} \geq 40|\text{data}) > 0.25$. Moreover, excursion sets can be calculated using the approach of [25].

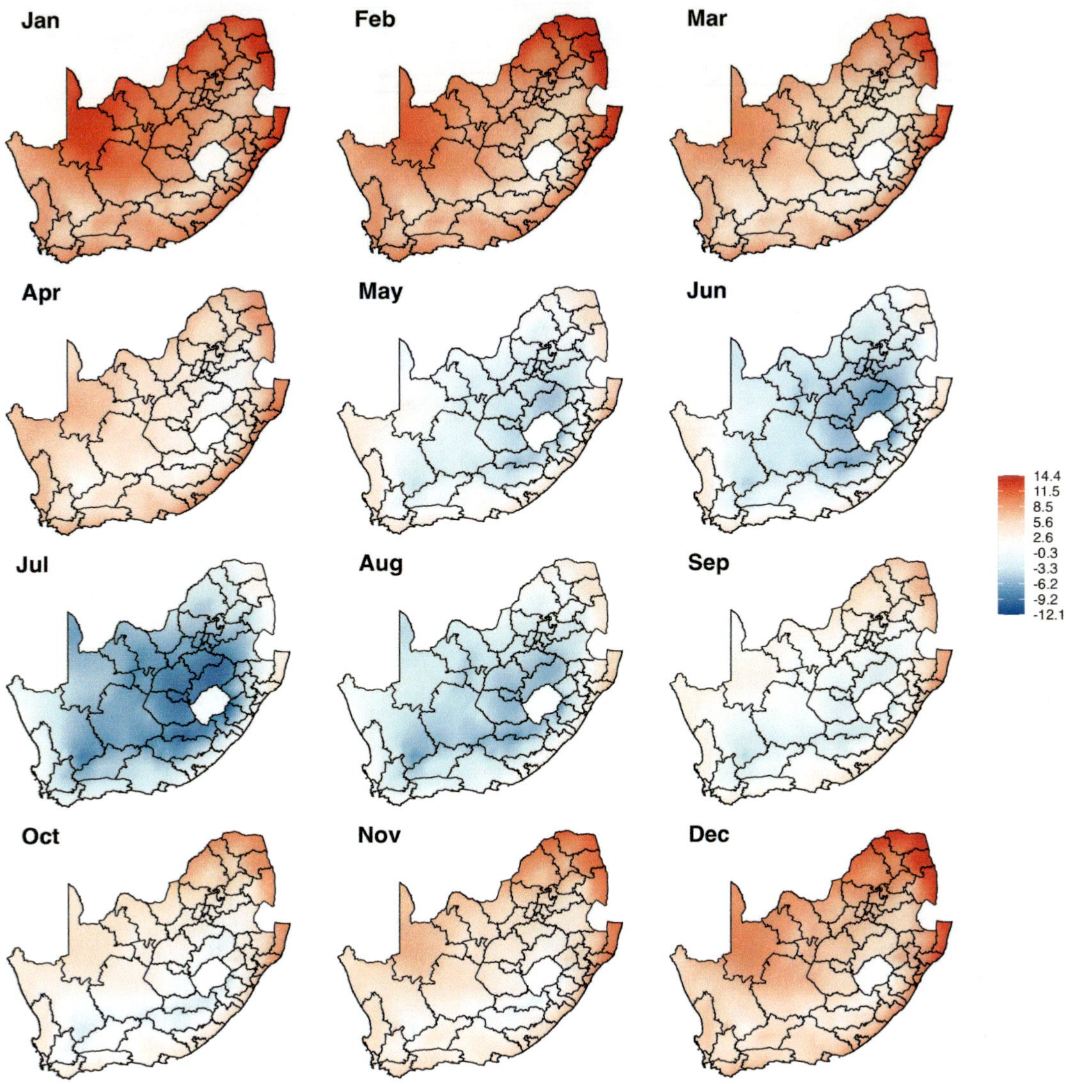

FIGURE 14.4
Spatio-temporal latent process $\boldsymbol{u}_{min}(\boldsymbol{s}, \boldsymbol{t})$ for the minimum temperatures for 2021 from Section 14.5.1.1

Comparing January 2021, 2022, and 2023 (first row in Figure 14.5) we note the usual desert-type areas in the Northern Cape indicating high risk of temperatures above 40. However, there are other areas that emerge in 2023 such as a part of Namaqualand, Beaufort-Wes, and the area around Graaff-Reinet in the Eastern Cape province. Moreover, most of the Northern Cape, parts of the Free State, North West, and the Kruger Park has a high risk of excessive temperatures in January 2023 compared to January 2021. This phenomena is notable also for November and December (rows 4 and 5 in Figure 14.5). This warning framework can support intervention programs by identifying the high risk communities at a finer level than municipalities or towns for effective roll out among the most vulnerable.

FIGURE 14.5
Posterior probability of temperature higher than 40 degrees Celsius for the warmer months—January, February, March, November, and December (from top to bottom)—of 2021, 2022 and 2023 (from left to right). The contours start at $P(\text{max\_t} \geq 40|\text{data}) = 0.25$.

TABLE 14.1
Posterior inference of the minimum, maximum, and joint temperature models (Runtime: 217, 241, and 1614 seconds, respectively)

	Individual models—Section 14.5.1.1	**Joint model—Section 14.5.1.2**
	Mode (95% CI)	**Mode (95% CI)**
	Fixed effects	
β_0^{min}	19.493 (4.821 ; 34.165)	16.558 (3.172 ; 29.945)
β_1^{min}	1.171 (−2.359 ; 4.700)	1.508 (−1.713 ; 4.730)
β_2^{min}	2.969 (−1.982 ; 7.920)	3.726 (−0.790 ; 8.242)
β_0^{max}	21.692 (0.392 ; 42.993)	21.463 (2.847 ; 40.080)
β_1^{max}	1.591 (−2.392 ; 5.574)	1.830 (−1.547 ; 5.206)
β_2^{max}	3.660 (−1.941 ; 9.260)	4.200 (−0.559 ; 8.960)
	Hyperparameters	
$1/\sigma^2_{min}$	107.08 (83.441 ; 136.530)	335.792 (156.56 ; 837.938)
r^{min}	25.70 (22.984 ; 28.867)	21.803 (1.389 ; 25.307)
σ_u^{min}	10.34 (9.163 ; 11.764)	5.134 (0.347 ; 6.034)
ρ_u^{min}	0.98 (0.977 ; 0.982)	0.957 (0.003 ; 0.963)
$1/\sigma^2_{max}$	130.241 (105.801 ; 157.363)	136.265 (33.514 ; 224.415)
r^{max}	29.415 (25.884 ; 33.849)	26.339 (1.888 ; 32.144)
σ_u^{max}	13.586 (11.830 ; 15.852)	12.258 (1.208 ; 16.219)
ρ_u^{max}	0.987 (0.986 ; 0.989)	0.988 (0.987; 0.990)
γ	—	0.644 (0.18; 0.681)

14.6 Summary and Extensions

This chapter showed how INLA can be used to efficiently fit complex spatio-temporal models for point-level data. There are various extensions available and possible. For point-level data, non-stationarity in the variance can be implemented using various covariance non-stationary models as describe in [19]. Although various ad-hoc approaches can be used, like subdivisions of the study area, these models define the non-stationarity through the SPDE and thus provide a principled approach to handle such covariance non-stationarity. Moreover, spatially varying coefficient models are also implemented in R-INLA to accommodate for covariates having different effects over space. Differing spatial and temporal observation densities can be accounted for using coregionalization spatio-temporal models, which is a trivial extension of the model proposed in Section 14.5.1.2. Spatial values at different levels of aggregation can be combined using the approach proposed by [26].

Various climate variables exhibit complexities for modeling. One such variable is precipitation, since it can only be measured if it is raining. There can thus be many zeros in precipitation data, although these zeros are not "real" zeros but rather structural. To accommodate this feature, we can use a hurdle model [16] by combining the information about the event (raining) and the measurements of the variable (precipitation) as follows:

$$p(s_i, t_j) = \begin{cases} \text{observed precipitation} & r(s_i, t_j) = 1 \\ \text{NA} & \text{otherwise} \end{cases},$$

where

$$r(s_i, t_j) = \begin{cases} 1 & \text{raining at location } s_i \text{ at timepoint } t_j \\ 0 & \text{otherwise} \end{cases}.$$

This model fits in our proposed framework since this accommodation impacts only the likelihood, and the spatio-temporal processes can thus still be formulated as GMRFs. Data fusion using geographical variables such as land use and elevation from other databases could be included in the model. Various sources from weather stations and numerical model outcome can be included in a data integration model using INLA, since these models are just specific joint models.

Various advances and extension have been and are being included in the INLA methodology and the R-INLA software, such as point process models, non-separable, non-stationary and fractional spatio-temporal models, generalized extreme value models, survival models, and many more likelihoods, while the *inlabru* R library [27] provides a user-friendly interface for spatial (and other) modeling with INLA.

The INLA methodology and the accompanying R-INLA interface allows efficient Bayesian inference of various complex models, including spatio-temporal processes, in a reasonable computational budget as shown in Section 14.5. This remarkable computational efficiency allows the practitioner to update their models regularly (daily) to enable near real-time prediction of processes, which cannot be achieved by using other inference approaches on this data size and model complexity.

Bibliography

[1] Andrew Gelman, John B Carlin, Hal S Stern, and Donald B Rubin. *Bayesian data analysis.* Chapman and Hall/CRC, 1995.

[2] Richard McElreath. *Statistical rethinking: A Bayesian course with examples in R and Stan.* Chapman and Hall/CRC, 2018.

[3] Peter McCullagh. *Generalized linear models.* Routledge, 2019.

[4] Arnold Zellner. Optimal information processing and Bayes's theorem. *The American Statistician*, 42(4):278–280, 1988.

[5] Håvard Rue and Leonhard Held. *Gaussian Markov random fields: Theory and applications.* CRC Press, 2005.

[6] Håvard Rue, Sara Martino, and Nicolas Chopin. Approximate Bayesian inference for latent Gaussian models by using integrated nested Laplace approximations. *Journal of the Royal Statistical Society Series B: Statistical Methodology*, 71(2):319–392, 2009.

[7] Janet Van Niekerk, Elias Krainski, Denis Rustand, and Håvard Rue. A new avenue for Bayesian inference with INLA. *Computational Statistics & Data Analysis*, 181:107692, 2023.

[8] Luke Tierney and Joseph B Kadane. Accurate approximations for posterior moments and marginal densities. *Journal of the American Statistical Association*, 81(393):82–86, 1986.

[9] Janet van Niekerk and Håvard Rue. Low-rank variational Bayes correction to the Laplace method. *Journal of Machine Learning Research*, 25(62):1–25, 2024.

[10] Shourya Dutta, Janet van Niekerk, and Håvard Rue. Scalable skewed Bayesian inference for latent Gaussian models. *arXiv preprint arXiv:2502.19083*, 2025.

[11] Lisa Gaedke-Merzhäuser, Janet van Niekerk, Olaf Schenk, and Håvard Rue. Parallelized integrated nested Laplace approximations for fast Bayesian inference. *Statistics and Computing*, 33(1):25, 2023.

[12] David Bolin, Alexandre B Simas, and Zhen Xiong. Covariance–based rational approximations of fractional SPDEs for computationally efficient Bayesian inference. *Journal of Computational and Graphical Statistics*, 33(1):64–74, 2024.

[13] Daniel Simpson, Håvard Rue, Andrea Riebler, Thiago G Martins, and Sigrunn H Sørbye. Penalising model component complexity: A principled, practical approach to constructing priors. *Statistical Science*, 32(1):1–28, 2017.

[14] Geir-Arne Fuglstad, Daniel Simpson, Finn Lindgren, and Håvard Rue. Constructing priors that penalize the complexity of Gaussian random fields. *Journal of the American Statistical Association*, 114(525):445–452, 2019.

[15] Finn Lindgren, Håvard Rue, and Johan Lindström. An explicit link between Gaussian fields and Gaussian Markov random fields: the stochastic partial differential equation approach. *Journal of the Royal Statistical Society Series B: Statistical Methodology*, 73(4):423–498, 2011.

[16] Elias Krainski, Virgilio Gómez-Rubio, Haakon Bakka, Amanda Lenzi, Daniela Castro-Camilo, Daniel Simpson, Finn Lindgren, and Håvard Rue. *Advanced spatial modeling with stochastic partial differential equations using R and INLA*. Chapman and Hall/CRC, 2018.

[17] Haakon Bakka, Håvard Rue, Geir-Arne Fuglstad, Andrea Riebler, David Bolin, Janine Illian, Elias Krainski, Daniel Simpson, and Finn Lindgren. Spatial modeling with R-INLA: A review. *Wiley Interdisciplinary Reviews: Computational Statistics*, 10(6):e1443, 2018.

[18] Haakon Bakka, Jarno Vanhatalo, Janine B Illian, Daniel Simpson, and Håvard Rue. Non-stationary Gaussian models with physical barriers. *Spatial Statistics*, 29:268–288, 2019.

[19] F. Lindgren, H. Bakka, D. Bolin, E. T. Krainski, and H. Rue. The diffusion-based extension of the Matérn field to space-time (invited article with discussion). *SORT*, 48(1):3–66, 2024.

[20] Finn Lindgren, David Bolin, and Håvard Rue. The SPDE approach for gaussian and non-gaussian fields: 10 years and still running. *Spatial Statistics*, 50:100599, 2022. Special Issue: The Impact of Spatial Statistics.

[21] Sigrunn Holbek Sørbye and Håvard Rue. Penalised complexity priors for stationary autoregressive processes. *Journal of Time Series Analysis*, 38(6):923–935, 2017.

[22] Janet Van Niekerk, Haakon Bakka, and Håvard Rue. Stable non-linear generalized Bayesian joint models for survival-longitudinal data. *Sankhya A*, 85(1):102–128, 2023.

[23] Finn Lindgren and Håvard Rue. On the second-order random walk model for irregular locations. *Scandinavian Journal of Statistics*, 35(4):691–700, 2008.

[24] Leonhard Knorr-Held. Bayesian modelling of inseparable space-time variation in disease risk. *Statistics in Medicine*, 19(17-18):2555–2567, 2000.

[25] David Bolin and Finn Lindgren. Calculating probabilistic excursion sets and related quantities using excursions. *Journal of Statistical Software*, 86(5):1–20, 2018.

[26] Paula Moraga, Susanna M Cramb, Kerrie L Mengersen, and Marcello Pagano. A geostatistical model for combined analysis of point-level and area-level data using INLA and SPDE. *Spatial Statistics*, 21:27–41, 2017.

[27] Fabian E Bachl, Finn Lindgren, David L Borchers, and Janine B Illian. inlabru: An R package for Bayesian spatial modelling from ecological survey data. *Methods in Ecology and Evolution*, 10(6):760–766, 2019.

15

Real-time Forecasting of Fire Front Propagation Using the Level Set Method and Echo State Networks

Myungsoo Yoo, Likun Zhang, and Christopher K. Wikle

15.1 Introduction

Despite ongoing efforts to mitigate the devastating impact of wildfires, we continue to experience their destructive consequences every year. For example, more than one hundred deaths were reported from the Maui wildfires in 2023 with economic losses estimated to be in the billions of dollars [Holliday et al., 2024]. More recently, the Palisades Fire and the Eaton Fire, which occurred in Los Angeles County, California in January 2025, burned 23,707 acres and 14,021 acres, respectively, and ranked among the most destructive California wildfires on record. The effect of massive wildfires can be catastrophic. Beyond loss of life and economic losses, they also undermine ecological resilience, accelerate land erosion, and significantly impact air quality, potentially offsetting the benefits of strategies designed to reduce premature mortality from air pollution [Carreras-Sospedra et al., 2024]. Given these impacts, fire prediction models can be essential for mitigating wildfire risks and for saving lives [Albini, 1984].

One major interest in studying wildfire behavior is predicting the probability of fire occurrence [see Taylor et al., 2013, for a review of prediction tasks in wildfire management and the statistical approaches used to address them]. To achieve this, factors such as vegetation type, fuel moisture, and wind velocity are typically used to estimate the fire occurrence probability [Jafari Goldarag et al., 2016]. For example, Opitz et al. [2020] used spatio-temporal log-Gaussian Cox process models to indirectly quantify the human influence on wildfire occurrence in the French Mediterranean basin. Koh et al. [2023] introduced a joint model for wildfire occurrence intensity and burnt area using log-Gaussian Cox process models and extreme value theory. Neural network models also provide promising approaches for this purpose. For example, "Fire-Net" by Seydi et al. [2022] uses satellite data as input for a neural network model based on multiscale residual convolution layers, with applications to forest fires in many different regions. Similarly, Bergado et al. [2021] use convolutional neural network models along with covariates to predict ignition probability in Australia. For a review of fire occurrence prediction using machine learning algorithms, see Abid [2021].

Despite the effectiveness of wildfire occurrence models, it is unrealistic to assume that wildfire ignition can be completely prevented. Therefore, fire risk management should also

DOI: 10.1201/9781003564423-15

incorporate strategies to mitigate fire spread and minimize its impact. This involves modeling the rate of spread through equations [e.g., model proposed by Rothermel, 1972] and forecasting the growth of the fire front through time. Over the years, various methods have been developed for this purpose, ranging from physics-based simulators to data-driven models. An early model by Wagner [1969] represents the fire front as an ellipse and describes its evolution based on the rate of spread at the head, flanks, and rear. More recently, the fire growth model "FARSITE" incorporates spatial data, including wind, weather, fuel, and elevation, to simulate long-term fire growth [Finney, 2004]. "FARSITE" is used for fire prediction in national parks in the USA [Finney and Ryan, 1995] and has also been combined with the ensemble Kalman filter to assimilate data in near real time [Srivas et al., 2016]. Grieshop and Wikle [2024a] consider a stochastic cellular automata model in which the neighborhood is learned based on Rothermel's equation and propagation rules are learned to describe unburned, burning, and burned wildfire states. Similarly, Grieshop and Wikle [2024b] consider a binary cellular automata model that includes an echo state network (ESN) and symbolic regression to learn propagation rules for wildfire front prediction. In addition to these methods, other approaches for fire front propagation consider Lagrangian movement of tracer particles [Clark et al., 2004] and the method of markers [Baptiste Filippi et al., 2009]. This remains an active area of research.

Arguably, the level set method (or level set equation) proposed by Osher and Sethian [1988] is among the most ubiquitous models for wildfire fire front propagation. The level set method is based on the partial differential system that describes the evolution of an implicit function. For example, a signed distance function, which is an example of an implicit function, is well-suited for representing a fire front, as it can accommodate complex boundaries and effectively describe topological changes. The key parameter in the level set equation is often parametrized by the model for the rate of spread. For example, Rochoux et al. [2013] and Mallet et al. [2009] relied on the rate of spread models proposed by Rothermel [1972] and Fendell and Wolff [2001], respectively. Emphasizing the importance of uncertainty quantification, Yoo and Wikle [2024] incorporated the level set equation into a Bayesian hierarchical model, where the rate of spread is approximated with low-rank basis functions and learned from the data. As an alternative to facilitate inference and incorporate nonlinear dynamics, Yoo and Wikle [2023] used an ESN model to learn the rate of spread rate in the level set equation. Instead of employing numerical schemes such as finite differencing or upwind differencing to solve the level set equation, Dabrowski et al. [2023] applied a physics-informed neural network to model the evolution of the fire front and demonstrated its performance in a controlled burn under homogeneous environmental conditions. One potential limitation of this approach is the assumption of a homogeneous propagation speed across space and time, which may not accurately reflect real-world wildfire dynamics.

Our goal in this chapter is to develop an efficient forecasting model capable of predicting fire front evolution with updates as data become available, opening a new gateway to near real-time forecasts and decision-making support in wildfire management. A wildfire front forecasting model should be accurate, computationally efficient, and capable of quantifying uncertainty. Moreover, the model should be able to update its parameters dynamically as new data become available [see, e.g., Jain et al., 2014, for a review of online learning in neural network models], a feature particularly valuable for computationally expensive models. Re-fitting the model to the entire dataset each time new data arrive can be computationally demanding and impractical for real-time wildfire management. Historically, "data assimilation" approaches have employed sequential filtering techniques, including the Kalman filter and particle filter, to efficiently update model estimates [Wikle and Berliner, 2007]. The application of such approaches to the level set method has also been explored. For example, Dabrowski et al. [2022] employed the ensemble Kalman filter (EnKF) to update the parameters in the Rothermel model, which was then used to represent the rate of spread in the level set method. However, they assumed that key parameters (e.g., evolution speed) was were known and constant across the domain, which may not be realistic.

Our approach builds upon the model by Yoo and Wikle [2023] but extends it to allow online updating to aid real-time wildfire front forecasting. Yoo and Wikle [2023] estimated spatially-varying propagation speed in the level set equation using the flexible reservoir computing neural ESN model. However, it still requires re-fitting the model to the entire dataset as new data become available, which is not suited for real-time forecasting. We address this by using recursive regularized least squares estimation, enabling efficient sequential updates as new data become available. This approach offers several key contributions. First, our model enables real-time wildfire forecasting without assuming known parameters for propagation speed. Relaxing this assumption is crucial in practice, as wildfire evolution speed can vary across space and time. Second, the flexibility of the ESN allows the model to capture temporal non-linearity that may arise in fire front propagation. Accounting for non-linearity is important, as it is frequently observed in dynamic systems [Chen and Billings, 1992], yet quite difficult to parameterize efficiently [Cressie and Wikle, 2011, Wikle and Hooten, 2010]. We show real-time forecast performance using a real wildfire event.

This chapter is organized as follows. Section 15.1 provides a brief overview of the level set method and the ESN model. Section 15.3 introduces the forecast model that integrates the level set method with ESN in Yoo and Wikle [2023], along with the algorithm for online updating. Section 15.4 presents the model's performance on real wild fire data. Finally, we conclude with a discussion and future study in Section 15.5.

15.2 Background

This section provides a brief overview of the level set method and the ESN model, which serve as the two main building blocks of the hybrid model in Section 15.3.

15.2.1 The Level Set Method

The level set method (or level set equation) is a partial differential equation that describes the evolution of an implicit function in any dimension defined as:

$$\frac{\partial \phi(\boldsymbol{s},t)}{\partial t} + \boldsymbol{v}(\boldsymbol{s},t) \cdot \nabla \phi(\boldsymbol{s},t) = 0, \tag{15.1}$$

where $\phi(\boldsymbol{s},t)$ denotes an implicit function evaluated at location $\boldsymbol{s} \in \mathbb{R}^d$ and time $t \in \mathbb{R}$, and $\nabla \phi = (\partial \phi/\partial s_1, \ldots, \partial \phi/\partial s_d)^\top$ denotes the gradient of ϕ over space. Equation (15.1) can be interpreted as a deterministic model describing how $\phi(\boldsymbol{s},t)$ varies across space at a fixed time and evolves over time, governed by its gradient $\nabla \phi$ and temporal derivative $\partial \phi/\partial t$, respectively. Most importantly, the direction and speed of evolution are determined by the spatially and temporally varying velocity $\boldsymbol{v}(\boldsymbol{s},t)$. Indeed, modeling the velocity requires careful consideration to accurately capture the realistic evolution of the underlying implicit function. It should be noted that equation (15.1) has no closed-form solution, and so its solution requires a numerical method, such as the upwind differencing scheme, the Hamilton-Jacobi ENO method, or the Hamilton-Jacobi WENO method [Osher and Fedkiw, 2002].

In the context of wildfire modeling, it is commonly assumed that the evolution follows the normal direction to the boundary [e.g., Alessandri et al., 2021, Mallet et al., 2009, Yoo and Wikle, 2023, 2024], which further simplifies equation (15.1) to:

$$\frac{\partial \phi(\boldsymbol{s},t)}{\partial t} + v_n(\boldsymbol{s},t) \cdot \|\nabla \phi(\boldsymbol{s},t)\| = 0, \tag{15.2}$$

where the scalar velocity (speed) in the *normal* direction is denoted by $v_n(\boldsymbol{s},t)$ and $\|\cdot\|$ represents the Euclidean (L_2) norm. A signed distance function is often used as the implicit function $\boldsymbol{\phi}(\boldsymbol{s},t)$, for $\boldsymbol{s} \in \mathbb{R}^2$ [e.g., Alessandri et al., 2021, Mallet et al., 2009, Yoo and Wikle, 2023, 2024]: i.e.,

$$\phi(\boldsymbol{s}) = \begin{cases} -d(\boldsymbol{s}), & \boldsymbol{s} \in \Omega^- \\ 0, & \boldsymbol{s} \in \partial\Omega \\ d(\boldsymbol{s}), & \boldsymbol{s} \in \Omega^+ \end{cases}, \tag{15.3}$$

where the locations on the boundary $\partial\Omega$ have a signed distance 0, and those inside and outside the boundary are assigned the minimum distance from $\boldsymbol{s}$ to the boundary with negative and positive signs, respectively. In practice, $d(\boldsymbol{s})$ can be defined by the minimum Euclidean distance as

$$d(\boldsymbol{s}) = \min(\|\boldsymbol{s} - \boldsymbol{s}_I\|), \tag{15.4}$$

where the set of points on the boundary is denoted by $\boldsymbol{s}_I \in \partial\Omega$.

Figure 15.1 illustrates the representation of a fire front using the signed distance function. Intuitively, the fire front is depicted indirectly by the zero contour of the signed distance function, rather than by a set of points on the fire front. One of the main advantages of using a signed distance function to represent a wildfire boundary is its easy interpretation and flexibility in describing potentially complex boundaries. Note that any boundary shape and its evolution over time can be captured by the signed distance function. The details of the fire front can be preserved in the signed distance function representation if it is computed on a dense spatial grid.

An additional advantage of the signed-distance function representation lies in the simplification of equation (15.2) to:

$$\frac{\partial \phi(\boldsymbol{s},t)}{\partial t} + v_n(\boldsymbol{s},t) = 0, \tag{15.5}$$

where the key property of a signed distance function, $\|\nabla \phi(\boldsymbol{s},t)\| = 1$, is utilized. Although a numerical method is still required to approximate equation (15.5), the simplifications based on the normal flow assumption in (15.2) and the key signed distance function property allow the sole focus to be on modeling the speed $v_n(\boldsymbol{s},t)$. Therefore, the model for $v_n(\boldsymbol{s},t)$ should be highly flexible to capture realistic and extreme fire behavior. To achieve this, Yoo and Wikle [2024] considered a latent Gaussian spatio-temporal dynamic process and Yoo and Wikle [2023] utilized the flexibility of an ESN to model the speed. The ESN approach worked better in those examples, and so we consider it here.

15.2.2 Echo State Network

Deep learning neural network models have revolutionized many aspects of machine learning, offering substantial improvements in prediction and classification tasks when data have complex relationships. As with many statistical models, accounting for dependency in data is one of the key strengths of neural network models. In the case of sequential data (e.g., language, time series), recurrent neural networks (RNNs) provide a relatively simple yet

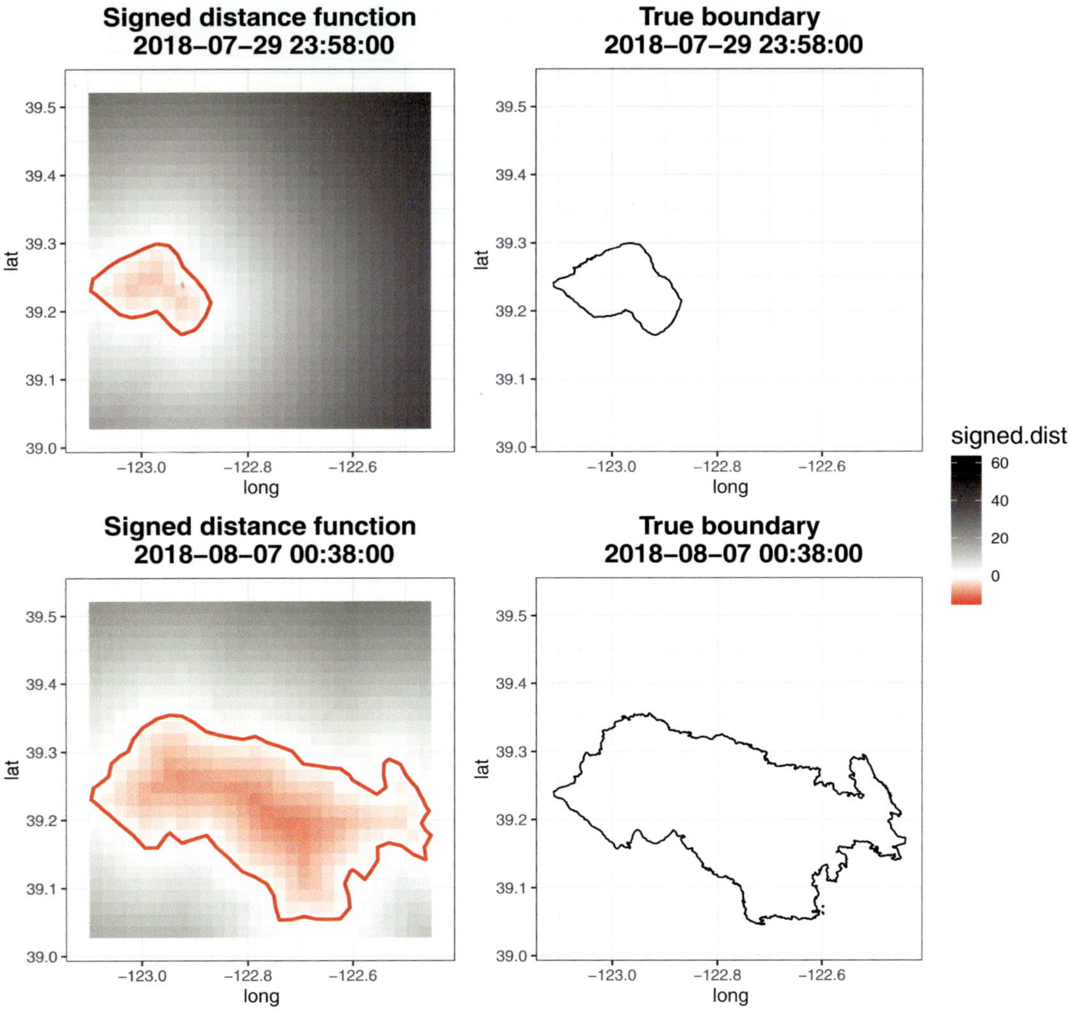

FIGURE 15.1
Representations of the Ranch fire front using the signed distance function. Left: The signed distance function, where the boundary, indicated by a solid red line, corresponds to the zero contour. Right: The true fire front. Note that the true fire front is represented by the set of points on the boundary itself.

powerful approach to capture serial dependence. Despite their flexibility, training such models via "backpropagation through time" [Werbos, 1990] can be challenging, often failing due to the vanishing/exploding gradient problem [Doya, 1992]. As an alternative, the ESN has received substantial attention [Lukoševičious, 2012] and has been widely utilized in recent years, particularly in cases where data volume is not sufficient to utilize gated recurrent neural networks [e.g., Long Short-Term Memory models; see Hochreiter and Schmidhuber, 1997].

The ESN is simply a classical RNN where the weights connecting the input to the hidden states and between hidden states are randomly initialized and remain fixed, rather than being estimated [Lukoševičious, 2012]. Such models with random weights are considered a type of reservoir computing. Despite its simplicity, the universal approximation property still holds for the ESN [Hart et al., 2021], meaning that, in theory, it can approximate any continuous function given a sufficiently large number of hidden states. The basic ESN model for time $t = 1, ..., T$ can be written as:

$$\begin{aligned}
\text{output}: \quad & \boldsymbol{y_t} = \boldsymbol{W}^{\text{out}} \boldsymbol{h}_t \\
\text{hidden state}: \quad & \boldsymbol{h}_t = (1 - \alpha_\ell) \boldsymbol{h}_{t-1} + \alpha_\ell \tilde{\boldsymbol{h}}_t \\
& \tilde{\boldsymbol{h}}_t = g_h \left(\frac{\nu}{|\lambda_w|} \boldsymbol{W}^{\text{in}} \boldsymbol{h}_{t-1} + \boldsymbol{U} \boldsymbol{x}_t \right), \\
\text{parameters}: \quad & \boldsymbol{W}^{\text{in}} = [w^{\text{in}}_{i,\ell}]_{i,\ell} : \gamma^w_{i,\ell} \text{Unif}(-a_w, a_w) + (1 - \gamma^w_{i,\ell}) \delta_0 \\
& \boldsymbol{U} = [u_{i,j}]_{i,j} : \gamma^u_{i,j} \text{Unif}(-a_u, a_u) + (1 - \gamma^u_{i,j}) \delta_0 \\
& \gamma^w_{i,\ell} \sim \text{Bernoulli}(\pi_w) \\
& \gamma^u_{i,j} \sim \text{Bernoulli}(\pi_u),
\end{aligned} \tag{15.6}$$

where $N \times 1$ vector $\boldsymbol{y}_t$, $J \times 1$ vector $\boldsymbol{h}_t$, and $n_u \times 1$ vector $\boldsymbol{x}_t$ represent the output, hidden states (reservoir), and inputs, respectively. Notably, by the nature of the reservoir computing framework, the $J \times J$ sparse matrix $\boldsymbol{W}^{\text{in}}$ and the $J \times n_u$ sparse matrix $\boldsymbol{U}$ are randomly initialized and fixed. For example, elements can be drawn from a uniform distribution with range parameters a_w and a_u. Parameters π_w and π_u control the sparsity of $\boldsymbol{W}^{\text{in}}$ and $\boldsymbol{U}$, with δ_0 representing a Dirac delta function at zero.

For an ESN to effectively approximate a nonlinear system, it must satisfy the so-called echo state property [Zhang et al., 2012]. Intuitively, the "echo state property" means that the hidden states $\boldsymbol{h}_t$ should not be sensitive to initial conditions, allowing the network to "echo" the influence of past inputs rather than initial transients. In practice, this condition is met when the spectral radius of $\boldsymbol{W}^{\text{in}}$ is less than 1 [Lukoševičious, 2012], which is often achieved by scaling $\boldsymbol{W}^{\text{in}}$ using a factor ν (with $0 < \nu < 1$), such that ν divided by the largest eigenvalue λ_w of $\boldsymbol{W}^{\text{in}}$ ensures the desired spectral radius. Additionally, the element-wise activation function $g_h(\cdot)$ (e.g., tanh or sigmoid functions) introduces the nonlinearity to the model. By construction, the information in $\boldsymbol{h}_{t-1}$ is propagated to $\boldsymbol{h}_t$, and the leaking rate α_ℓ (where $0 \leq \alpha_\ell \leq 1$) controls how much of the memory from $\boldsymbol{h}_{t-1}$ is retained in $\boldsymbol{h}_t$.

The appealing aspect of ESNs compared to RNNs is their low computational cost and training stability. Importantly, it should be emphasized that the $N \times J$ readout matrix $\boldsymbol{W}^{\text{out}}$ contains the only parameters that need to be estimated during training. With an ℓ_2 penalty added to the mean squared error in the loss function to prevent overfitting, $\boldsymbol{W}^{\text{out}}$ can be estimated explicitly without relying on gradient descent. That is,

$$\begin{aligned}
\widehat{\boldsymbol{W}}^{\text{out}} &= \underset{\boldsymbol{W}^{\text{out}}}{\arg\min} \left[\frac{1}{N} \sum_{t=1}^{T} \{ (\boldsymbol{y_t} - \boldsymbol{W}^{\text{out}} \boldsymbol{h_t})^\top (\boldsymbol{y_t} - \boldsymbol{W}^{\text{out}} \boldsymbol{h_t}) \} + \tau \sum_{j=1}^{J} \| \boldsymbol{w}^{\text{out}}_{\cdot j} \|^2 \right] \\
&= \boldsymbol{Y} \boldsymbol{H}^\top (\boldsymbol{H} \boldsymbol{H}^\top + \tau \boldsymbol{I})^{-1},
\end{aligned} \tag{15.7}$$

where $\boldsymbol{Y} = [\boldsymbol{y}_1, \cdots, \boldsymbol{y}_T]$ is the $N \times T$ matrix where each column corresponds to $\boldsymbol{y}_t$, for $t = 1, ..., T$, and $\boldsymbol{H} = [\boldsymbol{h}_1, \cdots, \boldsymbol{h}_T]$ is a $J \times T$ matrix. Additionally, $\boldsymbol{w}^{\text{out}}_{\cdot j}$ denotes j^{th} column of $\boldsymbol{W}^{\text{out}}$, $\| \cdot \|$ is the ℓ_2 norm, and the regularization parameter $\tau \in (0, \infty)$ controls the extent to which the estimate shrinks toward 0.

As with other neural network models, the predictive performance of an ESN is influenced by its hyperparameters. Lukoševičious [2012] suggested specifying small values for π_w and π_u and using a large reservoir size J to promote sparsity in $\boldsymbol{W}^{\text{in}}$ and $\boldsymbol{U}$ while effectively capturing the complex dynamics of the data. Other hyperparameters can be optimized using traditional cross-validation or validation approaches with grid search [McDermott, 2017], genetic algorithms [McDermott and Wikle, 2019], the backtracking search algorithm [Wang et al., 2019], and Bayesian optimization [Bai et al., 2023, Maat et al., 2018, Ribeiro et al., 2020]. Additionally, Racca and Magri [2021] introduced the "recycle validation" approach, which provides performance comparable to cross-validation at a lower computational cost. Rather than optimizing hyperparameters, Yoo and Wikle [2023] and Grieshop and Wikle [2024b] sampled them from a specified distribution. A review of hyperparameter optimization for ESN is provided in Matzner [2022].

As uncertainty quantification has gained significant attention in neural network models, various approaches have also been adopted for uncertainty quantification in ESN. One approach is the so-called ensemble method, where uncertainty is quantified using ensembles of forecasts generated from the reservoir [McDermott, 2017, Sheng et al., 2013, Yao et al., 2013]. Each forecast is associated with $\boldsymbol{W}^{\text{in}}$ and $\boldsymbol{U}$, which are independently drawn from the parameter distribution in equation (15.6). The mean or quantiles of the ensembles can then be obtained from the collection of forecasts. Other approaches include the Bayesian approach [Chatzis and Demiris, 2011, McDermott and Wikle, 2019] and the calibration method [Bonas and Castruccio, 2021].

The ESN presented in equation (15.6) is arguably the most basic ESN. Building upon this baseline, more advanced ESNs are studied in the literature for specific applications. To name a few, McDermott [2017] included the quadratic term $\boldsymbol{h}_t^2$ and the corresponding readout matrix in the output layer, highlighting the importance of quadratic interactions in many physical and biological processes. Deep ESNs have also been used to account for multiple scales in the dynamics [see, e.g., Bonas et al., 2024, Kim and King, 2020, McDermott and Wikle, 2019]. The deep ESN consists of L hidden states, where the output of the $(\ell-1)^{\text{th}}$ hidden state is fed into the ℓ^{in} hidden state as input, allowing it to capture more complex dynamic patterns and model intricate temporal dependencies. Furthermore, as physics-informed neural network models have introduced a new paradigm for incorporating physical laws into neural networks [Raissi et al., 2019], the same principle can be applied to the ESN. The physics-informed ESN by Doan et al. [2020] achieves this by regularizing the readout matrix to ensure that the predictions remain consistent with the physical laws of interest. A systematic review of ESN design is provided in Sun et al. [2024]. It should also be noted that an ESN is used for other types of data such as areal-level data [Wang et al., 2024] and count data [Wang et al., 2025]. These developments highlight the growing interest in both integrating ESNs with domain-specific structures and in developing new models based on ESNs.

15.3 Hybrid Level Set/ESN Model for Real-Time Forecasting

As indicated in Section 15.2.1, modeling the rate of spread $v_n(\boldsymbol{s}, t)$ across space and time is crucial, as it determines the propagation of the fire front. To this end, Yoo and Wikle [2023] employed the basic ESN in equation (15.6) to infer $v_n(\boldsymbol{s}, t)$ in equation (15.5) and demonstrated its effectiveness. Building on their approach, we develop an algorithm for the online updating of parameters. The algorithm introduced in this section can also be applied to other ESNs, provided that the readout matrix is estimated in a manner similar to equation (15.7).

The fire-spread model of Yoo and Wikle [2023] is given in vector notation as:

$$\begin{aligned}
\text{Response}: \quad & \boldsymbol{\phi}_t = \boldsymbol{\phi}_{t-\triangle t} - \boldsymbol{v}_{n,t-\triangle t}\triangle t + \boldsymbol{\epsilon}_t \\
\text{Rate of spread}: \quad & \boldsymbol{v}_{n,t-\triangle t} = \boldsymbol{W}^{\text{out}}\boldsymbol{h}_{t-\triangle t} \\
\text{Hidden state}: \quad & \boldsymbol{h}_{t-\triangle t} = (1-\alpha_\ell)\boldsymbol{h}_{t-2\triangle t} + \alpha_\ell \tilde{\boldsymbol{h}}_{t-\triangle t} \\
& \tilde{\boldsymbol{h}}_{t-\triangle t} = g_h\left(\frac{\nu}{|\lambda_w|}\boldsymbol{W}^{in}\boldsymbol{h}_{t-2\triangle t} + \boldsymbol{U}\boldsymbol{x}_{t-\triangle t}\right) \\
\text{Parameters}: \quad & \boldsymbol{W}^{\text{in}} = [w^{\text{in}}_{i,\ell}]_{i,\ell} : \gamma^w_{i,\ell} Unif(-a_w, a_w) + (1-\gamma^w_{i,\ell})\delta_0 \\
& \boldsymbol{U} = [u_{i,j}]_{i,j} : \gamma^u_{i,j} Unif(-a_u, a_u) + (1-\gamma^u_{i,j})\delta_0 \\
& \gamma^w_{i,\ell} \sim Bern(\pi_w) \\
& \gamma^u_{i,j} \sim Bern(\pi_u).
\end{aligned} \tag{15.8}$$

The response model approximates equation (15.5) using forward Euler discretization, where $\boldsymbol{\phi}_t = (\phi(\boldsymbol{s}_1, t), \ldots, \phi(\boldsymbol{s}_N, t))^\top$ represents the observed signed distance function for $\boldsymbol{s}_1, ..., \boldsymbol{s}_N$ at time t. The rate of spread in the normal direction to the boundary at $\boldsymbol{s}$ and $t-\triangle t$ is given by $\boldsymbol{v}_{n,t-\triangle t} = (v_n(\boldsymbol{s}_1, t-\triangle t), \ldots, v_n(\boldsymbol{s}_N, t-\triangle t))^\top$. Additionally, $\boldsymbol{\epsilon}_t = (\epsilon(\boldsymbol{s}_1, t), ..., \epsilon(\boldsymbol{s}_N, t))^\top$ denotes an error term, and $\triangle t$ is the time step size in the forward Euler discretization. The rate of spread $\boldsymbol{v}_{n,t-\triangle t}$ is parameterized by the hidden state $\boldsymbol{h}_{t-\triangle t}$ and the readout matrix $\boldsymbol{W}^{\text{out}}$, where $\boldsymbol{h}_{t-\triangle t}$ is generated based on the structure of an ESN. Other parameters in the hidden state are defined in Section 15.2.2. Then, the readout matrix $\boldsymbol{W}^{\text{out}}$ can be estimated by

$$\widehat{\boldsymbol{W}}^{\text{out}} = \left[\triangle t \boldsymbol{\Phi}_{T-\triangle t}\boldsymbol{H}^\top_{T-\triangle t} - \triangle t \boldsymbol{\Phi}_T \boldsymbol{H}^\top_{T-\triangle t}\right]\left[\triangle t^2 \boldsymbol{H}_{T-\triangle t}\boldsymbol{H}^\top_{T-\triangle t} + \tau \boldsymbol{I}\right]^{-1}, \tag{15.9}$$

where $\boldsymbol{\Phi}_T = [\boldsymbol{\phi}_{1+\triangle t}; \cdots; \boldsymbol{\phi}_T]$, $\boldsymbol{\Phi}_{T-\triangle t} = [\boldsymbol{\phi}_1; \cdots; \boldsymbol{\phi}_{T-\triangle t}]$, $\boldsymbol{H}_{T-\triangle t} = [\boldsymbol{h}_1; \cdots; \boldsymbol{h}_{T-\triangle t}]$, and $\tau \in (0, \infty)$ is the ridge penalty. We use the lagged values of the data for $\boldsymbol{x}_{t-\triangle t} = (\boldsymbol{\phi}^\top_{t-\triangle t}, \boldsymbol{\phi}^\top_{t-2\triangle t})^\top$, following the approach in McDermott [2017]. Intuitively, this allows inference on the rate of spread $\boldsymbol{v}_{n,t-\triangle t}$ at time $t-\triangle t$ using the observed signed distance at $t-\triangle t$ and the previous time $t-2\triangle t$. Once $\widehat{\boldsymbol{W}}^{\text{out}}$ is estimated, the forecast for $T+\triangle t$ is obtained by

$$\hat{\boldsymbol{\phi}}_{T+\triangle t} = \boldsymbol{\phi}_T - \widehat{\boldsymbol{W}}^{\text{out}}\boldsymbol{h}_T \triangle t. \tag{15.10}$$

Estimating the readout matrix via equation (15.9) corresponds to a scenario where the data is collected in advance and remains fixed, leading to an "off-line" update of the readout matrix. This approach may be impractical for real-time forecasting, where new data is continuously collected, as it requires refitting the model to the entire dataset. The limitation becomes particularly evident when model fitting is computationally expensive. While an ESN may appear less affected by this limitation, it still requires inverting a $J \times J$ matrix in equation (15.9), which can be computationally demanding, especially since J is generally set to the largest value computational power allows [Lukoševičious, 2012]. Consequently, in such cases, updating the readout matrix $\boldsymbol{W}^{\text{out}}$ in an "on-line" manner is preferable.

There have been attempts within the ESN framework to address this need. For example, Goswami [2023] proposed a sequential learning strategy for an ESN using the EnKF algorithm. Similarly, Chen et al. [2024] updated the readout matrix in an ESN recursively using the "Woodbury matrix identity" [Woodbury, 1950]. Our "online" learning algorithm is similar to that of Chen et al. [2024]. With some algebra, one can show that both the

Algorithm 1 On-line Update of a Hybrid Level Set/ESN Model

1: **Input** observed signed distance function for $t = 1, 1 + \triangle t$, $(\boldsymbol{\phi}_1,\ \boldsymbol{\phi}_{1+\triangle t})$, hyper parameters $(a_w,\ a_u,\ \pi_w,\ \pi_u,\ \alpha_\ell,\ J,$ and $\tau)$.
2: **Initial parameter estimation and forecasting**
3: Construct $\boldsymbol{W}^{\text{in}}$ and $\boldsymbol{U}$ using $a_w,\ a_u,\ \pi_u,$ and π_w by equation (15.8).
4: Construct $\boldsymbol{h}_1 = \boldsymbol{U}\widetilde{\boldsymbol{x}}_{1+\triangle t}$ where $\widetilde{\boldsymbol{x}}_{1+\triangle t} = [\boldsymbol{\phi}_{1+\triangle t}^\top, \boldsymbol{\phi}_1^\top]^\top$.
5: Estimate the initial readout matrix $\boldsymbol{W}^{\text{out},(1+\triangle t)}$ and covariance matrix $\boldsymbol{P}^{(1+\triangle t)}$using $\boldsymbol{h}_1$ and ridge penalty τ by equation (15.13).
6: Obtain the forecast for $t = 1+2\triangle t$ by $\hat{\boldsymbol{\phi}}_{1+2\triangle t} = \boldsymbol{\phi}_{1+\triangle t} - \boldsymbol{W}^{\text{out},(1+\triangle t)}\boldsymbol{h}_{1+\triangle t}\triangle t$ where $\boldsymbol{h}_{1+\triangle t}$ is obtained by equation (15.8) with $\tilde{\boldsymbol{x}}_{1+\triangle t}$
7: **Adaptive Parameter Updating and Forecasting with New Data** $\boldsymbol{\phi}_{1+2\triangle t}$
8: Construct $\boldsymbol{h}_{1+2\triangle t}$ by equation (15.8) with $\tilde{\boldsymbol{x}}_{1+2\triangle t} = [\boldsymbol{\phi}_{1+2\triangle t}^\top, \boldsymbol{\phi}_{1+\triangle t}^\top]^\top$.
9: Update the covariance and readout matrices by equations (15.11) and (15.12).
10: Obtain the forecast for $t = 1 + 3\triangle t$ by $\hat{\boldsymbol{\phi}}_{1+3\triangle t} = \boldsymbol{\phi}_{1+2\triangle t} - \boldsymbol{W}^{\text{out},(1+2\triangle t)}\boldsymbol{h}_{1+2\triangle t}\triangle t$.
11: **output** $\hat{\boldsymbol{\phi}}_{1+3\triangle t}$

covariance matrix and the readout matrix can be updated recursively as:

$$\boldsymbol{P}^{(t)} = \frac{\boldsymbol{P}^{(t-\triangle t)} - \boldsymbol{P}^{(t-\triangle t)}[\triangle t \cdot \boldsymbol{h}_{t-\triangle t}][\triangle t \cdot \boldsymbol{h}_{t-\triangle t}]^\top \boldsymbol{P}^{(t-\triangle t)}}{\left(1 + [\triangle t \cdot \boldsymbol{h}_{t-\triangle t}]^\top \boldsymbol{P}^{(t-\triangle t)}[\triangle t \cdot \boldsymbol{h}_{t-\triangle t}]\right)} \tag{15.11}$$

$$\widehat{\boldsymbol{W}}^{\text{out},(t)} = \widehat{\boldsymbol{W}}^{\text{out},(t-\triangle t)} + \left[\boldsymbol{\phi}_{t-\triangle t} - \boldsymbol{\phi}_t - \widehat{\boldsymbol{W}}^{\text{out},(t-\triangle t)}[\triangle t \cdot \boldsymbol{h}_{t-\triangle t}]\right][\triangle t \cdot \boldsymbol{h}_{t-\triangle t}]^\top (\boldsymbol{P}^{(t)})^\top, \tag{15.12}$$

where $\boldsymbol{P}^{(t)}$ and $\widehat{\boldsymbol{W}}^{\text{out},(t)}$ denote the estimated "covariance" and readout matrices based on data up to time t. Note that the recursive updating algorithm requires an initial estimate of $\boldsymbol{P}$ and $\boldsymbol{W}^{\text{out}}$ based on the initial data. We refer to the first two observations, $\boldsymbol{\phi}_1$ and $\boldsymbol{\phi}_{1+\triangle t}$, as the initial data, since the response model in equation (15.8) involves $\boldsymbol{\phi}_t$ and $\boldsymbol{\phi}_{t-\triangle t}$. Then, the initial estimates incorporating the ridge penalty are given by:

$$\boldsymbol{P}^{(1+\triangle t)} = [\triangle t^2 \boldsymbol{h}_1 \boldsymbol{h}_1^\top + \tau \boldsymbol{I}]^{-1} \tag{15.13}$$

$$\widehat{\boldsymbol{W}}^{\text{out},(1+\triangle t)} = [\triangle t \boldsymbol{\phi}_1 \boldsymbol{h}_1^\top - \triangle t \boldsymbol{\phi}_{1+\triangle t} \boldsymbol{h}_1^\top][\triangle t^2 \boldsymbol{h}_1 \boldsymbol{h}_1^\top + \tau \boldsymbol{I}]^{-1}, \tag{15.14}$$

where $\boldsymbol{h}_1$ is defined as $\boldsymbol{U}\tilde{\boldsymbol{x}}_{1+\triangle t}$ with $\tilde{\boldsymbol{x}}_{1+\triangle t} = [\boldsymbol{\phi}_{1+\triangle t}^\top, \boldsymbol{\phi}_1^\top]^\top$. As new data (e.g., $\boldsymbol{\phi}_{1+2\triangle t}$) become available, $\boldsymbol{P}^{(1+2\triangle t)}$ and $\widehat{\boldsymbol{W}}^{\text{out},(1+2\triangle t)}$ can be obtained recursively using equations (15.11) and (15.12), and the one-step-ahead forecast for $t = 1 + 3\triangle t$ can be obtained using equation (15.10). This procedure then continues as new observations become available.

Algorithm 1 outlines the steps for real-time fire front forecasting. Lines 3 to 6 involve estimating the initial parameters and obtaining the one step-ahead forecast. The one-step ahead forecast obtained from the initial data is typically not very accurate, as it is based only on the first two observations. Indeed, in our experiments, we observed that the initial one-step ahead forecast was not reliable when fitting the model with a large reservoir (i.e., large J). This is expected, as the initial readout matrix $\widehat{\boldsymbol{W}}^{\text{out},(1+\triangle t)}$, which maps the hidden states to the output, has dimensions $N \times J$, making it challenging to estimate with only two observations. Therefore, we recommend a weak learner with a relatively small J, as emphasized in McDermott [2017]. Lines 8 to 10 demonstrate the steps for updating the parameters and obtaining the forecast adaptively. Specifically, these steps are illustrated

Algorithm 2 Uncertainty quantification for Hybrid Level Set/ESN Model

1: **Input** The prior distributions of J, τ, ν, and α_ℓ denoted by $p(J)$, $p(\tau)$, $p(\nu)$, $p(\alpha_\ell)$, respectively. Fixed constants a_w, a_u, π_w, and π_u.
2: **for** $k = 1 : N_{\text{ensemble}}$ **do**
3: **Initial parameter estimation and forecasting**
4: Randomly draw $J^{(k)}$, $\tau^{(k)}$, $\nu^{(k)}$, and $\alpha_\ell^{(k)}$ from prior distributions.
5: Construct $\boldsymbol{W}^{\text{in},(k)}$, $\boldsymbol{U}^{(k)}$ using the sampled $\nu^{(k)}$ and the fixed parameters a_u, a_w, π_u, and π_w.
6: Obtain initial parameter estimates and forecasts with $J^{(k)}$, $\alpha_\ell^{(k)}$, and $\tau^{(k)}$ using Lines 3, 4, and 5 in Algorithm 1.
7: **Adaptive Parameter Updating and Forecasting with New Data** $\boldsymbol{\phi}_{1+2\triangle t}$
8: Update parameters and obtain forecasts with $J^{(k)}$, $\alpha_\ell^{(k)}$, and $\tau^{(k)}$ using Lines 8, 9, and 10 in Algorithm 1.
9: **end for**
10: Define the ensemble of forecasts: $\widehat{\boldsymbol{\phi}}_{\text{en},1+3\triangle t} = \{\widehat{\boldsymbol{\phi}}^{(i)}_{1+3\triangle t}\}_{i=1}^{N_{\text{ensemble}}}$
11: Ensemble forecast mean: $\widehat{\boldsymbol{\phi}}_{\text{mean},1+3\triangle t} = \frac{1}{N_{\text{ensemble}}} \sum_{i=1}^{N_{\text{ensemble}}} \widehat{\boldsymbol{\phi}}^{(i)}_{1+3\triangle t}$
12: Ensemble uncertainty: $(\hat{\ell}_{1+3\triangle t}, \hat{u}_{1+3\triangle t})$ obtained by equations (15.15) and (15.16) corresponding to the significance level α.
13: **output** $\widehat{\boldsymbol{\phi}}_{\text{mean},1+3\triangle t}$ and $(\hat{\ell}_{1+3\triangle t}, \hat{u}_{1+3\triangle t})$

for the case where new data $\boldsymbol{\phi}_{1+2\triangle t}$ become available. For subsequent time points, lines 8 to 10 can be applied iteratively as new observations are obtained, allowing for real-time forecasting.

15.3.1 Uncertainty Quantification

A wildfire forecasting model should be capable of appropriately quantifying uncertainty. We emphasize that this applies not only to our model but also to other models. This need arises from several factors, such as the limited understanding of wildfire behavior and measurement errors [Thompson and Calkin, 2011]. Therefore, we follow the approach of Yoo and Wikle [2023] to quantify forecast uncertainty, as their method has been shown to be effective in applications to two wildfires. The key concept is to use an ensemble of forecasts to quantify uncertainty, with each forecast obtained using different hyperparameters. The algorithm is summarized in Algorithm 2. For each ensemble, hyperparameters are randomly drawn from certain distributions and used to generate the input matrices $\boldsymbol{W}^{\text{in}}$ and $\boldsymbol{U}$ (Lines 4 and 5). The initial estimates and forecasts are then obtained for each ensemble using the sampled hyperparameters (Line 6). As new data is collected, the parameters are adaptively updated following the procedure described in Algorithm 1 (Line 8). Intuitively, this approach can be viewed as fitting N_{ensemble} independent forecasting models, with hyperparameters randomly drawn from distributions rather than optimized.

Then, analogous to conformal inference [Vovk et al., 2005],the $(1-\alpha)\%$ forecast interval at location $\boldsymbol{s}$ and time t is defined as

$$\left(\hat{\ell}_t(\boldsymbol{s}), \hat{u}_t(\boldsymbol{s})\right) = \underset{\ell_t(\boldsymbol{s}), u_t(\boldsymbol{s})}{\arg\min} \left(u_t(\boldsymbol{s}) - \ell_t(\boldsymbol{s})\right) \tag{15.15}$$

such that

$$\frac{1}{N_{\text{ensemble}}} \sum_{k=1}^{N_{\text{ensemble}}} I\left(\ell_t(\boldsymbol{s}) \leq \widehat{\phi}_t^{(k)}(\boldsymbol{s}) \leq u_t(\boldsymbol{s})\right) = 1 - \alpha, \tag{15.16}$$

where $\ell_t(\boldsymbol{s})$ and $u_t(\boldsymbol{s})$ are the upper and lower bounds and $\widehat{\phi}_t^{(k)}(\boldsymbol{s})$ is forecast on $\boldsymbol{s}$ at t from the kth ensemble member. Crucially, this method is appealing for our case as it does not require optimization of hyperparameters *a priori*. In contrast, other approaches typically tune hyperparameters by splitting the data into training and validation sets, which adds extra computational cost.

15.4 Real Data Analysis: The Ranch Fire

In this section, we evaluate the forecasting performance of our model using data from the so-called Ranch Fire, which was one of the largest wildfires in California's history (as part of the Mendocino Complex fire). This fire burned nearly 186,000 hectares [Brown et al., 2020].

The wildfire boundary data were retrieved from the Geomac database [Geospatial Multi-Agency Coordination Group, 2019] and the signed distance functions were calculated using the package `sf` [Pebesma and Bivand, 2023] in `R` [R Core Team, 2022]. We consider the first 11 observations, which are measured approximately every 24 hours. The left panel of Figure 15.2 illustrates the evolution of the Ranch Fire from July 28, 2018 to August 7, 2018, highlighting its extremely fast spread, especially during the first few days. Notably, the fire's progression is predominantly eastward.

We discretize the domain of interest into $N = 2500$ uniformly spaced locations within $[-123.1399, -122.4092] \times [39.0157, 39.3861]$, where we compute the signed distance function. To evaluate forecasting performance, we recursively update the model parameter and obtain one-step-ahead forecasts as new data becomes available, assuming the first two observations are given. Specifically, we obtain the forecast mean and 95% forecast intervals from 1000 ensembles, as described in Section 15.3. The distributional assumptions for the hyperparameters are given by

$$J \sim \mathrm{DU}(10, 30), \quad \alpha_\ell \sim \mathrm{U}(0, 1), \quad \nu \sim \mathrm{U}(0, 1), \quad \tau \sim \mathrm{U}(0.1, 30), \tag{15.17}$$

where $\mathrm{DU}(\cdot, \cdot)$ and $\mathrm{U}(\cdot, \cdot)$ denote the discrete and continuous uniform distributions, respectively, with lower and upper bounds. Moreover, we set $\pi_w = a_w = a_u = 0.3$ and $\pi_u = 0.5$.

Figure 15.3 shows the forecast results. The comparison between the true fire front (black solid line) and the forecast mean (blue solid line) suggests that the model effectively learns the eastward evolution, despite being trained with only a few time observations. In particular, the forecast intervals are narrower on the west side of the true fire front than on the east side, accounting for the predominant direction of evolution. Nonetheless, the extreme rate of spread to the east (e.g., the evolution between August 2 at 01:30 and 23:33) is not fully captured in the forecast mean. This is because the evolution speed observed in the new data is significantly faster than that in the data used to update the parameters. However, it is important to note that the 95% forecast intervals nearly cover the observed fire front. It emphasizes the importance of uncertainty quantification in forecasting, particularly in the early stages of a fire.

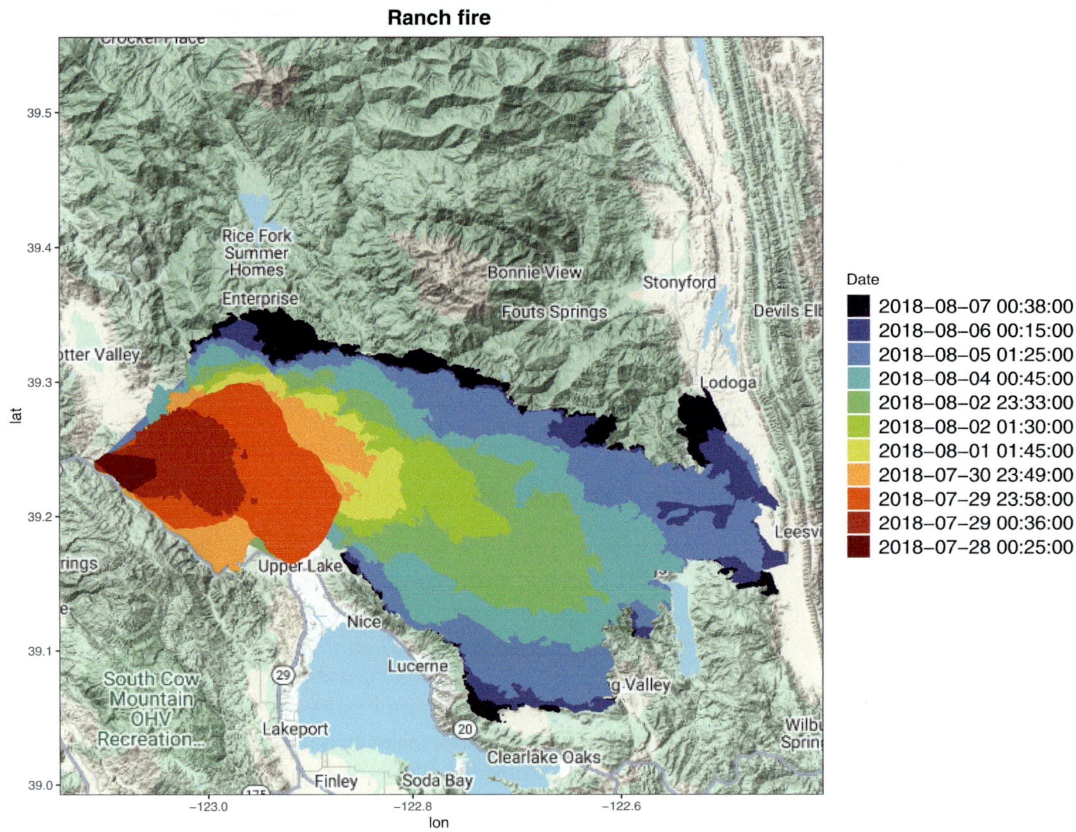

FIGURE 15.2
Ranch Fire progression from July 28, 2018 to August 10, 2018.

To more formally evaluate the forecasting performance, we use the threat score [TS; Wikle et al., 2019]:

$$\mathrm{TS} = \frac{\mathrm{A}_{11}}{\mathrm{A}_{11} + \mathrm{A}_{10} + \mathrm{A}_{01}}, \tag{15.18}$$

where A_{11} represents the area where the model correctly predicts an event that occurs, A_{10} corresponds to the area where the model predicts an event occurrence that does not occur, and A_{01} denotes the area where an event occurs but is not predicted by the model. The TS summary measure ranges from 0 to 1, with a value of 1 indicating a perfect forecast. As our primary focus is on the fire front, represented by the signed distance function $\phi(\boldsymbol{s}, t) = 0$, we consider the event to have occurred if $\phi(\boldsymbol{s}, t) \leq 0$. The TS summary measure, averaged over time with the hyperparameters specified in equation (15.18), is 0.794.

We also check the quality of the forecast intervals using the Interval Score [IS; Gneiting and Raftery, 2012]:

$$\mathrm{IS}_\alpha(\ell_t(\boldsymbol{s}), u_t(\boldsymbol{s}), \phi_t(\boldsymbol{s})) = \\ (u_t(\boldsymbol{s}) - \ell_t(\boldsymbol{s})) + \frac{2}{\alpha}(\ell_t(\boldsymbol{s}) - \phi_t(\boldsymbol{s}))I(\phi_t(\boldsymbol{s}) < \ell_t(\boldsymbol{s})) + \frac{2}{\alpha}(\phi_t(\boldsymbol{s}) - u_t(\boldsymbol{s}))I(\phi_t(\boldsymbol{s}) > u_t(\boldsymbol{s})), \tag{15.19}$$

where $\ell_t(\boldsymbol{s})$, $u_t(\boldsymbol{s})$, α, and $\phi_t(\boldsymbol{s})$ denote the lower bound, upper bound, significance level, and true signed distance function at $\boldsymbol{s}$ and time t, respectively. The IS is given by the

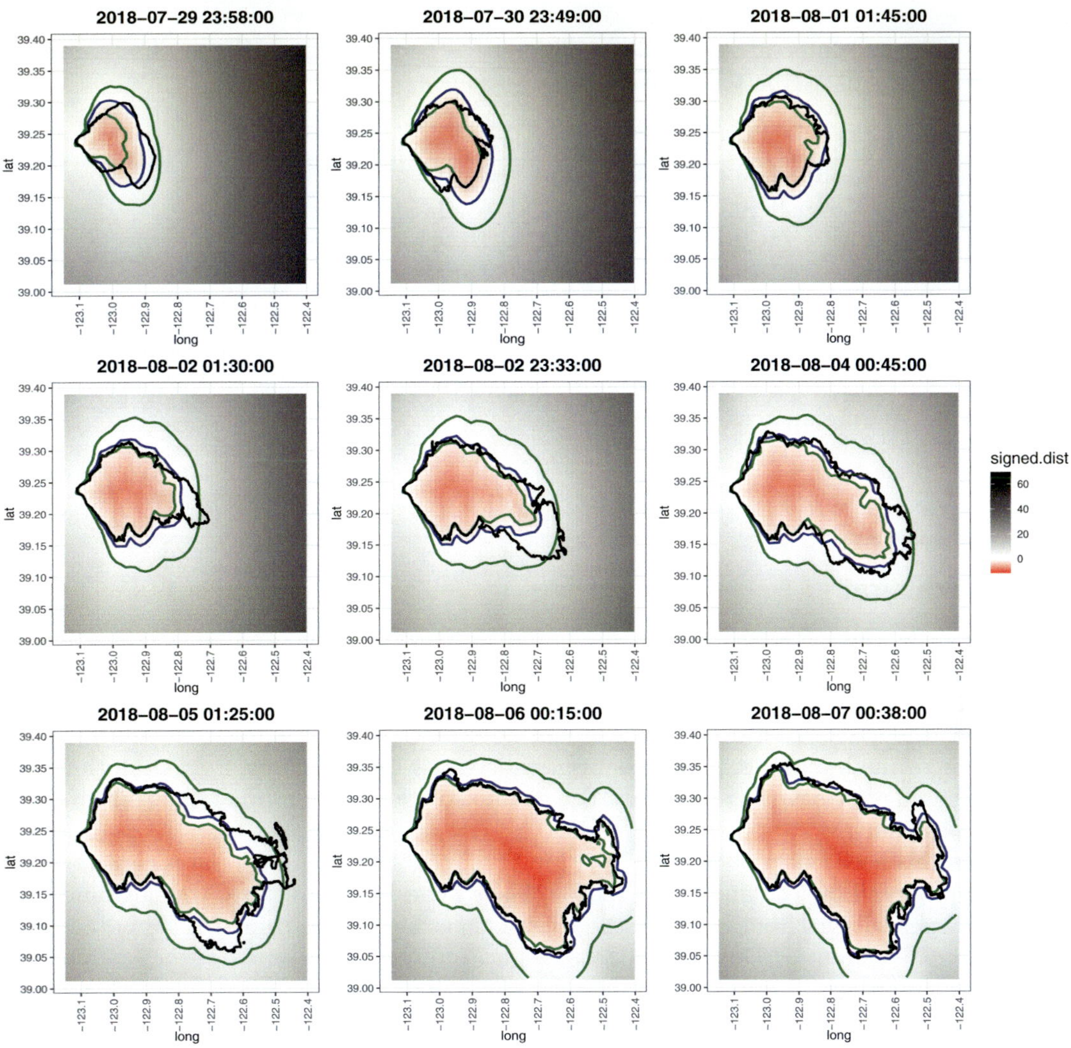

FIGURE 15.3
The comparison between the wildfire forecasts and the true fire front. Black and blue solid lines indicate the true fire front and the forecast mean, respectively. Green solid lines represent the 95% lower and upper bounds for the forecasts.

interval length $(u_t(\boldsymbol{s}) - \ell_t(\boldsymbol{s}))$ with a penalty term, $\frac{2}{\alpha}(\ell_t(\boldsymbol{s}) - \phi_t(\boldsymbol{s}))$ or $\frac{2}{\alpha}(\phi_t(\boldsymbol{s}) - u_t(\boldsymbol{s}))$, introduced when the forecast interval does not contain $\phi_t(\boldsymbol{s})$. A smaller IS indicates more accurate forecast intervals. We compute the point-wise IS only at locations within the true fire front and average it over space:

$$\mathrm{IS}(t) = \frac{\sum_{\boldsymbol{s}:\phi_t(\boldsymbol{s})\leq 0} \mathrm{IS}_\alpha(\ell_t(\boldsymbol{s}), u_t(\boldsymbol{s}), \phi_t(\boldsymbol{s}))}{|\{\boldsymbol{s} : \phi_t(\boldsymbol{s}) \leq 0\}|}, \tag{15.20}$$

where $|\cdot|$ indicates the cardinality of the set. The overall IS is then obtained by averaging $\mathrm{IS}(t)$ over time.

To examine the sensitivity to the hyperparameters, we evaluate forecasting performance using TS and IS for different values of a_w, a_u, π_w, π_u and distributions for J and τ. Table 15.1 presents the TS and IS for different hyperparameters and their distributions, with the

TABLE 15.1
Threat score (TS) and Interval score (IS) for different hyperparameters. Higher TS scores and lower IS scores indicate better forecasting performance.

$p(J)$	$p(\tau)$	a_w	a_u	π_w	π_u	TS	IS
DU(10,30)	U(0.1,30)	0.3	0.3	0.3	0.5	0.794	4.873
DU(50,80)	U(0.1,30)	0.3	0.3	0.3	0.5	0.789	5.179
DU(10,50)	U(0.1,10)	0.3	0.3	0.3	0.3	0.784	5.156
DU(50,80)	U(0.1,10)	0.3	0.3	0.3	0.3	0.790	5.420

best scores obtained using the hyperparameters in equation (15.17). Notably, a model with fewer parameters (i.e., a weak learner) performs better, though it remains relatively robust to the choice of hyperparameters.

15.5 Conclusion

The devastating effects of wildfires extend beyond property damage and economic loss, potentially impacting mental health and well-being [Eisenman and Galway, 2022]. While wildfire behavior remains uncertain, numerous studies have attempted to model it, employing approaches ranging from simulations and data-driven methods to neural network models. Although the level set method has been widely used to model fire front evolution, most models based on it have not been data-driven and so are less appropriate for real-time wildfire management.

To address this need, we propose an algorithm that recursively updates the parameters in the model proposed by Yoo and Wikle [2023]. Their hybrid model utilizes the flexibility of an ESN to infer wildfire front speed in the level set method. Then, using the Woodbury matrix identity [Woodbury, 1950], the proposed algorithm recursively updates the readout matrix in the ESN as new data becomes available. We evaluate the forecasting performance and demonstrate the feasibility of our algorithm for real-time forecasting on the infamous Ranch Fire. Notably, our results emphasize the importance of forecast intervals. While a weak learner is suitable in our application, we emphasize that a large reservoir may be beneficial for capturing complex dynamics, such as those in the Lorenz system. In such cases, the advantage of the online learning algorithm would become more apparent.

Building an accurate, efficient, and data-driven forecasting model that recursively updates its parameters can be challenging. One of the promising approaches is to integrate the underlying mechanistic laws into the forecasting model, as seen in the use of the level set method for modeling wildfire propagation. While equations (15.1), (15.2), and (15.5) provide a baseline model for fire front evolution, incorporating a curvature-dependent term or an external velocity field could improve its flexibility [Osher and Fedkiw, 2002], e.g.,

$$\frac{\partial \phi(\boldsymbol{s},t)}{\partial t} + v_n(\boldsymbol{s},t) \cdot \|\nabla \phi(\boldsymbol{s},t)\| + \boldsymbol{v}(\boldsymbol{s},t) \cdot \nabla \phi(\boldsymbol{s},t) = v_c(\boldsymbol{s},t) \cdot \|\nabla \phi(\boldsymbol{s},t)\|, \tag{15.21}$$

where $\boldsymbol{v}(\boldsymbol{s},t)$ and $v_c(\boldsymbol{s},t)$ represent the external velocity field and the curvature of the fire front, respectively. Equation (15.21) can be further simplified using the signed distance function property $\|\nabla \phi(\boldsymbol{s},t)\| = 1$. Therefore, one potential extension of our model and algorithm is to incorporate equation (15.21) into the response model.

Another extension is to incorporate covariates into the forecasting model, which may improve its performance by capturing additional factors influencing wildfire spread such as weather, fuels, and topography [Agee, 1996]. One straightforward approach to incorporating these factors is through the hidden states $\boldsymbol{h}_t$. For example, one can define $\tilde{\boldsymbol{x}}_t = (\boldsymbol{\phi}_t^\top, \boldsymbol{\phi}_{t-\triangle t}^\top, \boldsymbol{z}_t^\top)^\top$ where $\boldsymbol{z}_t$ denotes the covariates. However, Yoo and Wikle [2023] found this approach to be ineffective their examples, suggesting that an alternative integration strategy may be helpful.

Last, alternative loss functions for the ESN could capture the rapid wild fire evolution. The squared error loss function corresponds to maximizing the likelihood of a Gaussian model for the residuals, which may be inadequate for modeling processes with heavy tails, skewness, or abrupt changes. For example, the spread rate of the Ranch Fire increases rapidly between August 2 and August 4, 2018, a pattern not observed in the earlier stages of the fire. A loss function based on a heavy-tailed distribution (e.g., Cauchy loss) may be more suitable. Such a loss may not allow for a closed-form solution as in equation (15.6), preventing the application of the same recursive updating algorithm as used here, and hence is the subject of future research.

Bibliography

F. Abid. A survey of machine learning algorithms based forest fires prediction and detection systems. *Fire Technology*, 57(2):559–590, 2021. doi: 10.1007/s10694-020-01056-z.

J. K. Agee. The influence of forest structure on fire behavior. In *the 17th Annual Forest Vegetation Management Conference*, 1996.

F. A. Albini. Wildland fires: Predicting the behavior of wildland fires—among nature's most potent forces—can save lives, money, and natural resources. *American Scientist*, 72 (6):590–597, 1984.

A. Alessandri, P. Bagnerini, M. Gaggero, and L. Mantelli. Parameter estimation of fire propagation models using level set methods. *Applied Mathematical Modelling*, 92:731–747, 2021. `https://doi.org/10.1016/j.apm.2020.11.030`.

Y.-T. Bai, W. Jia, X.-B. Jin, T.-L. Su, J.-L. Kong, and Z.-G. Shi. Nonstationary time series prediction based on deep echo state network tuned by Bayesian optimization. *Mathematics*, 11(6), 2023. doi: 10.3390/math11061503.

J. Baptiste Filippi, F. Bosseur, C. Mari, C. Lac, P. Le Moigne, B. Cuenot, D. Veynante, D. Cariolle, and J.-H. Balbi. Coupled atmosphere-wildland fire modelling. *Journal of Advances in Modeling Earth Systems*, 1(4), 2009. doi: https://doi.org/10.3894/JAMES.2009.1.11.

J. R. Bergado, C. Persello, K. Reinke, and A. Stein. Predicting wildfire burns from big geodata using deep learning. *Safety Science*, 140:105276, 2021. doi: https://doi.org/10.1016/j.ssci.2021.105276.

M. Bonas and S. Castruccio. Calibration of spatial forecasts from citizen science urban air pollution data with sparse recurrent neural networks. *arXiv preprint arXiv:2105.02971*, 2021.

M. Bonas, C. K. Wikle, and S. Castruccio. Calibrated forecasts of quasi-periodic climate processes with deep echo state networks and penalized quantile regression. *Environmetrics*, 35(1):e2833, 2024. doi: https://doi.org/10.1002/env.2833.

T. Brown, S. Leach, B. Wachter, and B. Gardunio. The extreme 2018 northern California fire season. *Bulletin of the American Meteorological Society*, 101(1):S1–S4, 2020. doi: 10.1175/BAMS-D-19-0275.1.

M. Carreras-Sospedra, S. Zhu, M. MacKinnon, W. Lassman, J. D. Mirocha, M. Barbato, and D. Dabdub. Air quality and health impacts of the 2020 wildfires in California. *Fire Ecology*, 20(1):6, 2024.

S. P. Chatzis and Y. Demiris. Echo state Gaussian process. *IEEE Transactions on Neural Networks*, 22(9):1435–1445, 2011. doi: 10.1109/TNN.2011.2162109.

S. Chen and S. Billings. Neural networks for nonlinear dynamic system modelling and identification. *International Journal of Control*, 56(2):319–346, 1992. doi: https://doi.org/10.1080/00207179208934317.

X. Chen, L. Jin, and S. Li. An online learning strategy for echo state network. *IEEE Transactions on Systems, Man, and Cybernetics: Systems*, 54(1):644–655, 2024. doi: 10.1109/TSMC.2023.3319357.

T. Clark, J. Coen, and D. Latham. Description of a coupled atmosphere–fire model. *International Journal of Wildland Fire*, 13:49–63, 2004. `https://doi.org/10.1071/WF03043`.

N. Cressie and C. K. Wikle. *Statistics for spatio-temporal data.* John Wiley & Sons, 2011.

J. J. Dabrowski, C. Huston, J. Hilton, S. Mangeon, and P. Kuhnert. Towards data assimilation in level-set wildfire models using bayesian filtering. *arXiv preprint arXiv:2206.08501*, 2022.

J. J. Dabrowski, D. E. Pagendam, J. Hilton, C. Sanderson, D. MacKinlay, C. Huston, A. Bolt, and P. Kuhnert. Bayesian physics informed neural networks for data assimilation and spatio-temporal modelling of wildfires. *Spatial Statistics*, 55:100746, 2023. doi: https://doi.org/10.1016/j.spasta.2023.100746.

N. Doan, W. Polifke, and L. Magri. Physics-informed echo state networks. *Journal of Computational Science*, 47:101237, 2020. doi: https://doi.org/10.1016/j.jocs.2020.101237.

K. Doya. Bifurcations in the learning of recurrent neural networks. In *[Proceedings] 1992 IEEE International Symposium on Circuits and Systems*, volume 6, pages 2777–2780 vol.6, 1992. doi: 10.1109/ISCAS.1992.230622.

D. P. Eisenman and L. P. Galway. The mental health and well-being effects of wildfire smoke: a scoping review. *BMC Public Health*, 22(1):2274, 2022. doi: 10.1186/s12889-022-14662-z.

F. Fendell and M. Wolff. Wind-aided fire spread. In *Forest Fires: Behavior and Ecological Effects*, pages 171–223. Academic Press, 2001. `10.1016/B978-012386660-8/50008-8`.

M. A. Finney. FARSITE: Fire area simulator-model development and evaluation. *Res. Pap. RMRS-RP-4, Revised 2004, Ogden, UT: U.S. Department of Agriculture, Forest Service, Rocky Mountain Research Station*, 2004.

M. A. Finney and K. C. Ryan. Use of the FARSITE fire growth model for fire prediction in us national parks. In *Proceedings of the international emergency management and engineering conference*, pages 183–189. Citeseer, 1995.

Geospatial Multi-Agency Coordination Group. Historic perimeters combined 2000-2018 geomac. *Geospatial Multi-Agency Coordination Group*, 2019.

T. Gneiting and A. Raftery. Strictly proper scoring rules, prediction, and estimation. *Journal of the American Statistical Association*, 102:359–378, 2012. https://doi.org/10.1198/016214506000001437.

D. Goswami. Sequential learning from noisy data: Data-assimilation meets echo-state network. *arXiv preprint arXiv:2304.00198*, 2023.

N. Grieshop and C. K. Wikle. Data-driven modeling of wildfire spread with stochastic cellular automata and latent spatio-temporal dynamics. *Spatial Statistics*, 59:100794, 2024a. doi: https://doi.org/10.1016/j.spasta.2023.100794.

N. Grieshop and C. K. Wikle. Echo state network-enhanced symbolic regression for spatio-temporal binary stochastic cellular automata. *Spatial Statistics*, 60:100827, 2024b. doi: https://doi.org/10.1016/j.spasta.2024.100827.

A. G. Hart, J. L. Hook, and J. H. Dawes. Echo state networks trained by Tikhonov least squares are $L^2(\mu)$ approximators of ergodic dynamical systems. *Physica D: Nonlinear Phenomena*, 421:132882, 2021. doi: https://doi.org/10.1016/j.physd.2021.132882.

S. Hochreiter and J. Schmidhuber. Long short-term memory. *Neural Computation*, 9(8): 1735–1780, 1997.

R. Holliday, L. S. Krishnamurti, S. E. Jordan, M. A. Sia, L. A. Brenner, and L. L. Monteith. The health and social impacts of the Maui wildfires: Post-disaster care from a sociocultural lens. *Hawai'i Journal of Health & Social Welfare*, 83(3):85–87, 2024.

Y. Jafari Goldarag, A. Mohammadzadeh, and A. S. Ardakani. Fire risk assessment using neural network and logistic regression. *Journal of the Indian Society of Remote Sensing*, 44(6):885 – 894, 2016. doi: 10.1007/s12524-016-0557-6.

L. C. Jain, M. Seera, C. P. Lim, and P. Balasubramaniam. A review of online learning in supervised neural networks. *Neural Computing and Applications*, 25(3):491–509, 2014.

T. Kim and B. R. King. Time series prediction using deep echo state networks. *Neural Computing and Applications*, 32(23):17769–17787, 2020. doi: 10.1007/s00521-020-04948-x.

J. Koh, F. Pimont, J.-L. Dupuy, and T. Opitz. Spatiotemporal wildfire modeling through point processes with moderate and extreme marks. *The Annals of Applied Statistics*, 17 (1):560 – 582, 2023. doi: 10.1214/22-AOAS1642.

M. Lukoševičious. A practical guide to applying echo state networks. In *Neural Networks: Tricks of the Trade*, pages 659–686. Springer, 2012. https://doi.org/10.1007/978-3-642-35289-8_36.

J. R. Maat, N. Gianniotis, and P. Protopapas. Efficient optimization of echo state networks for time series datasets. In *2018 International Joint Conference on Neural Networks (IJCNN)*, pages 1–7, 2018. doi: 10.1109/IJCNN.2018.8489094.

V. Mallet, D. E. Keyes, and F. E. Fendell. Modeling wildland fire propagation with level set methods. *Computers and Mathematics with Applications*, 57(7):1089–1101, 2009. https://doi.org/10.1016/j.camwa.2008.10.089.

F. Matzner. Hyperparameter tuning in echo state networks. In *Proceedings of the Genetic and Evolutionary Computation Conference*, GECCO '22, pages 404–412, New York, NY, USA, 2022. Association for Computing Machinery. ISBN 9781450392372. doi: 10.1145/3512290.3528721.

P. L. McDermott and C. Wikle. An ensemble quadratic echo state network for non-linear spatio-temporal forecasting. *Stat*, 6:315–330, 2017. https://doi.org/10.1002/sta4.160.

P. McDermott and C. K. Wikle. Deep echo state networks with uncertainty quantification for spatio-temporal forecasting. *Environmetrics*, 30:e2553, 2019. https://doi.org/10.1002/env.2553.

T. Opitz, F. Bonneu, and E. Gabriel. Point-process based Bayesian modeling of space–time structures of forest fire occurrences in Mediterranean France. *Spatial Statistics*, 40:100429, 2020. doi: https://doi.org/10.1016/j.spasta.2020.100429.

S. Osher and R. Fedkiw. *Level Set Methods and Dynamic Implicit Surfaces*. Springer-Verlag, New York, 2002. https://doi.org/10.1007/b98879.

S. Osher and J. A. Sethian. Fronts propagating with curvature dependent speed: Algorithms based on Hamilton-Jacobi formulations. *Journal of Computational Physics*, 79(1):12–49, 1988. https://doi.org/10.1016/0021-9991(88)90002-2.

E. Pebesma and R. Bivand. *Spatial Data Science: With Applications in R*. Chapman and Hall/CRC, Boca Raton, 2023. doi: 10.1201/9780429459016.

R Core Team. *R: A Language and Environment for Statistical Computing*. R Foundation for Statistical Computing, Vienna, Austria, 2022.

A. Racca and L. Magri. Robust optimization and validation of echo state networks for learning chaotic dynamics. *Neural Networks*, 142:252–268, 2021. doi: https://doi.org/10.1016/j.neunet.2021.05.004.

M. Raissi, P. Perdikaris, and G. Karniadakis. Physics-informed neural networks: A deep learning framework for solving forward and inverse problems involving nonlinear partial differential equations. *Journal of Computational Physics*, 378:686–707, 2019. doi: https://doi.org/10.1016/j.jcp.2018.10.045.

G. Ribeiro, J. Sauer, N. Fraccanabbia, V. Mariani, and L. Coelho. Bayesian optimized echo state network applied to short-term load forecasting. *Energies*, 13:2390, 2020. https://doi.org/10.3390/en13092390.

M. Rochoux, B. Delmotte, B. Cuenot, S. Ricci, and A. Trouvé. Regional-scale simulations of wildland fire spread informed by real-time flame front observations. *Proceedings of the Combustion Institute*, pages 2641–2647, 2013. https://doi.org/10.1016/j.proci.2012.06.090.

R. C. Rothermel. A mathematical model for predicting fire spread in wildland fuels. *Res. Pap. INT-115. Ogden, UT: U.S. Department of Agriculture, Intermountain Forest and Range Experiment Station*, 1972.

S. T. Seydi, V. Saeidi, B. Kalantar, N. Ueda, and A. A. Halin. Fire-net: A deep learning framework for active forest fire detection. *Journal of Sensors*, 2022(1):8044390, 2022. doi: https://doi.org/10.1155/2022/8044390.

C. Sheng, J. Zhao, W. Wang, and H. Leung. Prediction intervals for a noisy nonlinear time series based on a bootstrapping reservoir computing network ensemble. *IEEE Transactions on Neural Networks and Learning Systems*, 24(7):1036–1048, 2013. 10.1109/TNNLS.2013.2250299.

T. Srivas, T. Artés, R. de Callafon, and I. Altintas. Wildfire spread prediction and assimilation for FARSITE using ensemble Kalman filtering. *Procedia Computer Science*, 80: 897–908, 2016. `https://doi.org/10.1016/j.procs.2016.05.328`.

C. Sun, M. Song, D. Cai, B. Zhang, S. Hong, and H. Li. A systematic review of echo state networks from design to application. *IEEE Transactions on Artificial Intelligence*, 5(1): 23–37, 2024. doi: 10.1109/TAI.2022.3225780.

S. W. Taylor, D. G. Woolford, C. B. Dean, and D. L. Martell. Wildfire Prediction to Inform Fire Management: Statistical Science Challenges. *Statistical Science*, 28(4):586 – 615, 2013. doi: 10.1214/13-STS451.

M. P. Thompson and D. E. Calkin. Uncertainty and risk in wildland fire management: A review. *Journal of Environmental Management*, 92(8):1895–1909, 2011. doi: https://doi.org/10.1016/j.jenvman.2011.03.015.

V. Vovk, A. Gammerman, and G. Shafer. *Algorithmic learning in a random world*, volume 29. Springer, 2005.

C. E. V. Wagner. A simple fire-growth model. *The Forestry Chronicle*, 45(2):103–104, 1969. doi: 10.5558/tfc45103-2.

Q. Wang, P. A. Parker, and R. B. Lund. Hierarchical count echo state network models with application to graduate student enrollments. *arXiv:2501.14698*, 2025.

Z. Wang, Y.-R. Zeng, S. Wang, and L. Wang. Optimizing echo state network with backtracking search optimization algorithm for time series forecasting. *Engineering Applications of Artificial Intelligence*, 81:117–132, 2019. doi: https://doi.org/10.1016/j.engappai.2019.02.009.

Z. Wang, S. H. Holan, and C. K. Wikle. Echo state networks for spatio-temporal area-level data. *arXiv preprint arXiv:2410.10641*, 2024.

P. Werbos. Backpropagation through time: what it does and how to do it. *Proceedings of the IEEE*, 78(10):1550–1560, 1990. doi: 10.1109/5.58337.

C. K. Wikle and L. M. Berliner. A Bayesian tutorial for data assimilation. *Physica D: Nonlinear Phenomena*, 230(1):1–16, 2007. doi: https://doi.org/10.1016/j.physd.2006.09.017.

C. K. Wikle and M. B. Hooten. A general science-based framework for dynamical spatio-temporal models. *Test*, 19:417–451, 2010.

C. K. Wikle, A. Zammit-Mangion, and N. Cressie. *Spatio-Temporal Statistics with R*. Chapman & Hall/CRC, Florida, 2019.

M. A. Woodbury. Inverting modified matrices. In *Memorandum Rept. 42, Statistical Research Group*, page 4. Princeton Univ., 1950.

W. Yao, Z. Zeng, C. Lian, and H. Tang. Ensembles of echo state networks for time series prediction. In *2013 Sixth International Conference on Advanced Computational Intelligence (ICACI)*, pages 299–304, 2013. `10.1109/ICACI.2013.6748520`.

M. Yoo and C. K. Wikle. Using echo state networks to inform physical models for fire front propagation. *Spatial Statistics*, 54:100732, 2023. doi: https://doi.org/10.1016/j.spasta.2023.100732.

M. Yoo and C. K. Wikle. A Bayesian spatio-temporal level set dynamic model and application to fire front propagation. *The Annals of Applied Statistics*, 18(1):404–423, 2024. doi: 10.1214/23-AOAS1794.

B. Zhang, D. J. Miller, and Y. Wang. Nonlinear system modeling with random matrices: Echo state networks revisited. *IEEE Transactions on Neural Networks and Learning Systems*, 23(1):175–182, 2012. 10.1109/TNNLS.2011.2178562.

16

Spatial Meta-Analysis for Finite Populations

Sudipto Banerjee

Abstract

This chapter presents a framework for Bayesian inference for a finite population of units by assimilating information from different studies. The chapter specifically attends to settings where the units in the finite population are spatially referenced and the values are posited to be spatially associated. The framework discussed here introduces spatial dependence by treating the units in the finite population as a partial realisation of a spatial process. Dependence is modelled using spatial covariance functions. Instead of focusing on the richest possible classes of hierarchical models, the chapter offers insights into some of the accessible probability distributions in conjugate Bayesian linear regression models and adapts them to carry out predictive inference for finite populations. The chapter concludes with some discussions on possible future investigations in this area including the adoption of emerging machine learning methods for pooling and assimilating information across different studies.

16.1 Introduction

The analysis of geographically referenced data plays a key role in understanding environmental processes. The fundamental concept underlying statistical inference for environmental data is a *stochastic process* (see, e.g., Gneiting and Guttorp, 2010). A stochastic process is a (possibly uncountable) collection of random variables indexed over space and/or time endowed with a probability law. Spatial processes refer to stochastic processes defined as an uncountable set $\{w(s) : s \in \mathcal{D}\}$, where s denotes a spatial location in a region $\mathcal{D}$ of interest. The location can be referenced as spherical or geographic coordinates (latitude and longitude) or as Euclidean coordinates from a map projection. For example, $w(s)$ may represent the level of temperature at s. Conceptually, the temperature exists throughout the domain. In practice, temperature measurements will be available to us only from a finite set of locations. We envision temperature as a spatial process and the data as a *partial realisation* of that process.

In this chapter, I will focus on situations where inference is sought for a finite population of spatially referenced variables. Common examples include data obtained from sample

DOI: 10.1201/9781003564423-16

surveys. Surveys are conducted to estimate scientific variables, such as biomass from a population of trees or the amount of contaminants in groundwater from the population of wells in a state. The inferential objectives for environmental survey data are somewhat different from those in customary geostatistics where we aim to learn about the underlying environmental process from realised data. In finite population contexts, we seek to estimate quantities associated with the finite population while acknowledging that the units in the finite population are themselves a partial realisation of the underlying process. Ver Hoef (2002) provides an excellent account of classical design-based sampling and geostatistical modelling for estimation of population quantities.

Finite population survey sampling has an extensive literature that is beyond the scope of a single article (see classic texts such as Cochran, 1977; Kish, 1965, emphasising design-based approaches). Bayesian inference for finite population survey sampling has been discussed from varied perspectives, for example, in Ericson (1969), Rao and Ghangurde (1972), Arora et al. (1997), Ghosh and Meeden (1997), Little and Rubin (2002), Gelman (2007), Gelman et al. (2014), and Banerjee (2024), while an assessment of classical and Bayesian approaches is offered by Rao (2011). Little (2004) offers insight into inference for finite populations in an illuminating broader discussion of model-based and design-based approaches in survey sampling.

In this chapter, I will focus on spatial meta-analysis for a finite population. The Bayesian approach to statistical inference quantifies uncertainty for unknown quantities of interest conditional on known (or seen) quantities. If $\mathcal{U}$ represents unknowns and $\mathcal{K}$ represents what is observed or known, inference proceeds from the posterior distribution $[\mathcal{U} \mid \mathcal{K}] \propto [\mathcal{U}, \mathcal{K}]$. Probabilistic inference specifies $[\mathcal{K} \mid \mathcal{U}]$ (the likelihood) and $[\mathcal{U}]$ (the prior). Bayesian data analysis from finite populations is appealing and conceptually simple. Here, $\mathcal{K}$ represents the sampled data and $\mathcal{U}$ represents all unknowns. The unknowns include potential values of measurements from population units not sampled and parameters in the probability law for $[\mathcal{U}]$.

The sheer volume of original research in this domain is massive and I do not intend to undertake a comprehensive review of Bayesian methods for survey sampling here with this article (see, e.g., Gelman et al., 2014; Kalton, 2019; Little and Rubin, 2002; Little, 1982; Tang et al., 2003, and references therein). Rather, I develop a Bayesian modelling framework that assimilates inference from different studies analysing samples from a finite population of interest whose units are appropriately modelled as a partial realisation of a continuous spatial process. I will review some preliminary distributions useful for Bayesian meta-analysis for finite populations. This is extended to an augmented Bayesian linear regression model that produces posterior inference from standard distributions for meta-analysis. Further insights are offered into how posterior inference from different samples can be assimilated into more general spatial meta-analysis frameworks.

16.2 Bayesian Hierarchical Modelling for the Finite Population Mean

Let $\mathcal{N} = \{1, 2, \ldots, N\}$ be a finite set of integers labelling the population units. Given a sample $\{i_1, i_2, \ldots, i_n\} \subset \mathcal{N}$ of size n, units in the population and sample are described by $\{(Y_t, Z_t) : t \in \mathcal{N}\}$, where Y_t is the measurement from the population unit labelled t and Z_t is a binary variable indicating if unit t is in the sample or not. The binary vector $Z = (Z_1, Z_2, \ldots, Z_N)^{\mathrm{T}}$ completely encodes the sampled units in the population and, analogously, $Y = (Y_1, Y_2, \ldots, Y_N)^{\mathrm{T}}$ denotes the values of all units in the finite population and $\mathcal{D}$ denotes the sampled data. Model-based inference (likelihood or Bayesian) assigns a probability law to the finite collection (Y, Z), which defines the sampling design.

In simple random sampling without replacement (SRSWOR) from a population of N units, $p(Y, Z) = p(Y)p(Z)$ and $p(Z) = 1/\binom{N}{n}$ whenever $\sum_{i=1}^{N} Z_i = n$ and 0 otherwise. The "prior" distribution for Y induces the distribution of the sampled data $p(\mathcal{D} \mid Y, Z)$. If the probability model assumes exchangeable population units, then we can reorder the labels to denote the sampled data $\mathcal{D}$ by $y = (y_1, y_2, \ldots, y_n)^{\mathrm{T}}$ and the unknown values from the remaining population units that are not sampled as $Y_u = (Y_{n+1}, \ldots, Y_N)^{\mathrm{T}}$.

The following presents a simple hierarchical model for population units:

$$Y_i \overset{iid}{\sim} N(\mu, \sigma^2),\ i = 1, 2, \ldots, N; \quad \mu \sim N(\theta, \sigma^2/n_0); \quad \sigma^2 \sim IG(a_0, b_0), \tag{16.1}$$

where θ, n_0, a_0, and b_0 are known scalars. The joint posterior distribution is

$$p(\mu, \sigma^2 \mid y) \propto \underbrace{IG(\sigma^2 \mid a_1, b_1)}_{p(\sigma^2 \mid y)} \times \underbrace{N\left(\mu \,\middle|\, \frac{n_0\theta + n\bar{y}}{n_0 + n}, \frac{\sigma^2}{n_0 + n}\right)}_{p(\mu \mid \sigma^2, y)},$$

where $\bar{y} = \sum_{i=1}^{n} y_i/n$ is the sample mean, $a_1 = a_0 + \frac{n}{2}$ and $b_1 = b_0 + \frac{f(\theta,\bar{y})}{2}$, $f(\theta, \bar{y}) = (n-1)s^2 + \frac{n_0 n}{n_0+n}(\bar{y} - \theta)^2$ and $s^2 = \sum_{i=1}^{n}(y_i - \bar{y})^2/(n-1)$ is the sample variance.

The model in (16.1) is represented as a linear model of the population units,

$$\begin{bmatrix} y \\ Y_u \end{bmatrix} = \begin{bmatrix} 1_n \\ 1_{N-n} \end{bmatrix} (\theta + \omega) + \begin{bmatrix} \epsilon_{1:n} \\ \epsilon_{(n+1):N} \end{bmatrix}, \quad \begin{bmatrix} \epsilon_{1:n} \\ \epsilon_{(n+1):N} \end{bmatrix} \sim N\left(\begin{bmatrix} 0_{n\times 1} \\ 0_{(N-n)\times 1} \end{bmatrix}, \sigma^2 \begin{bmatrix} I_n & O \\ O & I_{N-n} \end{bmatrix}\right)$$

where μ in (16.1) is substituted by the linear model $\mu = \theta + \omega$, $\omega \sim N(0, \sigma^2/n_0)$. The joint distribution is derived from the above model as

$$\begin{bmatrix} y \\ Y_u \end{bmatrix} \,\middle|\, \sigma^2 \sim N\left(\begin{bmatrix} 1_n \\ 1_{N-n} \end{bmatrix} \theta, \frac{\sigma^2}{n_0} \begin{bmatrix} 1_n 1_n^{\mathrm{T}} + n_0 I_n & 1_n 1_{N-n}^{\mathrm{T}} \\ 1_{N-n} 1_n^{\mathrm{T}} & 1_{N-n} 1_{N-n}^{\mathrm{T}} + n_0 I_{N-n} \end{bmatrix}\right). \tag{16.2}$$

The posterior distribution for the missing units conditional on σ^2 is derived from (16.2) as

$$p(Y_u \mid \sigma^2, y) = N\left(Y_u \,\middle|\, \frac{n_0\theta + n\bar{y}}{n + n_0} 1_{N-n}\,, \sigma^2\left(\frac{1}{n + n_0} 1_{N-n} 1_{N-n}^{\mathrm{T}} + \frac{1}{n} I_{N-n}\right)\right) \tag{16.3}$$

The joint density of the sampled values $\mathcal{D} = y$ and the population mean $\bar{Y}$ is

$$\begin{bmatrix} y \\ \bar{Y} \end{bmatrix} \,\middle|\, \sigma^2 \sim N_{n+1}\left(\begin{bmatrix} 1_n \\ 1 \end{bmatrix} \theta, \frac{\sigma^2}{n_0} \begin{bmatrix} 1_n 1_n^{\mathrm{T}} + n_0 I_n & (1 + f_0) 1_n \\ (1 + f_0) 1_n^{\mathrm{T}} & (1 + f_0) \end{bmatrix}\right), \tag{16.4}$$

where $f = n/N$ and $f_0 = n_0/N$. The likelihood $p(y \mid \bar{Y}, \sigma^2)$ derived from (16.4)

$$\begin{aligned} p(y \mid \bar{Y}, \sigma^2) &\propto \frac{1}{(\sigma^2)^{n/2}} \exp\left(-\frac{1}{2\sigma^2} (y - 1_n\bar{Y})^{\mathrm{T}} \left(I_n + \frac{1_n 1_n^{\mathrm{T}}}{N - n}\right) (y - 1_n\bar{Y})\right) \\ &\propto \frac{1}{(\sigma^2)^{n/2}} \exp\left(-\frac{1}{2\sigma^2}\left((n-1)s^2 + \frac{n}{1-f}(\bar{y} - \bar{Y})^2\right)\right) \\ &\propto \left(\frac{1}{\sigma^2}\right)^{\frac{n-1}{2}} \exp\left(-\frac{(n-1)s^2}{2\sigma^2}\right) \times \left(\frac{1}{\sigma^2}\right)^{\frac{1}{2}} \exp\left(-\frac{n(\bar{y} - \bar{Y})^2}{2\sigma^2(1-f)}\right). \end{aligned} \tag{16.5}$$

Therefore, the density $p(y \mid \bar{Y}, \sigma^2)$ is a function of only the mean of the sample and the variance of the sample. We have shown that the sample mean and sample variance contain "sufficient" information to describe the distribution of the finite sample given the finite population mean and the "super-population" variance parameter σ^2.

The posterior distribution of the population mean, $p(\bar{Y} \mid \sigma^2, y)$, is derived from (16.4) as

$$\bar{Y} \mid \sigma^2, y \sim N\left(f\bar{y} + (1-f)\frac{n_0\theta + n\bar{y}}{n_0 + n}, \sigma^2\frac{(1-f)(1+f_0)}{n+n_0}\right) . \tag{16.6}$$

If the prior distribution of μ becomes vague, i.e., $n_0 \to 0$ in (16.1), then the posterior mean and variance in (16.6) are $\bar{y}$ and $(1-f)\sigma^2/n$, respectively. As $n_0 \to 0$ and $a_0 \to -1/2$, which corresponds to a uniform prior on the real line for $(\mu, \log\sigma^2)$, the resulting posterior distribution is $p(\sigma^2 \mid y) \to IG((n-1)/2, (n-1)s^2/2)$ and the marginal posterior distribution for the population mean is $\left[\frac{\bar{Y}-\bar{y}}{s\sqrt{(1-f)/n}} \,\middle|\, y\right] \sim T_{n-1}$, where T_{n-1} denotes the Student's t distribution with $n-1$ degrees of freedom. This, numerically, reproduces inference from the classical theory of design-based estimators while retaining the appealing interpretation of direct probability statements about the population mean without recourse to the specialised Central Limit Theorems for SRSWOR sampling from finite populations. We also do not need to work with the density of the T_{n-1} distribution. Instead, we draw samples of $\sigma^2 \sim IG(a_1, b_1)$ and then draw a value of $\bar{Y}$ from $p(\bar{Y} \mid \sigma^2, y)$ given in (16.6) for each drawn value of σ^2. This produces samples from $p(\sigma^2, \bar{Y} \mid y)$.

16.3 Bayesian Linear Models for Finite Populations

We introduce covariates in (16.1) and construct a linear regression model

$$\begin{aligned} Y_i \mid \beta, \sigma^2 &\overset{ind}{\sim} N(x_i^{\mathrm{T}}\beta, \sigma^2);\; i = 1, 2, \ldots, N; \\ \beta \mid \sigma^2 &\sim N(M_0 m_0, \sigma^2 M_0); \quad \sigma^2 \sim IG(a_0, b_0) , \end{aligned} \tag{16.7}$$

where Y_i is our outcome of interest for unit i in the population, x_i is a $p \times 1$ vector of design variables assumed to be available for all units in the population, β is the corresponding $p \times 1$ vector of slopes and σ^2 is the residual variance. The prior for β is fully specified using a normal distribution with mean $M_0 m_0$ and variance $\sigma^2 M_0$, where M_0 is $p \times p$ positive definite and m_0 is $p \times 1$. There is no loss of generality in this parametric representation: a prior mean μ_β is mapped to $m_0 = M_0^{-1}\mu_\beta$.

The goal in finite population inference is to estimate the functions of the $N \times 1$ vector Y in (16.7) based on the $n \times 1$ vector of samples y. Since Y comprises y and the unknown values Y_u corresponding to the units that were not sampled, we seek to sample from the posterior predictive distribution $p(Y_u \mid y)$. This is distinct from other applications of Bayesian regression in which inference is primarily sought for β and σ^2. It is instructive to derive the required distribution theory using the linear model for y and Y_u corresponding to (16.7),

$$\begin{aligned} \begin{bmatrix} y \\ Y_u \end{bmatrix} = \begin{bmatrix} X_s \\ X_u \end{bmatrix}\beta + \begin{bmatrix} \epsilon_s \\ \epsilon_u \end{bmatrix} ; \quad \begin{bmatrix} \epsilon_s \\ \epsilon_u \end{bmatrix} \sim N\left(\begin{bmatrix} 0 \\ 0 \end{bmatrix}, \sigma^2\begin{bmatrix} I_n & O \\ O & I_{N-n} \end{bmatrix}\right) ; \\ \beta = M_0 m_0 + \eta \,; \quad \eta \sim N(0, \sigma^2 M_0) , \quad \sigma^2 \sim IG(a_0, b_0) . \end{aligned} \tag{16.8}$$

where $y = (y_1, y_2, \ldots, y_n)^{\mathrm{T}}$, X_s is $n \times p$ with rows x_i^{T} corresponding to sampled units, and X_u is $(N-n) \times p$ corresponding to non-sampled units. The posterior density is

$$p(\beta, \sigma^2 \mid y) = \underbrace{IG(\sigma^2 \mid a, b)}_{p(\sigma^2 \mid y)} \times \underbrace{N\left(\beta \mid Mm, \sigma^2 M\right)}_{p(\beta \mid \sigma^2, y)} , \tag{16.9}$$

where $m = m_0 + X_s^{\mathrm{T}}y$, $M^{-1} = M_0^{-1} + X_s^{\mathrm{T}}X_s$, $a = a_0 + \frac{n}{2}$, and $b = b_0 + \frac{1}{2}\left(y^{\mathrm{T}}y + m_0^{\mathrm{T}}M_0m_0 - m^{\mathrm{T}}Mm\right)$. Substituting the prior model for β into (16.8) produces the joint distribution of y and Y_u given σ^2, which corresponds to the linear model

$$\begin{bmatrix} y \\ Y_u \end{bmatrix} = \begin{bmatrix} X_s \\ X_u \end{bmatrix} M_0m_0 + \begin{bmatrix} \epsilon_s^* \\ \epsilon_u^* \end{bmatrix}; \quad \begin{bmatrix} \epsilon_s^* \\ \epsilon_u^* \end{bmatrix} \sim N\left(\begin{bmatrix} 0 \\ 0 \end{bmatrix}, \sigma^2 \begin{bmatrix} X_sM_0X_s^{\mathrm{T}} + I_n & X_sM_0X_u^{\mathrm{T}} \\ X_uM_0X_s^{\mathrm{T}} & X_uM_0X_u^{\mathrm{T}} + I_{N-n} \end{bmatrix}\right),$$

which implies that the conditional density $p(Y_u \mid y, \sigma^2)$ is also Gaussian with

$$\begin{aligned} &\mathbb{E}[Y_u \mid y, \sigma^2] = X_uM_0m_0 + X_uM_0X_s^{\mathrm{T}}\left(X_sM_0X_s^{\mathrm{T}} + I_n\right)^{-1}\left(y - X_sM_0m_0\right) \quad \text{and} \\ &\mathbb{V}(Y_u \mid y, \sigma^2) = \sigma^2\left\{I_{N-n} + X_uM_0X_u^{\mathrm{T}} - X_uM_0X_s^{\mathrm{T}}\left(X_sM_0X_s^{\mathrm{T}} + I_n\right)^{-1}X_sM_0X_u^{\mathrm{T}}\right\}. \end{aligned} \tag{16.10}$$

Further simplifications are possible with the expressions in (16.10). Thus,

$$\begin{aligned} \mathbb{V}(Y_u \mid y, \sigma^2) &= \sigma^2 I_{N-n} + \sigma^2 X_u \underbrace{\left\{M_0 - M_0X_s^{\mathrm{T}}\left(I_n + X_sM_0X_s^{\mathrm{T}}\right)^{-1}X_sM_0\right\}}_{\left(M_0^{-1}+X_s^{\mathrm{T}}X_s\right)^{-1}=M} X_u^{\mathrm{T}} \\ &= \sigma^2\left(I_{N-n} + X_uMX_u^{\mathrm{T}}\right), \end{aligned}$$

where the Woodbury matrix identity recognises the matrix between X_u and X_u^{T} as M. Also,

$$\begin{aligned} \mathbb{E}[Y_u \mid y, \sigma^2] &= X_uM_0\left\{m_0 + X_s^{\mathrm{T}}\underbrace{\left(I_n + X_sM_0X_s^{\mathrm{T}}\right)^{-1}}_{\left(I_n - X_sMX_s^{\mathrm{T}}\right)}\left(y - X_sM_0m_0\right)\right\} \\ &= X_uM_0\left\{m_0 + X_s^{\mathrm{T}}(y - X_sM_0m_0) - \underbrace{X_s^{\mathrm{T}}X_s}_{M^{-1}-M_0^{-1}} MX_s^{\mathrm{T}}(y - X_sM_0m_0)\right\} \\ &= X_uM_0m_0 + X_uMX_s^{\mathrm{T}}(y - X_sM_0m_0) \\ &= X_uM\left\{\underbrace{M^{-1}M_0}_{I_p+X_s^{\mathrm{T}}X_sM_0} m_0 + X_s^{\mathrm{T}}(y - X_sM_0m_0)\right\} = X_uM\underbrace{\left(m_0 + X_s^{\mathrm{T}}y\right)}_{m}. \end{aligned}$$

This summarises the predictive density as

$$p(Y_u \mid y, \sigma^2) = N_{N-n}\left(Y_u \mid X_uMm, \sigma^2\left(I_{N-n} + X_uMX_u^{\mathrm{T}}\right)\right). \tag{16.11}$$

Inference proceeds either by sampling directly from (16.11) or in two steps as follows: (i) Draw samples from the posterior distribution $p(\beta, \sigma^2 \mid y)$ and (ii) Draw one instance of Y_u for each drawn value of $\{\beta, \sigma^2\}$ from $Y_u \mid \beta, \sigma^2, y \sim N(Y_u \mid X_u\beta, \sigma^2 I_{N-n})$. The posterior samples Y_u and the observations y are then substituted into the desired functions of the population units to obtain the posterior samples of the desired functions. The posterior distribution of $\bar{Y}$ will be computed by calculating $f\bar{y} + (1 - f)\bar{Y}_u$ using each posterior sample of Y_u.

We are typically interested in estimating some linear function $a^{\mathrm{T}}Y$, where a is $N \times 1$ and fixed. The joint distribution for y and $a^{\mathrm{T}}Y$ is derived from (16.8) and is

$$\begin{bmatrix} y \\ a^{\mathrm{T}}Y \end{bmatrix} \Bigg| \sigma^2 \sim N_{n+1}\left(\begin{bmatrix} X_sM_0m_0 \\ a^{\mathrm{T}}XM_0m_0 \end{bmatrix}, \sigma^2\begin{bmatrix} X_sM_0X_s^{\mathrm{T}} + I_n & r \\ r^{\mathrm{T}} & v \end{bmatrix}\right), \tag{16.12}$$

where $r = a_s + X_s M_0 X^{\mathrm{T}} a$ is $n \times 1$, $v = a^{\mathrm{T}}(X M_0 X^{\mathrm{T}} + I_N) a$ is a positive scalar and $X = \begin{bmatrix} X_s \\ X_u \end{bmatrix}$. Both $p(y \mid a^{\mathrm{T}} Y, \sigma^2)$ and $p(a^{\mathrm{T}} Y \mid y, \sigma^2)$ can be derived from (16.12). The former represents the data likelihood and is Gaussian with

$$\begin{aligned} \mathbb{E}[y \mid a^{\mathrm{T}} Y, \sigma^2] &= X_s M_0 m_0 + \frac{a^{\mathrm{T}}(Y - X M_0 m_0)}{v} r \\ \text{and} \quad \mathbb{V}\left(y \mid a^{\mathrm{T}} Y, \sigma^2\right) &= \sigma^2 \left\{ X_s M_0 X_s^{\mathrm{T}} + I_n - \frac{r r^{\mathrm{T}}}{v} \right\} . \end{aligned} \tag{16.13}$$

Bayesian inference proceeds from $p(a^{\mathrm{T}} Y \mid y, \sigma^2)$, which is Gaussian with mean and variance $\mathbb{E}[a^{\mathrm{T}} Y \mid y, \sigma^2] = a_s^{\mathrm{T}} y + a_u^{\mathrm{T}} \mathbb{E}[Y_u \mid y, \sigma^2]$ and $\mathbb{V}(a^{\mathrm{T}} Y \mid y, \sigma^2) = a_u^{\mathrm{T}} \mathbb{V}(Y_u \mid y, \sigma^2) a_u$, where a_s and a_u are elements of a corresponding to sampled and unobserved units, and $\mathbb{E}[Y_u \mid y, \sigma^2]$ and $\mathbb{V}(Y_u \mid y, \sigma^2)$ are supplied in (16.10).

The Bayesian hierarchical model and linear regression framework is versatile and accommodates multi-stage cluster sampling designs (see, e.g., Banerjee, 2024; Malec and Sedransk, 1985; Scott and Smith, 1969). Ghosh and Sinha (1990), Little (2004), and Banerjee (2024) offer different perspectives on how the ubiquitous Horvitz-Thompson estimator (Horvitz and Thompson, 1952) in design-based survey sampling emerges as the posterior expectation of Bayesian models. Bayesian analysis for finite populations avoids asymptotic inference and delivers legitimate uncertainty estimates, thereby obviating issues with numerically unstable variance estimators.

16.4 Meta-analysis Using Survey Estimates

A common exercise in survey sampling is to provide pooled estimates for population quantities from different surveys. This is referred to as a meta-analysis for a finite population. For example, suppose that we are provided with $\mathcal{D}_k = \{n_k, \bar{y}_k, s_k^2\}$ consisting of sample size, sample mean, and sample variance from each of $k = 1, \ldots K$ different surveys. A straightforward model for pooling the survey estimates to infer the population mean assumes that the surveys are independent. The posterior distribution of interest is

$$p\left(\bar{Y}, \sigma^2 \mid \{\bar{y}_k, s_k^2\}_{k=1}^K\right) \propto IG(\sigma^2 \mid a_0, b_0) \times N\left(\bar{Y} \,\middle|\, \theta, \sigma^2 \frac{1+f_0}{n_0}\right) \times \prod_{k=1}^{K} p(\bar{y}_k, s_k^2 \mid \bar{Y}, \sigma^2) , \tag{16.14}$$

where each $p(\bar{y}_k, s_k^2 \mid \bar{Y}, \sigma^2)$ is of the form in (16.5). This simplifies to the pooled posterior

$$\begin{aligned} p\left(\bar{Y}, \sigma^2 \mid \mathcal{D}_{1:K}\right) &\propto IG(\sigma^2 \mid a_0, b_0) \times N\left(\bar{Y} \,\middle|\, \theta, \sigma^2 \frac{1+f_0}{n_0}\right) \\ &\quad \times \prod_{k=1}^{K} \left(\frac{1}{\sigma^2}\right)^{\frac{n_k-1}{2}} \exp\left(-\frac{(n_k-1)s_k^2}{2\sigma^2}\right) \times \left(\frac{1}{\sigma^2}\right)^{\frac{1}{2}} \exp\left(-\frac{n_k(\bar{y}_k - \bar{Y})^2}{2\sigma^2(1-f_k)}\right) \\ &\propto \underbrace{IG\left(\sigma^2 \,\middle|\, a_0 + \frac{\sum_{k=1}^K n_k}{2}, b_0 + \frac{q_K}{2}\right)}_{p(\sigma^2 \mid \mathcal{D}_{1:K})} \times \underbrace{N\left(\bar{Y} \,\middle|\, \frac{\lambda_0 \theta + \sum_{k=1}^K \lambda_k \bar{y}_k}{\lambda_0 + \sum_{k=1}^K \lambda_k}, \frac{\sigma^2}{\lambda_0 + \sum_{k=1}^K \lambda_k}\right)}_{p(\bar{Y} \mid \sigma^2, \mathcal{D}_{1:K})} , \end{aligned} \tag{16.15}$$

where $\lambda_0 = n_0/(1+f_0)$, $\lambda_k = n_k/(1-f_k)$ with $f_k = n_k/N$, $q_K = \sum_{t=1}^K (n_t - 1)s_t^2 + \sum_{t=0}^K \lambda_t (\bar{y}_t - \bar{y}_\cdot)^2$ with $\bar{y}_0 := \theta$ and $\bar{y}_\cdot = \sum_{t=0}^K \lambda_t \bar{y}_t / \left(\sum_{t=0}^K \lambda_t\right)$. We draw samples from

(16.15) by first sampling values of $\sigma^2 \sim p(\sigma^2 \mid \mathcal{D}_{1:K})$ and then sampling one value of $\bar{Y} \sim p(\bar{Y} \mid \sigma^2, \mathcal{D}_{1:K})$ for each drawn value of σ^2, thus delivering a pooled inference.

I will shortly turn to Bayesian meta-analysis with linear regression models, but first consider how information flows from posterior distributions based upon subsets of the data. Let $\mathcal{D}_k = \{n_k, y_k, X_k\}$ denote mutually exclusive (disjoint) subsets of the data for $k = 1, \ldots, K$, where each y_k is $n_k \times 1$, X_k is $n_k \times p$, and $\sum_{k=1}^{K} n_k = n$ is the sample size. Assuming that the surveys are independent, sequential Bayesian updating uses the simple observation that $p\left(\beta, \sigma^2 \mid \mathcal{D}_{1:k}\right) \propto p\left(\beta, \sigma^2 \mid \mathcal{D}_{1:(k-1)}\right) \times p\left(\mathcal{D}_k \mid \beta, \sigma^2\right)$. This yields the joint posterior for each $k = 1, 2, \ldots, K$ as

$$p(\beta, \sigma^2 \mid \mathcal{D}_{1:k}) = \underbrace{IG(\sigma^2 \mid a_k, b_k)}_{p\left(\sigma^2 \mid \mathcal{D}_{1:k}\right)} \times \underbrace{N\left(\beta \mid M_k m_k, \sigma^2 M_k\right)}_{p\left(\beta \mid \sigma^2, \mathcal{D}_{1:k}\right)}, \tag{16.16}$$

where $a_k = a_{k-1} + n_k/2$, $b_k = b_{k-1} + q_k/2$, $q_k = y_k^{\mathrm{T}} y_k + m_{k-1}^{\mathrm{T}} M_{k-1} m_{m-1} - m_k^{\mathrm{T}} M_k m_k$, $m_k = m_{k-1} + X_k^{\mathrm{T}} y_k$, and $M_k^{-1} = M_{k-1}^{-1} + X_k^{\mathrm{T}} X_k$. If $\mathcal{D} = \mathcal{D}_{1:K}$, then we obtain our desired posterior distribution at $k = K$ by applying (16.16) to each subset $\mathcal{D}_t$. This is computationally efficient when $\mathcal{D}$ is large because we do not need to store all the data and can overwrite the objects in memory as we move to the next subset. Straightforward calculations reveal that $M_K^{-1} = M_0^{-1} + \sum_{k=1}^{K} X_k^{\mathrm{T}} X_k$, $m_K = m_0 + \sum_{k=1}^{K} X_k^{\mathrm{T}} y_k$, $a = a_0 + \sum_{k=1}^{K} n_k/2$ and $b = b_0 + \frac{1}{2}\left(\sum_{k=1}^{K} \|y_k\|^2 + m_0^{\mathrm{T}} M_0 m_0 - m_K^{\mathrm{T}} M_K m_K\right)$. Hence, we are able to compute $p(\beta, \sigma^2 \mid \mathcal{D})$ sequentially from the subsets if each $\mathcal{D}_k = \{n_k, \|y_k\|^2, X_k^{\mathrm{T}} y_k, X_k^{\mathrm{T}} X_k\}$. Complete data $\{y_k, X_k\}$ used in each subset is not required. This principle has been extended to spatial statistics in the form of "meta-kriging" (Guhaniyogi and Banerjee, 2018).

Meta-analysis from different surveys requires different considerations when the objective is to estimate a finite population quantity such as $a^{\mathrm{T}} Y$. Here, (16.12) proves instructive. Analogous to (16.14) and (16.15), the desired posterior is

$$p(a^{\mathrm{T}} Y, \sigma^2 \mid \mathcal{D}) \propto \underbrace{IG(\sigma^2 \mid a_0, b_0)}_{p(\sigma^2)} \times \underbrace{N\left(a^{\mathrm{T}} Y \mid a^{\mathrm{T}} X M_0 m_0, \sigma^2 v\right)}_{p(a^{\mathrm{T}} Y \mid \sigma^2)} \times \prod_{k=1}^{K} \underbrace{N\left(y_k \mid \mu_k, \sigma^2 V_k\right)}_{p(y_k \mid a^{\mathrm{T}} Y, \sigma^2)}, \tag{16.17}$$

where μ_k and $\sigma^2 V_k$ are the mean and variance calculated in (16.13) using respective design matrices X_k over sampled units for each $k = 1, \ldots, K$. Letting r_k denote the $n_k \times 1$ vector evaluated in study k that corresponds to r in the error variance-covariance matrix of (16.12), we can write $\mu_k = \alpha_{0k} + \alpha_{1k}\left(a^{\mathrm{T}} Y\right)$, where $\alpha_{0k} = X_k M_0 m_0 - r_k v^{-1} a^{\mathrm{T}} X M_0 m_0$, $\alpha_{1k} = r_k v^{-1}$, and $V_k = X_k M_0 X_k^{\mathrm{T}} + I_n - r_k v^{-1} r_k^{\mathrm{T}}$. The resulting posterior is

$$p(a^{\mathrm{T}} Y, \sigma^2 \mid \mathcal{D}) = \underbrace{IG\left(\sigma^2 \mid a_1, b_1\right)}_{p(\sigma^2 \mid \mathcal{D})} \times \underbrace{N\left(a^{\mathrm{T}} Y \mid Mm, \sigma^2 M\right)}_{p(a^{\mathrm{T}} Y \mid \mathcal{D}, \sigma^2)}, \tag{16.18}$$

where $M^{-1} = v^{-1} + \sum_{k=1}^{K} \alpha_{1k}^{\mathrm{T}} V_k^{-1} \alpha_{1k}$, $m = v^{-1} a^{\mathrm{T}} X M_0 m_0 + \sum_{k=1}^{K} \alpha_{1k}^{\mathrm{T}} V_k^{-1}(y_k - \alpha_{0k})$, $a_1 = a_0 + \sum_{k=1}^{K} n_k/2$, $b_1 = b_0 + q/2$, and $q = \sum_{k=1}^{K} (y_k - \alpha_{0k})^{\mathrm{T}} V_k^{-1} (y_k - \alpha_{0k}) + \left(a^{\mathrm{T}} X M_0 m_0\right)^2 v^{-1} - Mm^2$. Pooled inference is performed by drawing $\sigma^2 \sim IG(a_1, b_1)$ and then sampling $a^{\mathrm{T}} Y \sim N(Mm, \sigma^2 M)$ for each drawn value of σ^2, which produces the desired samples from (16.18).

16.5 Spatial Process Models

Spatial data analysis is conveniently carried out by embedding a spatial process within the familiar hierarchical modelling paradigm,

$$[\text{data} \mid \text{process}] \times [\text{process} \mid \text{parameters}] \times [\text{parameters}] . \tag{16.19}$$

Point-referenced data are made up of variables referenced by points with coordinates (latitude-longitude, Easting-Northing, etc.). Modelling and analysing such data proceed from a random field that introduces dependence among any finite collection of random variables. The random field is a stochastic process defined as an uncountable set of random variables, say $\{w(\ell) : \ell \in \mathcal{L}\}$, in a domain of interest $\mathcal{L}$. This uncountable set is endowed with a probability law specifying the joint distribution for any finite subset of random variables. Spatial processes are usually constructed assuming $\mathcal{L} \subseteq \Re^d$ (usually $d = 2$ or 3) or perhaps as a subset of points on a sphere or ellipsoid. In spatiotemporal settings $\mathcal{L} = \mathcal{S} \times \mathcal{T}$, where $\mathcal{S} \subset \Re^d$ and $\mathcal{T} \subset [0, \infty)$ are the space and time domains, respectively, and $\ell = (s, t)$ is a space-time coordinate with spatial location $s \in \mathcal{S}$ and time point $t \in \mathcal{T}$ (see, e.g., Gneiting and Guttorp, 2010, for details).

The probability laws on random fields are specified with a covariance function $\text{cov}\{w(\ell), w(\ell')\} = K_\theta(\ell, \ell')$ for any two points ℓ and ℓ' in $\mathcal{L}$. If $\mathcal{U}$ and $\mathcal{V}$ are finite sets comprising n and m points in $\mathcal{L}$, respectively, then $K_\theta(\mathcal{U}, \mathcal{V})$ denotes the $n \times m$ matrix whose (i, j)-th element is evaluated using the covariance function $K_\theta(\cdot, \cdot)$ between the i-th point $\mathcal{U}$ and the j-th point in $\mathcal{V}$. If $\mathcal{U}$ or $\mathcal{V}$ comprises a single point, $K_\theta(\mathcal{U}, \mathcal{V})$ is a row or column vector, respectively. A valid spatiotemporal covariance function ensures that $K_\theta(\mathcal{U},\mathcal{U})$ is positive definite for any finite set $\mathcal{U}$, which we will denote simply as K_θ if the context is clear. A customary specification models $\{w(\ell) : \ell \in \mathcal{L}\}$ as a zero-centred Gaussian process, denoted as $w(\ell) \sim GP(0, K_\theta(\cdot, \cdot))$. For any finite collection $\mathcal{U} = \{\ell_1, \ell_2, \ldots, \ell_n\}$ in $\mathcal{L}$, the $n \times 1$ random vector $w_\mathcal{U} = (w(\ell_1)), w(\ell_2), \ldots, w(\ell_n))^\mathrm{T}$ is distributed as $N(0, K_\theta)$, where $K_\theta = K_\theta(\mathcal{U},\mathcal{U})$. Further details on valid spatial (and spatiotemporal) covariance functions can be found in Gneiting and Guttorp (2010), Cressie (1993), Stein (1999), Gelfand et al. (2010), Cressie and Wikle (2011), and Banerjee et al. (2014) and numerous references therein. The more common assumptions are *stationarity* and *isotropy*. The former assumes that $K_\theta(\ell, \ell') = K_\theta(\ell - \ell')$ depends upon the coordinates only through their separation vector, while isotropy assumes the covariance is a function of the distance between locations.

Several choices for the covariance function are available (see, e.g. Banerjee et al., 2014). The isotropic Matérn function, which models spatial covariance as a function of distance between spatial locations, while also introducing a parameter to control smoothness of process realisations, is defined as $K_\theta(d_{ab}) = \sigma^2 \frac{2^{1-\phi_2}}{\Gamma(\phi_2)} \left(\sqrt{2\phi_2} d_{ab} \phi_1\right)^{\phi_2} \mathcal{K}_{\phi_2}\left(\sqrt{2\phi_2} d_{ab} \phi_1\right)$, where $\mathcal{K}_{\phi_2}(\cdot)$ is the modified Bessel function and d_{ab} is the distance between two locations ℓ_a and ℓ_b. The parameters $\theta = \{\sigma^2, \phi = \{\phi_1, \phi_2\}\}$ capture variation due to spatial structure (σ^2), the rate of decline in spatial association (ϕ_1), and smoothness of the process (ϕ_2). The exponential function $K_\theta(d_{ab}) = \sigma^2 \exp(-\phi_1 d_{ab})$ is a special case with $\phi_2 = 1/2$.

If $y(\ell)$ is the value of a variable at ℓ, then a customary spatial regression model is

$$y(\ell) = x^\mathrm{T}(\ell)\beta + w(\ell) + \epsilon(\ell) , \tag{16.20}$$

where $x(\ell)$ is a $p \times 1$ ($p < n$) vector of spatially referenced predictors, β is the $p \times 1$ vector of slopes, and $w(\ell) \sim GP(0, K_\theta(\cdot, \cdot))$ is the spatial or spatiotemporal process and $\epsilon(\ell)$ is a white noise process modelling measurement error or fine scale variation attributed to disturbances at distances smaller than the minimum observed separations in space and/or time. We now embed the spatial process within the Bayesian hierarchical model

$$p(\theta, \beta, \tau) \times N(w \mid 0, K_\theta) \times N(y \mid X\beta + w, D_\tau) , \tag{16.21}$$

where $y = (y(\ell_1), y(\ell_2), \ldots, y(\ell_n))^{\mathrm{T}}$ is the $n \times 1$ vector of observed outcomes, X is the $n \times p$ matrix of predictors with i-th row $x^{\mathrm{T}}(\ell_i)$, and D_τ is the covariance matrix for $\epsilon(\ell)$ over $\{\ell_1, \ell_2, \ldots, \ell_n\}$. The hierarchy is completed by assigning prior distributions to β, θ, and τ.

Models for spatially dependent finite populations can be built by treating the finite population units as a partial realisation of the spatial process. In particular, we can integrate out w from (16.21) into a collapsed model $p(\theta, \beta, \tau) \times N(y \mid X\beta, K_\theta + D_\tau)$ (see, e.g., Banerjee, 2017, 2020, for computational considerations using collapsed models and augmented linear systems). A common specification is $D_\tau = \tau^2 I_n$, where τ^2 is the measurement error variance called the "nugget." A convenient parametric form is to write the spatial covariance function as $K_\theta(\ell, \ell') = \sigma^2 C_\phi(\ell, \ell')$, where $C_\phi(\ell, \ell')$ is the correlation between $w(\ell)$ and $w(\ell')$. The model has two variance components, σ^2 and τ^2. The former captures spatial variation and the latter estimates measurement error or other unexplained variability at fine scales. A convenient parametric form is to define and express the model in terms of one of the variance components and the ratio between the two variances.

For example, below we express our model in terms of σ^2 and $\delta^2 = \tau^2/\sigma^2$. The advantage of this parametric form is that exploratory data analysis can often inform about the underlying process and parameters such as δ^2 and ϕ can be fixed using tools such as variograms. This, in turn, allows us to exploit conjugate distributions and avoid computational bottlenecks (Banerjee, 2020; Zhang et al., 2019). In addition to fixing δ^2, variograms also inform about the rate of decay in spatial correlation as a function of distance and help us fix ϕ_1. If the Matérn covariance kernel is used, then the smoothness parameter ϕ_2 is not easily identified by the data and is fixed from mechanistic considerations. The parameters $\{\sigma^2, \phi\}$ are not consistently estimated from the data (Tang et al., 2021; Zhang, 2004) in a fixed-domain paradigm where we populate a fixed area with increasing number of locations generating observations. This implies that the effect of the prior distribution does not erode with increasing number of observations. Ideally, informative priors using scientific considerations should be considered. When such information is unavailable, one might as well fix these parameters and focus on spatial prediction of the latent process $w(\ell)$ and the outcome $y(\ell)$ at new locations that are still robust.

Inference for finite population units proceeds from regression models by modifying (16.8) to

$$\begin{bmatrix} y \\ Y_u \end{bmatrix} = \begin{bmatrix} X_s \\ X_u \end{bmatrix} \beta + \begin{bmatrix} \epsilon_s \\ \epsilon_u \end{bmatrix} ; \quad \begin{bmatrix} \epsilon_s \\ \epsilon_u \end{bmatrix} \sim N\left(\begin{bmatrix} 0 \\ 0 \end{bmatrix}, \sigma^2 \underbrace{\begin{bmatrix} H_s(\phi, \delta^2) & H_{su}(\phi) \\ H_{us}(\phi) & H_u(\phi, \delta^2) \end{bmatrix}}_{H(\phi, \delta^2)} \right) ; \tag{16.22}$$

$$\beta = M_0 m_0 + \eta ; \quad \eta \sim N(0, \sigma^2 M_0) ; \quad \sigma^2 \sim IG(a_0, b_0) ; \quad \phi \sim p(\phi) ,$$

where X_s and X_u are fixed design matrices corresponding to sampled and non-sampled units, respectively, $H(\phi, \delta^2) = \begin{bmatrix} H_s(\phi, \delta^2) & H_{su}(\phi) \\ H_{us}(\phi) & H_u(\phi, \delta^2) \end{bmatrix}$ is a positive definite correlation matrix partitioned accordingly, and M_0 and m_0 are assumed fixed.

The elements of the block covariance matrix $\sigma^2 H(\phi, \delta^2)$ in (16.22) are constructed using $C_\phi(d)$ evaluated using distances between the different locations indexed by rows and columns of $H(\phi, \delta^2)$ and the value of δ^2. In particular, all off-diagonal elements of $H_s(\phi, \delta^2)$ and $H_u(\phi, \delta^2)$ and all elements of $H_{su}(\phi)$ (hence also $H_{us}(\phi) = H_{su}(\phi)^{\mathrm{T}}$) are specified by $C_\phi(\ell_i, \ell_j)$ for all pairwise locations in the finite population. The diagonal elements of $H_s(\phi, \delta^2)$ and $H_u(\phi, \delta^2)$ are all equal to $1 + \delta^2$ implying that the variance of $y(\ell)$ at any location (sampled or not) is $\sigma^2 + \tau^2$.

For any fixed values of $\{\delta^2, \phi\}$, we draw samples from $p(Y_u \mid y, \delta^2, \phi)$ as follows: (i) Draw samples from the posterior distribution $p(\beta, \sigma^2 \mid y, \delta^2, \phi)$ and (ii) For each sampled value of $\{\beta, \sigma^2\}$, draw samples from $p(Y_u \mid \beta, \sigma^2, \delta^2, \phi, y)$. The resulting samples of Y_u are from the desired posterior predictive distribution $p(Y_u \mid y, \delta^2, \phi) = \mathbb{E}_{\beta,\sigma^2|y,\delta^2,\phi}\left[N\left(Y_u \mid X_u\beta + H_{us}H_s^{-1}(y - X_s\beta), \sigma^2\left(H_u - H_{us}H_s^{-1}H_{su}\right)^{-1}\right)\right]$. The posterior samples of Y_u and the observations y are then substituted into the desired functions of the population units to obtain the posterior samples of the desired functions. The posterior estimator obtained from (16.22) is $h(y; \beta, \sigma^2, \delta^2, \phi) = \mathbb{E}[Y_u \mid y, \beta, \sigma^2, \delta^2, \phi] = X_u\beta + H_{us}H_s^{-1}(y - X_s\beta)$ and $\mathbb{V}(Y_u \mid y, \beta, \sigma^2, \delta^2, \phi) = \sigma^2\left(H_u - H_{us}H_s^{-1}H_{su}\right)$. If the entire finite population is sampled, then $X_u = X_s$ and $H_{us} = H_s$ so $h(y; \beta, \sigma^2, \delta^2, \phi) = y$ and the conditional variance $\mathbb{V}(Y_u \mid y, \beta, \sigma^2, \delta^2, \phi) = 0$ for all values of $\{\beta, \sigma^2, \delta^2, \phi\}$. Hence, Bayesian inference is *consistent* in the sense that the posterior estimator equals the finite population quantity without uncertainty if the entire population is sampled.

Since $p(\beta, \sigma^2 \mid y, \delta^2, \phi)$ and $p(Y_u \mid \beta, \sigma^2, \delta^2, \phi, y)$ are both tractable distributions from which we can draw Monte Carlo samples, we are able to carry out Bayesian inference based on exact sampling from $p(Y_u \mid y, \delta^2, \phi)$ for any fixed $\{\delta^2, \phi\}$. Alternatively, we can carry out *full* Bayesian inference by assigning prior distributions to δ^2 and ϕ and devising iterative algorithms such as Markov chain Monte Carlo (MCMC; see, e.g., Robert and Casella, 2004) for posterior sampling or Integrated Nested Laplace Approximations (Rue et al., 2009) for quadrature-based posterior inference. A different approach, perhaps more in the spirit of machine learning methods, is to conduct Bayesian analysis using a weighted average ("stacking") of the exact conditional posterior distributions $p(Y_u \mid y, \delta^2, \phi)$ over a grid of values of δ^2 and ϕ. Bayesian predictive stacking (Wolpert, 1992; Yao et al., 2018) obtains weights to combine posterior distributions by using convex optimisation of a suitable score function or loss function. Once these weights have been obtained, inference on Y_u is carried out by sampling from each $p(Y_u \mid y, \delta^2, \phi)$ proportional to the weights. Recent developments in Bayesian geostatistics have shown the effectiveness of these methods to deliver rapid inference almost indistinguishable from more computationally expensive alternatives such as MCMC (Pan et al., in press; Presicce and Banerjee, 2025; Zhang et al., in press).

16.6 Spatial Survey Sampling Models in Complex Settings

More general frameworks for spatial survey sampling may also consider locations that yield "structural zeros", which represents all locations where sampling is precluded due to geographic considerations. For example, points situated over lakes and rivers may be precluded from the finite population of interest when one is interested in household economic surveys. Likewise, heavily urbanised metropolitan areas are precluded from studies concerning biomass interpolation from a finite population of trees in a forest (see, e.g., Finley et al., 2011, for data analysis accounting for structural zeros, although not in a finite population context). To infer about the process in (16.19) from the sampled observations, we use the spatial process to endow a probability law over the uncountable collection of random variables $\{(Y(\ell), Z(\ell)) : \ell \in \mathcal{L}\}$, where $\mathcal{L} \subset \mathbb{R}^2$ is our domain of interest. This induces the probability distribution for the finite population $\{(Y(\ell), Z(\ell)) : \ell \in \mathcal{L}_{FP}\}$, where $\mathcal{L}_{FP} = \{\ell_1, \ldots, \ell_N\}$ denotes a finite collection of N spatial locations. This set represents the units in our finite population and is made up of locations that are sampled, denoted by $\mathcal{L}_s$, and those that are not, denoted by $\mathcal{L}_u$; therefore, $\mathcal{L}_s \cup \mathcal{L}_u = \mathcal{L}_{FP}$.

Banerjee (2024) proposes a hierarchical modelling framework based on the following underlying stochastic processes generating the data: (i) a possibly vector-valued latent spatial process, $\{w(\ell) : \ell \in \mathcal{L}\}$, which models the variable(s) of interest at ℓ; (ii) a sampling indicator process, $\{Z(\ell) : \ell \in \mathcal{L}\}$, which indicates whether a location is sampled or not; and possibly (iii) a structural zero indicator process (partially observed), $\{u(\ell) : \ell \in \mathcal{L}\}$, indicating whether $\ell \in \mathcal{L}_{FP}$ or not. A joint hierarchical framework extends (16.19) to

$$[Y(\cdot) \mid w(\cdot), u(\cdot)] \times [Z(\cdot) \mid Y(\cdot), u(\cdot)] \times [w(\cdot) \mid u(\cdot)] \times [u(\cdot)] \,. \tag{16.23}$$

The above specification is rich and enriches our previous setting by allowing inference for non-ignorable sampling designs that allow for the sampling indicator process $Z(\cdot)$ to depend on the data generating process $Y(\cdot)$ (see, e.g., Gelman et al., 2014; Little and Rubin, 2002; Rubin, 1976, for details on non-ignorable designs).

It is worth remarking that model-based inference for complex survey data has a substantial body of literature that continues to evolve and expand. A comprehensive review is beyond the scope of a single article, but a few recent applications in the fields of forestry and ecology may serve as examples. Several applications have emerged from needs of data users in the United States National Forest Inventory (NFI) programs who seek higher spatial and temporal resolution forest status and change parameter estimates than can be provided by the traditional design-based estimation methods used by NFI programs (Betchold and Patterson, 2015; Tomppo et al., 2009; Westfall et al., 2022). In contrast to the focus on spatial process models in this chapter, the Fay-Herriot (FH) model (Fay and Herriot, 1979) is widely used in small area estimation (SAE) (Rao and Molina, 2015) for NFI data. The FH regression framework includes auxiliary design-based information as explanatory variables, while residual spatial (and temporal) dependence is captured using structured random effects based on the sampling design. It is customarily fit to small area direct estimates without using NFI plot locations.

Shannon et al. (2024) pointed out limitations of the FH model when design-based information are unavailable or unreliable due to small sample sizes. In a follow-up article, Shannon et al. (2025) devise a Bayesian spatial-temporal SAE model that efficiently leverages sparse NFI data to estimate status and trends for forest parameters. These authors also point out proposed model obviates the use of direct estimates and instead uses plot-level NFI measurements along with auxiliary data including remotely sensed tree canopy cover. These rich Bayesian hierarchical frameworks produce forest carbon estimates from the United States NFI over 14 years across the contiguous United States and achieve improved precision and accuracy over traditional estimation methods for county-level forest carbon dynamics. Other examples of model-based inference include Finley et al. (2024) who assimilate remote-sensed data and field measurements to estimate forest parameters when inventory data are sparse or resources limit the collection of enough data to achieve desired accuracy and precision using design-based methods.

16.7 Spatial Meta-analysis Using Bayesian Linear Models

Given the groundwork in the preceding sections, spatial meta-analysis using Bayesian linear regression is conceptually straightforward. As mentioned above, we collect data $\mathcal{D}_k = \{y_k, X_k\}$ for $k = 1, \ldots, K$ studies and let $\mathcal{D} = \mathcal{D}_{1:K}$. We also assume that the design matrix $X = \begin{bmatrix} X_s \\ X_u \end{bmatrix}$ is fully available to us and is fixed for all units in the finite population. To build a model for spatial meta-analysis, we use the following analogue of (16.12),

$$\begin{bmatrix} y \\ a^{\mathrm{T}}Y \end{bmatrix} \Bigg| \sigma^2 \sim N_{n+1}\left(\begin{bmatrix} X_s M_0 m_0 \\ a^{\mathrm{T}} X M_0 m_0 \end{bmatrix}, \sigma^2 \begin{bmatrix} X_s M_0 X_s^{\mathrm{T}} + H_s(\phi, \delta^2) & r(\phi, \delta^2) \\ r(\phi, \delta^2)^{\mathrm{T}} & v(\phi, \delta^2) \end{bmatrix}\right), \tag{16.24}$$

where $r(\phi, \delta^2) = H_s(\phi, \delta^2)a_s + X_s M_0 X^{\mathrm{T}} a$ is $n \times 1$ and $v(\phi, \delta^2) = a^{\mathrm{T}}\left(X M_0 X^{\mathrm{T}} + H(\phi, \delta^2)\right) a$ is a positive scalar. A spatial meta-analysis model assumes that the individual studies are independent of each other, so the joint distribution analogous to (16.14) is

$$p(a^{\mathrm{T}}Y, \theta \mid \mathcal{D}) \propto p(\theta) \times \underbrace{N\left(a^{\mathrm{T}}Y \mid a^{\mathrm{T}} X M_0 m_0, \sigma^2 v(\phi, \delta^2)\right)}_{p(a^{\mathrm{T}}Y \mid \sigma^2, \delta^2, \phi)} \times \prod_{k=1}^{K} \underbrace{N\left(y_k \mid \mu_k(\phi, \delta^2), \sigma^2 V_k(\phi, \delta^2)\right)}_{p(y_k \mid a^{\mathrm{T}}Y, \sigma^2, \delta^2, \phi)}, \tag{16.25}$$

where $\theta = \{\sigma^2, \delta^2, \phi\}$, $p(\theta) = p(\delta^2, \phi) \times \underbrace{IG(\sigma^2 \mid a_0, b_0)}_{p(\sigma^2)}$, $\mu_k(\theta) = X_k M_0 m_0 + r_k a^{\mathrm{T}}(Y - X M_0 m_0)/v$, $V_k = X_k M_0 X_k^{\mathrm{T}} + H_{s,k} - r_k r_k^{\mathrm{T}}/v$ with r_k, $H_{s,k}$ and $H_{su,k}$ being quantities (suppressing their dependence on $\{\phi, \delta^2\}$ for ease of notation) defined in the covariance matrix in (16.24) corresponding to study k, while v, as defined in (16.24), is computed using quantities for the entire finite population that remain invariant over the K individual studies.

Conditional on $\{\delta^2, \phi\}$, the posterior distribution $p(a^{\mathrm{T}}Y, \sigma^2 \mid \delta^2, \phi, \mathcal{D})$ is available in a closed form analogous to (16.18) with r_k, v, and V_k redefined as $r_k(\phi, \delta^2)$, $v(\phi, \delta^2)$, and $V_k(\phi, \delta^2)$. Hence, α_{0k} and α_{1k} are redefined as $\alpha_{0k}(\phi, \delta^2) = X_k M_0 m_0 - r_k(\phi, \delta^2) v^{-1} a^{\mathrm{T}} X M_0 m_0$ and $\alpha_{1k}(\phi, \delta^2) = r_k(\phi, \delta^2) v(\phi, \delta^2)^{-1}$, M and m are redefined as $M(\phi, \delta^2)^{-1} = v(\phi, \delta^2)^{-1} + \sum_{k=1}^{K} \alpha_{1k}(\phi, \delta^2)^{\mathrm{T}} V_k(\phi, \delta^2)^{-1} \alpha_{1k}(\phi, \delta^2)$ and $m(\phi, \delta^2) = v(\phi, \delta^2)^{-1} a^{\mathrm{T}} X M_0 m_0 + \sum_{k=1}^{K} \alpha_{1k}(\phi, \delta^2)^{\mathrm{T}} V_k(\phi, \delta^2)^{-1}(y_k - \alpha_{0k}(\phi, \delta^2))$, $a_1 = a_0 + \sum_{k=1}^{K} n_k/2$, $b_1 = b_0 + q(\phi, \delta^2)/2$, and $q(\phi, \delta^2) = \sum_{k=1}^{K}(y_k - \alpha_{0k}(\phi, \delta^2))^{\mathrm{T}} V_k(\phi, \delta^2)^{-1}(y_k - \alpha_{0k}(\phi, \delta^2)) + \left(a^{\mathrm{T}} X M_0 m_0\right)^2 v(\phi, \delta^2)^{-1} - M(\phi, \delta^2) m(\phi, \delta^2)^2$.

For any fixed value of $\{\delta^2, \phi\}$, pooled inference is performed by drawing $\sigma^2 \sim IG(a_1, b_1)$ and then sampling $a^{\mathrm{T}}Y \sim N(M(\phi, \delta^2) m(\phi, \delta^2), \sigma^2 M(\phi, \delta^2))$ for each drawn value of σ^2. This produces the desired samples from $p(a^{\mathrm{T}}Y, \sigma^2 \mid \delta^2, \phi, \mathcal{D})$. If $\{\delta^2, \phi\}$ are unknown, then one can assign informative prior distributions to these quantities and carry out full Bayesian inference using iterative algorithms such as MCMC (Robert and Casella, 2004) or Integrated Nested Laplace Approximations (Rue et al., 2009). Alternatively, Bayesian predictive stacking can be performed on a grid of values of $\{\delta^2, \phi\}$ (Pan et al., in press; Presicce and Banerjee, 2025; Zhang et al., in press) to carry out approximate inference from an optimal weighted posterior that will approximate $p(a^{\mathrm{T}}Y, \sigma^2 \mid \delta^2, \phi, \mathcal{D})$.

16.8 Illustrations

I illustrate spatial meta-analysis using two examples with the disclaimer that the numerical results should be treated as preliminary proofs of concepts that are expected to stoke further investigations and more scientifically robust findings. The first example is a study of groundwater nitrate levels in the population of California Central Valley wells located in the Tulare Lake Basin (TLB), where health officials seek estimates of mean nitrate levels. The data were extracted from the California Ambient Spatio-Temporal Information on Nitrate

TABLE 16.1
Results of California Nitrate data analysis.

Model	Finite Population Mean in mg/L (95% CI)	WAIC
(i) Independent	25.6 (19.2, 35.1)	470.6
(ii) Spatially dependent	32.1 (25.8, 33.7)	385.4

in Groundwater (CASTING) database, which is described in Harter et al. (2017) and Boyle et al. (2012). The spatial locations of these wells are available. The health implications of high nitrate levels in wells are dire. At high levels, infants and pregnant women are more susceptible to nitrate poisoning, making it more difficult for oxygen to be distributed to the body and can be fatal to infants under six months of age.

The population consists of 6117 unique wells scattered in 63 zip codes. Nitrate levels in the sampled wells were measured in milligrams per litre (milligrams per litre). Chan-Golston et al. (2020) analysed this dataset using a two-stage Bayesian cluster sampling design (Banerjee, 2024). Here, I consider a spatial meta-analysis of $K = 20$ studies, where each study estimated the total nitrate levels in the population of wells by treating the data as a simple random sample from the finite population.

Table 16.1 presents the posterior mean and 95% credible intervals for the finite population means obtained using the meta-analysis for 20 studies. The results are presented for the independent linear regression model in (16.17) and for the spatially dependent model described in (16.25). The last column of Table 16.1 presents the average value of the Watanabe-Akaike (or Widely Applicable) Information Criterion or WAIC calculated for the $K = 20$ studies (see, e.g., Vehtari et al., 2017; Watanabe, 2013). The logarithm of the posterior predictive density is evaluated at each data point in $\mathcal{D}_k$ and adjusts with a penalty term as

$$WAIC = -2\sum_{i\in\mathcal{S}} \log \mathbb{E}_{\beta,\theta|y}[p(y_i \mid \beta,\theta)] + 2\sum_{i\in\mathcal{S}} \mathbb{V}_{\beta,\theta|y}\left(\log p(y_i \mid \beta,\theta)\right) \;, \tag{16.26}$$

where θ indicates the collection of parameters in the model besides the regression coefficients β. For the independent linear regression model, $\theta = \{\sigma^2\}$, while for the spatial model $\theta = \{\sigma^2, \delta^2, \phi\}$ with δ^2 being the ratio of the variances from the spatial effects and the measurement error. Smaller values of $WAIC$ indicate better performance.

The spatially dependent model fits the data better as seen in the lower average WAIC score. Independent models do not perform as well and have an approximately 22% higher average WAIC score than spatial models. Therefore, the estimates of total nitrates in milligrammes per L from the spatial model should be trusted more.

I present a second example to demonstrate spatial meta-analysis for a very large dataset. Rather than storing all the data, which would be cumbersome, I perform a meta-analysis of $K = 50$ sampled subsets of the data. A more comprehensive analysis of the entire dataset is available in Zhang et al. (2019). This is a spatial dataset from NASA that includes observations of sea surface temperature (in degrees Celsius) in 2,827,252 spatial locations, of which approximately 90% (2,544,527) were used to fit the model and the rest were withheld for cross-validatory predictive assessment. Details of the dataset can be found in `http://modis-atmos.gsfc.nasa.gov/index.html` and details on the analysis can be found in Zhang et al. (2019).

Spatial meta-analysis is able to deliver full inference including the estimation of the spatial latent effects in about 3000 seconds. Since spatial meta-analysis is fast and directly samples from the posterior, hence there is no burn-in period for convergence, it was run

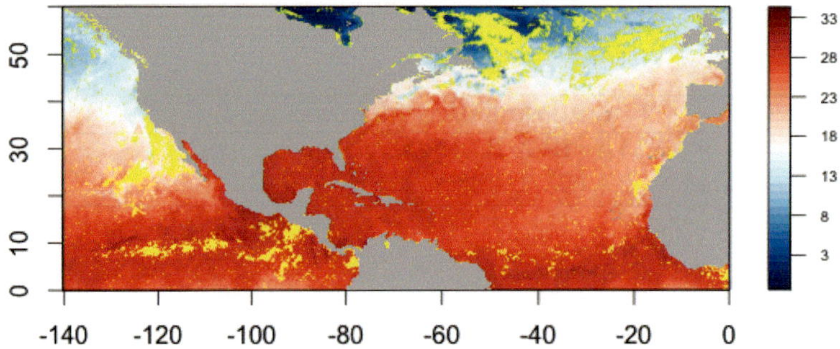

FIGURE 16.1
Posterior predictive maps of sea-surface temperature (in degree centigrade) from a spatial meta-analysis of 50 studies over the same region. The land is coloured in grey, locations in the ocean without observations are indicated in yellow.

over a grid of values of $\{\delta^2, \phi\}$. For each such value, a posterior predictive assessment was performed on the cross-validatory holdout set and the value of $\{\delta^2, \phi\}$ that produces the least RMSPE was selected as optimal input for which the estimates of $\{\gamma, \sigma^2\}$ were presented.

A crucial difference from the analysis in Zhang et al. (2019) is that spatial meta-analysis maps the outcome, which is sea surface temperature, by directly using the posterior distribution of $a^{\mathrm{T}}Y$ for appropriate choices of a. In particular, unlike the earlier example, we do not seek a posterior inference on some summary statistic, but the value of $Y(\ell)$ at a location on the grid. This is achieved in meta-analysis by taking $a = e_i$, where e_i is an appropriate vector that will draw out $Y(\ell)$ from Y. Repeating this for every ℓ on the grid produces Figure 16.1.

16.9 Concluding Remarks

This chapter has presented a Bayesian meta-analysis framework for spatially oriented data in the context of finite population inference using survey data. The framework developed here focuses upon pooled inference for the finite population units from different studies. The development intends to introduce preliminary notions and generate further investigations. The framework does not attempt pooled inference on the underlying spatial process itself.

Pooled inference of the spatial process is complicated since the subsets $\mathcal{D}_k$ may not sufficiently inform about the underlying spatial random field. Divide-and-conquer approaches for spatial random fields have been discussed in the context of spatial "BIG DATA" analysis such as meta-kriging (Guhaniyogi and Banerjee, 2018), dynamic transfer learning (Banerjee et al., 2025), and Bayesian predictive stacking (Presicce and Banerjee, 2025). While divide-and-conquer approaches for scaling statistical learning to massive volumes of data are gaining increasing attention, such developments for inferring spatial random fields and in carrying out spatial meta-analysis are still somewhat limited. The ideas and Bayesian distribution theory presented in this chapter may spur research in spatial meta-analysis and, in particular, investigate relative performances of model-based spatial meta-analysis as described in this chapter with emerging divide-and-conquer methods in machine learning.

Meta-analysis of spatially oriented survey data present opportunities for further research. Meta-analysis involving multiple survey variables that are associated among themselves as welly as spatially is of interest. Another direction can pursue richer spatial models than discussed in this chapter. For example, we can build on nonstationary spatial models in SAE (Chandra et al., 2015) and devise hierarchical nonstationary spatial process models. Here, dimension reduction using low-rank predictive processes or nearest-neighbour Gaussian processes can be used to reduce the parameter space (Coube-Sisqueille et al., 2025).

Acknowledgments

Sudipto Banerjee acknowledges funding from the National Science Foundation through DMS-1916349 and DMS-2113778; and funding from the National Institute of Health through NIEHS-R01ES027027, NIEHS-R01ES030210 and NIGMS-R01GM148761.

Bibliography

Arora, V., Lahiri, P., and Mukherjee, K. (1997). Empirical Bayes estimation of finite population means from complex surveys. *Journal of the American Statistical Association*, 92(440):1555–1562.

Banerjee, S. (2017). High-Dimensional Bayesian Geostatistics. *Bayesian Analysis*, 12(2):583 – 614.

Banerjee, S. (2020). Modeling massive spatial datasets using a conjugate Bayesian linear modeling framework. *Spatial Statistics*, 37:100417. Frontiers in Spatial and Spatio-temporal Research.

Banerjee, S. (2024). Finite population survey sampling: An unapologetic Bayesian perspective. *Sankhya A*, 86(1):95–124.

Banerjee, S., Carlin, B. P., and Gelfand, A. E. (2014). *Hierarchical Modeling and Analysis for Spatial Data.* Chapman & Hall/CRC, Boca Raton, FL, 2nd edition.

Banerjee, S., Chen, X., Frankenburg, I., and Zhou, D. (2025). Dynamic Bayesian learning for spatiotemporal mechanistic models.

Betchold, W. A. and Patterson, P. L. (2015). *The Enhanced Forest Inventory and Analysis Program – National Sampling Design and Estimation Procedures.* Gen. Tech. Rep. SRS-80. Asheville, NC: U.S. Department of Agriculture, Forest Service, Southern Research Station.

Boyle, D., King, A., Kourakos, G., Lockhart, K., Mayzelle, M., Fogg, G. E., and Harter, T. (2012). Groundwater Nitrate Occurrence, Technical Report 4. Technical report, Center for Watershed Sciences, University of California, Davis, Davis, CA.

Chan-Golston, A. M., Banerjee, S., and Handcock, M. S. (2020). Bayesian inference for finite populations under spatial process settings. *Environmetrics*, 31(3):e2606. e2606 env.2606.

Chandra, H., Salvati, N., and Chambers, R. (2015). A spatially nonstationary Fay-Herriot model for small area estimation. *Journal of Survey Statistics and Methodology*, 3(2):109–135.

Cochran, W. G. (1977). *Sampling Techniques.* John Wiley & Sons, Hoboken, NJ, 3rd edition.

Coube-Sisqueille, S., Banerjee, S., and Liquet, B. (2025). Nonstationary spatial process models with spatially varying covariance kernels. *Journal of Computational and Graphical Statistics*, pages 1–20.

Cressie, N. (1993). *Statistics for Spatial Data.* Wiley-Interscience, revised edition.

Cressie, N. and Wikle, C. K. (2011). *Statistics for Spatio-Temporal Data.* John Wiley & Sons, Hoboken, NJ.

Ericson, W. A. (1969). Subjective Bayesian models in sampling finite populations. *Journal of the Royal Statistical Society, Series B*, 31(2):195–233.

Fay, R. E. and Herriot, R. A. (1979). Estimates of income for small places: An application of james-stein procedures to census data. *Journal of the American Statistical Association*, 74(366).

Finley, A. O., Andersen, H.-E., Babcock, C., Cook, B. D., Morton, D. C., and Banerjee, S. (2024). Models to support forest inventory and small area estimation using sparsely sampled LIDAR: A case study involving G-LiHT lidar in Tanana, Alaska. *Journal of Agricultural, Biological and Environmental Statistics*, 29(4):695–722.

Finley, A. O., Banerjee, S., and MacFarlane, D. W. (2011). A hierarchical model for quantifying forest variables over large heterogeneous landscapes with uncertain forest areas. *Journal of the American Statistical Association*, 106(493):31–48. PMID: 26139950.

Gelfand, A., Diggle, P., Fuentes, M., and Guttorp, P. (2010). *Handbook of Spatial Statistics.* Boca Raton, FL: CRC Press.

Gelman, A. (2007). Struggles with Survey Weighting and Regression Modeling. *Statistical Science*, 22(2):153–164.

Gelman, A., Carlin, J. B., Stern, H. S., Dunson, D. B., Vehtari, A., and Rubin, D. B. (2014). *Bayesian Data Analysis.* Chapman & Hall/CRC, Boca Raton, FL, 3rd edition.

Ghosh, M. and Meeden, G. (1997). *Bayesian Methods for Finite Population Sampling.* Chapman & Hall, London.

Ghosh, M. and Sinha, B. K. (1990). On the consistency between model-and design-based estimators in survey sampling. *Communications in Statistics - Theory and Methods*, 19(2):689–702.

Gneiting, T. and Guttorp, P. (2010). Continuous-parameter spatio-temporal processes. In Gelfand, A., Diggle, P., Fuentes, M., and Guttorp, P., editors, *Handbook of Spatial Statistics*, pages 427–436. CRC Press, Boca Raton, FL.

Guhaniyogi, R. and Banerjee, S. (2018). Meta-kriging: Scalable Bayesian modeling and inference for massive spatial datasets. *Technometrics*, 60(4):430–444.

Harter, T., Dzurella, K., Kourakos, G., Hollander, A., Bell, A., Santos, N., Hart, Q., King, A., Quinn, J., Lampinen, G., Liptzin, D., Rosenstock, T., Zhang, M., Pettygrove, G., and Tomich, T. (2017). Nitrogen Fertilizer Loading to Groundwater in the Central Valley, Final Report to the Fertilizer Research Education Program Projects 11-0301 and 15-0454. Technical report, California Department of Food and Agriculture and University of California Davis, Davis, CA.

Horvitz, D. G. and Thompson, D. J. (1952). A Generalization of Sampling Without Replacement From a Finite Universe. *Journal of the American Statistical Association*, 47(260):663–685.

Kalton, G. (2019). Developments in survey research over the past 60 years: A personal perspective. *International Statistical Review*, 87(S1):S10–S30.

Kish, L. (1965). *Survey Sampling.* John Wiley & Sons, Inc., Hoboken, New Jersey.

Little, R. and Rubin, D. (2002). *Statistical Analysis with Missing Data.* John Wiley & Sons, Inc., Hoboken, New Jersey.

Little, R. J. (1982). Models for nonresponse in sample surveys. *Journal of the American statistical Association*, 77(378):237–250.

Little, R. J. (2004). To Model or Not To Model? Competing Modes of Inference for Finite Population Sampling. *Journal of the American Statistical Association*, 99(466):546–556.

Malec, D. and Sedransk, J. (1985). Bayesian Inference for Finite Population Parameters in Multistage Cluster Sampling. *Journal of the American Statistical Association*, 80(392):897–902.

Pan, S., Zhang, L., Bradley, J. R., and Banerjee, S. (in press). Bayesian inference for spatial-temporal non-gaussian data using predictive stacking. Bayesian Analysis. https://doi.org/10.1214/25-BA1582.

Presicce, L. and Banerjee, S. (2025). Bayesian transfer learning for artificially intelligent geospatial systems: A predictive stacking approach. arXiv:2410.09504v4.

Rao, J. N. K. (2011). Impact of Frequentist and Bayesian Methods on Survey Sampling Practice: A Selective Appraisal. *Statistical Science*, 26(2):240 – 256.

Rao, J. N. K. and Ghangurde, P. D. (1972). Bayesian optimization in sampling finite populations. *Journal of the American Statistical Association*, 67(338):439–443.

Rao, J. N. K. and Molina, I. (2015). *Small Area Estimation.* John Wiley & Sons, Hoboken, NJ, 2nd edition.

Robert, C. P. and Casella, G. (2004). *Monte Carlo Statistical Methods.* Springer, New York, NY, 2nd edition.

Rubin, D. B. (1976). Inference and missing data. *Biometrika*, 63(3):581–592.

Rue, H., Martino, S., and Chopin, N. (2009). Approximate Bayesian inference for latent Gaussian models by using integrated nested Laplace approximations. *Journal of the Royal Statistical Society: Series B (Statistical Methodology)*, 71(2):319–392.

Scott, A. and Smith, T. M. F. (1969). Estimation in Multi-Stage Surveys. *Journal of the American Statistical Association*, 64(327):830–840.

Shannon, E. S., Finley, A. O., Domke, G. M., May, P. B., Andersen, H.-E., Gaines III, G. C., and Banerjee, S. (2024). Toward spatio-temporal models to support national-scale forest carbon monitoring and reporting. *Environmental Research Letters*, 20(1):014052.

Shannon, E. S., Finley, A. O., May, P. B., Domke, G. M., Andersen, H.-E., George, C. III. G., Nothdurft, A., and Banerjee, S. (2025). Leveraging national forest inventory data to estimate forest carbon density status and trends for small areas.

Stein, M. L. (1999). *Interpolation of Spatial Data: Some Theory for Kriging*. Springer, first edition.

Tang, G., Little, R. J., and Raghunathan, T. E. (2003). Analysis of multivariate missing data with nonignorable nonresponse. *Biometrika*, 90(4):747–764.

Tang, W., Zhang, L., and Banerjee, S. (2021). On identifiability and consistency of the nugget in Gaussian spatial process models. *Journal of the Royal Statistical Society: Series B (Statistical Methodology)*, 83(5):1044–1070.

Tomppo, E., Schadauer, K., McRoberts, R. E., Gschwantner, T., Gabler, K., and Ståhl, G. (2009). *National forest inventories: pathways for common reporting (Vol. 1)*. Springer.

Vehtari, A., Gelman, A., and Gabry, J. (2017). Practical Bayesian model evaluation using leave-one-out cross-validation and WAIC. *Statistics and Computing*, 27(5):1413–1432.

Ver Hoef, J. (2002). Sampling and geostatistics for spatial data. *Écoscience*, 9(2):152–161.

Watanabe, S. (2013). A widely applicable Bayesian information criterion. *Journal of Machine Learning Research*, 14(27):867–897.

Westfall, J. A., Coulston, J. W., Moisen, G. G., and Andersen, H. E., editors (2022). *Sampling and estimation documentation for the Enhanced Forest Inventory and Analysis Program: 2022*. General Technical Report NRS-207, U.S. Department of Agriculture, Forest Service, Northern Research Station, Madison, WI.

Wolpert, D. H. (1992). Stacked generalization. *Neural Networks*, 5(2):241–259.

Yao, Y., Vehtari, A., Simpson, D., and Gelman, A. (2018). Using stacking to average Bayesian predictive distributions (with discussion). *Bayesian Analysis*, 13(3):917–1007.

Zhang, H. (2004). Inconsistent estimation and asymptotically equal interpolations in model-based geostatistics. *Journal of the American Statistical Association*, 99(465):250–261.

Zhang, L., Datta, A., and Banerjee, S. (2019). Practical Bayesian modeling and inference for massive spatial datasets on modest computing environments. *Statistical Analysis and Data Mining: The ASA Data Science Journal*, 12(3):197–209.

Zhang, L., Tang, W., and Banerjee, S. (in press). Bayesian geostatistics using predictive stacking. *Journal of the American Statistical Association*. https://doi.org/10.1080/01621459.2025.2566449.

17

A Review of Applications of Extreme Value Theory to Environmental Risk Assessment

Luis Gimeno-Sotelo, Jordan Richards, Arnab Hazra, Linda Mhalla, and Patrícia de Zea Bermudez

Assessing the risk of extreme environmental events is crucial for addressing the challenges brought forth by a world undergoing drastic climate change. The persistent increase in average global temperatures has induced a very strong impact on the environment, through an increase in the frequency and severity of devastating extreme events such as heatwaves or heavy rain storms. The consequences of such extreme events are far-reaching, affecting human lives through property damage and deaths. The field of statistical extreme value theory (EVT) encompasses the methodological tools developed to model such events, and the use of EVT in risk assessments is becoming increasingly pertinent. Predicting the likelihood and severity of future environmental extreme events allows us to prepare for future challenges imposed by our rapidly changing world and ultimately mitigate the risk of exposure to environmental extremes. This chapter revises the most important classical results and models of univariate EVT and reviews their use in applied settings. New contributions to the choice of the threshold in the peaks-over-threshold approach will be presented, alongside discussions of Bayesian, multivariate, and spatial approaches to modelling extreme environmental events. Emerging topics in the field of EVT, such as extensions of classical models, machine learning, and extremal causal inference, will be detailed alongside a review of their usage in recent environmental applications.

17.1 Motivation

The increasing occurrence of extreme environmental events, ranging from devastating floods and prolonged droughts to record-breaking heatwaves, underscores the urgent need for robust methods to assess and manage the associated risks. These events are not only a direct consequence of natural variability in the climate system, but are also exacerbated by anthropogenic factors such as greenhouse gas emissions. The impacts are multifaceted, affecting public health, infrastructure, ecosystems, and economies, with long-term implications for sustainable development and global resilience.

Extreme Value Theory or EVT, provides a statistical framework uniquely suited to understanding these phenomena by focusing on the tail behavior of probability distributions.

DOI: 10.1201/9781003564423-17

Unlike traditional methods that focus on average tendencies, EVT allows to accurately estimate the probability and magnitude of extremes, such as unprecedented rainfall leading to flash floods or extended heatwaves overwhelming energy and healthcare systems. The complexity of environmental systems characterized by interactions across spatial and temporal scales further emphasizes the need for EVT. Extreme events often exhibit dependencies over space and time, as seen in regional flood events influenced by upstream hydrological conditions or heatwaves driven by persistent atmospheric patterns. Climate change introduces additional challenges, as it alters the frequency, intensity, and distribution of extreme events, violating traditional stationarity assumptions. EVT's flexibility in addressing non-stationary processes makes it a crucial tool in a future climate context. By leveraging EVT, researchers and policymakers can better anticipate future extremes, evaluate their potential impacts, and implement evidence-based strategies for resilience and adaptation.

This chapter provides a review of EVT's applications in environmental risk assessment, from the traditional univariate methods to the most recent advances. Special attention will be given to multivariate extremes, which capture dependencies among variables critical for understanding compound events, such as simultaneous rainfall and wind extremes. The discussion extends to spatial extremes, emphasizing models that incorporate spatial dependence to assess regional risks and inform localized decision-making. The chapter will also explore the integration of Bayesian methods, which offer a flexible framework for incorporating prior knowledge and quantifying uncertainty in EVT models. Furthermore, it will delve into new developments, including extended models and distributions that better represent the complexities of real-world phenomena, enabling more accurate and nuanced risk assessments. Emerging approaches, such as causal inference, will be highlighted for their potential to disentangle the underlying drivers of extremes, facilitating targeted mitigation strategies. Additionally, the intersection of EVT with machine learning will be examined, showcasing how advanced algorithms can enhance the modeling of extremes. By synthesizing these topics, the chapter aims to illuminate the evolving landscape of EVT applications to environmental risk assessment, providing insights into how these tools are advancing our understanding of environmental phenomena.

17.2 Classical univariate Extreme Value Theory

EVT provides a statistical framework for modeling and analyzing extreme values (e.g., maximum or minimum observations) of datasets (see, e.g., [1]). Its core idea is to focus on the tails of probability distributions, which are most relevant when assessing extreme events. It has been widely applied to several areas, such as meteorology, hydrology, and engineering (see, e.g., [2, 3, 4, 5, 6, 7, 8]). The two classical univariate approaches for modeling extreme values are the Block Maxima and the Peaks-Over-Threshold (POT) approaches.

Block maxima approach

In the block maxima method, the data are divided into non-overlapping blocks (e.g., annual maxima of daily temperature records). The maximum observation from each block is then extracted, and the statistical behavior of these maxima is modeled. The theoretical foundation of the block maxima method is the **Fisher-Tippett-Gnedenko Theorem** [9, 10]. It states that, if the limit exists, appropriately normalized sample maxima must converge in distribution to the generalized Extreme Value (GEV) distribution with distribution function given by:

$$G(z;\mu,\sigma,\xi) = \begin{cases} \exp\left\{-\left[1+\xi\left(\frac{z-\mu}{\sigma}\right)\right]^{-1/\xi}\right\}, & \text{if} \quad \xi \neq 0, \\ \exp\{-\exp\left(-\frac{z-\mu}{\sigma}\right)\}, & \text{if} \quad \xi = 0, \end{cases} \tag{17.1}$$

for $1 + \xi(z - \mu)/\sigma > 0$, where $\mu \in \mathbb{R}$ is the location, $\sigma > 0$ is the scale, and $\xi \in \mathbb{R}$ is the shape parameter. The GEV reduces to one of three *extremal* types of distribution according to the value of ξ:

- $\xi > 0$: Fréchet distribution (heavy-tailed),
- $\xi < 0$: Weibull distribution (bounded upper tail),
- $\xi \to 0$: Gumbel distribution (exponential-type tail).

POT approach

The POT method uses all observations that exceed a certain high threshold u. The **Pickands-Balkema-de Haan Theorem** [11, 12] provides the theoretical basis to assume that the threshold excesses $Y = X - u$ can be modeled using the **generalized Pareto distribution (GPD)** with distribution function given by:

$$H(y;\sigma_u,\xi) = 1 - \left(1 + \xi\frac{y}{\sigma_u}\right)^{-1/\xi}, \quad \text{for } 1 + \xi\frac{y}{\sigma_u} > 0, \tag{17.2}$$

where $\sigma_u > 0$ is the scale parameter associated with the threshold u, and $\xi \in \mathbb{R}$ is the same shape parameter as defined in (17.1). In fact, the GPD and GEV distributions are intrinsically linked: if threshold excesses of a random variable have a GPD with shape parameter ξ, the corresponding maxima distribution is GEV with the same shape parameter. The GPD is particularly useful when data scarcity at the extreme tail makes block maxima approaches less practical.

Selecting an appropriate u is critical for application of the POT method [13], and this is usually difficult to achieve in practice. If u is too low, the assumptions of the GPD limit may fail; if u is too high, data scarcity can increase estimation uncertainty. Some frequently used methods for threshold selection are the mean excess function and parameter stability inspection. The **mean excess function** (also known as the mean residual life plot) is defined as $e(u) = E(X - u \mid X > u)$. In practice, the empirical mean excess function is plotted over different thresholds and a stable linear relationship indicates an appropriate threshold [14]. The **parameter stability** inspection consists of estimating the GPD parameters (σ_u, ξ) for a sequence of thresholds u, and choosing the threshold for which the parameter estimates stabilize. Moreover, diagnostic plots, such as quantile-quantile (QQ) plots, can be used to visually assess the suitability of the threshold.

Return periods and return levels

Return periods and return levels are critical tools in the management of environmental risk. They are crucial for designing infrastructure, preparing for natural disasters, and mitigating risks associated with extreme climate events. The **return period** T is the expected time interval between the occurrences of extreme events that exceed a specified value z. It is defined as $T = 1/p$, where p is the probability that z is exceeded. The **return level** represents the magnitude of an extreme event that is expected to be exceeded with a certain probability within a specified period. Return levels are often used to describe events such as 100-year floods, extreme temperature events, or extreme wind. The return level quantifies how extreme an event is likely to be, considering its probability of occurrence. For example, a return level corresponding to 100 years describes an event that, on average, is exceeded once every 100 years. For a given exceedance probability p, the return level z_p for a GEV distribution may be obtained as follows:

$$z_p = \mu + \frac{\sigma}{\xi}\left[(-\log(1-p))^{-\xi} - 1\right], \quad \xi \neq 0, \tag{17.3}$$

where μ is the location parameter, σ is the scale parameter, and ξ is the shape parameter of the GEV distribution. If the shape parameter $\xi \to 0$ (Gumbel case), the return level simplifies to:

$$z_p = \mu - \sigma \log\left(-\log(1-p)\right). \tag{17.4}$$

In the POT method, return levels are calculated using the GPD. For a given return period T (e.g., 100 years), the return level z_T is defined as the level expected to be exceeded once in T observations. The formula for the return level z_T under the GPD is

$$z_T = u + \frac{\sigma_u}{\xi}\left[(T\lambda)^{\xi} - 1\right], \quad \text{for } \xi \neq 0, \tag{17.5}$$

where u is the threshold, σ_u is the scale parameter at threshold u, ξ is the shape parameter, and $\lambda = n_u/N$ is the exceedance rate, where n_u is the number of exceedances above the threshold u in N observations. If the shape parameter $\xi \to 0$, the GPD reduces to the exponential distribution, with return level:

$$z_T = u + \sigma_u \log(T\lambda). \tag{17.6}$$

Non-stationarity in the classical univariate setting

Classical EVT assumes stationarity, such that the underlying stochastic process governing extremes does not change over time. However, real-world phenomena are often subject to non-stationary effects due to economic shifts, evolving engineering standards, and other dynamic influences. In the context of climate science, non-stationarity can arise from a variety of sources, including internal climate variability (such as the El Niño–Southern Oscillation) and external forcings (such as anthropogenic greenhouse gas emissions). Non-stationarity refers to changes in the statistical properties of a system over time or across different conditions. A common method to address this is to allow the parameters of extreme value distributions (such as the GEV distribution or GPD) to vary with time or other covariates [1]. For example, apart from time, non-stationary extreme value models for temperature and precipitation may also be fitted using other covariates such as indices representing large-scale atmospheric patterns, e.g., the western Pacific subtropical high or the East Asian summer monsoon [15, 16]. Recent approaches are based on the use of Generalized Additive Models (GAMs) in order to allow flexible, non-linear relationships between covariates and the parameters of the extreme value distributions [17, 18]. Moreover, advances in applications of machine learning make it possible to model non-stationarity in situations where there are many covariates or when there exists complex non-linear interactions between covariates (see Subsection 17.5.2 for further details).

Statistical methods for fitting distributions, goodness-of-fit procedures, and available software

When fitting univariate extreme value distributions, such as the GEV distribution or GPD, two prominent methods are Maximum Likelihood Estimation (MLE) and Probability Weighted Moments (PWMs) [19, 20].

MLE is a widely used approach that estimates distribution parameters by maximizing the likelihood function based on observed data. MLE is asymptotically efficient, meaning it provides optimal parameter estimates as sample size increases. However, it can be sensitive to small sample sizes, model misspecification, or the presence of outliers, which can affect the reliability of the estimates.

PWMs, on the other hand, are robust alternatives that rely on weighted averages of the data, emphasizing specific regions of the distribution, such as the tails. PWMs are particularly advantageous when dealing with extreme value distributions, as they allow for a straightforward estimation of parameters while being less sensitive to outliers when compared to MLE.

To assess the fit of EVT models, goodness-of-fit statistical tests, such as the Cramér-von Mises and the Anderson-Darling tests, are commonly used (see, e.g., [21]). Results of statistical tests can be complemented by graphical diagnostic procedures such as the probability and quantile plots, as well as return level plots.

Statistical programming languages, such as `R`, are typically used for conducting extreme value analyzes, and a large variety of `R` packages related to EVT methods are available, such as the `ismev`, `evd`, `eva`, `mev`, `evir`, or `extRemes` packages (see [22] for a review).

Practical application

To illustrate the classical univariate framework presented in this section, we analyze the annual maximum daily precipitation records of Lisbon (Portugal), spanning the period from 1864 to 2017 [23, 24].

Using the `R` packages `ismev` [25] and `extRemes` [26], we first fit a GEV distribution to the annual maxima assuming stationarity (see equation 17.1), using maximum-likelihood estimation. The following estimated parameters (standard errors in parentheses) are obtained: $\hat{\mu} = 42.75\,(1.16)$, $\hat{\sigma} = 12.57\,(0.90)$, and $\hat{\xi} = 0.12\,(0.07)$. The estimated shape parameter (positive) suggests a Fréchet class, but the Gumbel case is also plausible at 95% confidence level, which indicates that the upper tail is unbounded and slightly heavy. To assess the fit of the GEV distribution, two goodness-of-fit tests are applied using the `R` package `gnFit` [27]: the Cramér-von Mises and Anderson-Darling tests, which suggest a very good fit in both cases (with p-values of 0.993 and 0.953, respectively). Return levels are obtained using the fitted model, indicating the values of precipitation maxima that are exceeded, on average, every T years (see equation 17.3). For example, for $T = 50$ and $T = 150$ years, the values of the estimated return levels (normal-approximated 95% confidence intervals in parentheses) are 105.55 mm (87.24, 123.86) and 129.49 mm (97.22, 161.76), respectively.

However, a careful inspection of the data suggests that non-stationarity may be present in the annual precipitation maxima, both in its magnitude and variability. Thus, a non-stationary GEV model, for which its location and scale parameters are expressed as a linear function of time (t), is fitted to the annual maxima. The following estimated parameters are obtained (standard errors in parentheses): $\hat{\mu} = 38.61\,(1.82) + 0.06\,(0.02)\,t$, $\hat{\sigma} = 8.73\,(1.47) + 0.05\,(0.02)\,t$, $\hat{\xi} = 0.11\,(0.07)$, with $t = 1, ...154$. The estimated parameters, which also suggest a Fréchet class (plausibly a Gumbel class at 95% confidence level), point to a positive trend both in the magnitude and variability of the annual precipitation maxima. Using that fitted non-stationary model, we estimate conditional return levels for each year (see Figure 17.1). In our example, it can be seen that the estimated 50- and 150-year return levels for the precipitation maxima in Lisbon are higher in recent years than at the beginning of the study period, which is consistent with an increase in extreme precipitation during that period.

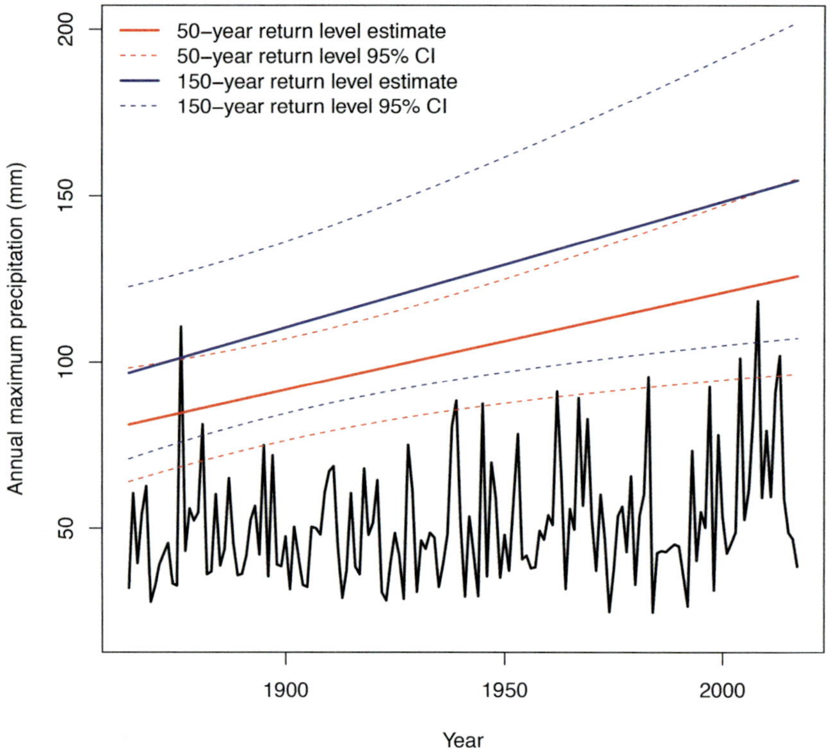

FIGURE 17.1
Annual precipitation maxima (mm) in Lisbon (Portugal) from 1864 to 2017, together with the estimated non-stationary 50- and 150-year return levels and their normal-approximated 95% confidence intervals. Non-stationary return levels are estimated using a non-stationary GEV model with both the location and scale parameters expressed as a linear function of time.

17.3 Bayesian Extreme Value Theory

Similar to most of the areas of statistics, frequentist inference is also prevalent in extreme value analysis, and most discussions in standard textbooks like [1] focus on such classical approaches. However, the Bayesian paradigm is well-known for several theoretical and computational advantages, including providing better uncertainty quantification and handling computations for hierarchical models in a straightforward manner using Markov chain Monte Carlo (MCMC) algorithms. Among the earliest known works on Bayesian extreme value analysis for environmental data, [28] analyzed annual maximum rainfall data obtained from a weather station in southwest England for 54 years between 1932 and 1988, ignoring a few years of data due to excessive missingness. The authors discussed the potential advantages of a Bayesian analysis over a classical one. Subsequently, [29] and [30] analyzed hydrological and precipitation extremes using Bayesian approaches.

The general setup of Bayesian inference is as follows. Suppose $\boldsymbol{Z} = (Z_1, \ldots, Z_n)'$ denotes the vector of possibly dependent but identically distributed observations from some distribution with density function $f(z; \boldsymbol{\theta})$ where $\boldsymbol{\theta}$ denote the vector of parameters in the model $f(\cdot)$. Here, we use f as a generic notation for the density function irrespective of the argument being scalar or vector. For block maxima, $f(\cdot)$ denotes the density of the GEV distribution, obtained from the distribution function $G(\cdot)$ in (17.1), and $\boldsymbol{\theta} = (\mu, \sigma, \xi)'$. Similarly, for threshold exceedances, $f(\cdot)$ denotes the GPD density function, obtained from the distribution function $H(\cdot)$ in (17.2), and $\boldsymbol{\theta} = (\sigma_u, \xi)'$.

Bayesian inference relies on two types of uncertainties: aleatoric uncertainty (also called statistical uncertainty) and epistemic uncertainty (also called systematic uncertainty). Here, the first type refers to the inherent randomness in a process, and the chosen model, i.e., $f(\cdot)$, represents this uncertainty. It arises due to natural variability in data. The second type derives from a lack of knowledge about a system. Here, we do not know the true $\boldsymbol{\theta}$. Thus, although the actual model has specific fixed parameter values, we assume they are random and assign some prior distribution to them. Depending on whether we have background information about the possible values of the parameter, we can quantify the epistemic uncertainty by choosing strong priors under a subjective Bayesian analysis. Here, expert knowledge [28], historical data [31], and physical-model-based predictions [32] are among the possible sources of background information. Otherwise, if we do not have enough information about the parameter, we prefer to select a weakly informative or objective prior. While there is no universal or common choice of priors for the GEV and GPD parameters, the principles of prior selection can be guided by [33]. Several common objective priors, including the Jeffreys prior, Berger-Bernardo reference prior, and maximal data information prior (MDIP), have been discussed in the literature [32, 34, 35, 36, 37, 38]. Some of these priors are not defined for all values of ξ; e.g., the Jeffreys priors exist for both GEV and GPD only if $\xi > -1/2$ [37]. Suppose we denote the parameter space by Θ. For the prior density $f(\boldsymbol{\theta}), \boldsymbol{\theta} \in \Theta$, using Bayes theorem, the posterior density is given by

$$f(\boldsymbol{\theta}|\boldsymbol{Z}) = \frac{f(\boldsymbol{Z}|\boldsymbol{\theta})f(\boldsymbol{\theta})}{\int_{\Theta} f(\boldsymbol{Z}|\boldsymbol{\phi})f(\boldsymbol{\phi})d\boldsymbol{\phi}} \propto f(\boldsymbol{Z}|\boldsymbol{\theta})f(\boldsymbol{\theta}); \tag{17.7}$$

in some cases, the posteriors can be improper unless a minimum number of observations are available. At the same time, some prior choices lead to improper posteriors irrespective of the choice of the parameter space [37, 38]; hence, the prior selection is crucial for valid posterior inferences.

In Bayesian estimation, our main aim lies in obtaining the posterior means, posterior variances, and credible intervals of the parameters of interest. For example, in the GEV case, we often focus on posterior inferences about μ, σ, ξ, and the return level z_p in (17.3). Similarly, in the GPD case, σ_u, ξ, and the return level z_p in (17.5) are generally the key parameters of interest. In the first case, with $\boldsymbol{\theta} = (\mu, \sigma, \xi)'$, the posterior mean of μ, i.e., $E[\mu|\boldsymbol{Z}]$, is given by

$$E[\mu|\boldsymbol{Z}] = \int \mu f(\mu|\boldsymbol{Z})\mathrm{d}\mu = \iiint \mu f((\mu, \sigma, \xi)'|\boldsymbol{Z})\mathrm{d}\mu\mathrm{d}\sigma\mathrm{d}\xi, \tag{17.8}$$

and similarly for the other measures of posterior inferences and other parameters of interest. A key advantage of a Bayesian analysis over a frequentist one lies in its ability to accurately quantify uncertainty. Given $\boldsymbol{Z}$, when we are interested in predicting Z_{n+1}, we use the predictive density $f(Z_{n+1}|\hat{\boldsymbol{\theta}}, \boldsymbol{Z})$, where $\hat{\boldsymbol{\theta}}$ is usually the maximum likelihood estimate of $\boldsymbol{\theta}$ based on $\boldsymbol{Z}$, i.e., we only consider the model uncertainty, and ignore the uncertainty involved in $\boldsymbol{\theta}$. On the other hand, the uncertainty in predicting Z_{n+1} given $\boldsymbol{Z}$ in a Bayesian approach is assessed through the posterior predictive distribution given by

$$f(Z_{n+1}|\boldsymbol{Z}) = \int f(Z_{n+1}, \boldsymbol{\theta}|\boldsymbol{Z})\mathrm{d}\boldsymbol{\theta} = \int f(Z_{n+1}|\boldsymbol{\theta}, \boldsymbol{Z})f(\boldsymbol{\theta}|\boldsymbol{Z})\mathrm{d}\boldsymbol{\theta}; \tag{17.9}$$

i.e., instead of a fixed parameter value of $\boldsymbol{\theta} = \hat{\boldsymbol{\theta}}$ in the frequentist setting, we average out

$f(Z_{n+1}|\boldsymbol{\theta}, \boldsymbol{Z})$ with weights given by $f(\boldsymbol{\theta}|\boldsymbol{Z})$. As a result, the uncertainty in the posterior distribution is correctly incorporated in the analysis [32, 39].

The computation of $f(\boldsymbol{\theta}|\boldsymbol{Z})$ is an integral part of a Bayesian analysis. Conjugate priors are popular choices in Bayesian inference; they allow mathematical pairing between the likelihood function and the prior density, and thus, the posterior distribution can be obtained in a closed form. For example, when Z_i are independently and identically distributed (IID) observations from a Bernoulli(θ) distribution, if we assume a prior $\theta \sim$ Beta(a, b), using straightforward algebra we can show that $\theta|\boldsymbol{Z} \sim$ Beta$(a + \sum_{i=1}^{n} Z_i, b + n - \sum_{i=1}^{n} Z_i)$. Thus, $E(\theta|\boldsymbol{Z}) = (a + \sum_{i=1}^{n} Z_i)/(a + b + n)$. However, in our case, the expressions in (17.8) and (17.9) are not analytically tractable because any closed-form expression of $f((\mu, \sigma, \xi)'|\boldsymbol{Z})$ is not known. The case of GPD is similar. As a result, we use the Monte Carlo approximation of $E[\mu|\boldsymbol{Z}]$ and other measures. In practice, we prefer MCMC algorithms; among various MCMC algorithms, the most common one is the Metropolis-Hastings (MH) algorithm. A detailed discussion on the MH algorithm is presented in Chapter 3 of [39]. The function `posterior` from the `R` package `evdbayes` [40] implements the MH algorithm for the GEV and GPD distributions. The mixing and convergence are usually monitored using trace plots, and the quality of the samples is judged using certain diagnostic checks, including effective sample size (ESS). Because the MCMC samples are autocorrelated, they carry less information than the same number of independent samples, and ESS is a measure to check the effective information carried by the autocorrelated samples. For each model parameter, the ESS value should be higher than 1000, as suggested in [39]. Furthermore, we can calculate the ESS using the function `effectiveSize` from the `R` package `coda` [41].

We illustrate the Bayesian extreme value analysis of annual maximum wind gust data for 42 years between 1971 and 2012 at a weather station at Twente in the Netherlands. The data is available in the `R` package `SpatialExtremes` [42]. We run the MCMC sampler using the `posterior` function for 10,000 iterations with different starting values of $\boldsymbol{\theta}$, and the trace plots are presented in Figure 17.2. For the GEV distribution, under the setting of weakly informative (non-objective) priors, some possible prior choices are $\mu \sim \text{Normal}(0, s_\mu^2)$, $\sigma^* = \log[\sigma] \sim \text{Normal}(0, s_\sigma^2)$, and $\xi \sim \text{Normal}(0, s_\xi^2)$, where s_μ^2, s_σ^2, and s_ξ^2 are usually chosen to be some high values like 10^2 or 100^2 and these choices can also be problem-specific. For the GPD case, we can specify $\sigma_u^* = \log[\sigma_u] \sim \text{Normal}(0, s_{\sigma^*}^2)$, and $\xi \sim \text{Normal}(0, s_\xi^2)$ for some high $s_{\sigma^*}^2$ and s_ξ^2. We choose independent priors with $s_\mu = 100$, $s_\sigma = 100$, and $s_\xi = 10$ following [1] and implement using the function `prior.norm` from the `R` package `evdbayes`. While [37] proved that uniform priors satisfy the posterior propriety for four or more observations, the authors also claim our vague normal priors to be "meaningful". We choose three different starting points of $\boldsymbol{\theta} = (\mu, \sigma, \xi)'$ with $\boldsymbol{\theta}^{(0)} = (200, 50, 0.5)'$, $\boldsymbol{\theta}^{(0)} = (100, 50, 0.5)'$, and $\boldsymbol{\theta}^{(0)} = (400, 50, -0.5)'$. In all cases, the chains converge to the stationary distribution; however, the burn-in period varies across different choices of $\boldsymbol{\theta}^{(0)}$. In practice, although MCMC theoretically converges to the stationary distribution as the number of iterations increases to infinity, different choices of the starting values can impact the length of the burn-in period. For $\boldsymbol{\theta}^{(0)} = (200, 50, 0.5)'$, the chains appear to converge around 250th iteration, and after discarding the burn-in period, the effective sample sizes for μ, σ, and ξ turn out to be approximately 67, 322, and 375 indicating the need for longer chains for drawing posterior inferences.

While the above analysis using the `evdbayes` package and similar vague normal prior choices for ξ in more complex modeling setups involving GEV and GPD are standard in practice [43, 44], [37], and [38] suggest using Jeffreys/MDIP/uniform and reference priors, respectively, as they satisfy posterior propriety.

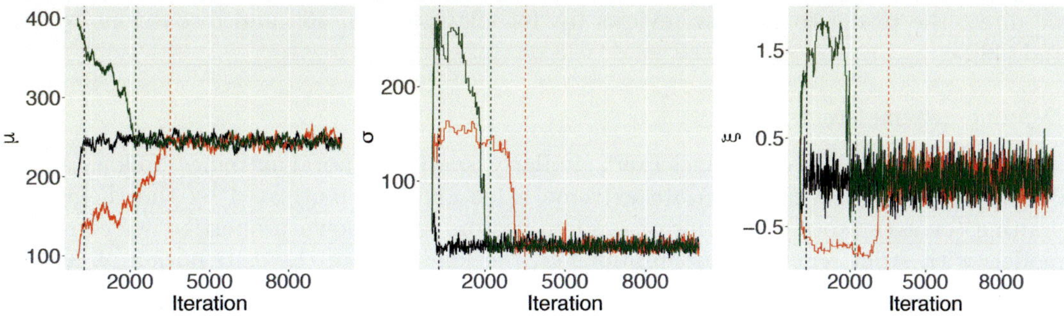

FIGURE 17.2
MCMC chains for the GEV parameters for the annual maximum wind gust data at Twente, the Netherlands. Here, different colors indicate the MCMC chains of μ, σ, and ξ, with different starting values $(\mu^{(0)}, \sigma^{(0)}, \xi^{(0)})'$. The black lines represent the starting value $(200, 50, 0.5)'$, the red lines for the starting value $(100, 50, 0.5)'$, and the green lines for the starting value $(400, 50, -0.5)'$.

17.4 Extremal Dependence and Modeling of Spatio-temporal Extremes

Overview

Robust environmental risk analysis requires not only an understanding of the extremal behavior of univariate random variables, but also the behavior of multivariate, spatio-temporal, and compound (or cascading) extreme events. The most devastating and impactful environmental events are not typically caused by extremes of a single variable, but are instead caused by concurrent extremes across multiple (potentially different) environmental variables, or prolonged extremes across a spatial region or time period. Take, for example, surface water flooding; this is not often a result of extreme rainfall at a single spatial location, but instead it is caused by sustained extreme rainfall over a spatial catchment area. The chance of surface flooding increases greatly with extreme lows of soil moisture in the days or weeks leading up to intense rainfall events, and low soil moisture can be a consequence of heatwaves or droughts; these, themselves, are a consequence of prolonged temperature extremes. Thus, any practitioners interested in mitigating the risk of surface flooding should first concern themselves with learning about the joint extremal behavior of rainfall, soil moisture, and temperatures, and how their extremal/tail dependence changes over space and time.

Multivariate extreme value theory is the sub-field of EVT for understanding the extremal behavior of a multivariate random vector, say $\mathbf{X} = (X_1, \ldots, X_d)^\top \in \mathbb{R}^d$, where d is the number of variables of interest. Multivariate EVT provides the foundations for the modeling of spatio-temporal extreme events; many of the modeling approaches for multivariate extremes that we describe below have been extended to a spatio-temporal setting by considering a stochastic process in place of $\mathbf{X}$, e.g., a spatio-temporal process $\{X(s,t) : s \in \mathcal{S}, t \in \mathbb{R}\}$ indexed by some spatial location $s \in \mathcal{S} \subset \mathbb{R}^p$ (typically $p = 2$ or $p = 3$) and time $t \in \mathbb{R}$. Due to the breadth and mathematical complexity of available approaches for modeling multivariate and spatio-temporal extremes, we provide here only a surface level description of some popular models of extremal dependence, and review their usage in environmental studies. We note that this is by no means an exhaustive list and, for more comprehensive details, we direct the reader to popular reviews by, for example, [45, 46], and discussions by [47] and [48].

Defining extreme events

Modeling of multivariate extremes (and, similarly, spatio-temporal extremes) follows a similar approach to that of univariate extreme value analysis: motivated by the theoretical limiting (or asymptotic) behavior of random vectors with known properties, we construct models with strict regularity assumptions on the joint tails of the distribution of $\mathbf{X}$, i.e., how the probability density or distribution function decays as $\mathbf{X}$ becomes more extreme. Then, as a result of the imposed regularity conditions, these models provide a theoretically-justified way of extrapolating beyond the range of observations to estimate extremely low probabilities. The first difficulty that multivariate EVT tries to overcome is defining what it means to be "extreme" and "in the tails" in a multivariate setting. Whilst univariate data have a natural ordering (we can define extreme univariate events as those being relatively large or small in magnitude; see Section 17.2), multivariate data have no such ordering [49], and there is no consensus in the literature on a definition of multivariate extreme events. Common approaches, to be discussed, include modeling componentwise maxima of $\mathbf{X}$, conditioning on at least one component of $\mathbf{X}$ being large or defining multivariate extremes via exceedances of an aggregate $A(\mathbf{X}) \in \mathbb{R}$ above some high threshold, i.e., $\mathbf{X} \mid A(\mathbf{X}) > u$ (popular choices for A include the maximum, minimum, or some other norm). We illustrate some of these approaches in Figure 17.3. There is no one-size-fits-all approach to modeling multivariate extremes, and different models are appropriate for different characterizations of multivariate extreme events.

Asymptotic models

We first discuss traditional modeling approaches for multivariate extremes, which arise as the natural extensions of the block-maxima and peaks-over-threshold approaches from the univariate to the multivariate setting. Asymptotically-motivated models include the multivariate max-stable and multivariate GPD (or, in the spatio-temporal setting, processes), which arise as the only non-degenerate limits of componentwise maxima and multivariate threshold exceedances, respectively. Max-stable distributions (processes) have seen abundant use in many environmental applications. For brevity, we do not provide an exhaustive list but instead identify some key application areas and references; see reviews by [45] and [51]. Multivariate (spatio-temporal) max-stable models have been used to model, for example, precipitation [52, 53], snowfall depths [54], forest fire weather indices [55], temperature extremes [56, 57], wind gusts [58], air pollution [59], sea levels [60], and river discharge [61]. They have also been used to model extremal dependence between different types of environmental variables, for example, rainfall and storm surge [62], wind speed and dryness [51], and air pollutants [63]. Multivariate and stochastic process extensions of the (univariate) GPD (see, e.g., [64]) have also seen extensive use by environmental statisticians. [65] and [66], and [67] model rainfall extremes in Burkina Faso, Florida, and Switzerland, respectively; [68] perform a landslide risk analysis using Swedish rainfall data; [69] model extremes of air pollutants in Milan, Italy; [70] model European windstorms; [71] and [72] consider the impacts of a changing climate on temperature extremes and precipitation extremes, respectively; [73] consider flood risk analysis in the Danube basin by modeling river discharge extremes.

Multivariate max-stable and Pareto models are *asymptotic*, in the sense that they describe the limiting dependence structure within multivariate extreme events (see [46]). As such, they are suitable models only for data that exhibit asymptotic dependence. A common measure for quantifying the strength of asymptotic dependence is the upper tail dependence

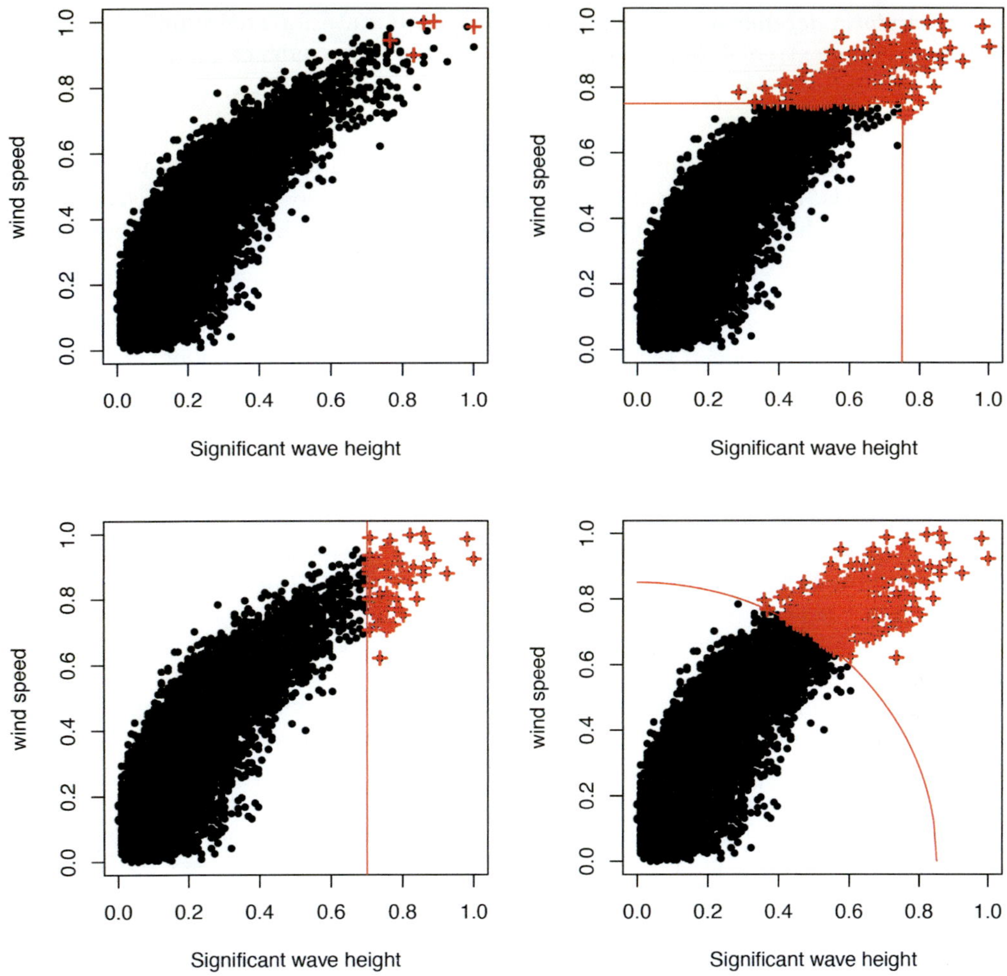

FIGURE 17.3
A pedagogical illustration of different definitions for multivariate extremes. Here, the observed data (black) are three-hourly significant wave heights and wind speeds recorded in the North Sea, UK, over a five year period (see [50]); these data have been scaled to lie within the unit interval $[0, 1]$. The black points denote observations of the random vector (X_1, X_2) and red crosses denote multivariate extreme events due to one of four definitions: (top-left) componentwise annual maxima; (top-right) $(X_1, X_2) \mid \max(X_1, X_2) > 0.75$; (bottom-left) $(X_1, X_2) \mid X_1 > 0.7$; and (bottom-right) $(X_1, X_2) \mid \sqrt{X_1^2 + X_2^2} > 0.85$. Note that the five componentwise annual maxima do not correspond to actual observations of (X_1, X_2).

coefficient, χ. For two random variables X and Y, we define χ as

$$\chi := \lim_{q \to 1} \Pr(F_X(X) > q | F_Y(Y) > q) = \lim_{q \to 1} \Pr(F_Y(Y) > q | F_X(X) > q),$$

where F_X and F_y denote the distribution functions of X and Y, respectively. If $\chi > 0$, we say that X and Y are *asymptotically dependent*, i.e., in the limit, as X (or Y) grows large, there is a non-zero probability that Y (or X) will also be extreme. Whilst models for asymptotic dependence ($\chi > 0$) are mathematically elegant, there have been recent studies arguing against their use in environmental applications (e.g., [74]). We now discuss a class of *sub-asymptotic dependence* models which are appropriate for environmental data that exhibit weakening extremal dependence, or asymptotic independence ($\chi = 0$).

Sub-asymptotic models

Many sub-asymptotic models can be seen as extensions or adaptations of the aforementioned asymptotic models. For example, a popular sub-asymptotic dependence model for maxima are max-infinitely divisible (max-id) distributions/processes, which extend classical max-stable distributions/processes. Examples of their usage in environmental studies include: [75], who model extreme sea levels along the coast of the UK; [76], who model Dutch wind gusts; [77], who model extreme North American precipitation; and [78], who model trends in the spatial extent of European temperature extremes. An alternative model is the inverted max-stable process, which was first developed by [79] and used to model significant wave heights in the North Sea. They can be constructed as the pointwise reciprocal of max-stable processes, and they offer a flexible class of models for asymptotically-independent environmental data (see, e.g., [80] and [81], who both modeled temperature and precipitation extremes). Mixtures of max-stable and inverted max-stable processes, so-called max-mixture processes [79], have also been used to model environmental extremes; for example, [82] used max-mixture process models to study rainfall extremes in Australia.

Max-mixture processes are *hybrid* extremal dependence models, as they can bridge asymptotic dependence and asymptotic independence; they are typically characterized by parameter(s) that determine whether or not the model captures asymptotic dependence or asymptotic independence. Such models are quite popular in environmental applications, as they do not require prior knowledge of the type of extremal dependence exhibited by the data. [83] used a hybrid process model to investigate precipitation extremes in Southern France; [84] used Laplace random fields to model extreme wind gusts in the Netherlands; [85] modeled oceanographic extremes using a hybrid bivariate extremal dependence model.

Inference for max-stable, Pareto, max-id, and max-mixture process models can be computationally-infeasible, or at least very expensive, when observed at a moderate number of spatial locations [74]. Without complex inference procedures (see discussions by [86] and [74]), typical applications of these models are limited to 100s or even 10s, of sampling locations. These limitations have motivated recent interest in another type of sub-asymptotic model, random scale, and location mixtures, which offer a more flexible characterization of extremal dependence (than asymptotic models) by capturing both asymptotic dependence and asymptotic independence [87]. Random scale and locations mixtures are often built upon Gaussian processes, and so applications of these models can benefit from the high-dimensional modeling techniques of classical geostatistics (see, e.g., the review by [88]). For example, by using sparse approximations of Gaussian processes, both [89] and [90] illustrated inference for random scale mixtures observed at 1000s of spatial locations. Spatial applications of random scale and location mixtures include: wind speeds [91, 92], temperatures [93, 94, 95], precipitation [89, 96, 97, 98], significant wave heights [99], streamflow [90], and ozone concentrations [100]. Moreover, space-time extensions of random scale mixture process models have been created by [101], who model Dutch precipitation extremes.

Conditional extremes

Another popular modeling framework for studying extremal dependence, first proposed by [102] and used to model UK air quality extremes, considers the behavior of a random vector (or process) conditional on at least one component of the random vector (or the process observed at a single location or time) being extreme. The multivariate *conditional extremes* framework, and its spatio-temporal extensions (see, e.g., [103, 104]), can flexibly capture both asymptotic dependence and asymptotic independence. These models have thus seen extensive use in environmental risk applications where asymptotic independence is prevalent, including the modeling of river flow extremes (e.g., [105]), flood risk (e.g., [106, 107]), rainfall extremes (e.g., [108, 109]), sea surface temperature extremes (e.g., [110] and [111]), metocean extremes (e.g., [112, 113, 114]), concurrent climate extremes (e.g., [115]), and heatwaves (e.g., [116]).

Geometric modeling of multivariate extremes

The geometric representation for multivariate extremes, known colloquially as *geometric extremes*, is becoming an increasingly popular tool for studying extremal dependence of environmental data. Through the study of the limiting shape of scaled sample clouds, the geometric framework links together several popular extremal dependence modeling frameworks (e.g., conditional extremes and regular variation; see [47]). Environmental applications in this field, and the related field of radial-angular modeling, consider extremes of air pollution [117, 118], fire risk [119], river flow [48, 120], wind speeds [121], and metocean characteristics [50, 122, 123, 124].

Bayesian modeling of spatio-temporal extremes

Our primary goal for extremes data collected across space and time is to estimate the return levels at unobserved locations. Among the initial works, [30], [43], and [125] ignored the spatial dependence of the precipitation data conditioning on spatially-varying coefficients of the GPD and GEV distributions and then model the coefficients using Gaussian process priors. Later, [44] proposed a Gaussian copula for rainfall extremes where the observations are dependent along with the similar latent Gaussian model framework for spatially-varying coefficients. While none of these works assume a max-stable process for the spatial extremes, [53] proposed a hierarchical Bayesian model, namely the Reich-Shaby model, for precipitation extremes that can be expressed hierarchically as a product of independent extremes and a nonlinear function of positive-stable random effects, allowing feasible inference for a large number of spatial locations. [126] proposed a Bayesian composite likelihood method for max-stable processes and used for analyzing precipitation extremes. [127] later extended the Reich-Shaby model for gridded temperature extremes data. When the number of time points is large, [128] explored an approximate Bayesian inference technique, max-and-smooth, for fast computation using stochastic partial differential equations.

A comment on multivariate and spatial non-stationarity

Considering non-stationarity is a crucial aspect of multivariate modeling. Non-stationary multivariate extremes refer to extreme events involving multiple variables whose dependence structure may change over time or depend on other covariates. Recent advances, based on regression-type models for angular measures and Pickands' dependence functions [129, 130, 131, 132] or Bayesian smoothing approaches [133], succeed in capturing covariate-dependent extremal dependence in both asymptotically dependent and asymptotically independent situations. These models allow for more accurate and realistic risk assessment, especially in fields such as climate science, where the joint behavior of extreme events can evolve considerably over time.

Regarding the spatial modeling, many of the aforementioned spatial extremal dependence models make the simplifying assumption that the process is stationary and isotropic, that is, that the strength of extremal dependence is a function of the Euclidean distance between locations. This assumption makes model inference much easier (than in the non-stationary setting), but it is unlikely to be met by data observed over large or topographically-complex spatial domains; this misspecification of the extremal dependence structure can lead to poor estimation of the spatial extent and intensity of extreme events [134]. Non-stationary spatial extremal dependence models have been constructed in one of two ways: i) by deforming the spatial domain or ii) by imposing non-stationarity directly into the extremal dependence structure. In the first setting, the spatial domain is warped into a latent space where one can reasonably assume stationarity and isotropy (proposed by [135] in classical geostatistics). For example, [136] used Gaussian processes, [137] used max-stable processes, and [138] used r-Pareto processes alongside spatial deformations to model rainfall extremes; also, [81] used inverted max-stable processes and [104] used spatial conditional extremes to model Australian temperature extremes. Instead of deforming the domain, one can impose non-stationarity directly into the extremal dependence structure by constructing a non-stationary variogram or covariance function. For example, [134] and [139] included covariates in a max-stable process variogram for non-stationary modeling of Colorado temperature extremes and German rainfall extremes, respectively; a similar approach was used by [78] to construct max-id processes for modeling changes in the spatial extent of European temperature extremes. [140], [141], and [57] used a regionalization of the spatial domain to construct non-stationary max-stable processes to model rainfall and temperature extremes. Non-stationary random scale mixture models have been constructed by, e.g., [96], [98], and [89] for modeling of rainfall extremes, and by [142] for modeling of extreme streamflow.

Moreover, dynamic space-time models provide robust frameworks by incorporating realistic representations of atmospheric systems. One way of dealing with the complexity of these systems is through hierarchical Bayesian dynamic models (see [143, 144] for applications to tropical Pacific sea surface temperatures and cloud intensity, respectively).

Practical application

To illustrate the classical bivariate extreme value analysis, we consider analyzing the annual significant wave height maxima and wind speed maxima recorded in the North Sea, off the north-east coast of the UK, for 52 years between 1958 and 2009. The original data are from three-hourly wave fields derived from the NORA10 hindcast gridded data set, which provides fields over the Norwegian Sea, the North Sea, and the Barents Sea at a spatial resolution of 10 km [145]. We take the wind speed and significant wave heights at one site located between the British Isles and Iceland. A part of this dataset is illustrated in Figure 17.3 to explain different definitions of multivariate extremes. Fitting a nonstationary GEV model individually to these time series, it turns out that no significant trend is visible at a 1% level based on a z-test for both variables. Hence, we fit a bivariate max-stable distribution with a logistic dependence structure and with GEV marginal distributions using the `fbvevd` function from the `R` package `evd` [146]. The estimated parameters for the marginal GEV location, scale, and shape parameters are: for wave height, 9.40, 0.99, and 0.04, respectively; for wind speed, 24.2, 1.47, and -0.20, respectively. Here, the estimated shape parameter values for the two variables are positive and negative, respectively, indicating Fréchet and Weibull classes. However, the corresponding standard errors are 0.10 and 0.09, respectively, and hence the wave height can be safely modeled using a Gumbel distribution as well. In either case, the support of the distribution of extreme wave heights is not bounded above, while it is bounded above for extreme wind speeds. The estimated dependence parameter value is 0.585 with a standard error of 0.076. The logistic dependence parameter varies between zero and one, with zero indicating complete dependence and one indicating independence; thus, we observe a mid-range dependence strength here. We calculate Gaussian-approximated confidence intervals of return level for different return periods using delta method. For low through high return levels, we demonstrate the estimated return levels based on univariate and bivariate GEV distributions in the left and middle panels of

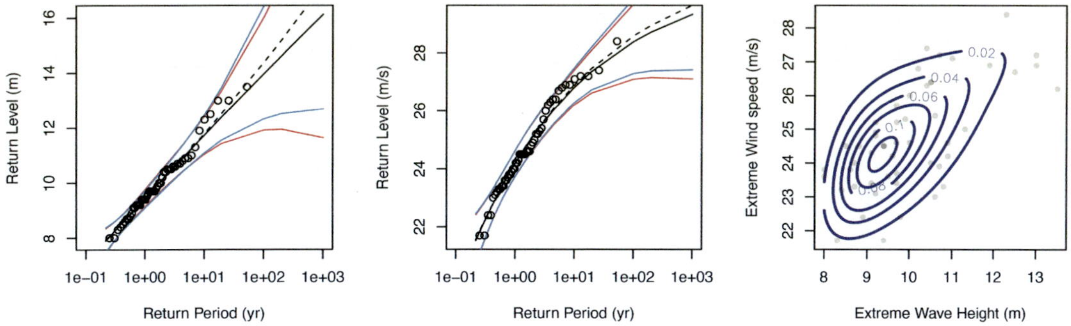

FIGURE 17.4
Left and Middle: Return level estimates for low through high return periods, along with 95% Gaussian-approximated confidence intervals, for significant wave height (in m) and wind speed (in m/s), estimated based on annual maxima recorded in the North Sea, UK, for 52 years between 1958 and 2009, using a bivariate GEV distribution with a logistic dependence function. Here, the black dots denote the empirical return levels, while the black solid and dashed lines represent the estimated return levels based on univariate models and bivariate models, respectively. The red and blue lines represent the 95% approximate confidence intervals based on univariate models and bivariate models, respectively. Right: The scatter plot of annual maxima for both variables, with the contour plots based on the fitted bivariate max-stable distribution.

Figure 17.4. The right panel of Figure 17.4 demonstrates the contours based on the fitted bivariate GEV distribution. The contours are non-elliptical, skewed, oriented diagonally upward from the lower left to the upper right, indicating a positive extremal dependence between the two variables; here the estimated χ-measure is $\hat{\chi} = 2 - 2^{0.585} = 0.500$. Borrowing information across variables, the uncertainty in return level estimation is generally lower (narrower confidence intervals) for both variables, demonstrating the advantage of bivariate modeling over fitting univariate models. The logistic dependence function can be replaced with, for example, an asymmetric logistic, Hüsler-Reiss or negative logistic type dependence functions, and the model selection can be based on information criteria.

17.5 Developments and Advances

17.5.1 Novel Approaches for Threshold Selection and New Distributions

Overview

As choosing an adequate threshold is a key aspect of the POT approach, the development of methods to choose u appropriately is a very active area of research. Moreover, although the traditional approaches to model extreme values, i.e., the block maxima method and the POT approach, have been widely used in environmental applications, different extensions to the GEV and the GPD have been recently proposed to overcome the intrinsic deficiencies of these distributions. Here, we review some of the most relevant advances.

Advances in threshold selection methods

For a classical review of methods for choosing the threshold, see the contributions of [147] and [148]. The difficulty of the problem is such that most methods available in the literature do not provide good answers when applied to real data. Subjectivity in the choice of the threshold is one of the most common characteristics that the practitioner must address. As such, the research is moving toward the development of methods that provide automatic answers and liberate the user from the ordeal of making the final decision.

One contribution is due to [149] and is extensively used in the literature. The authors address the choice of the threshold jointly with the assessment of goodness-of-fit. A set of ordered thresholds is considered and multiple testing is performed, automatically taking into account the false discovery rate. [150] proposed an automatic method to select the threshold that uses L-moments. This approach is very fast to provide an answer and outperforms other recent methods. Their proposal is a relevant improvement of the user selection approach provided by the L-moment Ratio Diagram implemented in the R-package `POT` [151]. The search for a satisfactory automatic method is addressed in a very recent paper by [152]. The proposal is based on the approach developed by [153] and uses bootstrap sampling and QQ-plots considering a grid of thresholds, and has the benefit of taking into account the uncertainty of both the choice of the threshold and the estimation of the corresponding GPD parameters. The authors show by means of a simulation study that the method is very robust to the choice of the set of thresholds. Moreover, it is easy to use in applications.

Some Bayesian contributions are also relevant to mention. [154] proposed a method that uses a common measure of surprise, the p-value. Measures of surprise enable us to determine the consistency between data and a null hypothesis, without considering an alternative hypothesis. The approach is very interesting but it is not automatic and has the significant disadvantage of being very time consuming. [155] resort to cross validation and Bayesian model averaging, considering various thresholds (instead of just one), with the aim of reducing the sensitivity that is transferred to the estimation of the GPD parameters. It is also an interesting method but requires the user intervention which may not be an easy task.

The most recent contribution to the topic of threshold choice seems to be that proposed by [156]. The method is an automatic extended version of the method proposed by [148].

Going beyond the GEV distribution

The existence of lower and upper bounds for the GEV distribution, in the case of non-zero shape parameters, is often unrealistic. In order to enable the GEV to have an unbounded support, [157] proposed a blended model (blended GEV), which combines the Fréchet distribution (right tail) and the Gumbel distribution (left tail). This extended model avoids the lower bound constraint for the GEV distribution with positive shape parameter. It has been successfully used for modeling air pollution [157], precipitation maxima [158, 159], and wildfires [160]. A version of that blended model for the case of the GEV with negative shape parameter has also been developed and applied to temperature and sea level maxima [161]. It combines the Weibull distribution (left tail) and the Gumbel distribution (right tail), thus avoiding the upper bound constraint.

As the GEV distribution may not be appropriate to model extreme events in some cases, other classes of distributions have been developed as generalizations of the GEV. For the sake of finding models which are more suitable for skewed data than the GEV distribution, generalizations, such as the dual gamma generalized extreme value (GGEV), the exponentiated generalized extreme value (EGEV), and the transmuted generalized extreme value (TGEV) distributions, have been proposed [162, 163], with environmental applications such as snowfall, river quotas, and rainfall. For example, the TGEV distribution is especially useful in the case of heavier tails than the simple GEV [164]. Another generalization of the

GEV, the odd generalized extreme value distribution (OGEV), has proven to be especially relevant when analyzing bimodal data, as illustrated for temperature extremes [165].

With the aim of avoiding the asymptotic constraints of the usual block maxima method based on the GEV distribution, non-asymptotic approaches have been widely used in recent years. Among these methodologies, the Metastatistical Extreme Value (MEV) distribution and its variants have gained considerable attention in hydrological research (see e.g., [166, 167, 168, 169]). This metastatistical approach uses the distribution of the ordinary events to make inferences on the distribution of block maxima, therefore allowing to employ as much information as possible for that inference. However, the usefulness of this approach for practical purposes has been questioned because of its redundancy with respect to the parent process, as its knowledge or estimation is needed to determine the metastatistical model [170].

Extending the GPD

Considerable efforts have been made in developing models that fix the problems of using the simple GPD. For example, the hybrid Pareto models [171] are extensions of the GPD to the whole real axis that are constructed by stitching a Gaussian density for the bulk of the distribution and a GPD for the tail, which have been used for modeling environmental variables such as runoff and precipitation [172, 173]. Recently, [119] proposed a blended version of the GPD, which combines the upper tail of the GPD with a semi-parametric model for the bulk of the distribution; they use this to model wildfire extremes in the United States. Another relevant advance in GPD modeling refers to the development of the so-called extended GPD (EGPD) models [174, 175]. EGPD models allow modeling the full range of observations, not only the upper tail, avoiding the threshold selection problem [175]. These models, for which both the upper and lower tails are in accordance with EVT, have been mostly used for analyzing rainfall [175, 176, 177, 178, 179], wildfires [180], and landslides [181]. In a similar direction, [182, 183] developed new parametric classes of distributions for the cases where both lower and upper extremes are relevant, applying their methodological framework to temperature data.

17.5.2 Applications of Machine Learning

Overview

Traditional models for extremes were developed under the assumption that data available for modeling would be sparsely sampled in both space and time. Extremes, by their nature, are rare events, and classical environmental applications of extreme value models had to suffice with small sample sizes. Take, e.g., applications of the block-maxima approach to modeling annual maxima, as described in Section 17.2; classic observational datasets are unlikely to include much more than tens of yearly maxima observations and only at very few measuring stations. However, with modern advancements in data collection techniques via, for example, remote sensing, reanalysis, and high-fidelity climate models, many practitioners of extreme value theory are now inundated with massive datasets that require sophisticated computer algorithms to process. To handle the increasing size of modern datasets, recent advancements in EVT have seen the combination of theoretically-justified modeling approaches for environmental extremes with the scalable and highly flexible algorithms of machine learning. We here discuss a few ways in which machine learning algorithms have been used to facilitate fast and reliable modeling of environmental extremes.

Regression-type models

To learn about the drivers of extremes of environmental processes, we may consider extremes of the distribution of a random variable Y conditional on covariates $\mathbf{X}$. This often proceeds by making parametric assumptions about the tails of $Y \mid \mathbf{X} = \mathbf{x}$ and performing regression, e.g., we assume that the upper-tails of $Y \mid \mathbf{X} = \mathbf{x}$ follow a GPD with scale parameter $\sigma(\mathbf{x}) > 0$ and shape parameter $\xi(\mathbf{x})$, which are functions of the realized covariates $\mathbf{x}$. When the dimension of the covariates becomes large, or the functions $\sigma(\mathbf{x})$ and $\xi(\mathbf{x})$ are expected to be very complex, we might model and estimate the parameter functions using techniques from the machine learning toolbox, e.g., decision trees, gradient boosting, or deep neural networks; see, e.g., [159] for further details. Machine-learning-based extreme value regression models have previously been used for environmental risk analyzes: gradient boosting was used by [184] and [185] to model US wild fire risk and Dutch precipitation extremes, respectively; decision trees were used by [186] to predict the cost of flooding events in France; [187] use random forests for post-processing extreme rainfall forecasts, and neural networks were used by: [160, 188], [180], and [189], to model wildfire risk across the United States, Australia, and Europe; [190], to estimate flood risk in the Swiss Danube river basin; [172], to model rainfall-runoff in France; [191], to model landslide hazard in the Himalayas; [173, 192, 193], and [159], to model precipitation extremes.

Fast inference

Deep learning models can also speed-up the modeling of environmental extremes. Many of the aforementioned models for multivariate and spatio-temporal extreme events (in Section 17.4) can be computationally difficult to estimate, particularly when the number of observations or sampling locations/variables (i.e., d) grows large. However, the computational issues related to the estimation of extremal dependence models can often be circumvented using simulation-based inference and deep learning; in brief, one trains a neural network model (on a vast amount of simulated training data examples) to map from data to important statistical objects, e.g., parameters, likelihood functions, or posterior densities/summary statistics (see review by [194]). The use of neural networks and the accessible software packages that facilitate their fast evaluation, means that simulation-based approaches to fitting extremal dependence model are orders-of-magnitude faster than classical inference approaches, and large scale studies of spatial environmental extremes are now computationally feasible. For example, simulation-based inference approaches were adopted by: [86] to model Arabian air pollution extremes; [90, 142, 195] and [196] to model US temperature extremes; [197] to model solar activity; [101] to model Dutch space-time rainfall extremes; and [198] to model Red Sea surface temperatures.

Extremal dependence modeling

While neural networks can be used to hasten the modeling of multivariate or spatio-temporal environmental extremes, they can also be used to increase the flexibility of models. For example, [50] and [124] used neural networks in the geometric framework to model ocean storm risk off the coast of the UK; [199] used neural networks to construct flexible multivariate extreme value distributions, and model extreme ozone levels and wind gusts; [138] use deep models to build non-stationary Pareto processes for UK precipitation. Moreover, some recent works have shown that generative AI models, when combined with rigorous methods for modeling extremal dependence, can improve inference on environmental extreme events: [200] use generative adversarial networks to model precipitation extremes and [201] also model temperature extremes; [202] and [203] generate river discharge and surface temperature extremes, respectively, using variational auto-encoders; [121] couch normalizing flows in the geometric extremes framework to model US wind speed extremes.

Further use cases

Finally, we note that machine learning methods can also help to aid in the decision-making process for practitioners of environmental risk modeling. For example, [204] and [205] used neural networks to classify environmental data as asymptotically dependent (or not), which helps modelers to choose an appropriate extremal dependence model (see Section 17.4) for their data. [206] perform extreme conformal prediction in an application to flood risk forecasting. [207] use Kolmogorov-Arnold networks to model the global triggering probability of tsunamis and tropical cyclones. Unsupervised machine learning algorithms, e.g., clustering, can also be used to identify patterns in environmental datasets, such as spatial regions that exhibit similar extremal characteristics: [208] cluster air pollution extremes, [209] and [141] cluster precipitation extremes, [210] cluster temperature extremes, and [211] and [212] cluster joint precipitation and wind extremes.

17.5.3 Causal Inference

Overview

When dealing with extreme events, a causal perspective is essential to distinguish mere associations from actual drivers, assess how interventions or external forces alter the likelihood of extremes, and make meaningful counterfactual claims about whether an event could have occurred under different conditions. This perspective is crucial for tackling pressing environmental problems such as attributing record-breaking heatwaves to anthropogenic emissions or understanding the impact of land-use change on flood risks. Until only recently, the literature on causal inference focused on average or quantile treatment causal effects and causal effects on extreme events and hence the tail distribution was not considered. [213] explored the problem of understanding causal effects of a binary treatment on a heavy-tailed variable of interest. The authors build on previous work on quantile treatment effects and perform tail extrapolation using a Hill-type estimator of the shape parameter that takes into account the presence of observed confounding through propensity score adjustments. The work of [213] bridges the potential outcome framework with extreme value theory in an elegant manner. While the authors discuss the application of their methodology to study the effect of college education on wages, one can think of environmental applications where effects of anthropogenic forcings on the tail of a climate or hydrological variable are studied. However, the developed methodology is tailored for univariate heavy-tailed outcomes and is thus inadequate for settings where causal effects on the joint tail distribution of a random vector are of interest. For such settings, the distributional links between the variables in the system of interest should be investigated through modeling of the joint tail behavior using multivariate EVT and formalizing the generating mechanism using structural causal models (SCMs). SCMs were suggested by [214] as a tool to mathematically encode assumptions on how variables are generated. For instance, if the system of interest is a network of d catchments where daily discharges $V_1, \ldots, V_d$ are observed, we assume that there exist unknown measurable functions $f_1, \ldots, f_d$ such that the observed variables satisfy

$$V_i := f_i(PA_i, \varepsilon_i) \quad i = 1, \ldots, d,$$

where PA_i is the set of random variables that are parents of V_i and $\varepsilon_0, \varepsilon_1, \ldots$ are random variables sometimes called exogenous variables or error terms. SCMs are defined such that the resulting graph is a directed acyclic graph (DAG), i.e., cycles are not allowed. The source node that is the variable with no parents, can represent the most upstream catchment in the network. Although the structural function f_i need only be measurable, when the interest is in the causal generating mechanisms of the variables when the system is under extreme conditions, specific choices for f_i were considered in the literature. For instance, extreme events in one variable of the system could be the result of extreme shocks in the noise variables propagating deterministically through the network or the result of an extreme isolated noise variable. Such characterization gives rise to the so-called recursive max-linear models that were introduced by [215] and defined as

$$X_i := \max_{j \in PA(i)} \max(\gamma_{ji} X_j, \varepsilon_i), \quad i = 1, \dots, d, \tag{17.10}$$

where the noise variables are assumed IID and regularly varying with index $\alpha > 0$ [216]. [217] proposed another way of generating extreme nodes. The authors assume that an extreme node is the result of either an extreme external noise variable or an extreme weighted sum of observations from its parents in the system. The resulting model, termed linear structural causal model (LSCM) is defined as

$$X_i := \sum_{j \in PA(i)} \beta_{ji} X_j + \varepsilon_i, \quad i = 1, \dots, d, \tag{17.11}$$

where ε_i are defined as in (17.10). The random vector $\mathbf{X} = (X_1, \dots, X_d)^\top$ as defined in (17.10) and (17.11) is regularly varying and its joint tail distribution converges to a multivariate max-stable distribution. The dependence structure of the limiting distribution is discrete with mass and weights described by the coefficients of the generating SCM. While, in both cases, extremes are described by multivariate max-stable models that do not possess a density, these form a dense subset of asymptotic multivariate extreme value distributions [218]. Such a property makes the heavy-tailed SCMs attractive as they can approximate any dependence structure between extremes arbitrarily well and possess an intuitive description of the extremal causal links through their regression-like formulation.

Heavy-tailed SCMs are well studied in the literature in terms of their identifiability and learning their structure from observational data. Here, identifiability refers to the problem of expressing the structural causal links f_i as a function of observable quantities only, i.e., no interventional quantities. We refer the reader to [219] for a recent review of identifiability results of heavy-tailed SCMs.

Causal structure learning

The task of structure learning consists of identifying the variables of the system (nodes in a DAG) that are causally related and the direction of their edges. In a multivariate setup, this boils down to estimating the topological order of the graph, which is the order of the nodes such that parents appear before their children. Most of the learning methods for extremes rely on a notion of asymmetry in the tail behavior that is induced by the causal link. For instance, one expects that extreme observations in the parent node propagate to their children but the relation in the other way does not necessarily hold. This is clearly the case in a network of catchments with upstream catchments being parents of some downstream catchments. As an illustration, the papers discussed below address the problem of uncovering causal links between extremes events in water discharges of basins. Such an application is unarguably important from a water and disaster management point of view but is also relevant from a statistical viewpoint as the limiting models for multivariate extremes are adequate for extreme water discharges [61] and the network structure driven by physical mechanisms reflects the ground truth about the causal connections. Beyond hydrological applications, heavy-tailed SCMs have also been employed in financial [220] and nutritional contexts [221], demonstrating their flexibility in modeling extremal causal mechanisms. [222] construct a bivariate tail score that exploits the asymmetry in the information in the causal and anti-causal directions to uncover causal relations in the joint tail of an asymptotically dependent random vector. Building on similar ideas but tightly related to the LSCM (17.11), [217] develop a causal discovery method using a so-called causal tail coefficient. The authors show consistency of their algorithm and discuss its robustness to the presence of confounders. [223] rely on the causal tail coefficient to propose testing for (direct) causal effects while accounting for confounding through a regression setup. A different line of work assumes that the recursive max-linear model holds and relies on the regular variation of its noise variables to express causal links as asymmetries in measures of dependence in the limiting max-stable distribution. Such an approach is proposed in [224, 225], with extensions to settings with hidden nodes in [221] and settings with propagating noise in [226]. Finally, [227] propose a general framework for studying structural causal models for extremes by defining the notion of extremal interventions and identifying the SCM emerging in the tails from the SCM describing the data generating process. Additionally, the authors extend a popular causal structure learning method in the bulk, namely the PC algorithm [228], by constructing a test for extremal conditional independence. The PC algorithm is also used for testing conditional tail independence in the framework of [229], who construct linear structural causal models using regularly-varying random vectors.

Extreme event attribution

A different take on causal inference for extreme climate and weather events is addressed in a recent line of work on extreme event attribution (EEA). This research provides important advances in using EVT, causal inference, and multivariate statistics to make counterfactual claims about the influence of external forcing on extreme events. [230] constitutes a notable methodological contribution by proposing a counterfactual attribution framework based on multivariate peaks-over-threshold models combined with probabilistic quantities of necessary and sufficient causation. [231] use Bayesian networks to assess the causal influence of different drivers, such as anthropogenic climate signals, on flood peak variability. While most EEA frameworks focus on exceedances above fixed thresholds, [232] offer an innovative approach by re-framing attribution in terms of record-breaking events. Their work introduces a new record-based fraction of attributable risk metric as well as parametric and non-parametric estimators of such quantities. Applied to records of yearly maxima of temperature where historical runs of a numerical climate model are compared to projections under a specific SSP scenario, they attribute increasing record frequency to anthropogenic forcing, especially in the tropics.

17.6 Summary and Conclusions

The increasing frequency and severity of climate-driven extreme events, such as catastrophic floods, prolonged droughts, and record-breaking heatwaves, have created an urgent demand for statistical tools that can support reliable risk assessment and informed decision-making. EVT provides a rigorous framework tailored to modeling events that lie far from typical environmental conditions. This chapter presents a structured overview of EVT key contributions to environmental risk assessment. It commences with classical univariate approaches which form the foundation of the theory, before introducing Bayesian methods that allow for incorporating prior knowledge and provide improved uncertainty quantification. The discussion then progresses to multivariate and spatio-temporal methodologies, which are crucial for analyzing data that exhibit dependence across variables and throughout space and time. Recent advancements in threshold selection and the development of extensions of the traditional extreme value models, which more effectively capture the complexity of environmental data, are also examined. The chapter concludes by examining emerging approaches which are at the intersection of EVT and data science, including the application of machine learning methods and the growing field of extremal causal inference. By integrating foundational theory with state-of-the-art methods, this chapter aims to provide both a comprehensive reference and a forward-looking perspective on how EVT continues to evolve to meet the challenges of environmental extremes in a changing climate.

Acknowledgments

The work by Luis Gimeno-Sotelo and Patrícia de Zea Bermudez is funded by national funds through FCT – Fundação para a Ciência e a Tecnologia, I.P., under CEAUL Research Unit, UID/00006/2025DOI: https://doi.org/10.54499/UID/00006/2025, and by the European Union – NextGenerationEU through the project UID/PRR/00006/2025, DOI:https://doi.org/10.54499/UID/PRR/00006/2025. They also acknowledge the INTEGRISK project, funded by CIÊNCIAS and FCiências.ID's own resources, under the CIÊNCIAS Internal Call for Research Projects, 2025 Edition. Arnab Hazra was partially supported by the Startup Research Grant, Science and Engineering Research Board, Department of Science and Technology, Government of India, with grant number SERB/MATH/2023632.

Bibliography

[1] S. Coles. *An Introduction to Statistical Modeling of Extreme Values.* Springer, 2001.

[2] K. Engeland, H. Hisdal, and A. Frigessi. Practical extreme value modelling of hydrological floods and droughts: a case study. *Extremes*, 7:5–30, 2004.

[3] O. Grigg and J. Tawn. Threshold models for river flow extremes. *Environmetrics*, 23(4):295–305, 2012.

[4] A. M. Alonso, P. de Zea Bermudez, and M. G. Scotto. Comparing generalized Pareto models fitted to extreme observations: an application to the largest temperatures in Spain. *Stochastic Environmental Research and Risk Assessment*, 28:1221–1233, 2014.

[5] G. D'Amico, F. Petroni, and F. Prattico. Wind speed prediction for wind farm applications by extreme value theory and copulas. *Journal of Wind Engineering and Industrial Aerodynamics*, 145:229–236, 2015.

[6] W. K. Huang, M. L. Stein, D. J. McInerney, S. Sun, and E. J. Moyer. Estimating changes in temperature extremes from millennial-scale climate simulations using generalized extreme value (GEV) distributions. *Advances in Statistical Climatology, Meteorology and Oceanography*, 2(1):79–103, 2016.

[7] L. Gimeno-Sotelo and L. Gimeno. Where does the link between atmospheric moisture transport and extreme precipitation matter? *Weather and Climate Extremes*, 39:100536, 2023.

[8] T. R. Kjeldsen and I. Prosdocimi. Use of peak over threshold data for flood frequency estimation: An application at the UK national scale. *Journal of Hydrology*, 626:130235, 2023.

[9] R. A. Fisher and L. H. C. Tippett. Limiting forms of the frequency distribution of the largest or smallest member of a sample. *Mathematical Proceedings of the Cambridge Philosophical Society*, 24(2):180–190, April 1928.

[10] B. Gnedenko. Sur la distribution limite du terme maximum d'une série aléatoire. *Annals of Mathematics*, 44(3):423–453, 1943.

[11] A. A. Balkema and L. De Haan. Residual life time at great age. *The Annals of Probability*, 2(5):792–804, 1974.

[12] J. Pickands III. Statistical inference using extreme order statistics. *The Annals of Statistics*, pages 119–131, 1975.

[13] S. Beguería. Uncertainties in partial duration series modelling of extremes related to the choice of the threshold value. *Journal of Hydrology*, 303(1-4):215–230, 2005.

[14] A. C. Davison and R. L. Smith. Models for exceedances over high thresholds. *Journal of the Royal Statistical Society Series B: Statistical Methodology*, 52(3):393–425, 1990.

[15] M. Gao and H. Zheng. Nonstationary extreme value analysis of temperature extremes in China. *Stochastic Environmental Research and Risk Assessment*, 32:1299–1315, 2018.

[16] M. Gao, D. Mo, and X. Wu. Nonstationary modeling of extreme precipitation in China. *Atmospheric Research*, 182:1–9, 2016.

[17] B. D. Youngman. Generalized additive models for exceedances of high thresholds with an application to return level estimation for US wind gusts. *Journal of the American Statistical Association*, 114(528):1865–1879, 2019.

[18] B. D. Youngman. evgam: An R package for generalized additive extreme value models. *Journal of Statistical Software*, 103:1–26, 2022.

[19] P. de Zea Bermudez and S. Kotz. Parameter estimation of the generalized Pareto distribution—part I. *Journal of Statistical Planning and Inference*, 140(6):1353–1373, 2010.

[20] A. Ferreira and L. de Haan. On the block maxima method in extreme value theory: PWM estimators. *Annals of Statistics*, 43(1):276–298, 2015.

[21] V. Choulakian and M. A. Stephens. Goodness-of-fit tests for the generalized Pareto distribution. *Technometrics*, 43(4):478–484, 2001.

[22] L. R. Belzile, C. Dutang, P. J. Northrop, and T. Opitz. A modeler's guide to extreme value software. *Extremes*, 26(4):595–638, 2023.

[23] M. A. Valente. Digitalização de Dados Históricos de Meteorologia do Instituto Geofísico do Infante D. Luís. Relatório de Estágio da Carreira Técnica Superior, Instituto Dom Luiz, Universidade de Lisboa, August 2007. 53pp.

[24] M. C. Gallego, R. M. Trigo, J. M. Vaquero, M. Brunet, J. A. García, J. Sigró, and M. A. Valente. Trends in frequency indices of daily precipitation over the Iberian Peninsula during the last century. *Journal of Geophysical Research: Atmospheres*, 116(D2), 2011.

[25] J. E. Heffernan and A. G. Stephenson. *ismev: An Introduction to Statistical Modeling of Extreme Values*, 2018. R package version 1.42.

[26] E. Gilleland and R. W. Katz. extRemes 2.0: An Extreme Value Analysis Package in R. *Journal of Statistical Software*, 72(8):1–39, 2016.

[27] A. Saeb. *gnFit: Goodness of Fit Test for Continuous Distribution Functions*, 2018. R package version 0.2.0.

[28] S. G. Coles and J. A. Tawn. A Bayesian analysis of extreme rainfall data. *Journal of the Royal Statistical Society Series C: Applied Statistics*, 45(4):463–478, 1996.

[29] R. W. Katz, M. B. Parlange, and P. Naveau. Statistics of extremes in hydrology. *Advances in Water Resources*, 25(8-12):1287–1304, 2002.

[30] D. Cooley, D. Nychka, and P. Naveau. Bayesian spatial modeling of extreme precipitation return levels. *Journal of the American Statistical Association*, 102(479):824–840, 2007.

[31] J. B. Elsner and T. H. Jagger. A hierarchical Bayesian approach to seasonal hurricane modeling. *Journal of Climate*, 17(14):2813–2827, 2004.

[32] N. Bousquet. Bayesian extreme value theory. In N. Bousquet and P. Bernardara, editors, *Extreme Value Theory with Applications to Natural Hazards: From Statistical Theory to Industrial Practice*, pages 271–325. Springer International Publishing, Cham, Switzerland, 2021.

[33] J. O. Berger, D. Ríos Insua, and F. Ruggeri, editors. *Robust Bayesian Analysis*, volume 152 of *Lecture Notes in Statistics*. Springer-Verlag, New York, NY, 2000.

[34] A. Zellner. Models, prior information, and Bayesian analysis. *Journal of Econometrics*, 75(1):51–68, 1996.

[35] S. Kotz and S. Nadarajah. *Extreme Value Distributions: Theory and Applications*. Imperial College Press / World Scientific, London / River Edge, NJ, 2000.

[36] M. E. Castellanos and S. Cabras. A default Bayesian procedure for the generalized Pareto distribution. *Journal of Statistical Planning and Inference*, 137(2):473–483, 2007.

[37] P. J. Northrop and N. Attalides. Posterior propriety in Bayesian extreme value analyses using reference priors. *Statistica Sinica*, 26(2):721–743, 2016.

[38] L. Zhang and B. A. Shaby. Reference priors for the generalized extreme value distribution. *Statistica Sinica*, 33(3):2185–2208, 2024.

[39] B. J. Reich and S. K. Ghosh. *Bayesian Statistical Methods*. Chapman and Hall/CRC, 2019.

[40] A. Stephenson and M. Ribatet. *evdbayes: Bayesian Analysis in Extreme Value Theory*, 2023. R package version 1.1-3.

[41] M. Plummer, N. Best, K. Cowles, and K. Vines. CODA: Convergence Diagnosis and Output Analysis for MCMC. *R News*, 6(1):7–11, 2006.

[42] M. Ribatet. *SpatialExtremes: Modelling Spatial Extremes*, 2022. R package version 2.1-0.

[43] H. Sang and A. E. Gelfand. Hierarchical modeling for extreme values observed over space and time. *Environmental and Ecological Statistics*, 16(3):407–426, 2009.

[44] H. Sang and A. E. Gelfand. Continuous spatial process models for spatial extreme values. *Journal of Agricultural, Biological, and Environmental Statistics*, 15:49–65, 2010.

[45] A. Davison and R. Huser. Statistics of extremes. *Annual Review of Statistics and Its Application*, 2:203–235, 4 2015.

[46] R. Huser and J. L. Wadsworth. Advances in statistical modeling of spatial extremes. *Wiley Interdisciplinary Reviews: Computational Statistics*, 14(1):e1537, 2022.

[47] N. Nolde and J. L. Wadsworth. Linking representations for multivariate extremes via a limit set. *Advances in Applied Probability*, 54(3):688–717, 2022.

[48] I. Papastathopoulos, L. De Monte, R. Campbell, and H. Rue. Statistical inference for radially-stable generalized Pareto distributions and return level-sets in geometric extremes. *preprint arXiv:2310.06130*, 2023.

[49] V. Barnett. The ordering of multivariate data. *Journal of the Royal Statistical Society. Series A (General)*, 139:318–344, 1976.

[50] C. J. Murphy-Barltrop, R. Majumder, and J. Richards. Deep learning of multivariate extremes via a geometric representation. *preprint arXiv:2406.19936*, 2024.

[51] D. Cooley, B. D. Hunter, and R. L. Smith. Univariate and multivariate extremes for the environmental sciences. In *Handbook of environmental and ecological statistics*, pages 153–180. Chapman and Hall/CRC, 2019.

[52] S. A. Padoan, M. Ribatet, and S. A. Sisson. Likelihood-based inference for max-stable processes. *Journal of the American Statistical Association*, 105(489):263–277, 2010.

[53] B. J. Reich and B. A. Shaby. A hierarchical max-stable spatial model for extreme precipitation. *The Annals of Applied Statistics*, 6(4):1430, 2012.

[54] J. Blanchet and A. C. Davison. Spatial modeling of extreme snow depth. *The Annals of Applied Statistics*, 5(3):1699–1725, 2011.

[55] A. G. Stephenson, B. A. Shaby, B. J. Reich, and A. L. Sullivan. Estimating spatially varying severity thresholds of a forest fire danger rating system using max-stable extreme-event modeling. *Journal of Applied Meteorology and Climatology*, 54(2):395–407, 2015.

[56] E. Thibaud, J. Aalto, D. S. Cooley, A. C. Davison, and J. Heikkinen. Bayesian inference for the Brown-Resnick process, with an application to extreme low temperatures. *The Annals of Applied Statistics*, pages 2303–2324, 2016.

[57] X. Shao, A. Hazra, J. Richards, and R. Huser. Flexible modeling of nonstationary extremal dependence using spatially fused LASSO and ridge penalties. *Technometrics*, 67(1):97–111, 2024.

[58] M. Oesting, M. Schlather, and P. Friederichs. Statistical post-processing of forecasts for extremes using bivariate Brown-Resnick processes with an application to wind gusts. *Extremes*, 20:309–332, 2017.

[59] S. Vettori, R. Huser, and M. G. Genton. Bayesian modeling of air pollution extremes using nested multivariate max-stable processes. *Biometrics*, 75(3):831–841, 2019.

[60] A. G. Stephenson. High-dimensional parametric modelling of multivariate extreme events. *Australian & New Zealand Journal of Statistics*, 51(1):77–88, 2009.

[61] P. Asadi, A. C. Davison, and S. Engelke. Extremes on river networks. *The Annals of Applied Statistics*, 9(4):2023 – 2050, 2015.

[62] F. Zheng, S. Westra, M. Leonard, and S. A. Sisson. Modeling dependence between extreme rainfall and storm surge to estimate coastal flooding risk. *Water Resources Research*, 50(3):2050–2071, 2014.

[63] B. Beranger and S. Padoan. Extreme dependence models. In *Extreme Value Modeling and Risk Analysis*, pages 345–372. Chapman and Hall/CRC, 2016.

[64] A. Ferreira and L. de Haan. The generalized Pareto process; with a view towards application and simulation. *Bernoulli*, 20(4):1717–1737, 2014.

[65] E. Thibaud and T. Opitz. Efficient inference and simulation for elliptical Pareto processes. *Biometrika*, 102(4):855–870, 2015.

[66] R. de Fondeville and A. C. Davison. High-dimensional peaks-over-threshold inference. *Biometrika*, 105(3):575–592, 2018.

[67] S. Béwentaoré and D. Barro. Space-time trend detection and dependence modeling in extreme event approaches by functional peaks-over-thresholds: Application to precipitation in Burkina Faso. *International Journal of Mathematics and Mathematical Sciences*, 2022(1):2608270, 2022.

[68] A. Kiriliouk, H. Rootzén, J. Segers, and J. L. Wadsworth. Peaks over thresholds modeling with multivariate generalized Pareto distributions. *Technometrics*, 61(1):123–135, 2019.

[69] M. Falk, S. A. Padoan, and F. Wisheckel. Generalized Pareto copulas: A key to multivariate extremes. *Journal of Multivariate Analysis*, 174:104538, 2019.

[70] R. de Fondeville and A. C. Davison. Functional peaks-over-threshold analysis. *Journal of the Royal Statistical Society Series B: Statistical Methodology*, 84(4):1392–1422, 2022.

[71] D. Healy, J. Tawn, P. Thorne, and A. Parnell. Inference for extreme spatial temperature events in a changing climate with application to Ireland. *Journal of the Royal Statistical Society Series C: Applied Statistics*, page qlae047, 2024.

[72] P. Zhong, M. Brunner, T. Opitz, and R. Huser. Spatial modeling and future projection of extreme precipitation extents. *Journal of the American Statistical Association*, 2024. To appear.

[73] S. Engelke and A. S. Hitz. Graphical models for extremes. *Journal of the Royal Statistical Society Series B: Statistical Methodology*, 82(4):871–932, 2020.

[74] R. Huser, T. Opitz, and J. Wadsworth. Modeling of spatial extremes in environmental data science: Time to move away from max-stable processes. *Environmental Data Science*, 2024. To appear.

[75] S. A. Padoan. Extreme dependence models based on event magnitude. *Journal of Multivariate Analysis*, 122:1–19, 2013.

[76] R. Huser, T. Opitz, and E. Thibaud. Max-infinitely divisible models and inference for spatial extremes. *Scandinavian Journal of Statistics*, 48(1):321–348, 2021.

[77] G. P. Bopp, B. A. Shaby, and R. Huser. A hierarchical max-infinitely divisible spatial model for extreme precipitation. *Journal of the American Statistical Association*, 116(533):93–106, 2021.

[78] P. Zhong, R. Huser, and T. Opitz. Modeling nonstationary temperature maxima based on extremal dependence changing with event magnitude. *The Annals of Applied Statistics*, 16(1):272–299, 2022.

[79] J. L. Wadsworth and J. A. Tawn. Dependence modelling for spatial extremes. *Biometrika*, 99(2):253–272, 2012.

[80] A. C. Davison, R. Huser, and E. Thibaud. Geostatistics of dependent and asymptotically independent extremes. *Mathematical Geosciences*, 45:511–529, 2013.

[81] J. Richards and J. L. Wadsworth. Spatial deformation for nonstationary extremal dependence. *Environmetrics*, 32(5):e2671, 2021.

[82] J.-N. Bacro, C. Gaetan, and G. Toulemonde. A flexible dependence model for spatial extremes. *Journal of Statistical Planning and Inference*, 172:36–52, 2016.

[83] J.-N. Bacro, C. Gaetan, T. Opitz, and G. Toulemonde. Hierarchical space-time modeling of asymptotically independent exceedances with an application to precipitation data. *Journal of the American Statistical Association*, 115(530):555–569, 2020.

[84] T. Opitz. Modeling asymptotically independent spatial extremes based on Laplace random fields. *Spatial Statistics*, 16:1–18, 5 2016.

[85] J. Wadsworth, J. Tawn, A. Davison, and D. Elton. Modelling across extremal dependence classes. *Journal of the Royal Statistical Society Series B: Statistical Methodology*, 79(1):149–175, 2017.

[86] J. Richards, M. Sainsbury-Dale, A. Zammit-Mangion, and R. Huser. Likelihood-free neural Bayes estimators for censored inference with peaks-over-threshold models. *Journal of Machine Learning Research*, 25(390):1–49, 2024.

[87] S. Engelke, T. Opitz, and J. Wadsworth. Extremal dependence of random scale constructions. *Extremes*, 22(4):623–666, 2019.

[88] M. J. Heaton, A. Datta, A. O. Finley, R. Furrer, J. Guinness, R. Guhaniyogi, F. Gerber, R. B. Gramacy, D. Hammerling, M. Katzfuss, et al. A case study competition among methods for analyzing large spatial data. *Journal of agricultural, biological and environmental Statistics*, 24:398–425, 2019.

[89] M. Shi, L. Zhang, M. D. Risser, and B. A. Shaby. Spatial scale-aware tail dependence modeling for high-dimensional spatial extremes. *preprint arXiv:2412.07957*, 2024.

[90] R. Majumder, B. J. Reich, and B. A. Shaby. Modeling extremal streamflow using deep learning approximations and a flexible spatial process. *The Annals of Applied Statistics*, 18(2):1519–1542, 2024.

[91] P. Krupskii and M. G. Genton. Factor copula models for data with spatio-temporal dependence. *Spatial Statistics*, 22:180–195, 2017.

[92] R. Huser, T. Opitz, and E. Thibaud. Bridging asymptotic independence and dependence in spatial extremes using Gaussian scale mixtures. *Spatial Statistics*, 21:166–186, 2017.

[93] P. Krupskii, R. Huser, and M. G. Genton. Factor copula models for replicated spatial data. *Journal of the American Statistical Association*, 113(521):467–479, 2018.

[94] P. Krupskii and R. Huser. Modeling spatial tail dependence with Cauchy convolution processes. *Electronic Journal of Statistics*, 16(2):6135–6174, 2022.

[95] A. Hazra and R. Huser. Estimating high-resolution red sea surface temperature hotspots, using a low-rank semiparametric spatial model. *The Annals of Applied Statistics*, 15(2):572–596, 2021.

[96] D. Castro-Camilo and R. Huser. Local likelihood estimation of complex tail dependence structures, applied to US precipitation extremes. *Journal of the American Statistical Association*, 115(531):1037–1054, 2020.

[97] A. Hazra, B. J. Reich, and A.-M. Staicu. A multivariate spatial skew-t process for joint modeling of extreme precipitation indexes. *Environmetrics*, 31(3):e2602, 2020.

[98] A. Hazra, R. Huser, and D. Bolin. Efficient modeling of spatial extremes over large geographical domains. *Journal of Computational and Graphical Statistics*, pages 1–17, 2024.

[99] R. Huser and J. L. Wadsworth. Modeling spatial processes with unknown extremal dependence class. *Journal of the American Statistical Association*, 114(525):434–444, 2019.

[100] S. A. Morris, B. J. Reich, E. Thibaud, and D. Cooley. A space-time skew-t model for threshold exceedances. *Biometrics*, 73(3):749–758, 2017.

[101] L. Dell'Oro and C. Gaetan. Flexible space-time models for extreme data. *preprint arXiv:2411.19184*, 2024.

[102] J. E. Heffernan and J. A. Tawn. A conditional approach for multivariate extreme values (with discussion). *Journal of the Royal Statistical Society Series B: Statistical Methodology*, 66(3):497–546, 2004.

[103] E. S. Simpson and J. L. Wadsworth. Conditional modelling of spatio-temporal extremes for Red Sea surface temperatures. *Spatial Statistics*, 41:100482, 2021.

[104] J. L. Wadsworth and J. Tawn. Higher-dimensional spatial extremes via single-site conditioning. *Spatial Statistics*, 51:100677, 2022.

[105] E. Gilleland, B. G. Brown, and C. M. Ammann. Spatial extreme value analysis to project extremes of large-scale indicators for severe weather. *Environmetrics*, 24(6):418–432, 2013.

[106] B. Winter, K. Schneeberger, M. Huttenlau, and J. Stötter. Sources of uncertainty in a probabilistic flood risk model. *Natural Hazards*, 91(2):431–446, 2018.

[107] R. Jane, L. Cadavid, J. Obeysekera, and T. Wahl. Multivariate statistical modelling of the drivers of compound flood events in South Florida. *Natural Hazards and Earth System Sciences*, 20(10):2681–2699, 2020.

[108] J. Richards, J. A. Tawn, and S. Brown. Modelling extremes of spatial aggregates of precipitation using conditional methods. *The Annals of Applied Statistics*, 16(4):2693–2713, 2022.

[109] J. Richards, J. A. Tawn, and S. Brown. Joint estimation of extreme spatially aggregated precipitation at different scales through mixture modelling. *Spatial Statistics*, 53:100725, 2023.

[110] E. S. Simpson, T. Opitz, and J. L. Wadsworth. High-dimensional modeling of spatial and spatio-temporal conditional extremes using INLA and Gaussian Markov random fields. *Extremes*, 26(4):669–713, 2023.

[111] L. Kakampakou, E. S. Simpson, and J. L. Wadsworth. Spatial extremal modelling: A case study on the interplay between margins and dependence. *Stat*, 13(4):e70021, 2024.

[112] P. Jonathan, K. Ewans, and D. Randell. Non-stationary conditional extremes of northern North Sea storm characteristics. *Environmetrics*, 25(3):172–188, 2014.

[113] J. Tawn, R. Shooter, R. Towe, and R. Lamb. Modelling spatial extreme events with environmental applications. *Spatial Statistics*, 28:39–58, 2018.

[114] S. Tendijck, E. Eastoe, J. Tawn, D. Randell, and P. Jonathan. Modeling the extremes of bivariate mixture distributions with application to oceanographic data. *Journal of the American Statistical Association*, 118(542):1373–1384, 2023.

[115] W. K. Huang, A. H. Monahan, and F. W. Zwiers. Estimating concurrent climate extremes: A conditional approach. *Weather and Climate Extremes*, 33:100332, 2021.

[116] H. C. Winter and J. A. Tawn. Modelling heatwaves in central France: a case-study in extremal dependence. *Journal of the Royal Statistical Society Series C: Applied Statistics*, 65(3):345–365, 2016.

[117] R. Campbell and J. Wadsworth. Piecewise-linear modeling of multivariate geometric extremes. *preprint arXiv:2412.05195*, 2024.

[118] E. S. Simpson and J. A. Tawn. Inference for new environmental contours using extreme value analysis. *Journal of Agricultural, Biological and Environmental Statistics*, 67(2):215–224, 2025.

[119] R. Majumder, B. A. Shaby, B. J. Reich, and D. S. Cooley. Semiparametric estimation of the shape of the limiting bivariate point cloud. *Bayesian Analysis*, 1(1):1–27, 2025.

[120] J. L. Wadsworth and R. Campbell. Statistical inference for multivariate extremes via a geometric approach. *Journal of the Royal Statistical Society Series B: Statistical Methodology*, 86(5):1243–1265, 2024.

[121] L. De Monte, R. Huser, I. Papastathopoulos, and J. Richards. Generative modelling of multivariate geometric extremes using normalising flows. *preprint arXiv:2505.02957*, 2025.

[122] E. S. Simpson and J. A. Tawn. Estimating the limiting shape of bivariate scaled sample clouds: with additional benefits of self-consistent inference for existing extremal dependence properties. *Electronic Journal of Statistics*, 18(2):4582–4611, 2024.

[123] E. Mackay, C. Murphy-Barltrop, and P. Jonathan. The SPAR model: A new paradigm for multivariate extremes: Application to joint distributions of metocean variables. *Journal of Offshore Mechanics and Arctic Engineering*, 147(1):011205, 2025.

[124] E. Mackay, C. J. Murphy-Barltrop, J. Richards, and P. Jonathan. Deep learning of joint extremes of metocean variables using the SPAR model. *preprint arXiv:2412.15808*, 2025.

[125] D. Cooley and S. R. Sain. Spatial hierarchical modeling of precipitation extremes from a regional climate model. *Journal of Agricultural, Biological, and Environmental Statistics*, 15:381–402, 2010.

[126] M. Ribatet, D. Cooley, and A. C. Davison. Bayesian inference from composite likelihoods, with an application to spatial extremes. *Statistica Sinica*, pages 813–845, 2012.

[127] B. J. Reich and B. A. Shaby. A spatial Markov model for climate extremes. *Journal of Computational and Graphical Statistics*, 28(1):117–126, 2019.

[128] Á. V. Jóhannesson, S. Siegert, R. Huser, H. Bakka, and B. Hrafnkelsson. Approximate Bayesian inference for analysis of spatiotemporal flood frequency data. *The Annals of Applied Statistics*, 16(2):905–935, 2022.

[129] D. Castro Camilo and M. de Carvalho. Spectral density regression for bivariate extremes. *Stochastic Environmental Research and Risk Assessment*, 31(7):1603–1613, 2017.

[130] L. Mhalla, V. Chavez-Demoulin, and P. Naveau. Non-linear models for extremal dependence. *Journal of Multivariate Analysis*, 159:49–66, 2017.

[131] L. Mhalla, M. de Carvalho, and V. Chavez-Demoulin. Regression-type models for extremal dependence. *Scandinavian Journal of Statistics*, 46(4):1141–1167, 2019.

[132] L. Mhalla, T. Opitz, and V. Chavez-Demoulin. Exceedance-based nonlinear regression of tail dependence. *Extremes*, 22:523–552, 2019.

[133] J. Lee, M. de Carvalho, A. Rua, and J. Avila. Bayesian smoothing for time-varying extremal dependence. *Journal of the Royal Statistical Society: Series C (Applied Statistics)*, 73(3):581–597, 2024.

[134] R. Huser and M. G. Genton. Non-stationary dependence structures for spatial extremes. *Journal of Agricultural, Biological, and Environmental Statistics*, 21:470–491, 2016.

[135] P. D. Sampson and P. Guttorp. Nonparametric estimation of nonstationary spatial covariance structure. *Journal of the American Statistical Association*, 87(417):108–119, 1992.

[136] B. D. Youngman. Flexible models for nonstationary dependence: Methodology and examples. *preprint arXiv:2001.06642*, 2020.

[137] C. Chevalier, O. Martius, and D. Ginsbourger. Modeling nonstationary extreme dependence with stationary max-stable processes and multidimensional scaling. *Journal of Computational and Graphical Statistics*, 30(3):745–755, 2021.

[138] X. Shao, J. Richards, and R. Huser. Modeling nonstationary extremal dependence via deep spatial deformations. *preprint arXiv:2505.12548*, 2025.

[139] C. Forster and M. Oesting. Non-stationary max-stable models with an application to heavy rainfall data. *Extremes*, 28:523–556, 2025.

[140] S. Buhl and C. Klüppelberg. Anisotropic Brown-Resnick space-time processes: estimation and model assessment. *Extremes*, 19:627–660, 2016.

[141] K. Saunders, A. Stephenson, and D. Karoly. A regionalisation approach for rainfall based on extremal dependence. *Extremes*, 24:215–240, 2021.

[142] R. Majumder and B. J. Reich. A deep learning synthetic likelihood approximation of a non-stationary spatial model for extreme streamflow forecasting. *Spatial Statistics*, 55:100755, 2023.

[143] L. M. Berliner, C. K. Wikle, and N. Cressie. Long-lead prediction of Pacific SSTs via Bayesian dynamic modeling. *Journal of Climate*, 13(22):3953–3968, 2000.

[144] C. K. Wikle. A kernel-based spectral model for non-Gaussian spatio-temporal processes. *Statistical Modelling*, 2(4):299–314, 2002.

[145] M. Reistad, Øyvind Breivik, H. Haakenstad, O. J. Aarnes, B. R. Furevik, and J.-R. Bidlot. A high-resolution hindcast of wind and waves for the North Sea, the Norwegian Sea, and the Barents Sea. *Journal of Geophysical Research*, 116:C05019, 5 2011.

[146] A. G. Stephenson. *evd: Extreme Value Distributions*, 2023. R package version 2.3-6.

[147] C. Scarlett and A. MacDonald. A review of extreme value threshold estimation and uncertainty quantification. *REVSTAT – Statistical Journal*, 10(1):33–60, 2012.

[148] A. Langousis, A. Mamalakis, M. Puliga, and R. Roberto. Threshold detection for the generalized Pareto distribution: Review of representative methods and application to the NOAA NCDC daily rainfall database. *Water Resources Research*, 52:2659–2681, 2016.

[149] J. S. Bader, J. Yan, and X. Zhang. Automated threshold selection for extreme value analysis via ordered goodness-of-fit tests with adjustment for false discovery rate. *The Annals of Applied Statistics*, 12(1):310–329, 2018.

[150] J. S. Lomba and M. I. Fraga Alves. L-moments for automatic threshold selection in extreme value analysis. *Stochastic Environmental Research and Risk Assessment*, 34:465–491, 2020.

[151] M. Ribatet and C. Dutang. *POT: Generalized Pareto Distribution and Peaks Over Threshold*, 2024. R package version 1.1-11.

[152] C. Murphy, J. A. Tawn, and Z. Varty. Automated threshold selection and associated inference uncertainty for univariate extremes. *Technometrics*, 2024. To appear.

[153] Z. Varty, J. A. Tawn, P. M. Atkinson, and S. Bierman. Inference for extreme earthquake magnitudes accounting for a time-varying measurement process. *arXiv:2102.00884*, 2021.

[154] J. Lee, Y. Fan, and S. Sisson. Bayesian threshold selection on extremal models using measures of surprise. *Computational Statistics And Data Analysis*, 85:84–99, 2015.

[155] P. J. Northrop, N. Attalides, and P. Jonathan. Cross-validatory extreme value threshold selection and uncertainty with application to ocean storm severity. *Journal of the Royal Statistical Society. Series C*, 66(1):93–120, 2017.

[156] R. Mínguez. Automatic threshold selection for generalized Pareto and Pareto–Poisson distributions in rainfall analysis: A case study using the NOAA NCDC daily rainfall databases. *Atmosphere 2025, 16, 61*, 16(61), 2025.

[157] D. Castro-Camilo, R. Huser, and H. Rue. Practical strategies for generalized extreme value-based regression models for extremes. *Environmetrics*, 33(6):e2742, 2022.

[158] S. M. Vandeskog, S. Martino, D. Castro-Camilo, and H. Rue. Modelling sub-daily precipitation extremes with the blended generalised extreme value distribution. *Journal of Agricultural, Biological and Environmental Statistics*, 27(4):598–621, 2022.

[159] J. Richards and R. Huser. Extreme quantile regression with deep learning. In M. de Carvalho, R. Huser, P. Naveau, and B. J. Reich, editors, *Handbook on Statistics of Extremes*. Chapman & Hall / CRC, 2024.

[160] J. Richards and R. Huser. Regression modelling of spatiotemporal extreme US wildfires via partially-interpretable neural networks. *arXiv:2208.07581*, 2022.

[161] N. Y. Krakauer. Extending the blended generalized extreme value distribution. *Discover Civil Engineering*, 1(1):97, 2024.

[162] G. R. Aryal and C. P. Tsokos. On the transmuted extreme value distribution with application. *Nonlinear Analysis: Theory, Methods & Applications*, 71(12):e1401–e1407, 2009.

[163] F. Nascimento, M. Bourguignon, and J. Leão. Extended generalized extreme value distribution with applications in environmental data. *Hacettepe Journal of Mathematics and Statistics*, 45(6):1847–1864, 2016.

[164] C. E. Otiniano, B. S. De Paiva, and D. S. Neto. The transmuted GEV distribution: properties and application. *Communications for Statistical Applications and Methods*, 26(3):239–259, 2019.

[165] W. Gyasi and K. Cooray. New generalized extreme value distribution with applications to extreme temperature data. *Environmetrics*, 35(3):e2836, 2024.

[166] M. Marani and M. Ignaccolo. A metastatistical approach to rainfall extremes. *Advances in Water Resources*, 79:121–126, 2015.

[167] F. Marra, D. Zoccatelli, M. Armon, and E. Morin. A simplified MEV formulation to model extremes emerging from multiple nonstationary underlying processes. *Advances in Water Resources*, 127:280–290, 2019.

[168] M. A. Falkensteiner, H. Schellander, G. Ehrensperger, and T. Hell. Accounting for seasonality in the metastatistical extreme value distribution. *Weather and Climate Extremes*, 42:100601, 2023.

[169] C. T. Vidrio-Sahagún, J. He, and A. Pietroniro. Nonstationary hydrological frequency analysis using the metastatistical extreme value distribution. *Advances in Water Resources*, 176:104460, 2023.

[170] F. Serinaldi, F. Lombardo, and C. G. Kilsby. Non-asymptotic distributions of water extremes: much ado about what? *Hydrology and Earth System Sciences*, 29(4):1159–1181, 2025.

[171] J. Carreau and Y. Bengio. A hybrid Pareto model for asymmetric fat-tailed data: The univariate case. *Extremes*, 12:53–76, 2009.

[172] J. Carreau, P. Naveau, and E. Sauquet. A statistical rainfall-runoff mixture model with heavy-tailed components. *Water Resources Research*, 45(10), 2009.

[173] J. Carreau and M. Vrac. Stochastic downscaling of precipitation with neural network conditional mixture models. *Water Resources Research*, 47(10), 2011.

[174] I. Papastathopoulos and J. A. Tawn. Extended generalised pareto models for tail estimation. *Journal of Statistical Planning and Inference*, 143(1):131–143, 2013.

[175] P. Naveau, R. Huser, P. Ribereau, and A. Hannart. Modeling jointly low, moderate, and heavy rainfall intensities without a threshold selection. *Water Resources Research*, 52(4):2753–2769, 2016.

[176] P. Tencaliec, A. C. Favre, P. Naveau, C. Prieur, and G. Nicolet. Flexible semiparametric generalized Pareto modeling of the entire range of rainfall amount. *Environmetrics*, 31(2):e2582, 2020.

[177] P. Rivoire, O. Martius, and P. Naveau. A comparison of moderate and extreme era-5 daily precipitation with two observational data sets. *Earth and Space Science*, 8(4):e2020EA001633, 2021.

[178] M. de Carvalho, S. Pereira, P. Pereira, and P. de Zea Bermudez. An extreme value Bayesian Lasso for the conditional left and right tails. *Journal of Agricultural, Biological and Environmental Statistics*, 27(2):222–239, 2022.

[179] J. S. Nanditha, G. Villarini, H. Kim, and P. Naveau. Strong linkage between observed daily precipitation extremes and anthropogenic emissions across the contiguous united states. *Geophysical Research Letters*, 51(20):e2024GL109553, 2024.

[180] D. Cisneros, J. Richards, A. Dahal, L. Lombardo, and R. Huser. Deep graphical regression for jointly moderate and extreme Australian wildfires. *Spatial Statistics*, 59:100811, 2024.

[181] R. Yadav, R. Huser, T. Opitz, and L. Lombardo. Joint modelling of landslide counts and sizes using spatial marked point processes with sub-asymptotic mark distributions. *Journal of the Royal Statistical Society Series C: Applied Statistics*, 72(5):1139–1161, 2023.

[182] M. L. Stein. Parametric models for distributions when interest is in extremes with an application to daily temperature. *Extremes*, 24(2):293–323, 2021.

[183] M. L. Stein. A parametric model for distributions with flexible behavior in both tails. *Environmetrics*, 32(2):e2658, 2021.

[184] J. Koh. Gradient boosting with extreme-value theory for wildfire prediction. *Extremes*, 26(2):273–299, 2023.

[185] J. Velthoen, C. Dombry, J.-J. Cai, and S. Engelke. Gradient boosting for extreme quantile regression. *Extremes*, 26(4):639–667, 2023.

[186] S. Farkas, A. Heranval, O. Lopez, and M. Thomas. Generalized Pareto regression trees for extreme event analysis. *Extremes*, 27:1–41, 2024.

[187] M. Taillardat, A.-L. Fougères, P. Naveau, and O. Mestre. Forest-based and semiparametric methods for the postprocessing of rainfall ensemble forecasting. *Weather and Forecasting*, 34(3):617–634, 2019.

[188] R. Majumder and J. Richards. Semi-parametric bulk and tail regression using spline-based neural networks. *arXiv preprint arXiv:2504.19994*, 2025.

[189] J. Richards, R. Huser, E. Bevacqua, and J. Zscheischler. Insights into the drivers and spatiotemporal trends of extreme Mediterranean wildfires with statistical deep learning. *Artificial Intelligence for the Earth Systems*, 2(4):e220095, 2023.

[190] O. C. Pasche and S. Engelke. Neural networks for extreme quantile regression with an application to forecasting of flood risk. *The Annals of Applied Statistics*, 18(4):2818–2839, 2024.

[191] A. Dahal, R. Huser, and L. Lombardo. At the junction between deep learning and statistics of extremes: formalizing the landslide hazard definition. *Journal of Geophysical Research: Machine Learning and Computation*, 1(3):e2024JH000164, 2024.

[192] A. J. Cannon. A flexible nonlinear modelling framework for nonstationary generalized extreme value analysis in hydroclimatology. *Hydrological Processes: An International Journal*, 24(6):673–685, 2010.

[193] M. Allouche, S. Girard, and E. Gobet. Estimation of extreme quantiles from heavy-tailed distributions with neural networks. *Statistics and Computing*, 34(1):12, 2024.

[194] A. Zammit-Mangion, M. Sainsbury-Dale, and R. Huser. Neural methods for amortized inference. *Annual Review of Statistics and Its Application*, 12, 2024.

[195] A. Lenzi, J. Bessac, J. Rudi, and M. L. Stein. Neural networks for parameter estimation in intractable models. *Computational Statistics & Data Analysis*, 185:107762, 2023.

[196] S. Rai, A. Hoffman, S. Lahiri, D. W. Nychka, S. R. Sain, and S. Bandyopadhyay. Fast parameter estimation of generalized extreme value distribution using neural networks. *Environmetrics*, 35(3):e2845, 2024.

[197] L. M. André, J. L. Wadsworth, and R. Huser. Neural Bayes inference for complex bivariate extremal dependence models. *preprint arXiv:2503.23156*, 2025.

[198] M. Sainsbury-Dale, A. Zammit-Mangion, and R. Huser. Likelihood-free parameter estimation with neural Bayes estimators. *The American Statistician*, 78(1):1–14, 2024.

[199] A. Hasan, K. Elkhalil, Y. Ng, J. M. Pereira, S. Farsiu, J. Blanchet, and V. Tarokh. Modeling extremes with d-max-decreasing neural networks. In *Uncertainty in Artificial Intelligence*, pages 759–768. PMLR, 2022.

[200] S. Bhatia, A. Jain, and B. Hooi. ExGAN: Adversarial generation of extreme samples. *Proceedings of the AAAI Conference on Artificial Intelligence*, 35(8):6750–6758, 2021.

[201] Y. Boulaguiem, J. Zscheischler, E. Vignotto, K. van der Wiel, and S. Engelke. Modeling and simulating spatial extremes by combining extreme value theory with generative adversarial networks. *Environmental Data Science*, 1:e5, 2022.

[202] N. Lafon, P. Naveau, and R. Fablet. A VAE approach to sample multivariate extremes. *preprint arXiv:2306.10987*, 2023.

[203] L. Zhang, X. Ma, C. K. Wikle, and R. Huser. Flexible and efficient spatial extremes emulation via variational autoencoders. *preprint arXiv:2307.08079*, 2023.

[204] M. Ahmed, V. Maume-Deschamps, and P. Ribereau. Recognizing a spatial extreme dependence structure: A deep learning approach. *Environmetrics*, 33(4):e2714, 2022.

[205] T. P. Wixson and D. Cooley. Neural network for asymptotic dependence/ independence classification: A series of experiments. *preprint doi.org/10.21203/rs.3.rs-3994810/v1*, 2024.

[206] O. C. Pasche, H. Lam, and S. Engelke. Extreme conformal prediction: Reliable intervals for high-impact events. *preprint arXiv:2505.08578*, 2025.

[207] M. de Carvalho, C. Ferrer, and R. Vallejos. A Kolmogorov-Arnold neural model for cascading extremes. *preprint arXiv:2505.13370*, 2025.

[208] A. Janßen and P. Wan. k-means clustering of extremes. *Electronic Journal of Statistics*, 14(1):1211–1233, 2020.

[209] E. Bernard, P. Naveau, M. Vrac, and O. Mestre. Clustering of maxima: Spatial dependencies among heavy rainfall in France. *Journal of Climate*, 26(20):7929–7937, 2013.

[210] M. Bador, P. Naveau, E. Gilleland, M. Castellà, and T. Arivelo. Spatial clustering of summer temperature maxima from the CNRM-CM5 climate model ensembles & E-OBS over Europe. *Weather and Climate Extremes*, 9:17–24, 2015.

[211] E. Vignotto, S. Engelke, and J. Zscheischler. Clustering bivariate dependencies of compound precipitation and wind extremes over Great Britain and Ireland. *Weather and Climate Extremes*, 32:100318, 2021.

[212] A. Boulin, E. Di Bernardino, T. Laloë, and G. Toulemonde. Identifying regions of concomitant compound precipitation and wind speed extremes over Europe. *preprint arXiv:2311.11292*, 2023.

[213] D. Deuber, J. Li, S. Engelke, and M. H. Maathuis. Estimation and inference of extremal quantile treatment effects for heavy-tailed distributions. *Journal of the American Statistical Association*, 119(547), 2023.

[214] J. Pearl. *Causality. Models, Reasoning, and Inference.* Cambridge University Press, Cambridge, 2010.

[215] N. Gissibl and C. Klüppelberg. Max-linear models on directed acyclic graphs. *Bernoulli*, 24(4A):2693–2720, 2018.

[216] S. I. Resnick. *Heavy-Tail Phenomena: Probabilistic and Statistical Modeling.* Springer, New York, 2006.

[217] N. Gnecco, N. Meinshausen, J. Peters, and S. Engelke. Causal discovery in heavy-tailed models. *The Annals of Statistics*, 49(3):1755–1778, 2021.

[218] A.-L. Fougères, C. Mercadier, and J. P. Nolan. Dense classes of multivariate extreme value distributions. *Journal of Multivariate Analysis*, 116:109–129, 2013.

[219] V. Chavez-Demoulin and L. Mhalla. Causality and extremes. *preprint arXiv:2403.05331*, 2024.

[220] C. Klüppelberg and M. Krali. Causal analysis of extreme risk in a network of industry portfolios, 2025.

[221] M. Krali, A. C. Davison, and C. Klüppelberg. Heavy-tailed max-linear structural equation models in networks with hidden nodes. *preprint arXiv:2306.15356*, 2023.

[222] L. Mhalla, V. Chavez-Demoulin, and D. J. Dupuis. Causal mechanism of extreme river discharges in the upper Danube basin network. *Journal of the Royal Statistical Society Series C: Applied Statistics*, 69(4):741–764, 2020.

[223] O. C. Pasche, V. Chavez-Demoulin, and A. C. Davison. Causal modelling of heavy-tailed variables and confounders with application to river flow. *Extremes*, 26(3):573–594, 2023.

[224] C. Klüppelberg and M. Krali. Estimating an extreme Bayesian network via scalings. *Journal of Multivariate Analysis*, 181, 2021.

[225] M. Krali. Causal discovery in heavy-tailed linear structural equation models via scalings. *preprint arXiv:2502.13762*, 2025.

[226] N. M. Tran, J. Buck, and C. Klüppelberg. Estimating a directed tree for extremes. *Journal of the Royal Statistical Society Series C: Applied Statistics*, 86(3):771–792, 2024.

[227] S. Engelke, N. Gnecco, and F. Röttger. Extremes of structural causal models. *preprint arXiv:2503.06536*, 2025.

[228] P. Spirtes, C. Glymour, and R. Scheines. *Causation, Prediction, and Search.* Adaptive Computation and Machine Learning. MIT Press, 2nd edition, 2000.

[229] J. Jiang, J. Richards, R. Huser, and D. Bolin. Separation-based causal discovery for extremes. *preprint arXiv:2505.08008*, 2025.

[230] A. Kiriliouk and P. Naveau. Climate extreme event attribution using multivariate peaks-over-thresholds modeling and counterfactual theory. *Annals of Applied Statistics*, 14(3):1342–1358, 2020.

[231] J. S. Nanditha, G. Villarini, H. Kim, and P. Naveau. Causal attribution of the interannual variability in flood peaks through Bayesian networks. *Water Resources Research*, 61(6), 2025.

[232] J. Worms and P. Naveau. Record events attribution in climate studies. *Environmetrics*, 33(8), 2022.

18

A Nonstationary Spatial Count Regression Using Gamma-Count: A Case Study on Canadian Precipitation

Mahsa Nadifar, Andriëtte Bekker, and Mohammad Arashi

This chapter develops a nonstationary spatial count regression model utilizing the Gamma-Count (GC) distribution. The approach is designed to handle spatially distributed count data under nonstationary dependence structures. Our method employs both structured additive regression (STAR) models and the stochastic partial differential equation (SPDE) framework to describe how dependencies evolve over space in non-stationary situations. It also encompasses nonlinear covariate-response connections and addresses uneven dispersion in count data. We illustrate the concept utilizing monthly precipitation data from Canada, emphasizing the importance of regional heterogeneity in rainfall patterns. This methodology aligns with the Sustainable Development Goals (SDGs) by improving climate and environmental research statistical tools. The integrated nested Laplace approximation (INLA) has been used to make a reliable Bayesian inference framework run faster. The findings highlight the significant benefits of the nonstationary GC model for analyzing spatially structured count data. This work enhances existing Bayesian approaches for nonstationary settings, providing valuable new insights for researchers in spatial econometrics, environmental statistics, and climate science.

18.1 Introduction

Spatial count data are fundamental to understanding complex phenomena in environmental science, epidemiology, and climatology, where intricate spatial patterns, nonlinear covariate relationships, and nonstationary dependence structures pose significant modeling challenges. In Alberta, Canada, precipitation counts, defined as days with precipitation exceeding 1.0 mm, exhibit marked spatial heterogeneity driven by topographic variations, from the Rocky Mountains in the west to the flat eastern prairies [1, 2]. This variability, captured in a synthetic dataset representing December 2024 wet-day counts, highlights the need for advanced statistical frameworks that accommodate nonstationarity, flexible dispersion, and spatial interactions. Such models are vital for enhancing spatial interpolation in sparse datasets,

DOI: 10.1201/9781003564423-18

informing agricultural planning, and optimizing water resource management, aligning with Sustainable Development Goal 13 (Climate Action) [3, 4].

Traditional approaches to spatial count data, such as spatial generalized linear mixed models (SGLMMs; [5]), rely on stationary dependence assumptions, which are inadequate for regions like Alberta with diverse orography and climatic gradients. Early count models, like Poisson regression, assume equi-dispersion (mean equals variance; [6]), an assumption often violated by precipitation data exhibiting over-dispersion from clustered wet days or under-dispersion in arid regions with sparse events. To address these limitations, models such as the negative binomial (NB; [7]), generalized Poisson [8], and Conway-Maxwell-Poisson (COM-Poisson; [9]) introduced parameters to handle dispersion variability. The NB model, blending Poisson and gamma distributions, effectively captures over-dispersion, while the COM-Poisson offers flexibility for both over- and under-dispersion. However, these models struggle to cohesively address diverse dispersion patterns in spatially dependent contexts, particularly when dependence varies across space due to physical or climatic factors [10].

The gamma-count (GC) model, rooted in renewal theory [11], provides a robust solution by modeling counts as the number of events within a fixed interval, with inter-event waiting times following a gamma distribution. This approach flexibly captures both over- and under-dispersion, making it ideal for precipitation counts. Recent advancements have extended the GC model to spatial applications, notably through STAR frameworks with penalized complexity (PC) priors [12, 13], stabilizing inference for geographically referenced counts. For instance, Nadifar et al. [10] integrated spatial structures into GC models, enhancing their applicability to environmental data. However, these extensions assume stationary dependence, oversimplifying processes influenced by physical barriers or covariates like elevation, which modulate both mean counts and spatial covariance in Alberta's precipitation patterns.

Nonstationary spatial models offer a powerful alternative by allowing dependence structures to vary across space, particularly through SPDEs [14]. Early nonstationary methods, such as Sampson and Guttorp's coordinate deformation [15], were computationally impractical for large datasets. Subsequent approaches, including Higdon's convolution-based models [16] and Paciorek and Schervish's covariance adjustments [17], improved flexibility but reduced interpretability. The SPDE framework, linking Gaussian random fields (GRFs) to sparse Gaussian Markov random fields (GMRFs), revolutionized nonstationary modeling by enabling efficient computation with spatially varying parameters [18]. Notable applications include Bakka et al.'s Barrier model for obstacle-driven dependence [19] and elevation-adjusted models for precipitation in complex terrains [1, 20]. While these excel for Gaussian data, nonstationary modeling of non-Gaussian count data, such as Alberta's precipitation counts, remains underexplored, as stationary GRF models [21] and hierarchical NB approaches [22] fail to capture elevation-driven nonstationarity.

This study introduces a Bayesian nonstationary spatial gamma-count (NS-GC) model, uniquely integrating the GC distribution's dispersion flexibility with SPDE-based nonstationary GRFs. Unlike stationary GC models [10] or Gaussian-centric SPDEs [20], the NS-GC model captures count-specific dispersion and elevation-driven spatial dependence within a hierarchical Bayesian framework, using INLA for computational efficiency [23]. Applied to Alberta's synthetic December 2024 wet-day counts, it tests the model's ability to handle sparse observations and topographic effects, outperforming stationary benchmarks like NB in fit and predictive accuracy. This work advances spatial statistics, supporting data-driven environmental planning and climate resilience [3].

The paper is organized as follows: Section 18.2 describes the dataset and exploratory analysis, Section 18.3 details the NS-GC model, Section 18.4 presents results, and Section 18.5 provides discussion and conclusions.

18.2 Data Motivation: Precipitation Counts in Alberta

This study examines a monthly climate summary dataset from Alberta, Canada, concentrating on the count of days with precipitation surpassing 1.0 mm in December 2024. The dataset, obtained from Environment and Climate Change Canada (`https://climate.weather.gc.ca/prods_servs/cdn_climate_summary_e.html`), includes 169 stations with recorded precipitation counts, along with covariates such as elevation, mean temperature (`Tm`), and total precipitation (`TP`). The response variable is the count of rainy days, referred to as `Precip_Days`, with covariates comprising geographical coordinates (longitude and latitude), mean monthly temperature (`Tm`), and total precipitation (`TP`). This analysis examines the spatial distribution of precipitation events, guiding agricultural practices, water resource management, and climate adaption techniques in Alberta's varied landscape.

Alberta's orography promotes notable geographic heterogeneity in precipitation patterns [2]: from the Rocky Mountains in the west to the level plains in the east. A vital season for agriculture, December shows counts of rainy days shaped by elevation, temperature, and regional climatic dynamics. Comprising sparse observations from urban and rural sites, the dataset has missing values for some precipitation records, so strong modeling is necessary to manage partial data. As will be explained in Section 18.3, our Bayesian NS-GC model is designed to handle this variability by utilizing elevation to manage the spatial relationships.

Exploratory data analysis (EDA) clarifies the data's structure and interrelationships. Figure 18.1 displays a pairwise scatterplot matrix of `Precip_Days` (days with precipitation $\geq$ 1.0 mm), `Tm`, and `TP`. Diagonal panels illustrate variable distributions, lower panels present scatterplots, and upper panels convey correlation coefficients. The correlations suggest nonlinear interactions between `TP` and `Precip_Days`, as well as possible collinearity involving the temperature variable, which guides the selection of covariates.

Furthermore, Figure 18.2 on the left demonstrates that the count variable `Precip_Days` displays skewness, with a maximum of roughly 9 days. This pattern indicates a bounded observation window but is theoretically unbounded [6, 7, 24, 25], rendering the GC model suitable for encapsulating the observed dispersion attributes. The right panel presents a box plot of `Precip_Days`, `Tm`, and `TP`, facilitating comprehension of the distribution of these variables. The asymmetry of `Precip_Days` indicates an unequal distribution within the dataset, a crucial insight for our modeling methodology.

Figure 18.3 illustrates sampling locations and recorded counts (left), along with a triangular mesh for SPDE modeling (right). The spatial distribution indicates increased precipitation frequency in western mountainous areas, aligning with topographic influences. The mesh, convexly expanded to alleviate boundary effects [14], facilitates finite element approximations in our NS-GC model.

Exploratory analyses highlight the necessity for nonstationary spatial modeling of Alberta's precipitation counts, influenced by the region's topography and climatic variability. Figure 18.4 displays empirical and residual variograms, highlighting significant spatial trends. The empirical variogram (right) demonstrates an increase in semivariance with distance, indicating diminishing spatial autocorrelation characteristics of count data. However, its irregular form, devoid of a consistent sill, suggests regional variations in dependence, aligning with nonstationarity caused by Alberta's varied orography. The residual variogram (right), obtained from a linear model accounting for factors such as temperature, exhibits a more gradual increase, signifying a continuous spatial pattern that a stationary model inadequately captures. This indicates that spatial dependence fluctuates throughout the domain, requiring sophisticated solutions beyond conventional frameworks. In addition, Figure 18.5 presents a heat map (left) of precipitation counts, demonstrating marked clustering in the

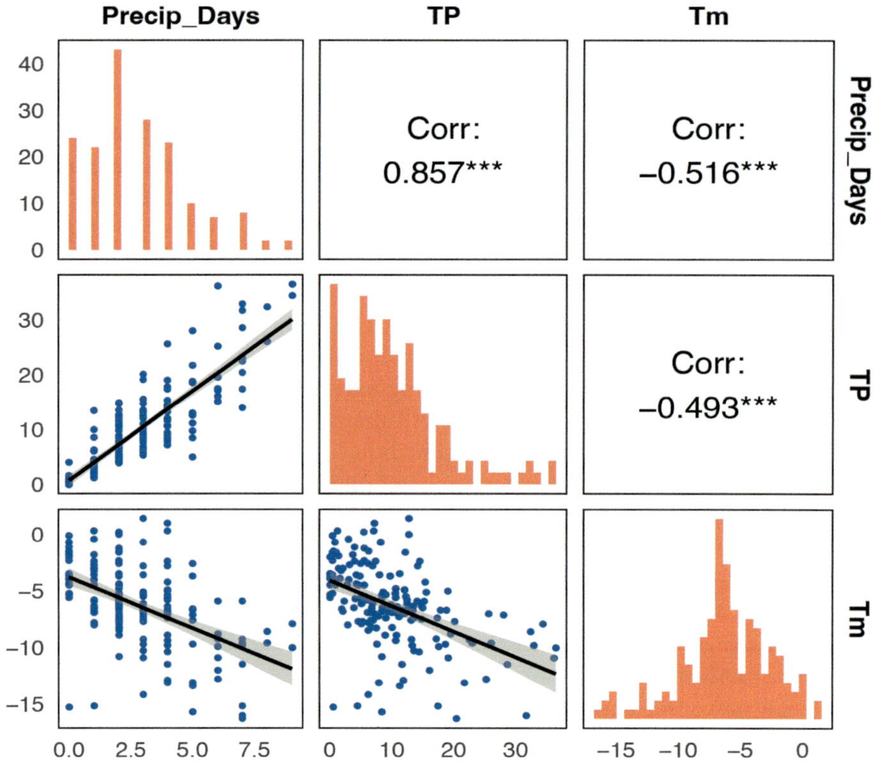

FIGURE 18.1
Pairwise scatterplot matrix of `Precip_Days`, `Tm`, and `TP`. Diagonal panels show distributions, lower panels display scatterplots, and upper panels report correlation coefficients.

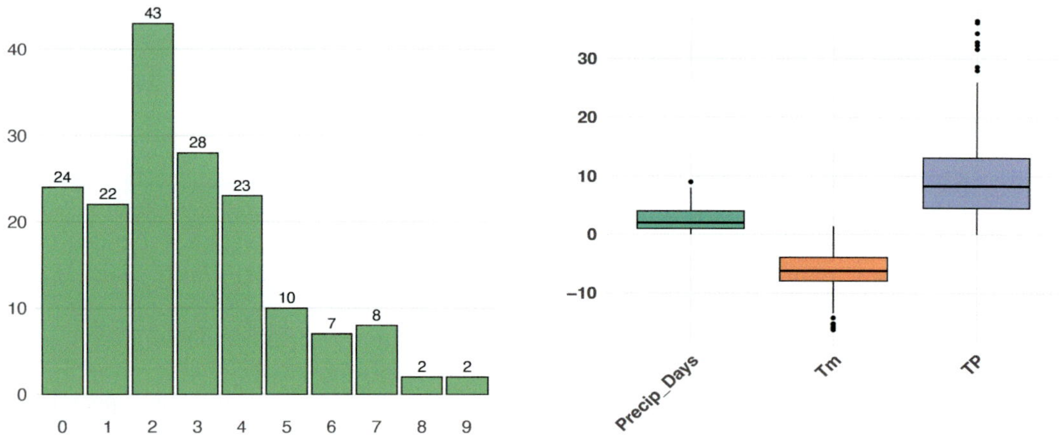

FIGURE 18.2
Bar plot of the response variable `Precip_Days` (left), illustrating its skewness and unbounded nature, alongside a box plot (right) displaying the distribution of `Precip_Days`, `Tm`, and `TP`.

mountainous western region compared to more sparse distributions in the eastern prairies, alongside a Moran's I scatterplot (right), which indicates a significant positive value (e.g., Moran's $I > 0$, $p < 0.05$). This validates global spatial autocorrelation; nevertheless, the regional differences depicted in the heat map, supported by exploratory covariate analyses

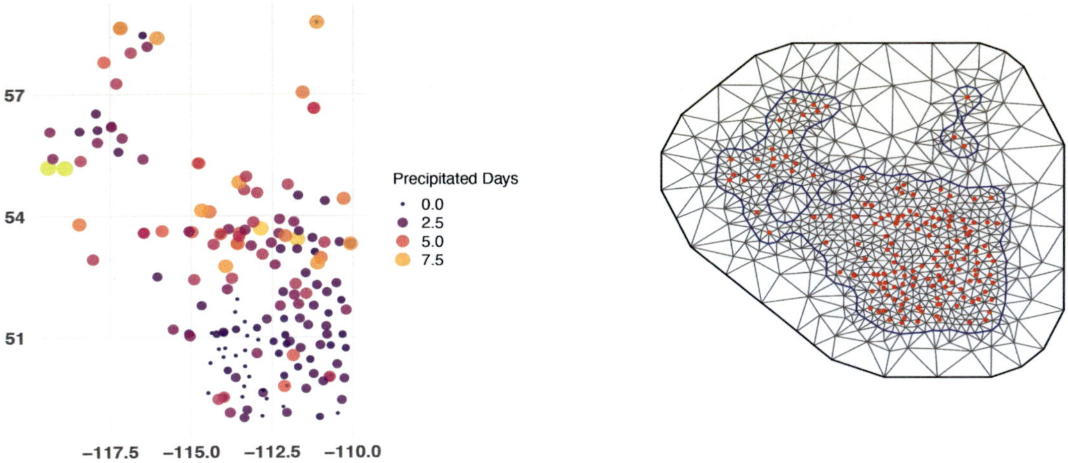

FIGURE 18.3
Sampling locations and days with precipitation $\geq$ 1.0 mm in Alberta, December 2024 (left), and triangular mesh for SPDE modeling (right).

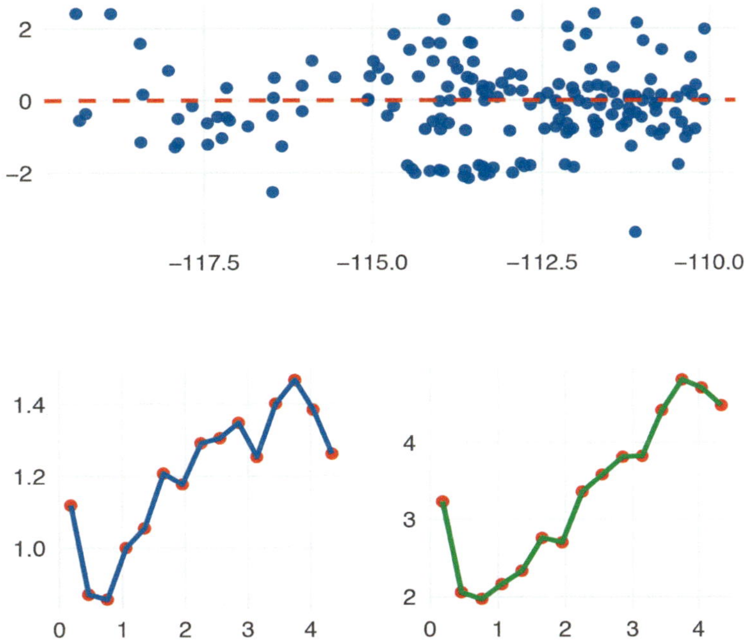

FIGURE 18.4
Residual of linear model (top), empirical variogram (down-right), and residual variogram (down-left) of precipitation counts, indicating spatial autocorrelation.

(Figure 18.1), indicate that this dependency varies with elevation and climatic gradients, exemplifying nonstationarity [20]. These findings corroborate prior evidence that stationary models, such as those predicated on isotropic Matérn covariance [26], inadequately represent dynamics in topographically intricate areas. Precipitation in Alberta, from the Rocky Mountains to the flat plains, illustrates this intricacy, since elevation influences both average amounts and their spatial covariance structure [2]. Conventional stationary methods

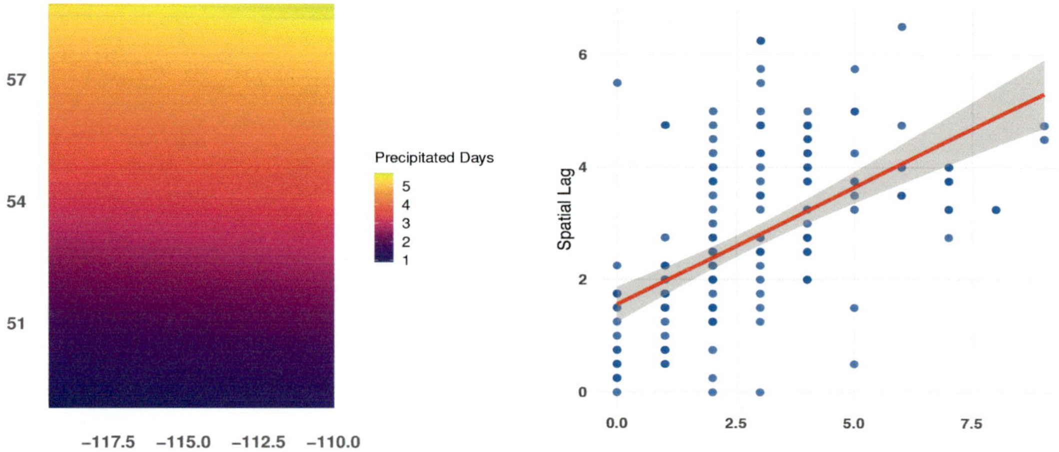

FIGURE 18.5
Heat map of precipitation counts (left) and Moran's I scatterplot (right), showing spatial clustering.

may distort these dynamics, so introducing bias into forecasts is essential for agricultural and water management. Consequently, a nonstationary spatial model that adjusts dependence based on local geographic characteristics is crucial for precise inference and effective environmental planning in this scenario.

These studies underscore the spatial heterogeneity, dispersion, and covariate interactions within the data set; hence, they validate the implementation of our NS-GC model in the next sections.

18.3 Bayesian Nonstationary Spatial Gamma-Count Regression

This section delineates the approach for a Bayesian nonstationary spatial gamma-count regression model (NS-GC), intended to describe spatially variable dependence structures and adaptable dispersion patterns in count data, specifically applied to precipitation counts in Alberta, Canada. Conventional spatial models frequently presume stationarity, which is insufficient for areas with topographic variability, such as Alberta, where precipitation patterns fluctuate due to elevation and regional climatic dynamics. Our objective is to incorporate the GC distribution, which addresses both over- and under-dispersion [11], into a nonstationary SPDE framework [14], permitting explanatory factors to affect the covariance structure as established by [20]. We utilize INLA for efficient Bayesian inference, capitalizing on its computing benefits compared to Markov Chain Monte Carlo (MCMC) approaches [23]. The methodology comprises two components: model development and Bayesian inference.

18.3.1 Nonstationary Spatial Gamma-Count Learning

Consider Y_i as the count of days with precipitation exceeding 1.0 mm at the location $\boldsymbol{s}_i = (s_{i1}, s_{i2})$ in Alberta during December 2024, for $i = 1, \dots, n$, paired with a covariate vector $\boldsymbol{x}_i = (1, x_{i1}, \dots, x_{ip})'$. We propose that Y_i follows a GC distribution, denoted $Y_i \sim$

GC(α, γ_i), where α governs dispersion and γ_i is a rate parameter shaped by covariates and spatial effects. Rooted in renewal theory [11], the GC model conceptualizes counts as the number of events within a fixed interval (here $t = 1$), with inter-event waiting times $v_{k_i} \sim \text{Gamma}(\alpha, \gamma_i)$. Its probability mass function is

$$P(Y_i = n) = G(n\alpha, \gamma_i) - G((n+1)\alpha, \gamma_i), \quad n = 0, 1, 2, \ldots, \tag{18.1}$$

where $G(k\alpha, \gamma_i) = \frac{1}{\Gamma(k\alpha)} \int_0^{\gamma_i} t^{k\alpha-1} e^{-t}\, dt$ denotes the incomplete gamma function. The mean, $\text{E}(Y_i) = \sum_{k=1}^{\infty} G(k\alpha, \gamma_i)$, lacks a closed form but inversely relates to the expected waiting time, $\text{E}(v_{k_i}) = \alpha/\gamma_i$.

For covariate effects, we adopt the GC regression framework of [10], modeling the log-expected waiting time as

$$\log(\text{E}(v_{k_i} | \boldsymbol{x}_i)) = \log\left(\frac{\alpha}{\gamma_i}\right) = -\eta_i, \tag{18.2}$$

where the linear predictor $\eta_i = \boldsymbol{x}_i'\boldsymbol{\beta}$, and $\boldsymbol{\beta} = (\beta_0, \beta_1, \ldots, \beta_p)'$ are regression coefficients. Thus, $\gamma_i = \alpha \exp(\eta_i)$, such that an increase in η_i reduces the expected count, reflecting the renewal process's intrinsic logic.

Spatial effects demand a sophisticated approach, as Alberta's precipitation exhibits pronounced variability tied to its topography. Stationary spatial models, such as the Matérn GRF, provide a baseline. Defined as the solution to:

$$(\kappa^2 - \Delta)^{\zeta/2}(\tau u(\boldsymbol{s})) = \mathcal{W}(\boldsymbol{s}), \tag{18.3}$$

where Δ is the Laplacian, $\mathcal{W}(\boldsymbol{s})$ is Gaussian white noise, κ determines the spatial range, and τ scales precision, the Matérn GRF assumes stationarity and isotropy. Here, the range $\rho = \sqrt{8\nu}/\kappa$ and variance $\sigma^2 \propto 1/(\kappa^{2\nu}\tau^2)$ are constant across space, with smoothness $\nu = 1$ (thus $\zeta = 2$) balancing realism and computation [14]. While widely used for [26] flexibility, it is insufficient for Alberta because of the spatially heterogeneous dependence created by the different terrains of the Rocky Mountains and prairies, as elaborated in Section 18.2.

To capture this nonstationarity, we extend the model using the SPDE approach, a paradigm shift from stationary constraints. Introduced by [14], the SPDE framework redefines GRFs as solutions to differential equations, linking them to sparse GMRFs for computational efficiency. In its stationary form, κ and τ are fixed scalars, enforcing a uniform covariance structure, e.g., a single range of approximately 100–200 km and consistent variance. However, *nonstationary modeling* relaxes these assumptions, allowing spatial parameters to vary across the domain, a critical adaptation for topographically diverse regions. Building on [20], we specify:

$$\log(\tau(\boldsymbol{s}_i)) = \theta_{\tau 0} + \theta_{\tau 1} z_i, \quad \log(\kappa(\boldsymbol{s}_i)) = \theta_{\kappa 0} + \theta_{\kappa 1} z_i, \tag{18.4}$$

where z_i is elevation at $\boldsymbol{s}_i$, and $\boldsymbol{\theta} = (\theta_{\tau 0}, \theta_{\tau 1}, \theta_{\kappa 0}, \theta_{\kappa 1})'$ are hyperparameters. *This nonstationary formulation* enables the spatial range $\rho(\boldsymbol{s}_i) = \sqrt{8\nu}/\kappa(\boldsymbol{s}_i)$ and marginal variance $\sigma^2(\boldsymbol{s}_i) \propto 1/(\kappa(\boldsymbol{s}_i)^{2\nu}\tau(\boldsymbol{s}_i)^2)$ to adapt locally, e.g., shorter ranges (tens of kilometers) in rugged highlands where precipitation patterns shift rapidly, and longer ranges (hundreds of kilometers) in the flatter prairies where dependence extends further. This flexibility reflects Alberta's physical reality, where elevation not only alters mean precipitation but also its spatial covariance, as supported by empirical variograms and heat maps (Section 18.2) and prior studies [1].

The SPDE's theoretical foundation lies in its ability to transform a continuous GRF into a discrete GMRF via the finite element method (FEM), applied over a triangular mesh (see Figure 18.3). This discretization approximates $u(\boldsymbol{s}) = \sum_{k=1}^{K} u_k \psi_k(\boldsymbol{s})$, where $\psi_k(\boldsymbol{s})$ are piecewise linear basis functions, and $\boldsymbol{u} = (u_1, \ldots, u_K)'$ follows a GMRF with precision matrix:

$$\mathbf{Q}(\boldsymbol{\theta}) = \mathbf{T}(\mathbf{K}^2\mathbf{C}\mathbf{K}^2 + \mathbf{K}^2\mathbf{G} + \mathbf{G}^\top\mathbf{K}^2 + \mathbf{G}\mathbf{C}^{-1}\mathbf{G})\mathbf{T}, \tag{18.5}$$

where $T_{kk} = \tau(\boldsymbol{s}_k)$, $K_{kk} = \kappa(\boldsymbol{s}_k)$, and $\mathbf{C}$ and $\mathbf{G}$ are mesh-derived matrices encoding spatial relationships [14]. Thus, $\boldsymbol{u}|\boldsymbol{\theta} \sim \mathcal{N}(\mathbf{0}, \mathbf{Q}(\boldsymbol{\theta})^{-1})$. *Unlike stationary GRFs*, where $\mathbf{Q}$ is uniform, this nonstationary precision matrix dynamically adjusts to local conditions, leveraging sparsity to scale efficiently with large datasets, a key advantage over traditional covariance-based methods [23]. For Alberta, with its sparse observation network and topographic extremes, *this nonstationary SPDE approach* ensures that spatial dependence aligns with geographic features, avoiding the oversimplification of isotropic models.

The full predictor integrates fixed and spatial effects:

$$\eta_i = \beta_0 + \sum_{j=1}^{p} f_j(x_{ji}) + u(\boldsymbol{s}_i), \tag{18.6}$$

where β_0 is the intercept, $f_j(x_{ji})$ are smooth nonlinear effects of covariates (e.g., temperature), modeled with second-order random walk priors for adaptability [21], and $u(\boldsymbol{s}_i)$ is the *nonstationary spatial field.* The resulting NS-GC model is

$$Y_i|\alpha, \beta_0, \boldsymbol{\theta}, \tau_x \sim \mathrm{GC}\left(\alpha, \alpha \exp\left(\beta_0 + \sum_{j=1}^{p} f_j(x_{ji}) + u(\boldsymbol{s}_i)\right)\right), \tag{18.7}$$

where τ_x governs the precision of nonlinear effects. This formulation seamlessly blends the GC likelihood's dispersion flexibility with a *nonstationary spatial structure*, tailored to Alberta's precipitation dynamics, offering a scientifically grounded advance over stationary alternatives in complex environmental settings.

18.3.2 Bayesian Inference with INLA

Given the complexity of the likelihood in equation (18.7), we adopt a hierarchical Bayesian framework with INLA for inference. The parameter vector includes latent Gaussian variables $\boldsymbol{\psi} = (\beta_0, \boldsymbol{f}_x, \boldsymbol{u})'$, where $\boldsymbol{f}_x = (f_1(x_{1i}), \ldots, f_p(x_{pi}))_{i=1}^n$, and hyperparameters $\boldsymbol{\theta} = (\alpha, \tau_x, \theta_{\tau 0}, \theta_{\tau 1}, \theta_{\kappa 0}, \theta_{\kappa 1})'$. The joint posterior is:

$$\pi(\boldsymbol{\psi}, \boldsymbol{\theta}|\boldsymbol{y}) \propto \left[\prod_{i=1}^{n} P(Y_i|\eta_i, \alpha)\right] \pi(\boldsymbol{\psi}|\boldsymbol{\theta})\pi(\boldsymbol{\theta}), \tag{18.8}$$

where $P(Y_i|\eta_i, \alpha) = G(\alpha y_i, \alpha e^{\eta_i}) - G(\alpha(y_i + 1), \alpha e^{\eta_i})$, and $\boldsymbol{y} = (y_1, \ldots, y_n)'$.

Priors are assigned as follows:

- $\beta_0 \sim \mathcal{N}(0, 100)$, a vague Gaussian prior for the intercept.
- Second-order random walk (RW2) prior for $f_j(x_{ji})$ due to their balance of smoothness and flexibility for nonlinear trends [21],

  $$f_j(x_{ji})|f_j(x_{j,i-1}), f_j(x_{j,i-2}) \sim \mathcal{N}(2f_j(x_{j,i-1}) - f_j(x_{j,i-2}), \tau_x^{-1}),$$

  with $\tau_x \sim$ PC.prec$(1, 0.01)$ a choice to penalize complexity conservatively [13].
- $$\boldsymbol{u}|\boldsymbol{\theta} \sim \mathcal{N}(\mathbf{0}, \mathbf{Q}(\boldsymbol{\theta})^{-1}),$$

  with $\theta_{\tau 0}, \theta_{\tau 1}, \theta_{\kappa 0}, \theta_{\kappa 1} \sim \mathcal{N}(0, 1)$, following [20].

- α uses a PC prior as in [12], balancing flexibility and simplicity:

$$\pi(\alpha) \quad = \quad \lambda e^{-\lambda\sqrt{-2\log\Gamma(\alpha)+2(\alpha-1)\psi(\alpha)+2}} \left| \frac{(\alpha-1)\frac{d}{d\alpha}\psi(\alpha)}{\sqrt{-2\log\Gamma(\alpha)+2(\alpha-1)\psi(\alpha)}} \right|, \tag{18.9}$$

where $\psi(\cdot)$ is the digamma function and λ is tuned based on prior knowledge [13]. Here, we consider the PC prior parameter as $\lambda = 3$, a result of [12].

INLA approximates the marginal posteriors:

$$\tilde{\pi}(\psi_j|\boldsymbol{y}) = \int \tilde{\pi}(\psi_j|\boldsymbol{\theta}, \boldsymbol{y})\tilde{\pi}(\boldsymbol{\theta}|\boldsymbol{y})\, d\boldsymbol{\theta}, \quad \tilde{\pi}(\theta_k|\boldsymbol{y}) = \int \tilde{\pi}(\boldsymbol{\theta}|\boldsymbol{y})\, d\boldsymbol{\theta}_{-k}, \tag{18.10}$$

using Laplace approximations and numerical integration [23]. The `gammacount` family in `R-INLA` facilitates this, enabling fast computation of posterior estimates, deviance information criterion (DIC) and Watanabe-Akaike information criterion (WAIC) for model comparison [27, 28]. This approach ensures efficient inference for the NS-GC model, suitable for Alberta's sparse precipitation data.

18.4 Data Fitting Results

The Bayesian NS-GC model was applied to the Alberta precipitation dataset for December 2024, utilizing the `R-INLA` package with the `gammacount` family, as detailed in Section 18.3. The model included variables, `Tm` and `TP`, with elevation influencing the spatial dependency structure in a nonstationary field through the SPDE framework [20]. The triangular mesh seen in Figure 18.3 (right) enabled finite element approximations, providing effective spatial interpolation throughout the varied topography of Alberta. Our analysis uses the following predictors:

$$\eta_i = f_{Tm}(\texttt{Tm}_i) + f_{TP}(\texttt{TP}_i) + u(s_{1i}, s_{2i}), \quad i = 1, \ldots, 169, \tag{18.11}$$

where $f_{Tm}(\texttt{Tm}_i)$ and $f_{TP}(\texttt{TP}_i)$ represent the unknown nonlinear effects of the monthly temperature mean and total precipitation, respectively. Additionally, $u(\cdot)$ denotes the spatial effect at a point with a longitudinal coordinate s_1 and a latitudinal coordinate s_2.

We assess the effectiveness of the proposed flexible nonstationary spatial count model, utilizing the latest data set on the frequency of days with at least 1.0 mm of precipitation in Alberta, Canada, during December 2024. This case study demonstrates that the model performs effectively in irregular and nonstationary spatial domains and is pertinent to those domains in comparison to alternative count models. To assess the merits of the new technique, we also analyze established alternatives, including the Poisson (Pois), generalized Poisson (GP; [8]), and negative binomial (NB) models. Table 18.1 provides a summary of the model goodness of fit, which includes DIC and WAIC. The GC model consistently surpasses competitors across all performance criteria, showcasing its robust predictive capability and accurate alignment with the dataset. The capability of our suggested model to accurately forecast outcomes and align exactly with data illustrates its effective performance.

Figure 18.6 displays the posterior mean estimates (points) and 95% credible intervals (lines) for the precision of fixed and spatial effects, in conjunction with the dispersion parameter of the NS-GC model. The initial two lines delineate the nonstationary SPDE

TABLE 18.1
Model criteria based on computed DIC and WAIC for the proposed model and its alternatives.

Model	DIC	WAIC
GC	**399.50**	**396.96**
Pois	505.55	499.35
NB	506.91	500.53
GP	508.47	501.99

parameters, whereby $\theta_{\tau_1} > 0$ signifies the elevation coefficient for $\log(\tau(\mathbf{s}_i))$ and $\theta_{\kappa_1} < 0$ denotes the elevation coefficient for $\log(\kappa(\mathbf{s}_i))$. A positive θ_{τ_1} indicates that spatial accuracy $\tau(\mathbf{s}_i)$ enhances with elevation, hence diminishing the marginal variance of the spatial field $u(\mathbf{s}_i)$ at elevated elevations. This indicates reduced local variability in precipitation frequency in Alberta's mountainous areas, possibly indicative of stable orographic precipitation patterns [1]. Conversely, a negative θ_{κ_1} signifies that $\kappa(\mathbf{s}_i)$ diminishes with elevation, hence increasing the spatial range $\rho(\mathbf{s}_i) = \sqrt{8\nu}/\kappa(\mathbf{s}_i)$. The overall consistency at elevated altitudes corresponds with extensive topographic effects, exemplified by the Rockies, in contrast to the more localized variability found in the grasslands [20]. The third line illustrates the accuracy parameter for the nonlinear influence of `TP`, demonstrating a low posterior mean, which signifies large variation. This considerable fluctuation highlights the pronounced nonlinearity of `TP`'s effect on wet-day counts, aligning with the structured additive framework's adaptability in representing Alberta's precipitation dynamics [21]. The fourth line indicates that the GC dispersion parameter, α, has a posterior mean significantly greater than five. The elevated value of the dispersion parameter, $\alpha > 1$, indicates under-dispersion, likely attributable to stable winter weather systems in December that diminish variability in precipitation days. This observation, wherein variance is less than the mean, supports the NS-GC model's enhanced fit compared to alternatives such as Poisson, as demonstrated by the model diagnostics in Table 18.1 [11]. The last result pertains to the precision parameter for the nonlinear impact of `Tm`, indicating a substantial posterior mean accompanied by an extensive credible interval. This high accuracy (low variance) indicates a rather constant contribution of `Tm` to the predictor. The extensive range indicates heightened uncertainty in the estimation, perhaps attributable to the intricate interplay between temperature and elevation across Alberta's many climate zones.

Figure 18.7 depicts the fitted smooth curves representing the nonlinear influences of `Tm` and `TP` on the anticipated $\texttt{Percip\_Days}_i$, as modeled inside the NS-GC framework. The curves illustrate the posterior mean estimates of the smooth functions $f_.(\cdot)$ in the predictor (18.11), obtained by second-order random walk priors [21]. The influence of `Tm` demonstrates a significant negative nonlinear trend, wherein elevated temperatures correlate with a diminished anticipated frequency of days with precipitation beyond 1.0 mm. The inverse association indicates that elevated temperatures in December 2024 reduce precipitation frequency in Alberta, likely due to less frontal activity or snowmelt transitions at higher temperatures, especially in the plains areas. In contrast, the influence of `TP` demonstrates a robust positive nonlinear correlation, wherein increases in total monthly precipitation are associated with a greater frequency of wet days, stabilizing at higher `TP` levels. This pattern is meteorologically consistent, as increased cumulative precipitation inherently results in more days exceeding the 1.0 mm threshold; however, the nonlinearity suggests overload effects, where additional precipitation grows on already wet days instead of increasing their frequency, a phenomenon intensified by Alberta's topographic variability. These findings highlight the NS-GC model's ability to encapsulate intricate covariate connections,

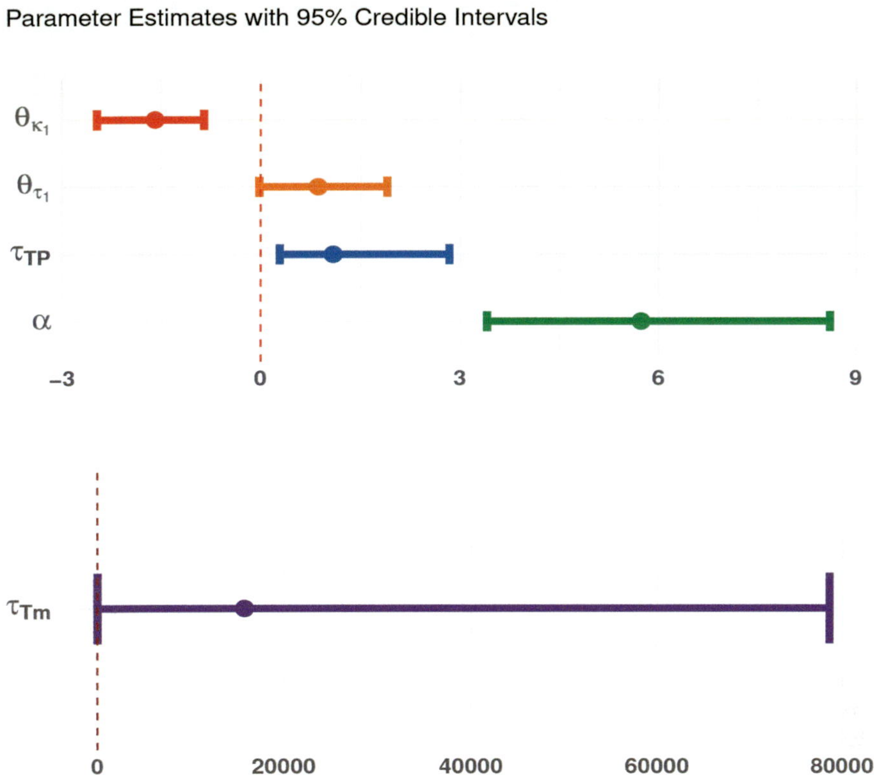

FIGURE 18.6
Posterior mean estimates (point) and 95% credible intervals (lines) of parameters and hyperparameters for the proposed model and its alternatives.

consistent with exploratory investigations (Section 18.2) and affirming its relevance to Alberta's varied precipitation regime.

Figure 18.8 illustrates the posterior mean, standard deviation, and first and third quartiles of the spatial effect $u(\mathbf{s}_i)$ in the NS-GC model, reflecting spatial variance in the anticipated number of wet days in Alberta for December 2024. The posterior mean indicates a significant spatial gradient, with diminished effects in the southeast, aligned with the grassland regions, in contrast to the center and northern parts, where topographic influences from the Rocky Mountains augment precipitation frequency. This pattern corresponds with the nonstationary SPDE framework, wherein elevation-induced fluctuations in $\tau(\mathbf{s}_i)$ and $\kappa(\mathbf{s}_i)$ influence spatial dependence (Section 18.3). The posterior standard deviation reveals increased uncertainty within Alberta's borders, especially in the western and northern peripheries, a conclusion supported by the broader range between the first and third quartiles in these areas. This increased variability indicates the intricate interaction of localized weather systems and limited observational coverage near boundaries. In contrast, center areas display narrower quartile ranges, indicating more reliable estimations. The NS-GC model's spatial mapping indicates a greater frequency of precipitation days near the domain's center, likely influenced by orographic effects, whereas peripheral areas, particularly the southeastern prairies, encounter fewer wet days, aligning with exploratory heat maps (Section 18.2).

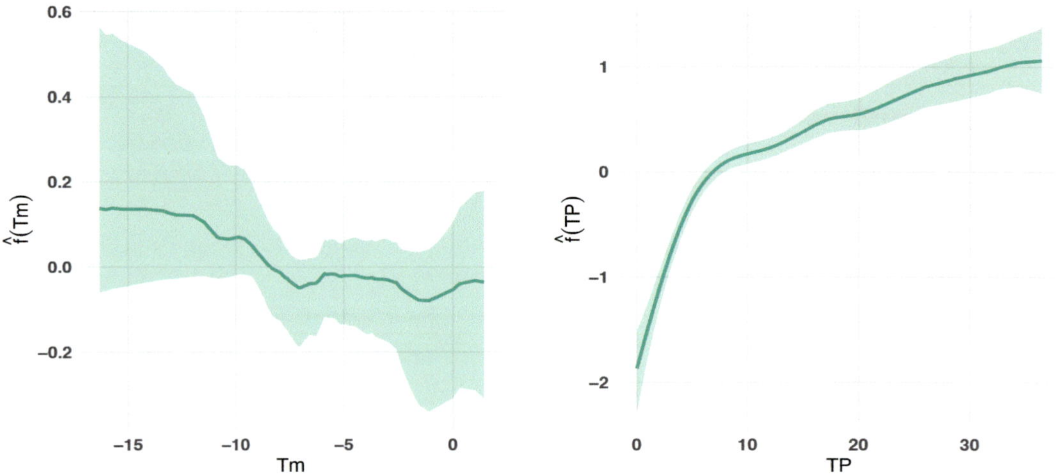

FIGURE 18.7
Fitted curve for nonlinear fixed effects including mean temperature (left), and total precipitation (right) for the NS-GC model.

18.5 Discussion and Conclusion

The nonstationary spatial gamma-count (NS-GC) model well represents Alberta's precipitation patterns in December 2024. Figure 18.8 illustrates the posterior mean of the spatial effect $u(\mathbf{s}_i)$, emphasizing increased wet-day frequency in the western mountainous areas compared to the arid southeastern prairies, in accordance with the topographic influences discussed in Section 18.2 [1]. This identification of high-precipitation areas facilitates winter water management by pinpointing essential reserves for spring, which is crucial for Alberta's agricultural planning. Probabilistic outputs, such as standard deviations and quartiles (Figure 18.8), quantify uncertainty, facilitating risk-informed decisions that correspond with SDGs for climate resilience.

The NS-GC model's enhanced fit, as indicated in Table 18.1, arises from the incorporation of the gamma-count distribution [11] within a nonstationary SPDE framework [14], effectively addressing dispersion (high α, Figure 18.6) and spatially variable dependence (Section 18.3). This research enhances spatial count modeling by providing a robust instrument that connects nonstationary techniques with count-specific likelihoods. The effectiveness of Alberta's December precipitation indicates its relevance to other topographically different areas, improving environmental analysis amid winter climate fluctuation.

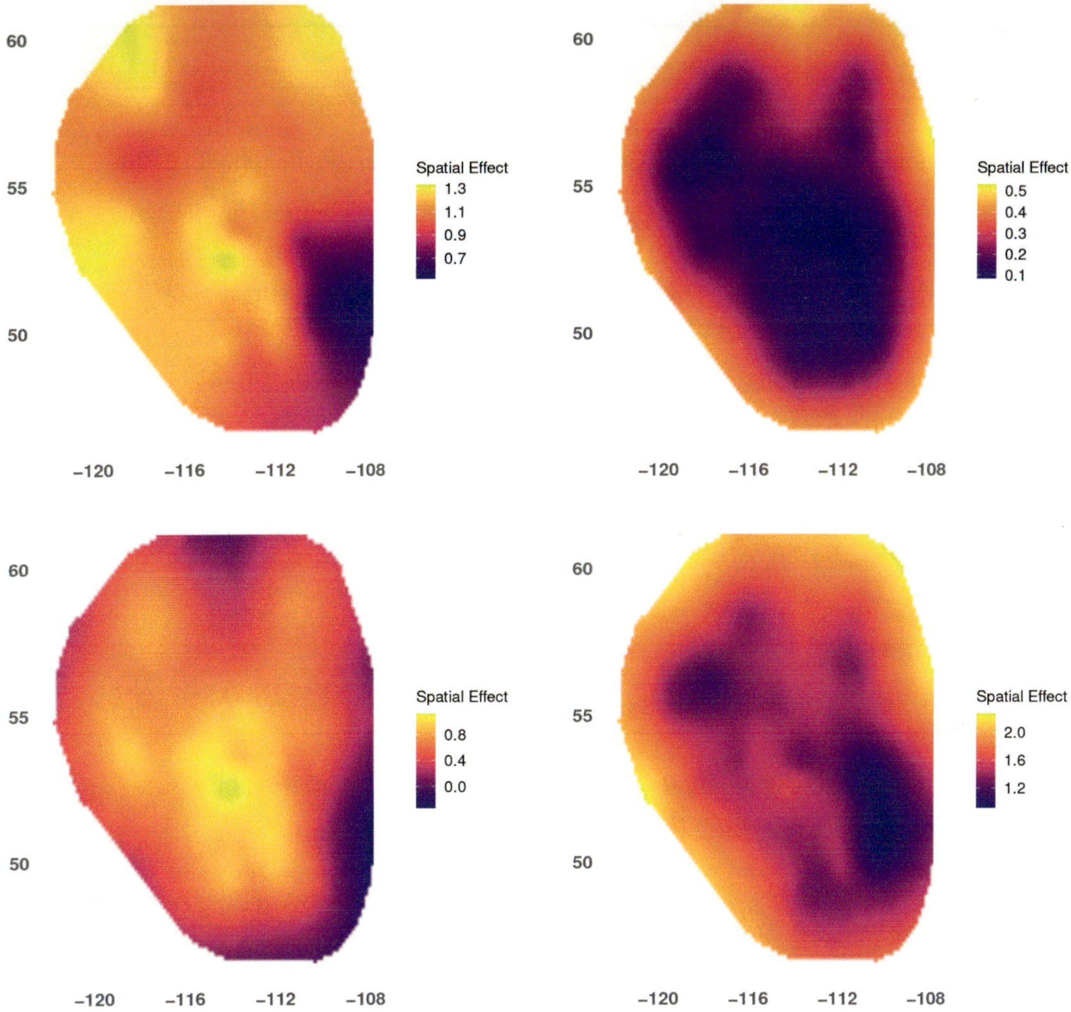

FIGURE 18.8
First row: posterior mean of the spatial effect (left) and its standard deviation (right); the second row presents the first (left) and third (right) quantiles of the posterior.

Acknowledgment

This work was based upon research supported in part by the National Research Foundation (NRF) of South Africa (SA), NRF ref. RA231117164450 and ref. RA211204653274, grant No. 151035, National Institute for Theoretical and Computational Sciences (NITheCS), as well as the Centre of Excellence in Mathematical and Statistical Sciences, based at the University of the Witwatersrand (SA). The opinions expressed and conclusions arrived at are those of the authors and are not necessarily to be attributed to the NRF. Mohammad Arashi's work is based on research supported in part by the Iran National Science Foundation (INSF) grant No. 4015320.

Bibliography

[1] Isa Marques, Nadja Klein, and Thomas Kneib. Non-stationary spatial regression for modelling monthly precipitation in Germany. *Spatial Statistics*, 40:100386, 2020.

[2] Kevin Shook and John Pomeroy. Prairie precipitation patterns and hydrological responses. *Journal of Hydrology*, 408(1-2):89–102, 2011.

[3] United Nations. Sustainable development goal 3: Ensure healthy lives and promote well-being for all at all ages, 2015. Accessed: Feb 23, 2025.

[4] Hossein Masoumi Karakani. Supporting the measurement of sustainable development goals in Africa: Geospatial sentiment data analysis. *IEEE Technology and Society Magazine*, 43(1):70–85, 2024.

[5] Hossein Baghishani and Mohsen Mohammadzadeh. A data cloning algorithm for computing maximum likelihood estimates in spatial generalized linear mixed models. *Computational Statistics & Data Analysis*, 55(4):1748–1759, 2011.

[6] A. Cameron and P. Trivedi. *Regression Analysis of Count Data*. Cambridge University Press, Cambridge, 2013.

[7] Joseph M. Hilbe. *Negative Binomial Regression*. Cambridge University Press, Cambridge, 2nd edition, 2011.

[8] Prem C. Consul and Felix Famoye. Generalized Poisson distribution. *Lagrangian Probability Distributions*, pages 165–190, 2006.

[9] D. Lord, S. Geedipally, and S. Guikema. Extension of the application of Conway-Maxwell-Poisson models: Analyzing traffic crash data exhibiting underdispersion. *Risk Analysis: An Official Publication of the Society for Risk Analysis*, 30:1268–1276, 2010.

[10] Mahsa Nadifar, Hossein Baghishani, and Afshin Fallah. A flexible generalized Poisson likelihood for spatial counts constructed by renewal theory, motivated by groundwater quality assessment. *Journal of Agricultural, Biological and Environmental Statistics*, 28(4):726–748, 2023.

[11] R. Winkelmann. Duration dependence and dispersion in count-data models. *Journal of Business & Economic Statistics*, 13(4):467–474, 1995.

[12] Mahsa Nadifar, Hossein Baghishani, Thomas Kneib, and Afshin Fallah. Flexible Bayesian modeling of counts: Constructing penalized complexity priors. *arXiv preprint arXiv:2105.08686*, 2021.

[13] Daniel Simpson, Håvard Rue, Andrea Riebler, Thiago G. Martins, and Sigrunn H. Sørbye. Penalising model component complexity: A principled, practical approach to constructing priors. *Statistical Science*, 32(1):1–28, 2017.

[14] Finn Lindgren, Håvard Rue, and Johan Lindström. An explicit link between Gaussian fields and Gaussian Markov random fields: The stochastic partial differential equation approach. *Journal of the Royal Statistical Society Series B: Statistical Methodology*, 73(4):423–498, 2011.

[15] Paul D. Sampson and Peter Guttorp. Nonparametric estimation of nonstationary spatial covariance structure. *Journal of the American Statistical Association*, 87(417):108–119, 1992.

[16] Dave Higdon. Non-stationary spatial modeling. In J. M. Bernardo, J. O. Berger, A. P. Dawid, and A. F. M. Smith, editors, *Bayesian Statistics 6*, pages 761–768. Oxford University Press, Oxford, 1999.

[17] Christopher J. Paciorek and Mark J. Schervish. Spatial modelling using a new class of nonstationary covariance functions. *Environmetrics*, 17(5):483–506, 2006.

[18] Geir-Arne Fuglstad, Finn Lindgren, Daniel Simpson, and Håvard Rue. Exploring a new class of non-stationary spatial Gaussian random fields with varying local anisotropy. *Statistica Sinica*, 25(1):115–133, 2015.

[19] Haakon Bakka, Jarno Vanhatalo, Janine B. Illian, Daniel Simpson, and Håvard Rue. Non-stationary Gaussian models with physical barriers. *Spatial Statistics*, 29:268–288, 2019.

[20] Rikke Ingebrigtsen, Finn Lindgren, and Ingelin Steinsland. Spatial models with explanatory variables in the dependence structure. *Spatial Statistics*, 8:20–38, 2014.

[21] H. Rue and L. Held. *Gaussian Markov Random Fields: Theory and Applications*. Chapman & Hall/CRC Press, London, 2005.

[22] William Kleiber, Richard W. Katz, and Balaji Rajagopalan. Daily spatiotemporal precipitation simulation using latent and transformed Gaussian processes. *Water Resources Research*, 48(1):W01523, 2012.

[23] H. Rue, S. Martino, and N. Chopin. Approximate Bayesian inference for latent Gaussian models by using integrated nested Laplace approximations. *Journal of the Royal Statistical Society Series B*, 71:319–392, 2009.

[24] Dominique Lord, Simon P. Washington, and John N. Ivan. Poisson, Poisson-gamma and zero-inflated regression models of motor vehicle crashes: Balancing statistical fit and theory. *Accident Analysis & Prevention*, 37(1):35–46, 2005.

[25] Rainer Winkelmann. *Econometric Analysis of Count Data*. Springer-Verlag, 2008.

[26] Michael L. Stein. *Interpolation of Spatial Data: Some Theory for Kriging*. Springer, New York, 1999.

[27] D. J. Spiegelhalter, N. Best, B. P. Carlin, and A. Linde. Bayesian measures of model complexity and fit (with discussion). *Journal of the Royal Statistical Society, Series B*, 64:1–34, 2002.

[28] A. Gelman, J. Hwang, and A. Vehtari. Understanding predictive information criteria for Bayesian models. *Statistics and Computing*, 24(6):997–1016, 2014.

19

More Explorations on a Parametric Model to Assess Segregation in Samples with Small Units

Mahdi Salehi, Angelo Mazza, and Antonio Punzo

Assessing segregation in populations with small units, such as establishments or classrooms is a crucial but challenging task. The use of conventional non-parametric segregation indexes in such contexts may lead to biased results due to the limited number of observations. To address this issue, a parametric approach has been recently introduced which has demonstrated superior performance compared to the most frequently used non-parametric solutions. This approach assumes a mixture of beta distributions as the prior distribution for the probability of an individual belonging to a minority group within a given unit. It has been claimed that one or two components for such a mixture model will be sufficient to get efficient estimators under the circumstances with small units. In this chapter, we intend to check whether increasing the number of components to three will potentially improve the proposed model. To this end, a numerical illustration as well as a simulation study with various data generating processes are carried out.

19.1 Introduction

Spatial segregation, as the residential separation of some groups within a certain society, is a phenomenon shaped by various influences, including socioeconomic mobility and acculturation [Ibraimovic et al. (2010); Massey (1985)]. Analysts commonly quantify segregation patterns using measures like the dissimilarity index originally defined by Duncan and Duncan (1955) and Gini segregation coefficient defined by Massey and Denton (1988). These metrics evaluate how unevenly two demographic groups (e.g., men and women) are distributed across neighborhoods, compared to a hypothetical "even distribution" where proportion of minority individuals is exactly the same in each unit. The indexes typically range from 0 to 1, where 0 indicates perfect integration (groups share all units proportionally) and 1 signifies complete segregation (groups occupy entirely separate units with no overlap).

The dissimilarity index, as the most frequently used measure of segregation, only represents one dimension of segregation while Massey and Denton (1988) argued that segregation is multidimensional comprising exposure, concentration, centralization, and clustering. Moreover, standard indexes (let's denote them by I) including the dissimilarity also reflect

DOI: 10.1201/9781003564423-19

randomness in the allocation of individuals to units; that is, they measure deviations from evenness, not deviations from randomness. But this fact is problematic in units with small unit sizes since it causes potentially large values for the conventional segregation indexes leading to upward biases. To overcome this issue, researchers have suggested several *non-parametric* adjustments. Among them, Carrington and Troske (1997) (CT) introduced an adjusted index which is the most frequently used in the applied literature [Rathelot (2012)]. CT developed a method that measures a sample's distance from *randomness* (the minority individuals are distributed randomly across units) rather than its distance from *evenness* which was mentioned earlier. More specifically, CT's proposed approach requires establishing a baseline segregation index I, e.g. the Gini, dissimilarity or Theil indexes, under random allocation. To calculate it, individuals are randomly reassigned (shuffled) to units while preserving original unit sizes. Then, the segregation index I is recalculated based on the new generated distributions of minority/majority groups obtained for each shuffling. The mean of these synthetic values of I (denoted by I^*) provides a reasonable estimation of the expected value of I under the random allocation. Given the above-mentioned I and I^*, the CT's adjusted index is defined as

$$\hat{I}_{\mathsf{CT}} = \begin{cases} \frac{I-I^*}{1-I^*}, & I \geq I^*, \\ \\ \frac{I-I^*}{I^*}, & I < I^*, \end{cases} \tag{19.1}$$

which ranges from -1 to 1. For the interpretation of $\hat{I}_{\mathsf{CT}}$ given by (19.1), one may refer to Figure 2 of Carrington and Troske (1997). They showed that I^* tends to I as unit size gets large. Another adjusting approach using a bootstrap procedure was proposed by Allen et al. (2009) as

$$\begin{aligned} \hat{I}_{\mathsf{ABW}} &= I - (\bar{I}_{\text{Boot}} - I) \\ &= 2I - \bar{I}_{\text{Boot}}, \end{aligned} \tag{19.2}$$

where $\bar{I}_{\text{Boot}} = \frac{1}{\mathcal{B}}\sum_{b=1}^{\mathcal{B}} I_{(b)}$ is the mean of $\mathcal{B}$ bootstrap observations of I. They showed that this bootstrap bias-correction works well provided both the minority proportion and unit size are not "very small". Furthermore, Mazza and Punzo (2015) proposed an analytically bias correction method for estimating the dissimilarity index that outperforms some bootstrap-based procedures including the one given by (19.2) in terms of both bias and MSE. Besides these essential nonparametric solutions to the mentioned issue, Rathelot (2012) proposed a *parametric* approach. More precisely, he assumed that the probability that an individual within a given unit belongs to the minority (denoted by p_a) is distributed as a mixture of Beta distributions, yielding the Beta-binomial mixture distribution. First, such a statistical model is fitted on the observed sample and then indexes are obtained accordingly. Through a simulation with seven scenarios for the DGPs, he showed that his approach performs well against $\hat{I}_{\mathsf{CT}}$ and $\hat{I}_{\mathsf{AWB}}$ techniques, even in most DGPs where a misspecification of the underlying population distribution happens. Moreover, according to the results of the mentioned simulation, he claimed that one or two number of components for such a mixture model will be sufficient to get efficient estimators under the circumstances with small unit sizes. In this chapter, we intend to check whether increasing the number of components to three will potentially improve the Rathelot's proposed model or not. In this regard, besides the likelihood ratio test (LRT), we will also employ Akaike's information criterion (AIC) and Bayesian information criterion (BIC) to choose which mixture model will outperform the other parametric competitors.

In this chapter, we will begin by reviewing Rathelot (2012)'s methodology, emphasizing its contributions and limitations in Section 19.2. We will then continue by doing a simulation study and a numerical illustration in Section 19.3. Finally, Section 19.4 concludes the chapter.

19.2 A Brief Overview of Rathelot's Approach

Suppose that the population is divided into two groups: a minority group and the rest of the population (a majority group), distributed across A units. In reality, groups can be defined by some characteristics such as gender, nationality, ethnicity, or social status, while units may represent entities like businesses, schools, neighborhoods, or districts. Assume further that the number of individuals in unit a ($a = 1, \ldots, A$) is denoted by M_a and the number of individuals among M_a persons who belong to the minority group is denoted by N_a, with $N_a \in \{0, 1, \ldots, M_a\}$. Typically, N_a is considered to be distributed as a Binomial $\text{Bin}(M_a, p_a)$, where p_a denotes the probability that an individual in unit a belongs to the minority group.

As discussed in the Section 19.1, measuring segregation when unit sizes (M_a's) are small is a challenging task. Conventional segregation indexes, such as the dissimilarity index, are biased under such a circumstance. This phenomenon arises because segregation indexes use the observed proportions $\pi_a = \frac{N_a}{M_a}$ to estimate the true but unobserved probabilities p_a, $a = 1, \ldots, A$. When units are small, π_a is an inconsistent estimate of p_a. As mentioned, this issue has been examined by several authors, including Carrington and Troske (1997), who introduced the adjusted index (19.1), defined using the Euclidean distance between the observed index and a simulation-based estimate of the index under the assumption of randomness. Their approach is the most widely employed in the applied literature. Another approach was proposed by Allen et al. (2009) who suggested using bootstrap techniques to adjust the index for the presence of a potential bias. Their idea is that the difference between the observed index and the average of the indexes computed on bootstrap samples is an estimator for the small-unit bias, and that adding this difference to the observed index provides the estimator given by (19.2) as an unbiased index.

Rathelot (2012) proposed a parametric method for measuring segregation to tackle the mentioned problem. In more details, under the assumption that A is large while M_a is small, Rathelot (2012) assumed that $p_1, \ldots, p_A$ are independent and identically distributed random variables following a mixture model with components having beta distributions. More specifically, the following hierarchical model holds

$$N_a \mid p_a \sim \text{Bin}(M_a, p_a), \qquad a = 1, \ldots, A, \tag{19.3}$$

$$p_a \sim f(p; \boldsymbol{\nu}_c) = \sum_{j=1}^{c} \lambda_j g(p; \alpha_j, \beta_j), \tag{19.4}$$

where $\boldsymbol{\nu}_c = \{(\alpha_j, \beta_j, \lambda_j),\ j = 1, \ldots, c\}$, c is the number of components, λ_j is the weight of the jth component ($\lambda_c = 1 - \sum_{j=1}^{c-1} \lambda_j$) and $g(\cdot; \alpha_j, \beta_j)$ is the density of the beta distribution with the shape parameters α_j and β_j. The marginal probability mass function (PMF) of N_a given $M_a = M$ is simply obtained as

$$P(N_a = n | M, \boldsymbol{\nu}_c) = \binom{M}{n} \sum_{j=1}^{c} \lambda_j \frac{\text{Beta}(\alpha_j + n, \beta_j + M - n)}{\text{Beta}(\alpha_j, \beta_j)}, \tag{19.5}$$

where Beta$(\cdot,\cdot)$ is the complete beta function. Equation (19.5) is indeed the probability that n individuals out of M persons belong to the minority group. Clearly, the random variables p_a are not directly observable, whereas the values of N_a are fully observed; let us denote their observed values by n_a. So, from the marginal distribution (19.5), the likelihood of the parameter $\boldsymbol{\nu}_c$ given the sample $\mathcal{N} := (n_1, \ldots, n_A)$ is derived as

$$\ell_M(\boldsymbol{\nu}_c) = \sum_{n=0}^{M} A_{nM} \log P(N_a = n|M, \boldsymbol{\nu}_c), \tag{19.6}$$

where $P(N_a = n|M, \boldsymbol{\nu}_c)$ is given by (19.5) and A_{nM} is the count of units with n minority individuals out of M total units. The function in (19.6) is the likelihood of $\boldsymbol{\nu}_c$ when all unit sizes are equal to M. In reality, where we have potentially different sizes of the units, say $\mathcal{M} := \{M_1, \ldots, M_r\}$, the maximum likelihood estimator (MLE) of $\boldsymbol{\nu}_c$ (let's say $\hat{\boldsymbol{\nu}}_c$) is obtained thorough a numerical maximization of $\ell(\boldsymbol{\nu}_c) = \sum_{k=1}^{r} \ell_{M_k}(\boldsymbol{\nu}_c)$ with $\ell_{M_k}(\cdot)$ given by (19.6). Once the parameters of this model are estimated, the segregation indexes can be computed accordingly.

Using a simulation study, Rathelot (2012) showed that estimators derived on the basis of the parametric model (19.3)–(19.4) have two essential advantages over traditional estimators as well as the adjusted methods suggested by CT and ABW. First, it provides a flexible and parsimonious way to model the distribution of p_a, which can be used to compute a variety of segregation indexes. Second, it allows to make an asymptotic confidence interval for the segregation indexes based on the asymptotic distribution of $\hat{\boldsymbol{\nu}}_c$ and then an application of the delta method which only needs an estimate of the variance-covariance matrix of $\hat{\boldsymbol{\nu}}_c$ and the gradient vector of the segregation index with respect to $\boldsymbol{\nu}_c$ (see Wasserman (2006) for more details on the delta method). Rathelot (2012) claimed that using the model (19.3)–(19.4) with at most two components ($c \leq 2$), "the segregation indexes corresponding to both continuous and discrete DGPs are accurately proxied by the mixtures". He employed the LRT to choose between the Beta-binomial mixture model (19.3)–(19.4) with $c = 1$ (denoted by BBM1) and $c = 2$ (denoted by BBM2).

However, we are seeking to add the third component as well to see if still there is any improvement. In this chapter, besides the LRT, the AIC and BIC metrics are added to select the best parametric models among BBM1, BBM2, and BBM3, where that last one stands for the Beta-binomial mixture model with three components.

19.3 Simulation Study and Numerical Illustrations

In this section, first, a numerical illustration is performed using a synthetic sample, then, a simulation study is carried out to see if Rathelot (2012)'s model given by (19.3)–(19.4) is still improved by using three components instead of 1 or 2 which were the cases used in his paper. For the simulation study, we use the same strategy of Rathelot (2012) by employing some continuous and discrete distributions as DGPs explained in Table 19.1.

As mentioned earlier in Section 19.2, for estimating indexes using the parametric method proposed by Rathelot (2012), we first need to obtain $\hat{\boldsymbol{\nu}}$ thorough a numerical maximization of the log-likelihood function (19.6). For $c = 1$, (19.6) reduces to the log-likelihood of the ordinary Beta-binomial model originally defined by Skellam (1948). In this case, we use the following method of moment estimators of the parameters $\boldsymbol{\nu}_1 = (\alpha, \beta)$ as reasonable initial

TABLE 19.1
Beta-binomial mixture (BBM) models with different numbers of components and data-generating processes (DGPs) used in the simulation. The DGPs are chosen such that $E(p_a) = 0.1$, yielding a small minority share.

DGP	$\mathrm{B}_c \sim f(p; \boldsymbol{\nu}_c)$, $c \in \{1,2,3\}$ (refer to (19.3)–(19.4))
	$\mathrm{D}_1 \sim P(D_1 = p) = 0.45I_{\{p=0\}} + 0.5I_{\{p=0.1\}} + 0.05I_{\{p=1\}}$,
	$\mathrm{D}_2 \sim P(D_2 = p) = 0.45I_{\{p=0.05\}} + 0.5I_{\{p=0.1\}} + 0.05I_{\{p=0.5\}}$
	$\mathrm{D}_3 \sim P(D_3 = p) = 0.2$, $p \in \{0.05, 0.1, 0.15, 0.2\}$
	$\mathrm{N} \stackrel{d}{=} p\|(0 < p < 1)$, where $p \sim \mathrm{N}(0.1, 0.05^2)$
	$\mathrm{W} \stackrel{d}{=} p\|(0 < p < 1)$, where $p \sim \mathrm{Weibull}(0.1, 1.1)$
Model	BBM1: The Beta-binomial mixture model (19.3)–(19.4) with $c = 1$
	BBM2: The Beta-binomial mixture model (19.3)–(19.4) with $c = 2$
	BBM3: The Beta-binomial mixture model (19.3)–(19.4) with $c = 3$
	BBM: The selected model among BBM1, BBM2 and BBM3 according to the LRT

* $I_{\{\cdot\}}$ stands for the indicator function.

guesses for the log-likelihood maximization process

$$\alpha_0 = \frac{Mm_1 - m_2}{M\left(\frac{m_2}{m_1} - m_1 - 1\right) + m_1}, \tag{19.7}$$

$$\beta_0 = \frac{(M - m_1)(M - \frac{m_2}{m_1})}{M\left(\frac{m_2}{m_1} - m_1 - 1\right) + m_1}, \tag{19.8}$$

where m_1 and m_2 are the first two raw moments of the observed sample $\mathcal{N}$. Then, a call to the R function `optim()` with `method = "L-BFGS-B"` and using the values given by (19.7) and (19.8) as starting points gives us $\hat{\boldsymbol{\nu}}_1 = (\hat{\alpha}, \hat{\beta})$ as the MLE of $\boldsymbol{\nu}_1$. For $c \in \{2, 3\}$, we use the global optimizer `Rsolnp::gosolnp()` with `n.restarts = 20` and `sim = 2000` to obtain desirable initial parameter estimates.

Before proceeding with the simulation, we investigate whether models BBM1 and BBM2 adequately fit all dgp's specified in Table 19.1, or whether we need the additional three degrees of freedom from BBM3. For illustration, we first generate p_a from DGP's in Table 19.1, then generate associated n_a values from $\mathrm{Bin}(10, p_a)$ for $a = 1, \ldots, 10000$, yielding the synthetic sample $\mathcal{N} = (n_1, \ldots, n_{10000})$. Figure 19.1 displays the models BBM1, BBM2, and BBM3 fitted on the generated sample $\mathcal{N}$ along with the relative frequencies of n_as, denoted by "Observed" in the plots. Intuition from the figure suggests that BBM1 performs as good as more complex models BBM2 and BBM3 on some DGPs e.g., B_1 and N, but it has not a good fit on the other DGP's specially the discrete ones D_2 and D_1 where the model BBM2 is significantly dominated by BBM3. These facts are also confirmed based on the total of the absolute differences between the observed probabilities and the corresponding estimated probabilities derived from the BBM1, BBM2, and BBM3 models in Figure 19.1. Thus, according to this illustration, it seems that there is a need to extend the parametric method by Rathelot (2012) to BBM3 under some specific DGPs. However, these conclusions are only based on one synthetic data. For making a more reliable decision, we need to do a Monte Carlo simulation where we have several samples enabling us to compare the mentioned parametric models using the average AIC and BIC values.

Now, Algorithm 1 is utilized for performing a simulation study. Table 19.2 reports the means and the medians of AIC and BIC values and Figures 19.2 and 19.3 display the distributions of AIC and BIC values obtained from the simulation.

It is evident from Table 19.2 and Figures 19.2 and 19.3 that in most cases BBM2 has outperformed the other two, but still there are some DGPs (e.g., the discrete distributions

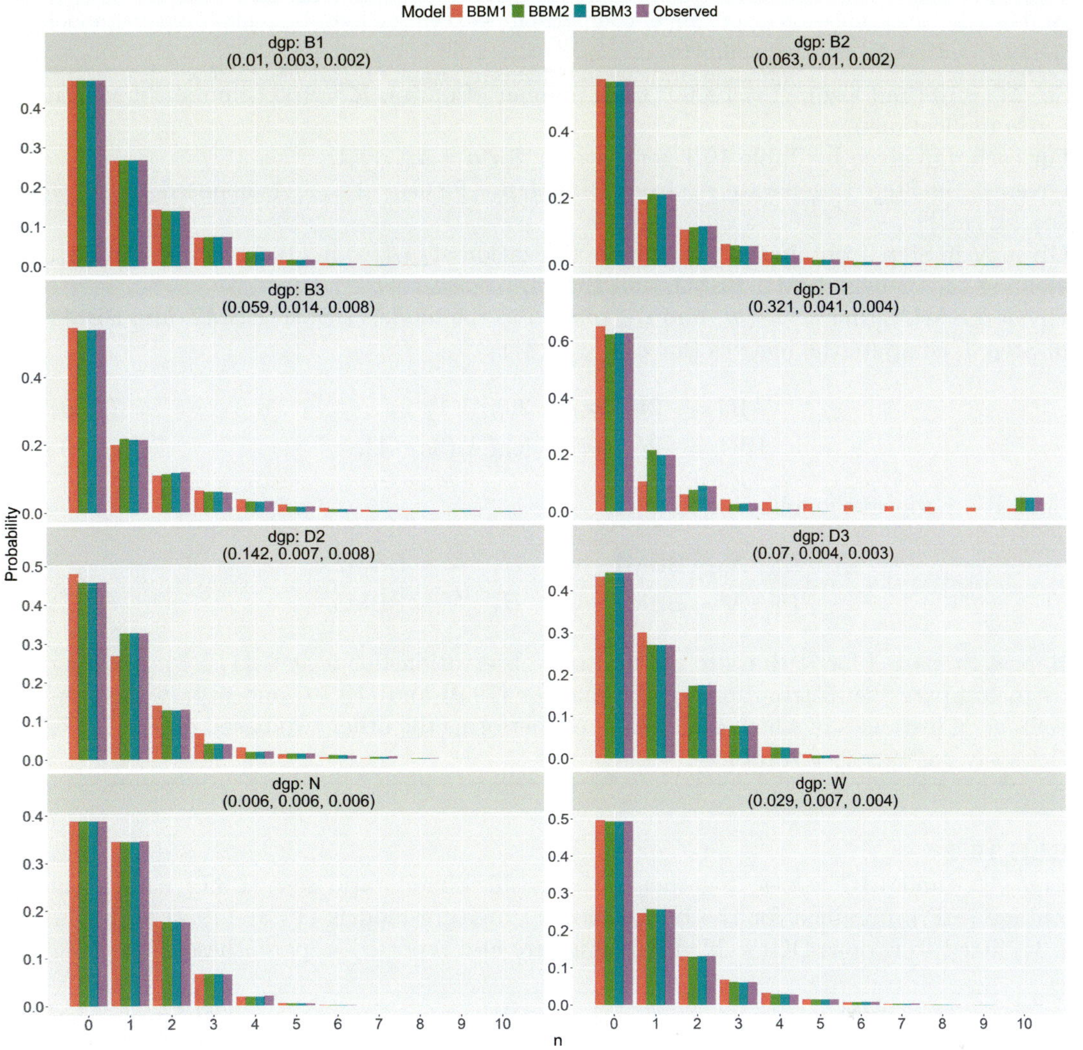

FIGURE 19.1
Fitting the distribution (19.5) with $c = 1$ component (the red bars), $c = 2$ components (the green bars), and $c = 3$ components (the blue bars) on synthetic samples drawn from the DGPs defined in Table 19.1. The purple bars represent the empirical distributions of the generated samples. The values shown in parentheses in the pane titles indicate the total of the absolute differences between the empirical probabilities and the corresponding estimated probabilities derived from the models BBM1, BBM2, BBM3, respectively.

D_1 and D_2) where BBM3 is the best. This fact is also confirmed by the sufficiency percentage of the LRT which indicates the percentage of cases in which each model outperforms the other two based on the LRT results. More precisely, it is seen that the model BBM3 has been the winner in all 100 synthetic samples generated from the DGP D_1 against its simpler competitors BBM1 and BBM2. This fact holds for the DGP D_2 as well, but for 78 cases out of 100. Now, on the basis of the three metrics AIC, BIC, and LRT, we can definitely conclude that the extension to the three components model BBM3 is fundamental under some circumstances. So, the previous results that we already obtained from Figure 19.1 are once again supported.

Algorithm 1 The simulation algorithm of comparison between the Beta-binomial mixture model (19.3)–(19.4) with some existing competitors.

Set $M = 10$ (unit size), $A = 10000$ (total number of units), $\mathcal{R} = 100$ (number of Monte Carlo replications).

1: For each dgp given by Table 19.1 generate p_a, for $a = 1, \ldots, A$.

2: Given p_a in Step 1, generate n_a from $\text{Bin}(M, p_a)$ for $a = 1, \ldots, A$, denoted by $\mathcal{N} = (n_1, \ldots, n_A)$

3: Given $\mathcal{N}$ in Step 2, obtain $\hat{\boldsymbol{\nu}}_c$ thorough maximization of $\ell_M(\boldsymbol{\nu}_c)$ in (19.6) for $c \in \{1, 2, 3\}$, yielding the models BBM1, BBM2, and BBM3, respectively.

4: Given the ML estimates $\hat{\boldsymbol{\nu}}_1$, $\hat{\boldsymbol{\nu}}_2$ and $\hat{\boldsymbol{\nu}}_3$, related to the models BBM1, BBM2, and BBM3 in Step 3, compute the metrics (for $c \in \{1, 2, 3\}$)

$$\text{AIC} = -2\ell_M(\hat{\boldsymbol{\nu}}_c) + 2(3c - 1), \tag{19.9}$$

$$\text{BIC} = -2\ell_M(\hat{\boldsymbol{\nu}}_c) + \log(A)(3c - 1). \tag{19.10}$$

and also perform the following LRTs

$$H_0 : \mathcal{N} \sim \text{BBM1} \quad \text{vs.} \quad H_1 : \mathcal{N} \sim \text{BBM2},$$
$$H_0 : \mathcal{N} \sim \text{BBM2} \quad \text{vs.} \quad H_1 : \mathcal{N} \sim \text{BBM3}.$$

5: Repeat Steps 1–4 for $\mathcal{R}$ times to get multiple observations for $\hat{\boldsymbol{\nu}}_c$, $c \in \{1, 2, 3\}$, allowing us to compare the distributions of the metrics (19.9) and (19.10) and obtain the percentage of instances in which each model outperforms the other two based on the LRTs (let's say *sufficiency percentage*).

TABLE 19.2
Average AIC and BIC values, together with the likelihood ratio test (LRT) sufficiency percentage (S.P.), obtained for the beta–binomial mixture models (19.3)–(19.4) with $c \in \{1, 2, 3\}$, fitted to various DGPs. Median values are also reported in parentheses.

	AIC			BIC			S.P.		
DGP	BBM1	BBM2	BBM3	BBM1	BBM2	BBM3	BBM1	BBM2	BBM3
B_1	**62801**	62805	62809	**62816**	62841	62867	**0.96**	0.04	0.01
	(**62833**)	(62839)	(62845)	(**62848**)	(62875)	(62903)			
B_2	57974	**57725**	57729	57988	**57761**	57787	0.00	**1.00**	0.00
	(57940)	(**57698**)	(57702)	(57954)	(**57734**)	(57759)			
B_3	58930	**58771**	58776	58944	**58807**	58833	0.00	**1.00**	0.00
	(58902)	(**58763**)	(58767)	(58916)	(**58799**)	(58825)			
D_1	46012	42301	**42203**	46027	42337	**42260**	0.00	0.00	**1.00**
	(46029)	(42332)	(**42216**)	(46044)	(42368)	(**42274**)			
D_2	61810	61247	**61240**	61825	**61283**	61297	0.00	0.22	**0.78**
	(61828)	(61268)	(**61263**)	(61843)	(**61304**)	(61321)			
D_3	63988	**63900**	63905	64003	**63936**	63963	0.00	**1.00**	0.00
	(64028)	(**63936**)	(63940)	(64042)	(**63973**)	(63998)			
N	**66037**	66039	66044	**66051**	66075	66102	**0.88**	0.12	0.00
	(**66104**)	(66109)	(66115)	(**66119**)	(66145)	(66172)			
W	62264	**62251**	62255	**62278**	62287	62313	0.07	**0.92**	0.01
	(62280)	(**62268**)	(62274)	(**62295**)	(62304)	(62332)			

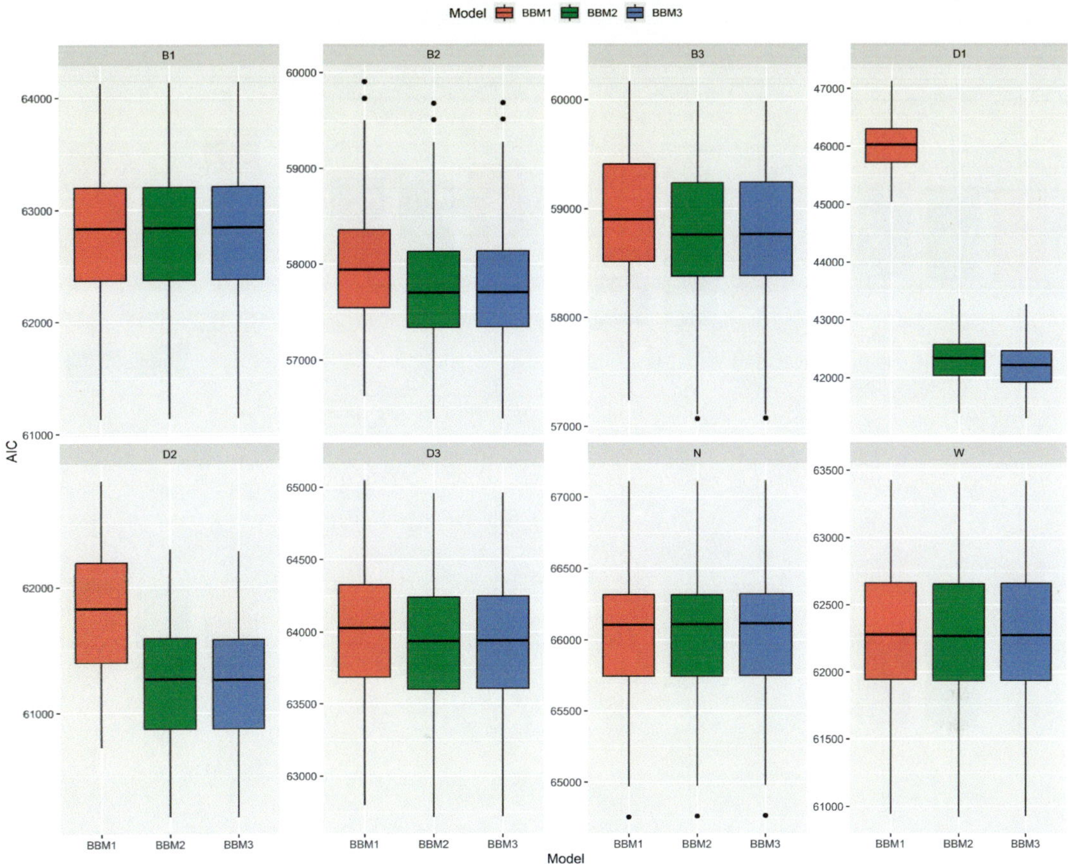

FIGURE 19.2
The AIC's obtained for Beta-binomial mixture model (19.3)–(19.4) with $c \in \{1, 2, 3\}$, fitted on various DGPs.

19.4 Conclusion

In this chapter, we addressed the issue of estimating segregation indexes when units have small sizes. Building on the parametric approach proposed by Rathelot (2012), we extended the idea by considering a three-component Beta-binomial mixture model (BBM3), as opposed to the one- (BBM1) and two-component (BBM2) models examined in Rathelot (2012). We conducted a simulation study to check the efficiency of this more complicated parametric model with those of the simpler ones. This comparison was done based on the LRT as well as the AIC and BIC metrics. Based on the numerical illustration and the simulation results we observed that there are still some specific scenarios and discrete DGPs with PMF's

$$P(D_1 = p) = 0.45I_{\{p=0\}} + 0.5I_{\{p=0.1\}} + 0.05I_{\{p=1\}}$$
$$P(D_2 = p) = 0.45I_{\{p=0.05\}} + 0.5I_{\{p=0.1\}} + 0.05I_{\{p=0.5\}}$$

that we strongly need to use BBM3 instead of BBM2 to gain more efficiency for the estimators proposed by Rathelot (2012).

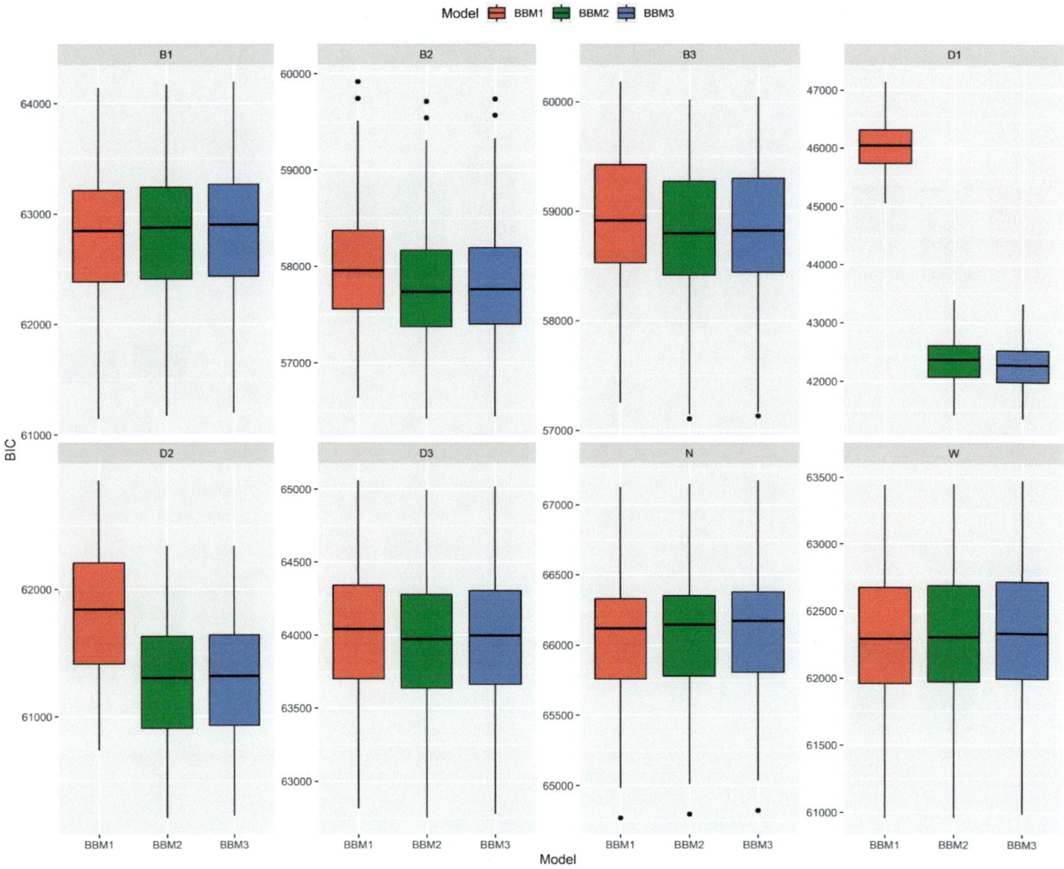

FIGURE 19.3
The BIC's obtained for Beta-binomial mixture models (19.3)–(19.4) with $c \in \{1, 2, 3\}$, fitted on various DGPs.

As mentioned in the introduction, Mazza and Punzo (2015) proposed an analytically bias correction method for the Direct estimator of dissimilarity index that outperforms some bootstrap-based procedures including ABW in terms of both bias and MSE. In future studies, we intend to compare the estimators explained in Table 19.1 with the solution of Mazza and Punzo (2015).

Acknowledgments

The authors are thankful to the reviewers for their comments that improved this chapter. This work has been partially supported by the PRIN 2022-PNRR research project "Foreign population and territory: integration processes, demographic imbalances, challenges and opportunities for the social and economic sustainability of the different local contexts (For.Pop.Ter)" [P2022WNLM7], funded by the European Union—Next Generation EU, component M4C2, Investment 1.1, CUP E53D23019200001 (Antonio Punzo). It has also been partially supported by the Italian Ministerial grant PRIN 2022 "Depopulation Risk: Experiences, Mobility and Subjective Well-being (DREAMS)" [2022RNKSEL], CUP E53D23010380006 (Mahdi Salehi and Angelo Mazza).

Bibliography

Allen, R., Burgess, S., and Windmeijer, F. (2009). More reliable inference for segregation indices. Technical Report 216, The Centre for Market and Public Organisation, University of Bristol, Bristol, UK.

Carrington, W. J. and Troske, K. R. (1997). On measuring segregation in samples with small units. *Journal of Business & Economic Statistics*, 15(4):402–409.

Duncan, O. D. and Duncan, B. (1955). A methodological analysis of segregation indexes. *American Sociological Review*, 20(2):210–217.

Ibraimovic, T., Lorenzo, M., and Stefano, S. (2010). Ethnic segregation and residential location choice of foreigners. In *Conference Paper STRC 2010*.

Massey, D. S. (1985). Ethnic residential segregation: A theoretical synthesis and empirical review. *Sociology and Social Research*, 69:315–350.

Massey, D. S. and Denton, N. A. (1988). The dimensions of residential segregation. *Social Forces*, 67(2):281–315.

Mazza, A. and Punzo, A. (2015). On the upward bias of the dissimilarity index and its corrections. *Sociological Methods & Research*, 44(1):80–107.

Rathelot, R. (2012). Measuring segregation when units are small: A parametric approach. *Journal of Business & Economic Statistics*, 30(4):546–553.

Skellam, J. G. (1948). A probability distribution derived from the binomial distribution by regarding the probability of success as variable between the sets of trials. *Journal of the Royal Statistical Society, Series B*, 10:257–261.

Wasserman, L. (2006). *All of Nonparametric Statistics*. Springer, New York.

Index

Note: Page numbers in **bold** and *italics* represent tables and figures, respectively.

For Product Safety Concerns and Information please contact our
EU representative GPSR@taylorandfrancis.com Taylor & Francis
Verlag GmbH, Kaufingerstraße 24, 80331 München, Germany